METALLIC AND INDUSTRIAL
MINERAL DEPOSITS

Metallic and Industrial Mineral Deposits

Carl A. Lamey
Professor Emeritus of Geology
The Ohio State University

McGraw-Hill Book Company
New York
St. Louis
San Francisco
Toronto
London
Sydney

This book is dedicated to my wife,
Mary Lucile, who rendered valuable assistance
in the task of checking the final manuscript,
and whose patience and understanding during
its writing added much to whatever merit
it may have.

Metallic and Industrial Mineral Deposits

Copyright © 1966 by McGraw-Hill, Inc. All Rights Reserved.
Printed in the United States of America. This book, or
parts thereof, may not be reproduced in any form without
permission of the publishers.
Library of Congress Catalog Card Number 65-28819

07-036091-X

3 4 5 6 7 8 9   KPKP   7 9 8 7 6 5 4

Preface

The general picture of the discovery, production, reserves, and utilization of metallic mineral deposits throughout the world has changed materially since 1945. Atomic bombs, nuclear missiles, and nuclear-powered submarines were developed; the uses of atomic energy for peaceful purposes were explored; and outer space was successfully invaded. All these developments emphasized the importance of certain metals, such as uranium, thorium, beryllium, and lithium, that previously had received scant notice in discussions of metallic mineral deposits. Moreover, industrial activity in general has increased tremendously. Many new products have been placed on the market, and this has stimulated the use of numerous nonmetallic or industrial materials, as well as metals.

Increased industrial demand has caused much exploration for new mineral deposits and has led to their discovery and exploitation and to the development of more efficient means of recovering material from low-grade deposits. In turn, this has given rise to a very great amount of periodical and other literature, which has become so extensive that the task of reading it is a formidable one. Consequently a great need has arisen for a condensation, in one volume, of some of the vast amount of information that has been published regarding individual deposits, the distribution, production, and reserves of mineral resources, and the more probable future developments. This book is an attempt to make such a condensation, except for the mineral fuels, and to list the more important references from which the condensation was made.

An attempt has been made to present a clear picture of world production by the use of graphs and tables based on the percentage of world production by various countries for the 15-year period 1945 to 1959 and for the single year 1962. These tables also show the actual average world production, in tons or other suitable units. Further, tables are included that show the world rank of the more productive countries for various materials during the period 1945 to 1959, so as to give the reader some concept of the importance of different countries. The past history of production from some deposits and of some countries, production trends, and possible future developments are discussed in the text where such information is available and is pertinent.

The book is organized in three parts, which deal respectively with (1) the characteristics and origin of mineral deposits in general, (2) the metallic deposits, and (3) the industrial deposits.

Part I is a discussion of the kinds of mineral deposits, their characteristics, localization, and origin. The formation of mineral deposits is considered under three principal categories: those formed by igneous activity, by metamorphic activity, and by surface agencies. This material is intended to be background information for those people who have not studied the origin and accumulation of mineral deposits and to serve as a review for those who have had an introduction to the subject but perhaps have been absorbed in the details of engineering or other work and need a readily available condensation of the principles that control the formation and accumulation of mineral deposits. The discussion is not, and is not intended to be, an advanced treatment of the origin of mineral deposits, but it does attempt to present a summary of the major concepts that have been stated in the literature.

Part II is a brief discussion of most of the more important metallic mineral deposits throughout the world and a summary of some features of production and reserves of various metals. An attempt has been made to present a concise sketch of those deposits that are now or have been important sources of world production or that give promise of future importance. The discussion tries to summarize, also, the factors that were of importance in bringing about the accumulation of individual deposits, and the major concepts of origin that have been suggested. No attempt has been made to state a definite conclusion in cases where the origin is a matter of sharp controversy, but rather the attempt has been to summarize the ideas that have been advanced and to cite literature covering the subject.

Part III is a summary of the industrial deposits, in which some of the more important materials are discussed briefly and the world-wide production picture is presented. These discussions are condensed more than are those of the metallic deposits, since several books dealing with industrial deposits were published in 1960 and 1961. The same general approach is used as that employed in Part II.

Throughout the book the purpose has been to make it as informative as possible within reasonable size limits. The hope is expressed that it can be read with understanding and profit not only by undergraduate students of geology but also by all people who have some knowledge of geology, mineralogy, and chemistry and by engineers, economists, and others interested in the mineral industry and that it may provide a suitable foundation for further reading from the wealth of literature now available.

A book of this type is, of necessity, a compilation from many sources. Free use has been made of published material, but care has been taken to show the sources of the information used. No attempt has been made to present complete bibliographies, but the references cited will provide the key to most of the important literature. A number of short references relating to new developments in mining and treatment of materials have been included for those who may wish to explore this field in more detail than is given in this book. Also, attention has been called to references relating to new deposits that give promise of being important in the future.

The writer is deeply indebted to all the geologists, mining engineers, and others who have investigated mineral deposits throughout the world and published the results of their studies. He is similarly indebted to all people who have compiled statistics of production. The writing of this book would have been impossible without the information thus made available. Further, he wishes to express grateful thanks to the many publishers who generously permitted the use of copyrighted material.

Carl A. Lamey

Contents

part **1**

The Characteristics and Origin of Mineral Deposits

chapter 1

Mineral

Resources

"The United States has been endowed with vast mineral resources, whose equal in quantity and variety has not yet been found in any other area of the world. Possession of these resources and ability to utilize them have made possible the preeminent industrial position of the United States and its unexcelled standard of living." (3, p. v)[1] This statement made in 1948 by J. A. Krug, then Secretary of the Interior of the United States, indicates clearly the value of mineral resources not only to the United States but also to any country. These resources, which include minerals and rocks of all types, unconsolidated masses of clay, sand, and gravel, deposits of peat, coal, petroleum, gas, and related materials, are second only to agricultural resources in the promotion of the welfare of a nation.

The history of the development of man is interwoven with the finding and utilization of mineral resources. Primitive man, just as modern man, needed food, a means of protection, and shelter; and, although not an absolute need, the desire for personal ornamentation seems to have existed from very early times. Man, the tool-making animal, at first used crude implements of wood and stone. Flint, rather widely distributed throughout the world and breaking with sharp edges, was at first rudely fashioned but later finished by grinding and polishing. It, and other stone materials, supplied scrapers, borers, adzes, arrowheads, and axes. With these man was enabled to carry on hunting, protect himself from his enemies, wage war, prepare skins for clothing, develop crude agriculture, and construct rude shelters. Ornamentation was furnished by such pieces of native copper, silver, and gold that might be found along streams or in surface gravels, and pigments for decoration were obtained from hematite, limonite, pyrolusite or wad, and cinnabar. Gradually metal implements and weapons—of copper, bronze, iron—replaced stone ones, and those who first obtained weapons of metal were able to overcome those who used weapons of stone. Wars were waged, kingdoms won and lost, and peoples enslaved for the possession of metals. This long and interesting story[2]

[1] Numbers in parentheses refer to references at the end of the chapter.
[2] See reference 4 for a detailed and interesting account of the part played by mineral resources in the development of nations. This reference has been used freely in the preparation of the material that follows.

cannot be recounted here, but a few features will be summarized in order to give a better understanding of the importance of mineral resources.

Early man moved about and bartered materials from one location for those of another. Flints from certain places were of particularly good quality, and these apparently were carried from one tribe to another and used in bartering. Copper in North America was carried from place to place by the Indians. Other metals also were used in bartering. Thus began early trade and the establishment of trade routes.[1]

Salt seems to have played an important role in this history. The salt of Palmyra apparently was an important factor in the commerce between Syrian ports and the Persian Gulf; salt oases were on the caravan route over the Libyan Desert in the time of Herodotus; salt pans on the estuary of the Dnieper River apparently were the cause of trade with people on the shore of the Aegean Sea and those living in southern Russia. The salt mines of India apparently were the center of widespread trade. The salt of Ostia was carried into the Sabine country over one of the oldest of Roman roads, the Via Salaria (from *sal,* Latin for *salt*). Salt was used instead of money in some places, and the Roman soldiers were given money with which to purchase salt.[2] During the Dark Ages, after the collapse of the Roman Empire, little mining of metals was done, but salt, because it was needed in the diet of most people, was still mined by some German tribes. Eventually the trade in salt led to the discovery of the silver mines of Freiburg in 1170, and this in turn brought about discovery of other important deposits of metals in the Erzgebirge (ore mountains) south of Freiburg.

Metals became of great importance after man learned to use fire for smelting ore and fashioning implements. The Phoenicians, a seafaring people, are reputed to have been the metal merchants of their time and to have carried on trade in tin, lead, iron, copper, silver, and gold. They are said to have sailed widely, not only in the Mediterranean Sea but also through the Strait of Gibraltar into the Atlantic Ocean, and to have reached England and France. The metals they used in trade were obtained from Africa, Spain, Greece, and the vicinity of Cornwall, England. Tin, which was highly prized in making bronze (an alloy of tin and copper), was obtained from the Cassiterides (Tin Islands) which apparently were in the vicinity of Cornwall. The sources of the metals, and the routes traveled, were kept secret, and for about two thousand years the Phoenicians dominated trade in metals.

Silver and gold also played a prominent part in world history, since they could be used as a highly prized medium of exchange and were, in time, converted into coins. A country that controlled much gold and silver was able to obtain the materials needed for industrial expansion and for war. The silver from the mines of Laurium, Greece, is said to have been used by the Athenians to enable them to build ships with which they defeated the Persians at Salamis in 480 B.C. The Romans regarded a mine as a prize of war and seized mines wherever they could. When they vanquished Carthage they obtained the silver of Spain. Much of the expansion of the Roman Empire depended on obtaining gold and silver, as well

---

[1] See reference 2, pp. 34–88, for an instructive discussion of the effect of mineral resources on trade.

[2] The word *salary* is derived from the Latin *salarium,* which was the money given to the Roman soldiers for the purchase of salt.

as other metals, and the decline of the empire was marked by the loss of gold and silver resulting from its use to import luxuries without balancing exportation of commodities, since the Romans were soldiers who seized mines rather than discovered and explored them, and thus production from the mines, commonly by slave labor, declined. The lure of gold is said to have resulted in the discovery of the New World. The Aztec country of Mexico was invaded by Spain because of gold and silver. This invasion, however, did not lead to commercial expansion in Mexico, since the aim of the Spaniards was to seize and remove the gold and silver.

The discovery of gold in California was of great importance in bringing about industrial expansion, not only in the United States, but also in Canada and Australia. The gold rush to California that started in 1848 brought in people from various parts of the world.

Edward H. Hargraves, an Australian, came to California in 1849. His experience in the California areas convinced him that gold should be present in some places that he had seen in New South Wales, because conditions appeared to be similar. He returned to New South Wales in 1851, and on February 12, 1851, discovered gold in the bed of a stream. This led to an intensive search for gold and the discovery on September 8, 1851, of gold in the area of Victoria that later developed into the rich Ballarat goldfield. "The nearness of the discoveries at Ballarat, Bendigo, and Castlemain rendered it easy for everybody to go thither in a hurry. . . . Sailors left their ships at Port Phillip; clerks jumped off their stools to decamp for the diggings; every able-bodied man shouldered his blankets and trudged through the bush to engage in the search for treasure" (4).[1] The great gold rush led eventually to the discovery of other metals in Australia— tin, copper, lead, zinc, and silver—a rapid increase in population, and industrial expansion. "Agriculture waited on mining, the need of foodstuffs stimulated farming; a complex civilization followed in the wake of the two fundamental industries" (4).[2]

Prospectors and miners commonly move from place to place, ever seeking new sources of valuable minerals; so the early gold prospectors of California moved northward through the present states of Oregon, Washington, and Idaho, into western Canada. This resulted in the finding of gold along the Fraser River in 1858 and the first great gold rush to British Columbia (1, p. 5). The Indians resisted the coming of prospectors, but this resistance was overcome by British troops. Prospectors moved up the Fraser and Thompson Rivers, reached the Cariboo country, and discovered some of the richest alluvial gold deposits of the world. The resulting gold rush caused the building of the first highway in western Canada.

In the train of all the various discoveries of gold . . . the foundations of agriculture, fruit growing, and ranching in British Columbia were laid. Soon there came a new outlook on the part of the people of British Columbia and with it a consciousness of the meaning and consequences of mineral development. The search for minerals other than gold was stimulated and great lead,

---

[1] From "Man and Metals" by T. A. Rickard, p. 751. Copyright 1932. McGraw-Hill Book Company. Used by permission.

[2] From "Man and Metals" by T. A. Rickard, p. 758. Copyright 1932. McGraw-Hill Book Company. Used by permission.

zinc, and copper deposits were discovered. For years British Columbia called herself the *mining province, and she had a right to the name.*[1]

It may seem strange that the great mineral wealth of Ontario and Quebec was not exploited before that of British Columbia, since eastern Canada was the scene of the early fur trade, the pioneer industry of Canada. Knowledge of mineral wealth in eastern Canada undoubtedly existed, but the policy of the fur traders was to discourage settlement and the development of mineral resources because of the probable adverse effect it would have on the fur supply. Eventually, however, the prospector succeeded the fur trader. The fabulous silver deposits of Cobalt, Ontario, were discovered in 1903, followed by the discovery of the world-famous gold areas of Porcupine and Kirkland Lake, Ontario.

Gold has always been a lure which took men into the remote corners of the earth. The value of the gold produced in Canada has always been greater than that of any other mineral product. However, the influence of gold has been vastly greater in the Canadian economy than mere production statistics would indicate. More than any other resource it has provided the stimulus for the opening-up of our northern regions. . . . The colonization of the clay belts of northern Quebec and Northern Ontario was made possible only because of markets for agricultural produce in the mining towns and the labour opportunities of the mines.[2]

With the discovery and development of the vast iron ore deposits of Labrador, a new chapter in the history of expansion began, not only in Canada but elsewhere. "As transportation facilities are established, power sites developed, and the mines brought into production, an industry of vast importance to North America, as well as to western Europe, will have been created. The story of that creation is a romance in itself."[3]

We have seen how the discovery of gold in California led to the development of Australia and Canada but have not yet considered its importance in the expansion of the United States. The early eastern colonists needed iron for tools and lead for bullets. These they found nearby in relatively small amounts, but these early supplies soon became insufficient. Hardy and adventurous pioneers, moving westward, discovered the iron deposits of Michigan in 1844 and the copper deposits of the same state in 1847. Coal was first discovered in the present state of Illinois in the seventeenth century. Small coal mines were opened in Virginia and Ohio early in the eighteenth century and in Pennsylvania about 1800. All these deposits, along with gold from Virginia, the Carolinas, and Georgia, aided early industrial expansion. The discovery of rich gold deposits in California, however, stimulated transcontinental travel and led to other important discoveries. As in Australia and Canada, food and shelter were needed for a rapidly expanding population. Surveys for transcontinental railway routes were made between 1850 and 1860, and these surveys included a search for other mineral deposits. They

---

[1] From Charles Camsell, p. 6, in "Out of the Earth" by G. B. Langford. Copyright 1954. University of Toronto Press. Used by permission.
[2] From Charles Camsell, p. 7, in "Out of the Earth" by G. B. Langford. Copyright 1954. University of Toronto Press. Used by permission.
[3] From Charles Camsell, p. 8, in "Out of the Earth" by G. B. Langford. Copyright 1954. University of Toronto Press. Used by permission.

were followed by the construction of four major railways, the Union Pacific, the Southern Pacific, the Santa Fe, and the Northern Pacific, which were completed, respectively, in 1869, 1881, 1882, and 1883. These railroads stimulated not only production from mineral deposits but also settlement and expansion of the country. Industrial expansion was stimulated not only in the western United States but also in the eastern United States, since the progress of railroad expansion required vastly increased supplies of iron ore and production of pig iron, gave employment to many people, and called for many supporting industries. The railroad expansion, and consequent additional production of pig iron and steel, are shown in Table 1-1.

TABLE 1-1.   Construction of Railroad and Production of Pig Iron and Steel, in the United States, 1830–1899*

Year	Railroad in operation, miles	Pig iron produced, tons	Steel produced, tons
1830	23	165,000	—
1840	2,818	286,903	—
1850	9,021	563,775	—
1860	30,626	821,223	—
1870	52,922	1,665,179	68,750
1880	92,296	3,835,191	1,247,335
1890	166,698	9,202,703	4,277,071
1899	189,295	11,255,000†	10,639,857

* Data from "Mineral Resources of the United States," 1900, U.S. Geological Survey, p. 81–89.
† Pig iron shipped, tons. S. H. Schurr and E. K. Vogeley, 1960, "Historical Statistics of Minerals in the United States," p. 17.

The development of nonmetallic resources commonly lags behind the production of gold, silver, iron ore, and some other metallic resources, because the nonmetallic resources have much lower unit value and generally must be used near the source of supply. Hence their utilization must await increase in nearby population and creation of local industries. As development proceeds, however, their production increases rapidly and may exceed that of the metallic resources. This is particularly true if sufficient supplies of mineral fuels are available to furnish power for manufacturing. These mineral fuels, themselves, may surpass all other mineral resources in value (Table 1-2).

TABLE 1-2.   Value of Mineral Resources Produced in the United States, 1880–1960*

Year	Metallic resources	Nonmetallic resources	Mineral fuels
1880	190,881,000	$ 56,341,000	$ 120,241,000
1900	514,232,000	188,328,000	406,376,000
1920	1,763,675,000	1,024,755,000	4,192,910,000
1940	1,678,600,000	818,800,000	3,116,500,000
1960	2,021,000,000	3,730,000,000	12,141,000,000

* Data from "Mineral Resources of the United States," 1880, 1900, 1920, U.S. Geological Survey, and *Minerals Yearbook*, 1940, 1960, U.S. Bureau of Mines.

Minerals were mined before the dawn of history. Throughout the ages they have afforded the means for commercial nations to gain dominion over nations whose economies were principally agricultural. . . .

The Industrial Revolution harnessed iron and coal to the service of man. As each of the 90-odd chemical elements that can be won from the ground has been graduated from the laboratory and put to work, civilization has leaped ahead until now virtually all sources of our national-income, even professions and personal-service businesses, depend more or less completely upon the free flow of mineral supply.[1]

SELECTED REFERENCES

1. G. B. Langford, 1954, "Out of the Earth," University of Toronto Press, Toronto, Canada, 125 pp. See especially pp. 3–9 and 109–114.
2. T. S. Lovering, 1944, "Minerals in World Affairs," Prentice-Hall, Inc., Englewood Cliffs, N.J., 394 pp. See especially pp. 15–88.
3. "Mineral Resources of the United States," 1948, Public Affairs Press, Washington, D.C., 212 pp. See especially pp. 21–27.
4. T. A. Rickard, 1932, "Man and Metals," McGraw-Hill Book Company, New York, 1068 pp.
5. P. M. Tyler, 1948, "From the Ground Up," McGraw-Hill Book Company, New York, 248 pp. See especially pp. 1–20.

[1] From "From the Ground Up," pp. 8–9, by P. M. Tyler. Copyright 1948. McGraw-Hill Book Company. Used by permission.

chapter **2**

General Features, Kinds, and Localization of Mineral Deposits

Mineral resources may be divided into groups designated the *metallic deposits,* the *nonmetallic* or *industrial deposits,* and the *mineral fuels.* Only the first two of these will be discussed, and they will be referred to collectively as *mineral deposits.* Since the history of the origin and accumulation of many mineral deposits is complex, it seems wise to present first a general survey of the characteristics and types of the various deposits and subsequently to discuss each type in more detail.

A mineral deposit is considered to be a natural accumulation of useful mineral materials that can be profitably extracted if conditions are favorable. Some accumulations of this type are designated *ore,* a term somewhat variously used. One definition (5, p. 475) states that an ore is "a natural mineral compound, of the elements of which one at least is a metal. The term is applied more loosely to all metalliferous rock, though it contains the metal in the free state, and occasionally to the compounds of non-metallic substances, as sulphur ore. . . . Also, materials mined and worked for nonmetals, as pyrite is an *ore* of sulphur. . . . A mineral of sufficient value as to quality and quantity which may be mined with profit. . . ." (See also references 6, p. 289, and 7, p. 205.) The original meaning of the word apparently was a lump of metal, probably copper or bronze. Various words from which the present term *ore* was derived referred to brass, copper, and wrought metal. Hence there has been a strong tendency to consider, as ore, material from which a metal can be obtained; but present usage by many includes materials such as sulfur, fluorite, and barite. Also, there is a strong tendency to regard profitable extraction as an important characteristic.

Almost every ore is composed of some minerals designated *ore minerals* and others designated *gangue minerals.* Further, many ores include more or less *country rock.*[1] If a metal is sought, an ore mineral is one from which one or more metals may be extracted, such as galena or sphalerite, but if the ore is considered to be an accumulation of fluorite, barite, or some other mineral not

[1] Many terms are defined and discussed in the glossaries listed at the end of this chapter.

used for the extraction of a metal, the ore mineral is the valuable mineral for which the deposit is mined. A gangue mineral generally is a worthless mineral in a particular deposit. In deposits mined for metals, some common gangue minerals are quartz, calcite, fluorite, and barite. If enough fluorite, for example, were present in a deposit mined for a metal, the fluorite could be obtained as a by-product, and thus it would be an ore mineral *in that particular deposit.* Thus the amount of a mineral present in a deposit may determine whether it is regarded as an ore mineral or a gangue mineral in that deposit. Country rock is the general mass of relatively sterile rock adjacent to or enclosing a mineral deposit, as granite, limestone, etc. In many deposits more or less country rock adheres to or is included with the ore as it is mined. *Gangue,* as distinct from gangue mineral, is used to designate the association of worthless minerals or enclosed rock, or both, that are present in a mineral deposit. The term *ore body* generally is used for a fairly continuous mass of ore, which may include low-grade material and waste, as well as high-grade material, but the form or character of which distinguish it from the adjoining country rock.

Deposits may be of *syngenetic* or of *epigenetic* origin. A syngenetic deposit is one that was formed essentially contemporaneously with the rock in which it is enclosed although it may be later than the underlying and earlier than the overlying rock, whereas an epigenetic deposit is one that was formed later than the rock in which it is enclosed. Also deposits may be *hypogene* or *supergene.* A hypogene deposit is one that was formed as a primary deposit by ascending solutions, generally conceived to have been of magmatic origin, whereas a supergene deposit is one that was formed by descending waters of meteoric origin, the action of which commonly resulted in the enrichment of a previously existing deposit.

Mineral deposits, except the mineral fuels, are generally separated into two groups, *metallic deposits* and *nonmetallic* or *industrial deposits.* The metallic deposits are those mined chiefly for the extraction of a metal or of metals, whereas the nonmetallic or industrial deposits are those generally used for other purposes. In the older writings, and in many of the modern ones, the term *nonmetallic deposits* is used exclusively. The term *industrial deposits,* which is more comprehensive, has come into use largely because various nonmetallics are utilized widely in the manufacture of glass, brick, tile, cement, and other products, and also in the steel industry. Moreover, not all materials used by industry are composed of nonmetallic minerals; chromite, for example, is used extensively in making refractory bricks; manganese ores are used for making dry batteries; and bauxite, the chief ore of aluminum, is used for making aluminous abrasives and aluminous refractories. Many rocks, such as limestone, slate, and granite, are nonmetallic deposits, but these are somewhat generally designated as bodies of the particular rock, as a limestone deposit, a slate deposit, etc.

Mineral deposits are characterized by various forms or shapes. Some occur as beds, some occur along beds as lenses or irregular bodies, many fill fractures of various types or follow other structural features, some make pipelike bodies, some are disseminated. These features are considered further in connection with the localization of deposits.

Many factors determine whether a deposit may be profitably extracted: the amount of valuable mineral present; the presence or lack of impurities; whether some materials can be obtained as by-products or co-products, such as gold or silver from ores of copper, lead, or zinc; whether material is at or very near the surface so that it can be removed by open-pit mining or whether it is some distance beneath the surface and must be removed by underground mining; favorable or unfavorable structural conditions for mining; the physical characteristics of certain materials; distance from market; current market price; ease or difficulty of metallurgical treatment of metallic ores; cost of labor; and many others. These are matters that fall in the fields of mining, milling, metallurgy, and mineral economics and cannot be treated here. One should realize, however, that many factors may determine the possibility of utilizing a particular deposit.

Mineral deposits are unevenly distributed throughout the world, and also throughout most countries in which they occur. Thus much of the world's molybdenum has been produced in the United States from a single deposit at Climax, Colorado; most of the diamonds have come from Africa; much platinum has come from Africa and also from Canada, the Canadian production chiefly from the deposits around Sudbury, Ontario. Further, certain kinds of deposits have been formed during certain geologic times, such as highly important gold deposits in the very old Precambrian rocks and important iron deposits in considerably younger Precambrian rocks throughout the world, or important deposits of copper, silver, lead, zinc, and a number of other metals during early Tertiary time. These are but a few examples. The reasons for the distribution of deposits in space and time are many, and the problem is complex. It has its roots in concepts of the origin of the earth, the ultimate source of the constituents of which mineral deposits are composed, and the manner of their redistribution—matters beyond the scope of an introductory treatment of mineral deposits.

Kinds of mineral deposits

The formation of a mineral deposit is an episode or a series of episodes in the geologic history of the region in which it occurs, and the kind of deposit formed is dependent on the geologic history of the region. The broad, general picture of the history of the continents, after initial formation, has been one of erosion of material in the hinterland and deposition in geosynclines, followed by orogeny with which, locally or regionally, there was associated metamorphism and igneous intrusion; and this may have been repeated several times. Mineral deposits may be grouped in three broad categories that are associated with geologic history of this kind: (1) those formed by igneous activity; (2) those formed by weathering, erosion, the action of groundwater, and sedimentary deposition, conveniently grouped together as deposits formed by surface agencies; and (3) those formed by metamorphism. The tabulations that follow are for the purpose of presenting concisely the various kinds of mineral deposits, their relationships, and the general manner in which they have formed. They are discussed more fully in succeeding chapters. Selected examples of the various kinds of deposits are listed in the Index under the specific headings, such as magmatic, contact metasomatic, placer, etc.

Deposits Formed by Igneous Activity

Genetic kind of deposit	Relationship to contemporaneous igneous activity	Major features	Some materials obtained commercially
Magmatic	Formed essentially during ordinary rock-forming stage	Form disseminations or segregations of valuable minerals within contemporaneous igneous body, or dikelike masses around margins of such body	Chromite, native platinum, some magnetite and ilmenite; diamond; corundum
Pegmatitic	Pegmatite bodies formed near the end of ordinary rock-forming stage, by low-melting and volatile constituents from magma chamber	Form small masses and dikes associated with contemporaneous igneous bodies; commonly of granitic composition; some minerals of very large size	Lithium-bearing minerals, columbite-tantalite, beryl, various minerals containing rare chemical elements; mica, feldspar, some quartz
Hydrothermal	Formed after consolidation of contemporaneous igneous rock, by solutions from deep magma chamber; temperature range is great	Fill fractures and various openings; also form replacements in consolidated contemporaneous igneous rock and in other rocks of the area	Source of many metals
Pneumatolytic or pneumotectic*	Form near end of ordinary rock-forming stage; may overlap in part pegmatitic and high-temperature hydrothermal stages	More generally associated with granitic rocks or pegmatites	Possibly some tin, tungsten, and molybdenum minerals
Carbonatites†	Apparently emplaced late in the history of certain alkalic rocks	Veinlike or dikelike masses and cores in contemporaneous volcanic plugs	Niobium (columbium); possibly minerals of various rare chemical elements

* Opinion is divided regarding the use of *pneumatolytic* or *pneumotectic* as a separate genetic kind of deposit. Generally *pneumatolytic* relates to materials formed wholly or in part from gaseous emanations, whereas *pneumotectic* relates to processes and products in which strictly magmatic material was modified and to some extent controlled by gaseous constituents. This is discussed further in Chap. 3.

† Carbonatites are carbonate-rich rocks genetically related to alkalic magma. Disagreement exists regarding their origin and exact place in the kinds of ore deposits. This is discussed further in Chap. 3.

Deposits Formed by Surface Agencies

Genetic kind of deposit	Origin	Major features	Some materials obtained commercially
DEPOSITS FORMED CHIEFLY BY WEATHERING AND EROSION, MECHANICAL OR CHEMICAL TRANSPORTATION, AND DEPOSITION IN STREAMS, IN BODIES OF STANDING WATER, OR ON THE SURFACE OF THE LAND			
Placers or alluvial deposits	Accumulation of heavy and chemically resistant material along stream valleys and beaches	Valuable minerals associated with sands and gravels; may be elevated or buried after formation	Much native gold; native platinum; cassiterite; some other metallic minerals; diamonds; monazite; several other industrial minerals
Sedimentary beds	Mechanical or chemical accumulation of material deposited as, or in, sedimentary beds at the time of their formation; includes evaporites	Occurrence as sedimentary beds or as lenses or disseminations in such beds if formed essentially at same time as deposition of beds	Sand, gravel, shale, some clay, limestone, phosphatic deposits, evaporites (gypsum, halite, potash salts, some others); important iron and manganese deposits; possibly copper and some other metals
DEPOSITS FORMED CHIEFLY BY WEATHERING AND THE ACTIVITY OF GROUNDWATER			
Residual	Residue from weathering of rocks, essentially *in situ*	Form blanket or pocket deposits at or very near present surface, or along buried unconformities	Bauxite and clay; iron and iron-nickel-cobalt; some nickel; manganese; some barite
Oxidation and supergene enrichment	Weathering and action of groundwater on previously existing deposit	May have residual capping, beneath which may be oxidized and enriched deposits and, beneath them, sulfide enriched deposits; not all divisions always present	Copper, silver, lead, zinc, and some others, as enrichments of previous deposits
Groundwater	Solution and transportation of material by groundwater, possibly for considerable distances, and subsequent deposition in favorable environment	Lenslike deposits in sedimentary beds, in places transecting bedding	Some uranium and vanadium

Deposits Formed by Metamorphic Activity

Genetic kind of deposit	Origin	Major features	Some materials obtained commercially
Contact meta-morphic and con-tact metasomatic	Metamorphism and metasomatism (replacement) around margins of contemporane-ous igneous in-trusion	Commonly replace limestone or some carbonate rock; much material added from mag-matic source	Tungsten; some copper, iron, and other metals; some nonmetals
Regional meta-morphic	Regional meta-morphism, usually associated with orogeny	Formed by rear-rangement and recombination of materials con-tained in rock that was meta-morphosed*	Various industrial deposits, as silli-manite, kyanite, slate, marble; possibly recon-centration of some metals

* Some disagreement exists regarding the cause of regional metamorphism and just how it is produced. Also, some people attribute the concentration of various metallic deposits to remobili-zation and further concentration of materials that were present in rocks affected by regional metamorphism. This is discussed further in Chap. 4.

Localization of mineral deposits

Within any particular region in which mineral deposits are formed, what determines just where the deposits will be localized and where, within a deposit, the higher values will occur? These questions have been the subject of investigation for many years, and although they are still not completely answered, many parts of the jigsaw puzzle have been put in place. Part of the difficulty is that almost every deposit seems to possess its own individuality; yet certain principles of general application have emerged. We now turn our attention to a consideration of those principles that apply to the localization of *epigenetic deposits* and to the localization of the richer parts of such deposits.

Emmons (19) stated that a great many, perhaps most, epigenetic deposits related to igneous activity were formed in connection with batholiths of general granitic composition. Further, since the deposits were formed by ascending solu-tions, they accumulated in the upper parts of the batholith—the hood area (Fig. 2-1)—especially in the small protuberances that formed stocks on solidification of the magma. These upward bulges, or cupolas, are therefore favorable places for ore deposition, and the deposits may be in the cupola or stock, or in fractures in the surrounding rock that composes the roof of the batholith.

More than fifty years ago Irving (33) wrote:

The fact that all portions of mineral veins or ore deposits are not equally rich and that the valuable or workable areas form but limited portions of the entire deposit is well recognized. . . . Where a segregation of higher

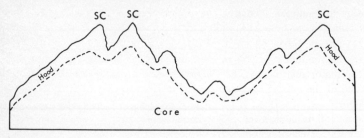

Fig. 2-1 Section through an ideal batholith. Mineral deposits tend to occur in the hood, especially in the summit cupolas (SC). After W. H. Emmons, "Ore Deposits of the Western States," p. 330. Copyright 1933. The American Institute of Mining and Metallurgical Engineers. Used by permission.

values in surrounding, or intervening, low-grade material is observable in an ore-body, the high-grade mass, if of sufficient size and sufficiently regular form, is usually termed a shoot.[1]

About ten years later Bateman (31) wrote:

Ore is where it is, not purely by chance, but by the result of definite processes which operated under certain conditions within the earth. . . . For example: Is a certain ore deposit located in a particular place because of a preexisting fissure, or a favorable rock formation, or its proximity to an igneous intrusion, or to a combination of all of them?[2]

Specific epigenetic deposits, and the ore shoots within them, are localized in one particular place rather than another because of the total physical and chemical environment, the structural features present, and the timing of events. Space for deposition of material is highly important. Space may result from original porosity of a rock, as in some sandstones and conglomerates, or from induced porosity resulting from earth movements, solution, alteration, or other causes. The chemical character of a rock determines the manner in which it will react as solutions come in contact with it, the ease or difficulty of its replacement by particular solutions, and the possibility of precipitation of material from solution. Consequently the order in which rocks are encountered by solutions, that is, the stratigraphic succession or the sequence of rocks of all types, is of considerable importance. Briefly, then, rock character, rock sequence, and structural features are the important factors that help localize mineral deposits, aside from the character, temperature, and rate of cooling of the ore-bearing solutions, and the length of time during which they pass through rocks. Here we will be concerned with rock character, rock sequence, and structural features. Although one of these could be the sole cause of localization of a deposit, commonly the localization results from a combination of two or all three of them.

[1] From R. D. Irving, *Economic Geology*, vol. 3, pp. 143–145. Copyright 1908. Economic Geology Publishing Company. Used by permission.
[2] From A. M. Bateman, *Economic Geology*, vol. 14, pp. 640–641. Copyright 1919. Economic Geology Publishing Company. Used by permission.

Much has been written about the localization of mineral deposits. Some of the material (17, 18, 21) deals with the localization of districts and the regional controls; some (19) deals with the localization around batholiths; but much of the material deals with the local controls of rock character and structure, especially the latter. Only the major features can be discussed here, but some useful general references are listed at the end of the chapter. These have been used freely in the following discussion.

Descriptions of many important mineralized areas show conclusively that a considerable amount of disturbance occurred at the general time of mineralization, commonly before, during, and after ore deposition. Newhouse stated (28):

> *It is the writer's belief that ore introduction goes on in the majority of ore deposits synchronously with certain stages of deformation which produce the structural features in which the ore occurs.*
>
> *The epigenetic ore deposits which have been closely studied appear rarely to have been introduced into a dead or static structural feature regardless of whether it is a fault or a fold, and whether the ore forms a filling or a replacement. The evidence for repeated faulting or intermineralization fracturing is widespread in the description of ore deposits . . . and has been emphasized widely in the literature.*[1]

Faults, fractures, and shear zones that come into existence during disturbances may be mineralized, and various types of veins formed.

A *vein,* for the purpose of description, may be considered to be "a tabular body of mineralized matter localized by a fracture or fracture zone, regardless of whether the vein matter was introduced by filling, by replacement or by a combination of the two processes."[2] (26)

Veins of numerous types are mentioned in the literature. Thus one encounters *fissure veins* (formed by fissure filling); *lode fissures* (formed by replacement); *linked veins* (adjacent roughly parallel veins connected by diagonal veins); *lenticular veins* (lens shaped); *ladder veins* (fillings of short transverse fractures in dikes); *pitches and flats* (fillings along steeply inclined joints, the pitches, and along flat or bedding fractures, the flats); *breccia veins* (vein matter filling spaces around fragments of wallrock or other rock); *saddle reefs* (fillings in spaces at the top of a fold). The general term *lode* usually refers to several more or less parallel mineralized veins that are spaced closely enough to be mined together rather than singly. Similarly, although not a type of vein, a *stockwork*[3] might be included here because commonly it is a mass of rock cut by many small interlacing veins.

Veins are localized in faults, shear zones, and fractures (Fig. 2-2). Commonly the veins are not in a major fault of a region, but are in smaller faults, shears, and fractures close to the major fault. Thus the mines of the great Kirkland Lake–

---

[1] From W. H. Newhouse (ed.), "Ore Deposits as Related to Structural Features," p. 34. Copyright 1942. Princeton University Press. Used by permission.

[2] From H. E. McKinstry, *Economic Geology,* Fiftieth Ann. Vol., pt. 1, p. 175. Copyright 1955. Economic Geology Publishing Company. Used by permission.

[3] A stockwork has been defined as "an ore deposit of such a form that it is worked in floors or stories. It may be a solid mass of ore, or a rock mass so interpenetrated by small veins of ore that the whole must be mined together" (5, p. 650). In some respects, stockworks resemble disseminated deposits.

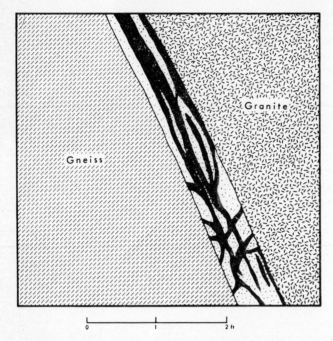

Fig. 2-2 Section to show sphalerite (solid black) filling fractures in crushed granite (stippled) and forming a vein in fault zone between gneiss and granite, Georgetown area, Colorado. After J. E. Spurr and G. H. Garrey, *U.S. Geol. Survey Prof. Paper* 63, p. 192.

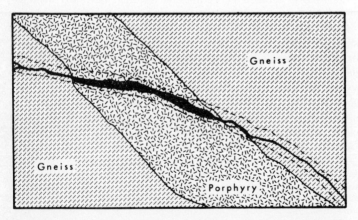

Fig. 2-3 Plan to show thickening of vein on passage from gneiss into porphyry, which fractured more readily than the gneiss, Georgetown area, Colorado. Boundaries of fracture zone shown by dashed line. After J. E. Spurr and G. H. Garrey, *U.S. Geol. Survey Prof. Paper* 63, p. 237.

Larder Lake gold area of Ontario, Canada, are close to major faults, and the copper deposits of northern Michigan are close to the large Keweenaw fault. Many writers have stressed the fact that ore bodies are related to minor faults and fractures that are associated with larger faults.

Localization of ore in particular places along a fault, and the formation of ore shoots, may result where faults cut across contacts of dissimilar rocks or pass through several rocks that are physically different or lie along contacts of dissimilar rocks. Faults commonly are deflected with passage from one rock to another, and openings made available for the localization of ore shoots. Deflection of this type is common in the Mother Lode system of California on passage from slate to greenstone or from greenstone into slate, where veins have been formed along reverse faults. Also, the width of a vein may change in passing from one rock type to another (Fig. 2-3), because one rock fractures readily whereas the other one tends to flow. At the Dunkin mine of the Breckenridge district, Colorado, the Dunkin vein, which formed along a fault of small vertical displacement, was productive where the vein is almost entirely within a thick monzonite porphyry sill, but nearly barren a short distance after it passed into shale (Fig. 2-4). In the same mine several lenticular ore shoots are separated by barren ground, and the vein passed into a tight and narrow seam of gouge. The contact of two unlike types of rock is one of the most common features with which ores are associated (28, p. 49). Such contacts commonly are weak zones that localize faulting or fracturing (Fig. 2-5).

Types of ore bodies, in addition to veins, are (1) *disseminated deposits,* (2) *concordant deposits,* and (3) *pipes* and *mantos.*

Disseminated deposits include "bulky bodies of no very uniform or consistent shape in which minor amounts of sulphide occur in scattered grains or along small closely spaced fractures within a large volume of rock."[1] (26) Many highly important copper deposits, known as "porphyry coppers," are of this type.

Concordant deposits are localized in certain beds or groups of beds. These include deposits in which pores or interstitial openings in the beds have been filled, those in which there has been some replacement, and those in which there has been extensive replacement. The deposits may be clearly related to hydrothermal activity, as certain deposits at Bingham, Utah, and at Leadville, Colorado; or they may belong to the great group of stratiform deposits in sedimentary rocks, some of which have been designated *stratabound deposits.* Much controversy exists about the origin of some of them, such as the Mississippi Valley and similar lead-zinc deposits, the deposits of the African copper belt, and others.

Pipes and mantos are similar but are differentiated because of position. Ore bodies that are roughly circular or oval in cross section are termed *pipes* or *chimneys* if they are vertical or steeply inclined. *Mantos* are generally more or less horizontal blanketlike ore bodies.[2] Pipes and mantos may be interconnected,

---

[1] From H. E. McKinstry, *Economic Geology,* Fiftieth Ann. Vol., pt. 1, pp. 190–191. Copyright 1955. Economic Geology Publishing Company. Used by permission.

[2] "*Manto* means blanket and should, literally speaking, be reserved for tabular bedded replacements. But in Santa Eulalia and other Mexican occurrences it is used for horizontal tubes along bedding planes in contradistinction to steep *chimneys.* . . ." (26) From H. E. McKinstry, *Economic Geology,* Fiftieth Ann. Vol., pt. 1, p. 207. Copyright 1955. Economic Geology Publishing Company. Used by permission.

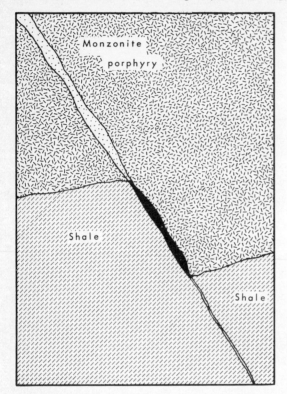

Fig. 2-4 Section to show rich gold ore (solid black) at contact between shale and monzonite porphyry, with a seam of quartz and pyrite in the shale below the gold ore, and oxidized lead ore (stippled) in the porphyry above the gold ore, Breckenridge district, Colorado. After T. S. Lovering, *U.S. Geol. Survey Prof. Paper* 176, p. 39.

especially in limestone, and the pipes appear to be feeders for the mantos. Some of these structures are large, the pipes as much as 300 ft in diameter and several hundred to several thousand feet in the direction of elongation, the mantos hundreds and even thousands of feet approximately along the bedding. Not all pipes are associated with mantos. Some are breccia pipes that have been interpreted as features formed by corrosive action of volcanic emanations followed by local cracking resulting from change in volume. Volcanic necks may become

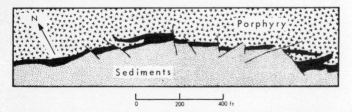

Fig. 2-5 Plan to show ore (solid black) concentrated along sheared and brecciated zone between porphyry and sediments, 350-ft level, Jerome mine, Ontario, Canada. Somewhat simplified after W. L. Brown, "Structural Geology of Canadian Ore Deposits," Jubilee Volume, p. 439. Copyright 1948. The Canadian Institute of Mining and Metallurgy. Used by permission.

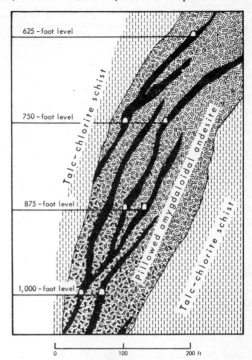

Fig. 2-6 Section to show veins (solid black) emplaced in andesite but not in schist, Aunor Gold Mine, Ontario, Canada. After B. S. W. Buffam, "Structural Geology of Canadian Ore Deposits," Jubilee Volume, p. 512. Copyright 1948. The Canadian Institute of Mining and Metallurgy. Used by permission.

mineralized, or ore bodies may form on or near their borders, making pipelike deposits, and some of these have been interpreted as resulting in part from collapse of material after faulting or corrosion.

Many attempts have been made to class various rocks as favorable or unfavorable for ore deposition, but this can be done only in a general way, because rocks that are favorable in one district may not be so in another one (25, pp. 279–289). Very commonly the matter of favorability depends on the types of rock that are in contact in a particular region. In a general way, competent, strong, brittle rocks break in such a manner that openings favorable for ore deposition occur (Fig. 2-6), whereas incompetent rocks tend to yield plastically or by flowage and thus do not provide openings favorable for ore deposition. Yet some incompetent rocks have yielded by shearing between competent beds and have been favorable places for ore deposition. In the Breckenridge district, Colorado, the quartzites were favorable for deposition because of the manner in which they fractured, whereas the shales were not (Fig. 2-7). Breccias, however they are formed, commonly are favorable sites for deposition (Fig. 2-8).

Limestones are somewhat generally considered to be favorable rocks, and certainly numerous ore deposits have been localized in limestones. This favorability is in part a result of the chemical character of limestone, which is easily replaceable and which may be porous because of solution; but porosity may be due also to its physical character, since it may act as a competent rock and fracture readily. The physical characteristics of favorable and unfavorable limestones and dolomites were investigated experimentally by Rove (34), who con-

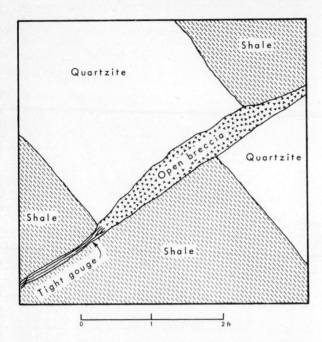

Fig. 2-7 Section to show open breccia where a fault is in quartzite and tight gouge where it is in shale, Breckenridge district, Colorado. After T. S. Lovering, *U.S. Geol. Survey Prof. Paper* 176, p. 58.

cluded that secondary deformational openings are more important than primary intergranular openings in causing permeability. Moreover, heating of limestone may cause permanent expansion and greatly increased permeability, as shown by experiments conducted at Princeton University (23). This increased permeability would permit entry of solutions and hence should greatly aid in replacement.

Localization of ore in fractures and in chemically favorable rocks may be aided by various types of impermeable barriers, which may be impervious rocks or fault gouge. In the Leadville district, Colorado (32, pp. 188–189),

> the blanket ores . . . nearly all lie at horizons where a readily replaceable rock is overlain by a relatively unreplaceable and impermeable rock. The replaceable rock is everywhere or nearly everywhere calcareous; the overlying rock may be shale, quartzite, or other sediment, but that above most of the important blankets is a sheet of porphyry, and the number of horizons favorable for ore varies principally with the number of porphyry sheets. The rocks at favorable horizons are, of course, not everywhere mineralized; suitable structure, and especially the presence of fissures through which the ore-bearing solutions may reach these horizons, are essential to the formation of large bed deposits.

The relationships in the Leadville district are well shown by Fig. 2-9. The effects of faults, shear zones, and an impermeable shale barrier in localizing ore in the East Shasta copper-zinc district, California, are shown by Fig. 2-10.

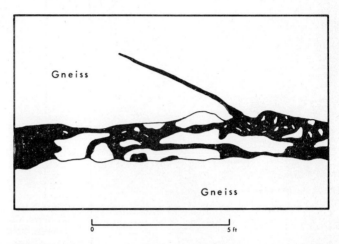

Fig. 2-8 Plan to show sphalerite (solid black) in a brecciated zone in gneiss, Georgetown area, Colorado. After J. E. Spurr and G. H. Garrey, *U.S. Geol. Survey Prof. Paper* 63, p. 228.

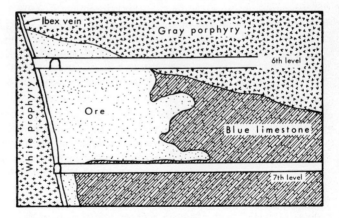

Fig. 2-9 Section to show replacement of limestone beneath porphyry, Leadville district, Colorado. After S. F. Emmons, J. D. Irving, and G. F. Loughlin, *U.S. Geol. Survey Prof. Paper* 148, p. 297.

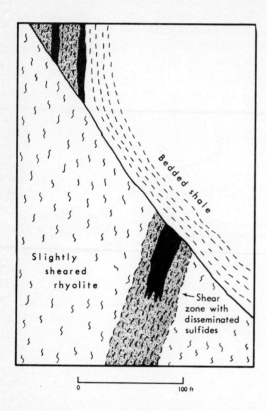

Fig. 2-10 Diagrammatic section to show ore bodies (solid black) concentrated in shear zone in rhyolite, Afterthought mine, East Shasta copper-zinc district, California. After J. P. Albers and J. F. Robertson, *U.S. Geol. Survey Prof. Paper* 338, p. 72.

Various other features related to faults and fractures, such as intersections of fractures, fractures ending against a premineral fault and impermeable rock, changes in strike or dip, or splitting of faults, all have contributed to the localization of ore at one place or another (Fig. 2-11). Structural features and rock character also cause localization of ore by the action of descending surface waters (Fig. 2-12).

Folds have been important in the localization of ore. In some districts, as in the Lake Superior iron region, synclines, commonly of complex, plunging type in which relatively impervious rocks underlie iron formation, have provided troughs in which downward-moving waters concentrated the iron ore (Fig. 2-13). In many other places, where ore was brought in during or after folding, faults and fractures have been an important factor in localizing ore (28, p. 39). This is well shown in some of the saddle reefs of the Ballarat and Bendigo gold districts, Australia (Figs. 2-14, 2-15).

The major features responsible for the localization of ore deposits and ore shoots discussed in the foregoing pages are (1) faults, fractures, and shear zones; (2) contacts between unlike rocks; (3) favorable and unfavorable rocks; (4) impermeable barriers; (5) folds. Various combinations of these features have accounted for the localization of many ore deposits and ore shoots.

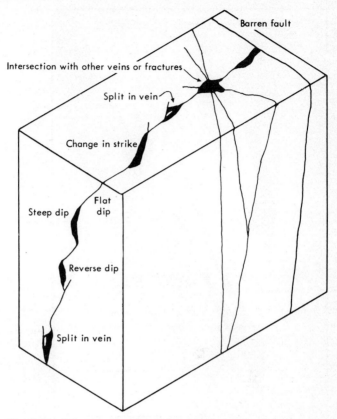

Fig. 2-11 Diagram showing influence of structural features on the development of ore shoots along veins, La Plata district, Colorado. After E. B. Eckel, *U.S. Geol. Survey Prof. Paper* 219, p. 78.

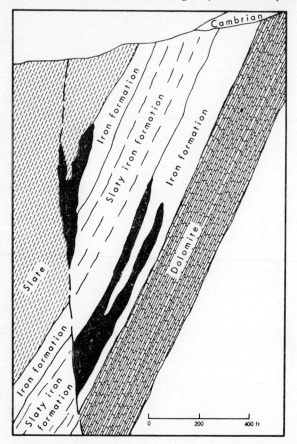

Fig. 2-12 Section to show concentration of iron ore against impermeable rock along a fault, Menominee iron district, Michigan. Slightly modified after Stephen Royce, "Ore Deposits as Related to Structural Features," p. 59. Copyright 1942. Princeton University Press. Used by permission.

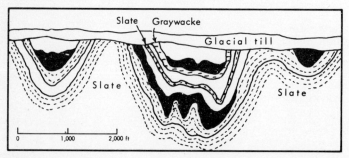

Fig. 2-13 Section to show influence of synclines and impervious rocks (slate and graywacke) in concentrating iron ore (solid black), Menominee iron district, Michigan. Slightly modified after Stephen Royce, "Ore Deposits as Related to Structural Features," p. 59. Copyright 1942. Princeton University Press. Used by permission.

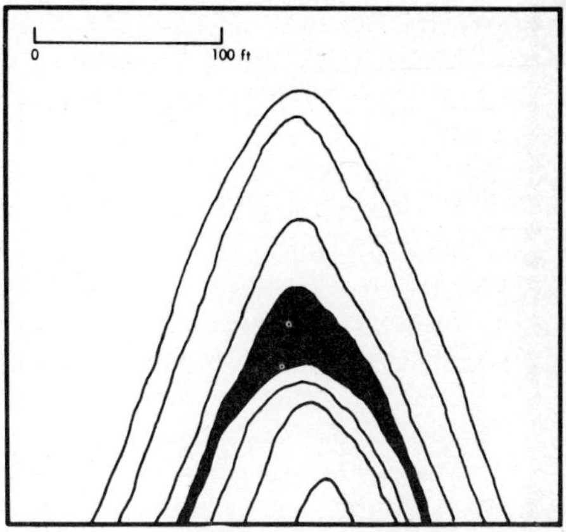

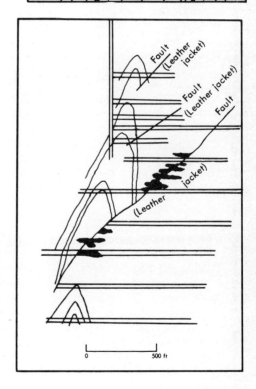

Fig. 2-14 Section to show saddle reef (solid black), Bendigo goldfield, Australia. Simplified after D. E. Thomas, 1953, in "Geology of Australian Ore Deposits," p. 1020, Fifth Empire Mining and Metallurgical Congress, Australia and New Zealand. Australasian Institute of Mining and Metallurgy.

Fig. 2-15 Section to show anticlinal structure and associated faults (leather jackets because of leathery character of accompanying crushed rock). Gold-bearing quartz (solid black) commonly was associated with the leather jackets. First Chance Mine, Ballarat gold field, Australia. Generalized and simplified after W. Baragwanath, 1953, in "Geology of Australian Ore Deposits," p. 998, Fifth Mining and Metallurgical Congress, Australia and New Zealand. Australasian Institute of Mining and Metallurgy.

GENERAL REFERENCES

1. A. M. Bateman, 1950, "Economic Mineral Deposits," 2d ed., John Wiley & Sons, Inc., New York, 916 pp.

2. R. L. Bates, 1960, "Geology of the Industrial Minerals and Rocks," Harper & Row, Publishers, Incorporated, New York, 441 pp.

3. F. W. Clarke, 1924, Data of geochemistry, *U.S. Geol. Survey Bull.* 770, 841 pp.

4. W. H. Emmons, 1940, "The Principles of Economic Geology," 2d ed., McGraw-Hill Book Company, New York, 529 pp.
5. A. H. Fay, 1920, A glossary of the mining and mineral industry, *U.S. Bur. Mines Bull.* 95, 754 pp.
6. A. A. G. Schieferdecker (ed.), 1959, "Geological Nomenclature," Royal Geological and Mining Society of the Netherlands, Gorinchem, Netherlands, 533 pp.
7. "Glossary of Geology and Related Sciences," 1957, The American Geological Institute, Washington, D.C., 325 pp.
8. "Glossary of Geology and Related Sciences, Supplement," 1960, The American Geological Institute, Washington, D.C., 72 pp.
9. "Industrial Minerals and Rocks," 3d ed., 1960, The American Institute of Mining, Metallurgical, and Petroleum Engineers, New York, 934 pp.
10. W. Lindgren, 1933, "Mineral Deposits," 4th ed., McGraw-Hill Book Company, New York, 930 pp.
11. B. Mason, 1958, "Principles of Geochemistry," 2d ed., John Wiley & Sons, Inc., New York, 310 pp.
12. H. E. McKinstry, 1948, "Mining Geology," Prentice-Hall, Inc., Englewood Cliffs, N.J., 680 pp.
13. F. J. Pettijohn, 1957, "Sedimentary Rocks," 2d ed., Harper & Row, Publishers, Incorporated, New York, 718 pp.
14. K. Rankama and T. G. Sahama, 1950, "Geochemistry," The University of Chicago Press, Chicago, 912 pp.
15. F. J. Turner and J. Verhoogen, 1960, "Igneous and Metamorphic Petrology," 2d ed., McGraw-Hill Book Company, New York, 694 pp.
16. H. Williams, F. J. Turner, and C. M. Gilbert, 1954, "Petrography," W. H. Freeman and Company, San Francisco, 406 pp.

SELECTED REFERENCES
REGARDING LOCALIZATION OF DEPOSITS

17. P. Billingsly and A. Locke, 1941, Structure of ore districts in the continental framework, *Am. Inst. Mining Engineers Trans.*, vol. 144, pp. 9–64.
18. B. S. Butler, 1933, Ore deposits as related to stratigraphic, structural and igneous geology in the western United States, in "Ore Deposits of the Western States" (Lindgren volume), pp. 198–240, The American Institute of Mining and Metallurgical Engineers, New York.
19. W. H. Emmons, 1933, On the mechanism of deposition of certain metalliferous lode systems associated with granite batholiths, in "Ore Deposits of the Western States" (Lindgren volume), pp. 327–340, The American Institute of Mining and Metallurgical Engineers, New York.
20. C. D. Hulin, 1929, Structural control of ore deposition, *Econ. Geology*, vol. 24, pp. 15–49.
21. C. D. Hulin, 1948, Factors in the localization of mineralized districts, *Am. Inst. Mining Engineers Trans.*, vol. 178, pp. 36–57.
22. R. A. Mackay, 1946, The control of impounding structures on ore deposition, *Econ. Geology*, vol. 41, pp. 13–46.

23. J. C. Maxwell and P. Verrall, 1953, Expansion and increase of permeability of carbonate rocks on heating, *Am. Geophys. Union Trans.*, vol. 34, pp. 101–106.

24. H. E. McKinstry, 1941, Structural control of ore deposition in fissure veins, *Am. Inst. Mining Engineers Trans.*, vol. 144, pp. 65–95.

25. H. E. McKinstry, 1948, "Mining Geology," Prentice-Hall, Inc., Englewood Cliffs, N.J., 680 pp. See pp. 277–289, Stratigraphic and lithologic guides; pp. 290–327, Fracture patterns as guides; pp. 328–342, Contacts and folds as guides.

26. H. E. McKinstry, 1955, Structure of hydrothermal ore deposits, *Econ. Geology,* Fiftieth Ann. Vol., pt. 1, pp. 170–225.

27. Various authors, L. C. Graton (chairman), 1941, Some observations in ore search, *Am. Inst. Mining Engineers Trans.*, vol. 144, pp. 111–146. This is a symposium that presents a discussion of three questions regarding (1) structural deformation, (2) chemical and physical characteristics of certain beds within a limestone formation, and (3) the probable bottom of an ore deposit.

28. Various authors, W. H. Newhouse (ed.), 1942, "Ore Deposits as Related to Structural Features," Princeton University Press, Princeton, N.J., 280 pp. See especially material by W. H. Newhouse, pp. 9–53, which is a summary of the structural features associated with the ore deposits described.

29. Various authors, M. E. Wilson (chairman), 1948, "Structural Geology of Canadian Ore Deposits," Canadian Institute of Mining and Metallurgy, Mercury Press, Montreal, 948 pp. This is a symposium dealing with the structure of various sections of Canada and the controls that caused the localization of many specific deposits.

30. E. Wisser, 1941, The environment of ore bodies, *Am. Inst. Mining Engineers Trans.*, vol. 144, pp. 96–110.

MISCELLANEOUS REFERENCES CITED

31. A. M. Bateman, 1919, Why ore is where it is, *Econ. Geology,* vol. 14, pp. 640–642.

32. S. F. Emmons, J. D. Irving, and G. F. Loughlin, 1926, Geology and ore deposits of the Leadville mining district, Colorado, *U.S. Geol. Survey Prof. Paper* 148, pp. 188–189.

33. R. D. Irving, 1908, The localization of values or occurrence of shoots in metalliferous deposits, *Econ. Geology,* vol. 3, pp. 143–145.

34. O. N. Rove, 1947, Some physical characteristics of certain favorable and unfavorable ore horizons, *Econ. Geology,* vol. 42, pp. 57–77, 161–193.

chapter 3

Deposits Formed by Igneous Activity

Many mineral deposits result from the crystallization and differentiation of magma; hence it is well to summarize those processes.

Modern concepts assume that two broad types of original magmas, basaltic and granitic, are sufficient for the development of practically all varieties of igneous rocks and that differentiation and contamination of such magmas will yield diverse products. The approximate composition of magmas of these two types is indicated by the rock analyses shown in Table 3-1, compiled from Daly (8). This table, of course, does not show the true composition of the magma

TABLE 3-1.   Approximate Composition of Basaltic and Granitic Rocks*

Oxide	Basaltic rocks, percent†	Granitic rocks, percent†
SiO_2	48.91	70.67
TiO_2	1.21	0.37
Al_2O_3	16.08	14.28
Fe_2O_3	4.99	1.54
FeO	6.29	1.61
MnO	0.27	0.11
MgO	6.39	0.78
CaO	9.29	1.83
Na_2O	3.01	3.45
K_2O	1.41	4.17
H_2O	1.59	0.95
P_2O_5	0.42	0.15

* Compiled from "Igneous Rocks and the Depths of the Earth," by R. A. Daly. Copyright 1933. McGraw-Hill Book Company. Used by permission.

† The composition of the basaltic rocks was obtained by using the weighted average of 198 basalts and 41 gabbros, analyses 58 and 57 of Daly (p. 17); the composition of the granitic rocks by using the weighted average of 546 granites and 126 rhyolites, analyses 4 and 5 of Daly (p. 9).

from which the rocks were formed, because of the loss of volatile constituents, but it does show the major differences that probably existed in the compositions of the two types of magmas. The outstanding contrasts between the composition of basaltic and granitic rocks, as shown by the table, are the higher percentages of Fe_2O_3, FeO, MgO, and CaO in the basaltic ones and the higher percentages of SiO_2 and K_2O in the granitic ones, which reflect outstanding differences in the two types of magma. The actual temperatures of the magmas from which rocks crystallize are unknown, but estimates indicate that the temperature of basaltic magma might be around 1100°C and the temperature of granitic magma around 700°C.

Certain minerals commonly begin to crystallize before others as the temperature of a magma falls, but two or more minerals may begin to crystallize at the same time. Bowen (4, p. 59), concluded that crystallization of magma is dominated by reactions of early-formed crystals with the uncrystallized magma and that, although certain minerals appear in a particular order during crystallization, they also disappear in a similar order. Consequently he formulated the *reaction series,* and stated that, as the temperature of a magma declines, certain minerals crystallize and then react, in whole or in part, with the magma to form others, which in turn may react to form still others. Thus from basaltic magma olivine might form, react wholly or partly to form magnesium pyroxene, which in turn might react to form magnesium-calcium pyroxene, and this would be followed by other reactions that would yield amphibole and finally biotite. Soon after the olivine began to form, calcic plagioclase would begin to crystallize. It would react continuously and, as temperature fell, the plagioclase that formed would be progressively more sodic. Bowen termed this plagioclase feldspar series a *continuous series* because it is an isomorphous series in which reaction is continuous as temperature falls, and the other series a *discontinuous series* because the change from one mineral to another is by steps at appropriate temperatures and compositions of the remaining magma. The two series together he designated the *reaction series.* The relations of the reaction series postulated by Bowen are shown by the diagram on page 30.

The diagram as presented by Bowen has been slightly modified, merely for the purpose of discussion, by placing certain mineral groups within numbered areas. Area 1 represents the condition in which crystallization had proceeded sufficiently that some olivine had reacted to form magnesium pyroxene, and a highly calcic plagioclase had reacted to form a somewhat less calcic one. If, at this stage, solidification of the magma could be very quickly accomplished, the remaining liquid might form a glass or a very fine-grained aggregate of minerals in which there would be somewhat larger crystals of olivine, magnesium pyroxene, and calcic plagioclase, the composition of some basalts and gabbros. Area 2 represents an assumed condition in which the composition of the original magma was such that all of any original olivine that may have formed had reacted to form magnesium pyroxene, and in turn all of the magnesium pyroxene had reacted to form magnesium-calcium pyroxene, but only part of that mineral had been converted into amphibole. The plagioclase had reacted continuously until a more alkalic plagioclase had formed. If very rapid cooling were then accomplished, the resulting product would consist of magnesium-calcium pyroxene,

The Reaction Series of N. L. Bowen*

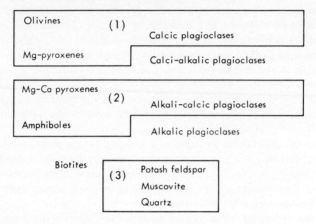

*Modified from N. L. Bowen, "The Evolution of the Igneous Rocks," p.60.
Copyright 1928. Princeton University Press. Used by permission.

amphibole, and alkali-calcic plagioclase in a glassy or fine-grained groundmass, the composition of some andesites and diorites. Similarly, products of crystallization might come to be amphibole, biotite, and alkalic plagioclase. All of this assumes, of course, proper starting compositions and exact control so that reaction could take place. *Area 3 is not part of either branch of the reaction series.* This should be emphasized, as confusion may exist regarding the minerals in that area. Bowen's original statement (4) is worth quoting.

> *It is to be noted that the minerals given at the end of the series after their convergence are on a somewhat different basis from the others. It can not, of course, be said that any liquid reacts with biotite and alkalic plagioclase to give potash feldspar. On the contrary these last minerals (principally potash feldspar and quartz) are those that form from the liquid, if any, which is left over at this final stage and are thus the result of the failure of this liquid to be used up in the reactions which produce biotite and alkalic plagioclase carrying some potash.*[1]

The minerals of area 3, therefore, are formed from liquid not used in the formation of minerals of the continuous and discontinuous branches of the series.

An important feature of the reaction series is that failure of a mineral to react with the remaining magma, either because the mineral sinks or is otherwise removed from contact with the magma, or because crystallization is too rapid, leads to the storing up of certain constituents in the final magma and also a lowering of the temperature of final crystallization. The net result of failure of reaction is a tendency to store up potash, soda, and silica, which would be used in forming muscovite, potash feldspar, albite, perthite,[2] and quartz.

[1] From N. L. Bowen, "The Evolution of the Igneous Rocks," p. 61. Copyright 1928. Princeton University Press. Used by permission.
[2] Perthite is an intergrowth of potash feldspar and soda feldspar.

Although the general course of crystallization of magma seems to be dominated by Bowen's reaction series, an alternative course appears to have been effective in the crystallization of some magmas of basaltic composition. The olivines and pyroxenes are not only members of Bowen's reaction series, but also members of isomorphous series of olivines and pyroxenes. Among the olivines, for example, one end member of the isomorphous series is forsterite, Mg_2SiO_4, which melts at 1890°C, and the other end member is fayalite, Fe_2SiO_4, which melts at 1205°C. Crystallization of a melt containing only the elements for these two end members, therefore, would proceed in the same manner as the crystallization of the plagioclase feldspars, but in the olivine series the first mineral to form would be rich in magnesia, and the reaction would produce olivines that contained progressively more iron. Failure to react would then tend to store up iron in the final product (3).

Magmatic differentiation[1] is brought about, at least in part, by failure of reaction during the crystallization of magma, which failure, in turn, may be promoted by the sinking of crystals in the magma, by earth movements that squeeze magma away from crystals that had formed (filter pressing), and in several other ways. (See Bowen, 5, for details.)

The processes that operate during ₁and immediately subsequent to the consolidation of a magma to form an igneous rock, and which bring about progressive changes in the character of some of the rock that crystallized early, and which lead to the formation of mineral deposits, have been discussed by many writers. Shand (25) has suggested certain stages and mineralogical events that are a useful means of summarizing the complex history of the consolidation of magma into igneous rock and the events that immediately follow the consolidation. His suggestions are shown by Table 3-2.

TABLE 3-2. Stages Reached during and subsequent to the Consolidation of Magma into Igneous Rock*

Stages	Events
MAGMATIC	
First magmatic	A stage when only anhydrous silicates (and possibly other anhydrous compounds) crystallize out of the magma
Second magmatic	A stage when hydroxyl-bearing silicates of low water content (mostly less than 5 percent) crystallize out of the magma in addition to anhydrous silicates

[1] Magmatic differentiation has been defined as "the process by which different types of igneous rocks are derived from a single parent magma, or by which different parts of a single molten mass assume different compositions and textures as it solidifies. . . ." (See reference 5, p. 413, of Chap. 2.) It has also been stated in such a manner that it includes the process by which magmatic ore deposits are formed and is regarded as the means of supplying solutions and vapors that form mineral veins. (See references 6, p. 295, and 7, p. 175, of Chap. 2.)

Table 3-2. Continued

Stages	Events
POSTMAGMATIC Deuteric† or high-temperature hydrothermal	A stage when the same hydroxyl-bearing silicates grow in the solid rock by replacement of earlier silicates
Low-temperature hydrothermal (or intermediate and low-temperature hydrothermal)	A final stage when all the previously formed silicates become unstable and are replaced by scaly, fibrous, or colloform decomposition products of higher water content (more than 10 percent) and by carbonates

* Reprinted from *Journal of Geology,* vol. 52, pp. 348–350, The Terminology of Late Magmatic and Post-magmatic Processes by S. J. Shand, by permission of The University of Chicago Press. Copyright 1944.

† The term *deuteric* relates to alterations produced in an igneous rock as a consequence of and during the later stages of magma consolidation. (See reference 7, p. 79, of Chap. 2.)

TABLE 3-3. Simplified Correlation of Stages of Magmatic and Postmagmatic Activity with Classes of Mineral Deposits*

Stages and subdivisions of igneous activity		Classes of mineral deposits	
Major stages	Subdivisions		
Magmatic	First magmatic	Magmatic	
	Second magmatic		
Postmagmatic	Deuteric or high-temperature hydrothermal	Pegmatitic	Pneumatolytic†
		High-temperature hydrothermal	
	Low-temperature hydrothermal (or intermediate and low-temperature hydrothermal)	Intermediate and low-temperature hydrothermal	

* Modified from E. E. Wahlstrom, "Introduction to Theoretical Igneous Petrology," p. 262. Copyright 1950. John Wiley and Sons. Used by permission.

† *Pneumatolytic* relates to the effects and agents of *pneumatolysis,* which has been stated to be a process whereby minerals are formed by emanation of gases and vapors from solidifying magma. Certain investigators, especially the Europeans, tend to limit pneumatolysis to temperatures between about 600 and 400°C. (See references 5, p. 524; 6, p. 297; and 7, p. 226, of Chap. 2.) The term *pneumotectic* has been used in connection with certain deposits similar to or the same as those designated *pneumatolytic* by some writers. The term was proposed by L. C. Graton and D. H. McLaughlin, 1918, *Econ. Geology,* vol. 13, p. 85.

This tabulation indicates that immediately after the consolidation of magma various postmagmatic changes take place as a continuation of the igneous history. During the postmagmatic stages emanations rise into and attack the previously formed rock. These emanations are thought to come from a deep magma chamber that contains the low-melting (hyperfusible) and volatile constituents that have accumulated as a result of prolonged magmatic differentiation. This material is designated *rest magma,* and from it certain mineral deposits may form. (See references 6, p. 296; and 7, p. 244, of Chap. 2.)

Attempts have been made to correlate the magmatic and postmagmatic stages with the stages of formation or classes of mineral deposits. Wahlstrom (26, p. 262) prepared a table showing various probable correlations, from which the simplifications shown in Table 3-3 was compiled.

Magmatic deposits

The magmatic mineral deposits form during the period of crystallization of the rock-forming minerals, that is, during the magmatic stage. In some of these deposits the ore minerals are more or less disseminated throughout the rock, in others they occur in rather definite layers or in dikelike masses. Magmatic deposits furnish but a small variety of metallic and industrial materials, but they are the chief original source of native metals of the platinum group, of chromite, diamond, and corundum, and also the source of some important iron ore, chiefly magnetite. The deposits of chromite and platinum metals occur in peridotite, pyroxenite, and anorthosite; the deposits of diamond in peridotite. The deposits of corundum and magnetite show more variable occurrences.

Four principal ideas have been proposed to account for the characteristics shown by magmatic deposits: (1) early crystallization of ore minerals, which either remained somewhat disseminated throughout the rock mass or became segregated into layers by settling; (2) segregation of ore material as a residual liquid or as an immiscible liquid, which crystallized in place and thus formed layers; (3) segregation of crystals of ore minerals into layers by settling, followed by melting and injection of that material as dikelike masses; (4) segregation of ore material as a residual or an immiscible liquid, followed by injection of that liquid. If the liquid was immiscible it was thought to consist of material that would form sulfides. In much of the earlier literature, magmatic deposits are designated *magmatic segregations,* and they were thought to have formed by settling of crystals of minerals such as chromite and magnetite. Observation of certain deposits, however, such as the magnetite deposits in the Kiruna area, Sweden, showed that at some places magnetite was injected, apparently as molten material. The idea was then proposed that the magnetite, after accumulation as a segregation of crystals, was remelted and intruded (15, p. 793). This concept was objected to, however, because the heat required to melt the magnetite seemed unlikely to have been available and, furthermore, if remelting had taken place the rock-forming minerals also should have melted. Consequently an alternative explanation was advanced that a fluid rich in iron remained as a residue after the crystallization of the usual rock-forming minerals. This fluid might have crystallized in the interstices of the rock-forming minerals and made disseminated deposits; it might have become segregated into layers and crystallized in place; or it might have been squeezed out and injected around the marginal parts of the rock mass.

Bateman (2) discussed the formation of magmatic deposits and proposed that they be classified as early-magmatic and late-magmatic deposits, the formation of which would correspond more or less to the first magmatic and second magmatic stages shown by Table 3-3. The early-magmatic deposits were considered to be accumulations of crystals, either as disseminations or as segregations, whereas the late-magmatic ones were considered to be accumulations of residual or immiscible liquids, which subsequently crystallized in place as segregations or were squeezed out and then crystallized as injections. Bateman admitted the possibility that some early-magmatic deposits might have been formed as injections of a crystal mush lubricated by some interstitial liquid, but he seemed inclined to think that all injections may belong in the late-magmatic group. The actual separation of magmatic deposits into these classes must be based on detailed field and laboratory study.

Pegmatitic deposits

The pegmatites, chiefly granite pegmatites, are important original sources of beryl, lithium minerals, columbite, tantalite, various minerals that yield rare chemical elements, feldspar, mica, and gem minerals. They also furnish, among other things, some cassiterite, wolframite, native bismuth and bismuthinite, molybdenite, and quartz.

The characteristics of pegmatites are somewhat variable. Commonly they form dikes and small, rather irregular pipelike or stocklike intrusions. Some are rather massive, others are zoned. They may be simple or complex, the simple type being composed chiefly of feldspar, quartz, and mica, with some accessory minerals, the complex type containing, in addition to the rock-forming minerals, many relatively uncommon to rare minerals. Many pegmatites are coarse-grained, individual crystals ranging in length from a few inches to almost 50 ft in exceptional cases. Many pegmatites, however, show much variability in grain size and arrangement. In some places the texture is that of graphic granite,[1] in others it is an admixture of graphic and granitoid textures, and in outcrops texture may change from relatively fine to relatively coarse in short distances. These textural variations have led to the concept that irregularity of grain size is one of the characteristics of pegmatites. Although granite pegmatite is the common type, most rock groups are represented among pegmatites. If the term *pegmatite* is used without modification, perhaps the usual intent is granite pegmatite. Confusion can be avoided by prefixing the name of the rock group, as syenite pegmatite, gabbro pegmatite, etc.

Many, many pages have been written about the origin of pegmatites. Jahns (12) recently discussed the study of pegmatites and cited 698 *selected* references

---

[1] The texture of graphic granite should be the typical texture of pegmatite if one adheres to the meaning of the name. The term *pegmatite* comes from *pegma,* alluding to something joined together, and relates to the type of texture caused by individuals that are mutually interpenetrating, which is the texture of graphic granite.

dealing with various studies that have been made. Only the most outstanding concepts can be summarized here.

An early concept ascribed the origin of pegmatites to segregation of material from surrounding country rock by groundwater,[1] but present concepts generally ascribe their origin either to igneous or to metamorphic activity. (See reference 14 for a review of theories proposed to 1933.) The metamorphic concept relates the formation of pegmatites to the process of granitization, and the growth of pegmatites in regionally metamorphosed rocks to a phase of granitization. According to this concept, transformation of other rocks into granite and pegmatite takes place without fusion of the rocks, magma is not formed, and the granite and pegmatite are essentially the end products of regional metamorphism. The igneous concept postulates that magma is formed, undergoes differentiation to such an extent that much volatile and low-melting material accumulates, and this material is injected into surrounding rocks and forms pegmatites.

Two major variations of the igneous concept are in vogue. One version is that the material was all injected into the rocks at one time and either crystallized soon afterward or remained fluid for some time, long enough to undergo further differentiation in place before complete solidification. This version assumes that essentially nothing was added from a deep magma chamber after the first injection of material and that crystallization proceeded, therefore, in a chemical system that was closed. A second version also assumes injection of differentiated material but postulates that, as crystallization proceeded, other material, usually more highly differentiated, was added from a deeper magma chamber and caused replacement, at least in part, of material that had crystallized earlier. *Simple pegmatites,* composed chiefly of feldspar, quartz, and mica, would represent material that had been injected and crystallized without further differentiation (first version) or without appreciable replacement (second version). *Complex pegmatites,* composed of notable amounts of less common minerals, would represent differentiated material that had been injected and then had undergone further differentiation in place during crystallization (first version) or much replacement of the material that crystallized early, as a result of addition of material (second version). Some replacement occurs in the outer parts of the pegmatites formed by further differentiation in place, by emanations from the central part of the chamber (first version), whereas replacement is of major importance in the second version.[2]

Zoned pegmatites are the most important source of mineral deposits. Cameron and others (7) described four types of zones in such pegmatites, which, from the margin inward, are (1) border zones, (2) wall zones, (3) intermediate zones, and (4) cores. Some pegmatites contain a number of intermediate zones. The zoning is not only lateral but also vertical, so that not all zones may be represented in a particular section through a pegmatite (Fig. 3-1). Further, not all zones are present in every pegmatite; zones pinch and swell, are asymmetrical or lenticular; and some cores are divided into segments. The various zones are characterized by somewhat distinctive mineral assemblages, as shown by Table 3-4.

[1] This was the process designated *lateral secretion* (7, p. 164, of Chap. 2).
[2] See reference 7, pp. 97–106, for a summary of pegmatites formed by the first version, and references 14, pp. 53–54, and 23 for summaries of pegmatites formed according to the second version.

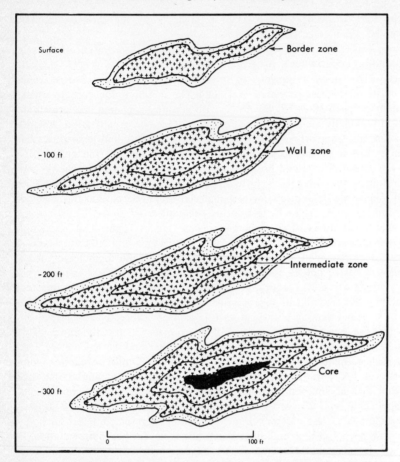

Fig. 3-1 Idealized horizontal sections of a zoned pegmatite, showing appearance at different levels. After E. N. Cameron and others, "Internal Structure of Granitic Pegmatites," *Econ. Geology Mon.* 2, p. 23. Copyright 1949. Economic Geology Publishing Company. Used by permission.

Although many valuable minerals are present in pegmatites, commercial production has come from only a relatively small number of bodies. Attempts to mine pegmatites have been discouraging in many cases, partly because of the small amount and scattered distribution of some valuable minerals, partly because of a lack of understanding of the distribution of minerals within a body. Work with zoned pegmatites now indicates that the estimation of grade and reserves[1] of at least some pegmatites is feasible (20), and lately attention has been given to methods of concentrating minerals scattered throughout pegmatites (1, 11, 18). Consequently some deposits previously thought not to be usable may come to be of commercial importance in the future.

[1] The ore reserve of a deposit generally is considered to be the tonnage of ore that can be calculated, although some of it may be inferred and not definitely proved to be present. See reference 16, pp. 470–478, for discussion.

TABLE 3-4.   Some Generalized Features of Zoned Pegmatites*

Zone	Thickness	Texture	Composition
Border	Usually a few inches; rarely 2 ft.	Aplitic to fine-grained pegmatitic.	Usually some combination of feldspar, quartz, and muscovite, possibly biotite, but not all may be present; more usual accessories are tourmaline, beryl, apatite, garnet.
Wall	Usually less than 10 ft, rarely more than 35 ft; some are asymmetric or discontinuous.	Generally coarser than border zone.	Common minerals are plagioclase, perthite, quartz, muscovite, tourmaline; less common minerals are biotite, apatite, columbite-tantalite, garnet, beryl. Important source of mica and beryl.
Intermediate (Number may be 0–5 or more.)	Single row of crystals to many feet, depending on size of body and thickness of wall zone; may be lenticular or merge.	Generally progressively coarser toward core.	*If no lithium present:* perthite, or combinations of plagioclase, quartz, perthite, muscovite, biotite. *If lithium present:* Above minerals plus one or more of amblygonite, spodumene, lepidolite.
Core	Depends on size of body and thickness of intermediate and wall zones; may be composed of segments.	Variable, but generally rather coarse; may contain some giant crystals.	Variable, but generally quartz and perthite; or quartz plus scattered perthite; or quartz, perthite, and plagioclase; tourmaline, beryl, or lithium-bearing minerals may be present.

* Compiled from E. N. Cameron and others, Internal structure of granitic pegmatites, *Economic Geology Mon.* 2. Copyright 1949. Economic Geology Publishing Company. Used by permission.

Hydrothermal deposits

The discussion of the crystallization and differentiation of magma in the first part of the present chapter indicates that hydrothermal deposits come into existence during the postmagmatic stage of an igneous intrusion as a result of expulsion of rest magma from a deep magma chamber. This material may enter fractures, pore spaces, and other openings, and form deposits as a result of cavity filling and replacement. Consequently the hydrothermal deposits are all of epigenetic origin and display a variety of forms, such as veins of many types, stockworks, pipes, mantos, concordant deposits, and disseminated deposits.

Hydrothermal deposits form throughout a temperature variation from about 50 to 500°C and a pressure variation from that existing at or almost at the surface of the earth to that several miles below the surface. Consequently, since both temperature and pressure influence deposition, the deposits formed contain a great variety of minerals. Nevertheless, a considerable amount of order exists, since some minerals that are deposited under conditions of high temperature and pressure, for example, would not form along with some others that are deposited at low temperature and pressure.

Attempts have been made to group together hydrothermal deposits formed under similar conditions of temperature and pressure, or depth beneath the earth's surface. These groups have been designated *depth-temperature* or *depth-zone* types of deposits. Lindgren (15, pp. 211–212) proposed the scheme shown in Table 3-5.

TABLE 3-5.   Some Types of Hydrothermal Deposits Proposed by Waldemar Lindgren*

Type of deposit	Depth of deposition and concentration	Temperature range, °C ±	Pressure
Epithermal	Slight	50–200	Moderate
Mesothermal	Intermediate	200–300	High
Hypothermal	Great†	300–500	Very high

* From "Mineral Deposits," 4th ed., by Waldemar Lindgren. Copyright 1933. McGraw-Hill Book Company. Used by permission.
† The modifying statement "or at high temperature and pressure" was added.

Graton (10, pp. 536–540) suggested two other types, the *leptothermal*[1] and the *telethermal*. The leptothermal deposits were to include those formed under conditions roughly intermediate between those needed for mesothermal and epithermal deposits, and the telethermal deposits were to include those formed near the surface by the very last energy available in the solutions.

How is one to know whether a hydrothermal deposit belongs in one of these various types? Lindgren (15) stated various minerals, textures, and structures that might be used as criteria for epithermal, mesothermal, and hypothermal deposits,

[1] *Leptothermal* is from the Greek *leptos,* implying moderate or subdued. *Telethermal* is from the Greek *tele;* far; these deposits have been referred to by some as the "far travelers," in allusion to the distance the solutions come from their source.

and these criteria, if used with proper discretion, are very helpful. Some minerals, however, form throughout a wide range of temperature and pressure, and some may owe their existence to the chemical environment in which they form rather than the temperature and pressure that exists. Descriptions of deposits in various parts of the world indicate, for example, that they were formed under conditions of high temperature but low pressure or at shallow depth, and thus do not appear to fit into the temperature-pressure scheme indicated in Table 3-5. This seems to have been the case in the highly important tin-silver deposits of Bolivia, in which there is evidence of crowding of minerals together, of telescoping of zones, and of tumultuous deposition. Regarding the telescoping of zones, Graton (10, p. 544) stated:

> *It seems probable that telescoping is especially likely to occur where the geothermal gradient is abnormally steep because of the near-surface cooling of hot effusive rocks that had but shortly preceded the introduction of the ores. In such instances the solutions would have moved upward for indefinite distances along "pre-heated" channelways, losing heat less rapidly than they would if passing through rocks where the usual thermal gradient prevails. In so far, therefore, as rate of heat-loss affects mineral precipitation, solutions under these circumstances would cause deposits characteristic of the deeper zones to be extended, or "stretched" as one might say, to abnormally shallow levels. But in close proximity to the surface, where the preceding effusives have become cold, the rate of heat-loss by the solutions will be so unusually rapid as to produce impetuous deposition which telescopes or crowds together kinds of minerals that should under more normal circumstances have been deposited successively at different places along the ascending pathway.*[1]

Deposits of this type that show telescoping of zones, Buddington (6, p. 209) designated *xenothermal,* "from the Greek *xeno* meaning strange, different, foreign, and hence suggestive of the peculiar textures for the normal high-temperature mineral assemblages involved, of the abnormal association of high temperature with shallow depth, and the 'telescoped' character of many of the deposits."[2]

The problem of the criteria that may be used to determine the particular hydrothermal type to which a deposit belongs is beyond the scope of the present discussion. Nevertheless, in a general way, deposits may be classed as probably low-temperature, moderate-temperature, or high-temperature types, as will be seen from subsequent discussions of various mineral deposits.

Because both physical and chemical conditions change from the source of mineralizing solutions outward and upward, some mineral deposits are zoned. Emmons (9) presented a "reconstructed vein system" from the surface downward. This was an ideal representation not seen anywhere but put together as a composite made from various deposits. Near the surface, and formed under conditions of relatively low temperature and pressure, were mercury, antimony, and certain gold-silver deposits; lower, in order of greater depth, certain deposits of silver, lead,

[1] L. C. Graton, *Econ. Geology,* vol. 28, p. 544. Copyright 1933. Economic Geology Publishing Company. Used by permission.
[2] A. F. Buddington, *Econ. Geology,* vol. 30, p. 209. Copyright 1933. Economic Geology Publishing Company. Used by permission.

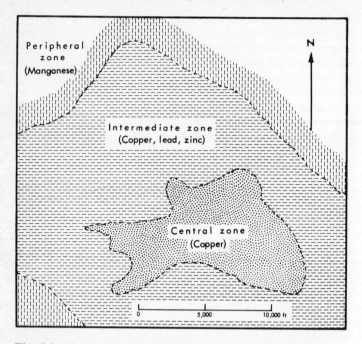

Fig. 3-2 Plan at elevation 4,600 ft, Butte district, Montana, to show zoning of the deposit. Generally silver increases in amount from the central to the peripheral zone. After Reno H. Sales, *Am. Inst. Min. Metall. Trans.*, vol. 46, p. 58. Copyright 1913. American Institute of Mining and Metallurgical Engineers. Used by permission.

zinc, and copper; and still lower, certain deposits of gold, arsenic, bismuth, tungsten, and tin.[1] This oversimplified the complex problem of mineral zoning. Nevertheless, certain parts of this succession appear in various deposits. Only two examples will be cited. The tin deposits of Cornwall, England, which have been mined since sometime before 1200, show a zoning from the deepest parts upward, which is (1) tin ore (cassiterite), (2) tin-copper ore, (3) copper-tin ore, (4) zinc-lead ore, with zinc more abundant than lead, and (5) lead-zinc ore, with lead more abundant than zinc, and silver ore possibly present (13, p. 105). The world-famous copper deposits of Butte, Montana, show a lateral zoning (Fig. 3-2) that consists of a central zone, an intermediate zone, and a peripheral zone. The central zone is characteristically a copper zone, in which some silver is present; the intermediate zone, although predominantly a copper zone, contains much more silver than the central zone and some lead, zinc, and manganese; the peripheral zone is one in which copper no longer is present in commercial amount and the values are chiefly in gold, silver, zinc, and manganese. Vertical zoning of the same type exists from deeper to shallower levels of the mines (22, pp. 10–11).

[1] Condensed from "The Principles of Economic Geology," by W. H. Emmons, 2d ed., p. 196. Copyright 1940. McGraw-Hill Book Company. Used by permission.

One further feature of hydrothermal deposits, wallrock alteration, deserves brief discussion. The rocks adjacent to mineralized veins and other types of hydrothermal deposits commonly show alteration in varying degrees. The alteration generally is most intense along the margins of veins or near the places of greatest ore mineralization and decreases away from such places. Further, the minerals formed in places of more intense alteration associated with a specific deposit may differ from those formed in places of less intense alteration. Thus at Butte, Montana, the mineralized veins, which occur in quartz monzonite, are surrounded by alteration envelopes derived from that rock. The ore veins were formed in fractures, and

> *the alteration minerals are arranged in zones parallel to the fracture. . . . Beginning with the vein itself and proceeding outward toward the unaltered quartz monzonite, one encounters successively zone A . . ., the sericitized zone, in which the predominant hydrothermal mineral is a white mica; and zone B, where clay minerals, principally of the kaolinite and montmorillonite groups, are dominant.*[1]

Moreover, the width of zones of alteration along vein walls generally increases with depth and laterally toward the central zone.

Several types of alteration are well known, and some of them are characteristically associated either with deposits of certain minerals or with deposits of certain hydrothermal types, such as hypothermal, mesothermal, and epithermal. Thus *greisenization*[2] is commonly associated with deposits of cassiterite and, to a less extent, deposits of wolframite. Further, *tourmalinization* may be common in such deposits, although it shows a wider range of association than does greisenization. Both are considered to be characteristic of relatively high-temperature deposits and are regarded by some to be hydrothermal alterations and by others to be pneumatolytic alterations. *Sericitization,* the process whereby minerals of a rock are replaced by sericite, is a very common type of alteration associated with mineral deposits, especially those formed at moderate temperatures. *Argillic alteration,* whereby the minerals of a rock are converted into various clay minerals, is also common at moderate to low temperatures. It may be associated with sericitization, as at Butte, Montana, where sericite is characteristic of more intense alteration and clay minerals are characteristic of less intense alteration. *Silicification* is also widespread and apparently takes place throughout a wide range of temperature. Replacement by quartz appears to be characteristic of higher-temperature silicification, replacement by chalcedony or opal, of lower-temperature silicification; but replacement by quartz may occur at low temperature also. A rock designated *jasperoid* is formed by silicification. It consists of very fine-grained quartz or

[1] From R. H. Sales and C. Meyer, *Am. Inst. Mining Engineers Trans.,* vol. 178, p. 12. Copyright 1948. The American Institute of Mining, Metallurgical, and Petroleum Engineers. Used by permission.

[2] *Greisen* is an altered granitic rock that generally consists largely of muscovite and quartz. Topaz is commonly present and locally may be the chief constituent. Lithium mica, tourmaline, and fluorite, rutile, cassiterite, or wolframite may be present. The term has been variously used and defined, and not uncommonly the origin is incorporated in the definition. (See references 6, p. 317, and 7, p. 131, of Chap. 2.) *Greisenization,* of course, is the process whereby rocks are converted into greisen.

chalcedony and generally was formed by replacement of limestone or dolomite. Other somewhat common types of alteration are *serpentinization,* most commonly affecting peridotites and pyroxenites; *chloritization,* characteristic of basalts, gabbros, and diorites; *albitization, alunitization,* and *zeolitization.* Albitization forms albite in a variety of igneous rocks. By alunitization various aluminous silicates are replaced by alunite, $KAl_3(SO_4)_2(OH)_6$, and by zeolitization various constituents of a rock are replaced by zeolites (stilbite, natrolite, and others). Another type of alteration that is characteristic of andesites and rocks of similar composition has been designated *propylitization.* This is the conversion of rocks into products composed chiefly of chlorite, sericite, epidote, carbonates, quartz, and disseminated pyrite. *Propylite,*[1] from which the term *propylitization* comes, was a name given to greenstonelike rocks in the Sierra Nevadas, which were at that time thought to be a new type of rock. Later it was shown that they are altered andesites, and the term *propylite* has rather generally been discarded, but *propylitization* is still used.

Why is a knowledge of wallrock alteration important in connection with a study of mineral deposits? The general concept is that, since it commonly is associated with hydrothermal deposits, it should furnish a guide to the location of ore bodies. Several geologists have studied this problem and offered suggestions regarding the use of wallrock alteration. Schwartz (24) stated:

> *Hydrothermal alteration because it is commonly pervasive in the vicinity of ore deposits furnishes a valuable guide in the exploration for new ore bodies. Epigenetic ores rarely occur in essentially unaltered country rock, and more or less zoning is characteristic of the alteration with the greatest intensity generally nearest the ore. Because of considerable differences in pressure, temperature, composition of solutions, composition of wall rocks, and the time factor, the use of hydrothermal alteration as a guide to ore is by no means simple.*[2]

Further discussion of this subject cannot be undertaken here, but even the student who is getting only an introduction to the study of mineral deposits cannot generally read descriptions of hydrothermal deposits without encountering at least some mention of wallrock alteration.

Pneumatolytic deposits

Discussion of pneumatolytic deposits has been deferred to the latter part of this chapter because, as indicated earlier, some confusion exists about what should be included in this category. Indeed, deposits that some desire to class as hydro-

---

[1] *Propylite* comes from the Greek *propolos,* referring to a servant who goes before one. The rock was so named because it was thought to be "the precursor of all other volcanic rocks, and its appearance on the surface inaugurated a grand revolutionary activity on the globe. It is this position, at the entrance as it were to a new era in the history of the earth, which has given rise to the name 'propylite.'" Reprinted from "A Descriptive Petrography of the Igneous Rocks," vol. 3, p. 177, by Albert Johannsen, by permission of The University of Chicago Press. Copyright 1937.

[2] From G. M. Schwartz, *Economic Geology,* Fiftieth Ann. Vol., pt. 1, p. 300. Copyright 1955. Economic Geology Publishing Company. Used by permission.

thermal, pegmatitic, or even magmatic, others wish to class as pneumatolytic. This is indicated by the position of the pneumatolytic deposits in Table 3-3.

The problem of the exact origin of deposits that have been considered to be pneumatolytic is beyond the scope of an introductory treatment of the origin of mineral deposits, but some comments are necessary for a better understanding of such deposits. Between the magmatic stage of ore formation, in which is included the formation of igneous rocks in the strict sense, and the hydrothermal stage, Niggli (19, pp. 2–6) placed the pegmatitic and pneumatolytic stage, which he regarded as a transition between the magmatic stage, characterized by silicate melts, and the hydrothermal stage, characterized by aqueous solutions. He discussed the separation of nonvolatile material from a magma, the vapor pressure as separation progressed, the igneous rock stage, the pegmatitic to pneumatolytic stage, and the hydrothermal stage (19, p. 3). Niggli postulated that, as crystallization progressed, a slight decrease in temperature would cause the separation of a large amount of nonvolatile material containing ordinary components of igneous rocks and that vapor pressure would rise. Subsequently during the progress of crystallization the volatile component would increase in amount, vapor pressure would increase to a maximum, and the material separating would be rich in volatile material. At this stage of maximum vapor pressure material would be expelled through any openings present, and pegmatites and pneumatolytic deposits would be formed. Following this the vapor pressure would fall and material rich in water would be expelled as hydrothermal deposits.

Various descriptions of cassiterite deposits and wolframite deposits indicate that their formation, as well as the formation of greisen and tourmaline associated with them, was caused by pneumatolysis.

Carbonatites

The carbonatites are carbonate bodies that form veinlike or dikelike masses and cores in volcanic plugs of alkalic rocks. The manner of their formation is not entirely clear. Pecora (21, p. 1549) stated that, among petrologists, four schools are represented in the ideas proposed for the origin of carbonatites: (1) magmatists, (2) hydrothermalists, (3) xenolithists, and (4) gas transferists. The magmatists postulate the formation of bodies from a carbonate-rich magma, and thus carbonatites would be a special type of igneous rock, just as are the pegmatites. The hydrothermalists think that the carbonatite bodies were formed by hydrothermal solutions and thus should be grouped with the hydrothermal mineral deposits. The xenolithists believe that the carbonatites are altered xenoliths of original carbonate-rich rock such as limestone. The gas transferists think that escaping gas from erupting volcanoes may account for the origin of the carbonatites, since many of them in East Africa occur in explosive vents.

One of the problems in connection with the origin of the carbonatites is the source of the large amount of CO_2 required for their formation. Pecora (21), from his own field investigations, concluded

> that the CO_2 is derived by a process of concentration of residual ingredients during crystallization of the silicate minerals. At elevated temperatures and

high confining pressures the residual fraction becomes increasingly richer in CO_2. *. . . As these hot solutions follow the pressure gradient toward the surface along intermittent fractures and conduits, the critical region is reached where* CO_2 *departs from the carbonatic solution as a gas and thereby provides conditions suitable for gas buildup, explosive phenomena, and precipitation of minerals.*[1]

Pecora considers the pressure factor to be more important than temperature, as he thinks the rate in fall of pressure is more rapid than is the rate in fall of temperature. According to this concept the CO_2 was in the alkalic magma, perhaps in place of or in addition to water, but the ultimate source of the CO_2 in the magma is a separate problem. Wyllie and Tuttle (27) determined experimentally that simplified carbonatite liquids exist at moderate temperatures throughout a wide range of pressure. This experimental discovery led them to state that there is little reason to doubt, if field evidence indicates an intrusive origin, that the material was emplaced as a liquid or largely liquid magma.

The carbonatites are extremely interesting rocks and are likely to prove highly important sources of niobium (columbium) and other rare constituents in various parts of the world—Canada, Brazil, Africa, and other places.

SELECTED REFERENCES

1. M. K. Banks, W. T. McDaniel, and P. N. Sales, 1953, A method for concentration of North Carolina spodumene ores, *Mining Eng.,* vol. 5, pp. 181–186.
2. A. M. Bateman, 1942, Magmas and ores, *Econ. Geology,* vol. 37, pp. 1–15.
3. A. M. Bateman, 1951, The formation of late magmatic oxide ores, *Econ. Geology,* vol. 46, pp. 404–426.
4. N. L. Bowen, 1928, "The Evolution of the Igneous Rocks," pp. 54–91, Princeton University Press, Princeton, N.J., 251 pp.
5. N. L. Bowen, 1933, The broader story of magmatic differentiation, briefly told, in "Ore Deposits of the Western States," (Lindgren volume), pp. 106–128, The American Institute of Mining and Metallurgical Engineers, New York.
6. A. F. Buddington, 1935, High-temperature mineral associations at shallow to moderate depths, *Econ. Geology,* vol. 30, pp. 205–222.
7. E. N. Cameron, R. H. Jahns, A. H. McNair, and L. R. Page, 1949, "Internal Structure of Granitic Pegmatites," *Econ. Geology Mon.* 2, 115 pp. Contains bibliography of 121 references.
8. R. A. Daly, 1933, "Igneous Rocks and the Depths of the Earth," pp. 9–25, McGraw-Hill Book Company, New York, 598 pp.
9. W. H. Emmons, 1940, "The Principles of Economic Geology," 2d ed., p. 196, McGraw-Hill Book Company, New York, 529 pp.
10. L. C. Graton, 1933, The depth zones in ore deposition, *Econ. Geology,* vol. 28, pp. 513–555.
11. B. L. Gunsallus, 1956, Feldspar, *U.S. Bur. Mines Bull.* 556, pp. 263–271.

[1] From W. T. Pecora, *Geol. Soc. America Bull.,* vol. 67, p. 1552. Copyright 1956. Geological Society of America. Used by permission.

12. R. H. Jahns, 1955, The study of pegmatites, *Econ. Geology*, Fiftieth Ann. Vol., pt. 2, pp. 1025–1130. Contains bibliography of 698 references.

13. W. R. Jones, 1925, "Tinfields of the World," p. 105, Mining Publications, Ltd., London, 423 pp.

14. K. K. Landes, 1933, Origin and classification of pegmatites, *Am. Mineralogist*, vol. 18, pp. 33–56, 95–103.

15. W. Lindgren, 1933, "Mineral Deposits," 4th ed., pp. 211–212, 793, McGraw-Hill Book Company, New York, 930 pp.

16. H. E. McKinstry, 1948, "Mining Geology," pp. 470–478, Prentice-Hall, Inc., Englewood Cliffs, N.J., 680 pp.

17. H. E. McKinstry, 1955, Structure of hydrothermal ore deposits, *Econ. Geology*, Fiftieth Ann. Vol., pt. 1, pp. 170–225.

18. G. A. Munson and F. F. Clarke, 1955, Mining and concentrating spodumene in the Black Hills, South Dakota, *Mining Eng.*, vol. 7, pp. 1041–1047.

19. P. Niggli, 1929, "Ore Deposits of Magmatic Origin," pp. 2–6, Thomas Murby and Company, London, 93 pp. (Translated from the original German by H. C. Boydell.)

20. J. J. Norton and L. R. Page, 1956, Methods used to determine grade and reserves of pegmatites, *Mining Eng.*, vol. 8, pp. 401–414.

21. W. T. Pecora, 1956, Carbonatites: A review, *Geol. Soc. America Bull.*, vol. 67, pp. 1537–1556.

22. R. H. Sales and C. Meyer, 1948, Wall rock alteration at Butte, Montana, *Am. Inst. Mining Engineers Trans.*, vol. 178, pp. 9–35.

23. W. T. Schaller, 1933, Pegmatites, in "Ore Deposits of the Western States," (Lindgren volume), pp. 144–151, The American Institute of Mining and Metallurgical Engineers, New York.

24. G. M. Schwartz, 1955, Hydrothermal alteration as a guide to ore, *Econ. Geology*, Fiftieth Ann. Vol. pt. 1, pp. 300–323.

25. S. J. Shand, 1944, The terminology of late magmatic and post-magmatic processes, *Jour. Geology*, vol. 52, pp. 342–350.

26. E. E. Wahlstrom, 1950, "Introduction to Theoretical Igneous Petrology," p. 262, John Wiley & Sons, Inc., New York, 365 pp.

27. P. J. Wyllie and O. F. Tuttle, 1960, Experimental verification for the magmatic origin of carbonatites, Internat. Geol. Cong., 21st, Copenhagen 1960, Petrographic Provinces, Igneous and Metamorphic Rocks, sec. 13, pt. 13, pp. 310–318.

chapter 4

Deposits Formed by Metamorphic Activity

The changes produced by metamorphism are those that take place as a result of reorganization and recrystallization of previously formed rock material without melting, but usually under conditions of increased temperature and pressure, and commonly in the presence of chemically active fluids. The variations of temperature and of local pressure during metamorphism are unknown, but estimates indicate a probable maximum variation of temperature from about 200 to nearly 1000°C, and of load pressure from about 100 bars or less to 10,000 bars. Most metamorphism, however, apparently takes place between temperatures of 300 and 750°C and load pressures between 2,000 and 8,000 bars (7).

Metamorphism, as used in this book, excludes alterations of igneous rocks by late-stage emanations from magma that formed the rocks in question (classed as autometamorphism or endometamorphism by some), normal wallrock alteration along the margins of mineral deposits, weathering, action of groundwater, and diagenesis. It is recognized, however, that diagenesis may grade insensibly into metamorphism and that at present we have no good means of distinguishing certain diagenetic reorganizations from metamorphic ones. Mineral assemblages, however, are reliable indicators if metamorphism has progressed beyond a very low degree of intensity.

Metamorphism, as used by the author and as presented in the most recent discussions of the subject, includes two major subdivisions which, if metamorphism has progressed to a very high degree of intensity, may grade into each other, although in many cases they are clearly separable. These are (1) *contact metamorphism* and (2) *regional metamorphism*. Turner and Verhoogen (7) include a third kind, *dislocation metamorphism*, as one of three main types associated with the occurrence of the common metamorphic rocks. This type is "restricted to zones of intense deformation, such as major faults and 'movement horizons.'"[1] Other kinds recognized by Turner and Verhoogen, usually restricted in occurrence, are

[1] From F. J. Turner and J. Verhoogen, "Igneous and Metamorphic Petrology," 2d ed., p. 452 Copyright 1960. McGraw-Hill Book Company. Used by permission.

pyrometamorphism (shown by xenoliths), *cataclastic metamorphism* (mechanical deformation without crystallization or chemical reaction), *metasomatic metamorphism* (involving substantial change in chemical composition), and *retrogressive metamorphism* (conversion of higher-temperature metamorphic minerals to lower-temperature ones, usually hydrous, that are more stable under the new conditions). Metasomatic metamorphism is, in the opinion of the writer, an integral part of much contact metamorphism and some regional metamorphism. Certainly some change in chemical composition takes place in almost any contact or regional metamorphism; this is clearly recognized by Turner and Verhoogen (7, pp. 450–451), and it is only the degree of change that gives rise to this separate classification. It seems, therefore, that the succeeding discussion will be much simplified if metasomatism[1] is included with metamorphism *if it takes place under metamorphic conditions.* The term *dynamic metamorphism* is not used by the author (nor by Turner and Verhoogen), because of lack of clarity regarding its exact meaning and various objections raised soon after Rosenbusch introduced the term, because dynamic metamorphism deformed but did not transform.[2]

Both contact metamorphism and regional metamorphism bring about the formation of mineral deposits. The deposits formed by contact metamorphism are chiefly those of metals, especially iron, copper, and tungsten, but also some nonmetals. The deposits formed by regional metamorphism are chiefly nonmetals, such as kyanite, sillimanite, some graphite, and various metamorphic rocks as, for example, slate, marble, and mica schist. Disagreement exists regarding the importance of regional metamorphism in the formation or concentration of metallic deposits.

Contact metasomatic deposits

Contact metamorphism takes place in the rocks around intrusive masses of magma and is always associated with igneous activity. During the formation of most mineral deposits produced by contact metamorphism, new minerals are formed in the intruded rock, in part by reorganization and recrystallization of constituents already present in that rock, in part by emanations from the magma. Where commercial metallic deposits have been formed (Fig. 4-1), such emanations contained large amounts of metallic constituents and, in some cases, also large amounts of nonmetallic constituents. The gangue minerals of such deposits are one of their characteristic features; hence their formation will be discussed briefly. Most contact-metasomatic deposits occur in metamorphosed limestone or dolomite, or, in more general terms, calcareous rocks, so the discussion will be limited to changes in such rocks.

The minerals formed by contact metamorphism of a limestone, or any other rock, as a result of reorganization and recrystallization without addition of constituents, depend on the minerals originally present in the rock, their size and

---

[1] *Metasomatism* has been defined in several ways. Generally it is used in the same sense as replacement. Since replacement, or metasomatism, will proceed throughout a great temperature range, only the metasomatism that is clearly a part of metamorphism should be included with it.
[2] See reference 6, pp. 1–43, for an excellent review of various concepts of metamorphism.

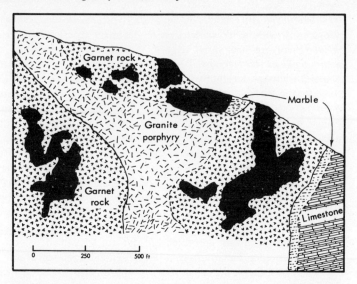

Fig. 4-1 Section showing contact-metasomatic copper deposits (solid black) and associated garnet rock, Empire Mine, Mackay region, Idaho. Large blocks of limestone were engulfed in the granite porphyry. After J. B. Umpleby, *U.S. Geol. Survey Prof. Paper* 97, p. 46.

state of admixture throughout the rock, that is, whether in close contact with one another or widely separated from one another, the temperature reached, and the rock pressure. Water, carbon dioxide, or both, may considerably influence the reactions that take place. Several possible reactions will be used as examples of the manner in which new minerals may be formed by metamorphism. The chemical reactions do not necessarily take place just as indicated by the chemical equations. The equations, however, are a useful means of showing why many minerals that are formed depend on the composition of the original rock.

A calcareous rock may be composed chiefly of calcite or chiefly of dolomite, but generally moderate amounts of other constituents are present. The most usual ones are chert, clay, and iron oxide. The simplest examples that can be used are those that assume the presence of silica only as an impurity in calcite or dolomite.

One reaction that has been carefully investigated in the laboratory is that involving calcite and silica. Such a combination of constituents would be the equivalent of a calcitic limestone that contained only chert as an impurity. If the pressure on such a rock is about 2,000 bars, reaction would take place at a temperature of about 800°C and a new mineral, wollastonite, would be formed. This can be put in the form of the following equation:

$$\underset{\text{calcite}}{CaCO_3} + \underset{\text{chert}}{SiO_2} \rightarrow \underset{\text{wollastonite}}{CaSiO_3} + \underset{\text{carbon dioxide}}{CO_2}$$

If more calcite were available than that needed to unite with the chert, it would simply recrystallize, probably with coarser grain. The carbon dioxide released in the formation of wollastonite is lost, and this reaction does not take place unless carbon dioxide can be evolved. In the event that carbon dioxide cannot be driven off, such as might result from too great pressure at a given temperature, simple recrystallization would result, the chert passing into interlocking quartz, the calcite into a more coarsely crystalline mosaic.

Much is known also about the formation of minerals from dolomite that contains silica as an impurity (1). A few examples will show some of the minerals that may form, depending on various temperatures and pressures and the amount of different constituents.

$$CaMg(CO_3)_2 + 2SiO_2 \rightarrow CaMgSi_2O_6 + 2CO_2$$

dolomite   chert   diopside, a pyroxene   carbon dioxide

$$2CaMg(CO_3)_2 + SiO_2 \rightarrow Mg_2SiO_4 + 2CaCO_3 + 2CO_2$$

dolomite   chert   forsterite, an olivine   calcite   carbon dioxide

$$5CaMg(CO_3)_2 + 8SiO_2 + H_2O \rightarrow$$

dolomite   chert   water

$$Ca_2Mg_5Si_8O_{22}(OH)_2 + 3CaCO_3 + 7CO_2$$

tremolite, an amphibole   calcite   carbon dioxide

These three equations show that three different minerals, diopside, forsterite, and tremolite, may be formed from dolomite and chert in different proportions but that water must be present if tremolite is formed, since that mineral contains hydroxyl. The reactions are controlled by temperature and pressure also, the details of which cannot be discussed here. Generally tremolite forms at lower temperatures than does diopside, and forsterite may form at temperatures intermediate between those two minerals if the content of silica is sufficiently low compared to the content of dolomite. It will be noted, also, that calcite is a mineral of the metamorphic suite (right-hand side of the equation) in two equations even though dolomite was the original mineral, since the magnesium of the dolomite is all used in forming forsterite or tremolite. Different proportions of constituents would yield somewhat different results.

If clay or iron oxide is present, as well as chert or quartz, garnet or epidote may form, the amphibole formed may be actinolite or anthophyllite, and the pyroxene formed may be hedenbergite or a pyroxene of the diopside-hedenbergite series; less commonly the amphibole and pyroxene formed will be aluminous varieties such as hornblende and augite. Some of the relationships among impurities in an original calcareous rock and the minerals that might form by metamorphism if various impurities are present are shown by Table 4-1.

The most common and characteristic gangue minerals formed by contact metamorphism of limestone and dolomite, based on the usual impurities present in those rocks, should be amphibole (tremolite, actinolite, anthophyllite), pyroxene (diopside or diopside-hedenbergite), forsterite, epidote, and garnet (grossularite-andradite). These, along with calcite and quartz, are, indeed, the common gangue minerals that occur in such deposits. Locally, however, minerals less easily explained on the basis of original composition are abundant in the gangue. Moreover, analyses of many calcareous rocks some distance from igneous intrusions have shown but little SiO_2, FeO, Fe_2O_3, and Al_2O_3, whereas the same rocks, as far as one can determine from field relations, near the intrusions have been converted into masses composed in part of amphibole, pyroxene, or garnet, along with other gangue minerals, and contain commercial amounts of various minerals, such as magnetite ($FeFe_2O_4$), chalcopyrite ($CuFeS_2$), and scheelite ($CaWO_4$). Furthermore, the characteristic pyroxene of many such commercial deposits is high in

TABLE 4-1. Some Relationships among Constituents of a Calcareous Rock and Some Common Minerals That Form by Metamorphism

Calcareous constituents	Impurities	Some common minerals that might form		
		Mineral group	*Mineral*	*Composition*
$CaCO_3$	SiO_2		Wollastonite	$CaSiO_3$
	$H_4Al_2Si_2O_9$ and Fe_2O_3	Epidote	Pistacite (common epidote)	$Ca_2(Al \cdot OH)(Al, Fe)_2(SiO_4)_3$
		Garnet	Grossularite-andradite series	$Ca_3(Al, Fe)_2(SiO_4)_3$
$CaMg(CO_3)_2$	SiO_2	Olivine	Forsterite	Mg_2SiO_4
		Amphibole	Tremolite	$Ca_2Mg_5Si_8O_{22}(OH)_2$
		Pyroxene	Diopside	$CaMgSi_2O_6$
	SiO_2 and FeO	Amphibole	Actinolite	$Ca_2(Mg, Fe)_5Si_8O_{22}(OH)_2$
			Anthophyllite	$(Mg, Fe)_7Si_8O_{22}(OH)_2$
		Pyroxene	Diopside-hedenbergite series	$Ca(Mg, Fe)Si_2O_6$

iron and may be nearly pure hedenbergite ($CaFeSi_2O_6$), and the characteristic garnet is also high in iron, nearly pure andradite [$Ca_3Fe_2(SiO_4)_3$]. These features, along with field and laboratory studies that show clear evidence of replacement, indicate that in many deposits not only the ore minerals were formed by metasomatism, but also many of the gangue minerals, the needed constituents having been supplied by emanations from the intrusive magma. Studies by Lindgren in the Clifton-Morenci district, Arizona (4), and at Bingham, Utah (5), and studies by Umpleby (8) in the Mackay region, Idaho, show that much material was added to the limestones. Studies by many other investigators have led to the same conclusion.

The gangue of certain Swedish contact-metasomatic deposits was early designated *skarn*.[1] The term *skarn* is widely used and is a convenient one as a designation for the gangue of many contact-metasomatic deposits. Recently, however, Geijer (2) subdivided skarn ores into two types, primary skarn ores and reaction

[1] *Skarn*, a Swedish mining term, has been used for the silicate gangue of some iron ore and other deposits of Archean age, especially those that occur in calcareous rocks, but it has been applied also to similar deposits of younger age. (See reference 6, p. 310, of Chap. 2.)

skarn ores. The primary skarn ores were formed by contact-metamorphic action, and large amounts of constituents were added. The reaction skarn ores were formed by metamorphism of original sedimentary or replacement deposits as a result of regional heating. The term *skarn* as used throughout the literature generally refers to primary skarn formed by contact metamorphism and metasomatism, and this is the manner in which it is used in this book. The writer prefers to designate the reaction skarns of Geijer, *metamorphosed deposits*. Such deposits are present in the Lake Superior and the Labrador iron regions as a result of metamorphism of sedimentary iron formations.

The term *tactite* has also been proposed for some but not all of the gangue of contact-metasomatic deposits. "Tactite may be defined as a rock of more or less complex mineralogy formed by the contact metamorphism of limestone, dolomite, or other soluble rocks into which foreign matter from the intruding magma has been introduced by hot solutions or gases. It does not include the inclosing zone of tremolite, wollastonite, and calcite."[1] (3) The discussion that accompanies this definition indicates that two zones of gangue minerals may surround contact-metasomatic deposits: an inner zone, designated tactite, which contains the ore minerals and the gangue minerals formed in part by addition of material; and an outer zone in which there are gangue minerals that were formed essentially without addition. The term *tactite*, then, has a meaning similar to that of *skarn* if the latter is extended to include deposits that are younger than Archean. *Skarn* is much more widely used than *tactite* in descriptions of contact-metasomatic deposits.

Contact-metasomatic deposits generally are of small size, and the distribution of the ore minerals in such deposits commonly is erratic. In some places the ore minerals are admixed with the gangue minerals; in other places they occur chiefly in recrystallized limestone that contains but few silicates. Some ore minerals appear to have formed essentially at the same time as the gangue minerals, others formed later and replaced gangue minerals. Lindgren (5, p. 534) stated, regarding metamorphism and metasomatism at Bingham, Utah, that some of the ore minerals formed practically simultaneously with the metamorphic silicates but that the important ore deposits, although they came into existence in part with the development of the silicates, continued to form later, after the temperature had fallen far below that needed for the formation of the silicates.

In some places limestone has been replaced (1) in part by ore minerals disseminated through it, (2) selectively along beds, or (3) almost entirely by massive ore. In some places the ore solutions entered fractures and filled cavities or replaced material along the margins of fractures. Various small openings, such as cleavage cracks in minerals and intergrain boundaries, were sites of deposition and replacement.

Regional metamorphic deposits

Regional metamorphism generally is conceived to take place as a result of increased temperature and pressure, especially stress, at some stage during the

[1] F. L. Hess, *American Journal of Science*, vol. 48, p. 378. Copyright 1918. Used by permission. The term *tactite* is derived from the Latin *tactus* (touching). Presumably it refers to the proximity of the combined gangue and ore material to the igneous intrusive, although this is not stated by Hess.

formation of folded mountain ranges. Constituents of the rocks affected are re-organized and recrystallized and, because of crystallization under stress, generally show preferred orientation, as in schists. Deformation of the rocks and movement of fluids through them appear to be highly important in assisting temperature and pressure to bring about metamorphism. Deformation apparently accelerates meta-morphic reactions (7, p. 477) and may bring about reorganizations that, in the absence of shear, might be so slow that results would not be observable. Yet some rocks that have been highly deformed are essentially unmetamorphosed. This could result because temperatures were too low to produce metamorphism, but another possibility is that fluids were not present in the rocks during metamorphism. Fluids, just as deformation, are thought to accelerate metamorphic reactions. Metamorphism might take place, therefore, in the presence of fluids, at a lower temperature than it would in their absence. Both fluids and deformation have come to be regarded as "catalysts" that aid increase of temperature in promoting metamorphism (10, p. 163). The source of the fluids is not always clear, but three possible sources will be discussed briefly.

Solutions held in the intergranular spaces of rocks certainly is one possible source. Water from this source could combine with constituents of the rock to form the common metamorphic minerals of the schists, such as chlorites, micas, and amphiboles, all of which contain hydroxyl. Metamorphic rocks containing those minerals generally were formed by metamorphism of low to moderate intensity.

Reorganization of hydroxyl-bearing minerals is a second possible source of obtaining fluids that might promote metamorphism. Minerals containing hydroxyl are essentially absent in many gneisses and granulites, which are characterized especially by feldspars, pyroxenes, and garnets. These rocks were formed by meta-morphism of greater intensity than that required to form schists. During this in-creased intensity, the micas in the rocks undergoing metamorphism were reorga-nized and feldspars and other minerals were formed from them. Thus muscovite, $KAl_3Si_3O_{10}(OH)_2$, may have been changed to orthoclase, $KAlSi_3O_8$. In order to accomplish this change, reorganization among other minerals must take place, since orthoclase contains less aluminum than muscovite, less oxygen, and no hydroxyl. Not uncommonly the excess aluminum is used to form kyanite or sillimanite, Al_2SiO_5, and the hydroxyl is driven off, just as carbon dioxide is driven off in re-organizations caused by metamorphism of limestone. Thus hydroxyl may be driven from some rocks as the intensity of metamorphism increases into others farther from the center of metamorphic activity, there to be used to accelerate meta-morphism of those rocks.

A third possible source is emanations derived from magmas that were intruded into the central part of the region undergoing orogeny or that were generated by melting or partial melting of rocks in the central part of a geosyncline during orog-eny. In many folded mountains the degree of regional metamorphism, as indicated by the rocks that were formed, increased from the flanks toward the core of the mountains. If magma were generated deep in the central part of a geosyncline that was undergoing folding, emanations might pass outward long distances into rocks and thus accelerate metamorphism there.

Various concepts of metamorphic reorganization, with the formation of gneisses and migmatites, have led to the concept that granitization is an end stage of metamorphism or, as expressed by some people, that granites are a product of ultrametamorphism without melting of rock material.[1] According to the more usual concept of granitization, emanations are important in promoting metamorphism outward from a core region, but such emanations do not come from magmas. This concept was well expressed by Read (6), who stated: "Out from the central theatre of granitization there pass waves of metasomatizing solutions, changing in composition and in temperature as they become more distant from the core and promoting thereby the formation of zones of metamorphism about it."[2]

Mineral deposits formed by regional metamorphism have generally been thought to have formed by reorganization and recrystallization of constituents contained within the original rock. Thus mica schist, used as a source of ground mica, was probably formed from some type of shale, since the constituents needed for its formation are contained in shale. Similarly, deposits of kyanite and sillimanite could be formed from a shale that contained a considerable amount of alumina. Most mineral deposits formed by regional metamorphism, therefore, form an integral part of the metamorphosed rocks of a region and follow the general trend of those rocks. Further, deposits generally follow specific stratigraphic horizons, because the composition of the original rocks was of major importance in determining the composition of the deposit. The concept of granitization as applied to the formation of mineral deposits by regional metamosphism, however, is used by some people to account for the migration and concentration of metallic constituents. In many cases the concept is that the metallic constituents had previously accumulated in sedimentary beds, and were reorganized and moved about during metamorphism, but remained essentially in the same stratigraphic horizon as originally, or very nearly so. Some investigators, however, believe that material was driven from previous igneous or metamorphic rocks rather long distances and reconcentrated in favorable locations.

SELECTED REFERENCES

1. N. L. Bowen, 1940, Progressive metamorphism of siliceous limestone and dolomite, *Jour. Geology*, vol. 48, pp. 225–274.

2. Per Geijer, 1959, Some aspects of the skarn ore problems in central Sweden, *Geol. Foren. Stockholm, Förh.*, vol. 81, pt. 3, no. 498, pp. 514–534. Abstract in English, pp. 514–515.

3. F. L. Hess, 1918, Tactite, the product of contact metamorphism, *Am. Jour. Sci.*, ser. 4, vol. 48, pp. 377–378.

4. W. Lindgren, 1905, The copper deposits of the Clifton-Morenci district, Arizona, *U.S. Geol. Survey Prof. Paper* 43, pp. 123–164. An early discussion of metamorphism and metasomatism as applied to ore deposition.

5. W. Lindgren, 1924, Contact metamorphism at Bingham, Utah, *Geol. Soc. America Bull.*, vol. 35, pp. 507–534. Shows that much material was added during metamorphism and metasomatism.

6. H. H. Read, 1957, "The Granite Controversy," Interscience Publishers, New York, 430 pp. See pp. 1–43 for a review and discussion of metamorphism

---

[1] See Walton (9) for a good summary of granitization.

[2] From "The Granite Controversy," by H. H. Read, p. 355. Copyright 1957. John Wiley and Sons. Used by permission.

in general and pp. 354–359 for a discussion of regional metamorphism as viewed by Read.

7. F. J. Turner and J. Verhoogen, 1960, "Igneous and Metamorphic Petrology," 2d ed., McGraw-Hill Book Company, New York, 694 pp. See pp. 450–452 for statement of kinds of metamorphism and p. 477 for discussion of the function of deformation in metamorphism.

8. J. B. Umpleby, 1917, Geology and ore deposits of the Mackay region, Idaho, *U.S. Geol. Survey Prof. Paper* 97, pp. 70–79. Discusses addition of material during metamorphism.

9. Matt Walton, 1960, Granite problems, *Science,* vol. 131, pp. 635–646.

10. H. Williams, F. J. Turner, and C. M. Gilbert, 1954, "Petrography," W. H. Freeman and Company, San Francisco, 406 pp. See pp. 161–171 for a good, uncomplicated discussion of metamorphism.

chapter 5

Deposits Formed by Surface Agencies

Surface agencies, as used here, include weathering, erosion, the work of ground-water, transportation and subsequent deposition of material by agents other than those involved in metamorphism and igneous action, the activity of organisms in connection with the various processes, and diagenesis. The work of surface agencies, therefore, covers an extremely broad field of activity and embraces the formation of all mineral deposits not formed by igneous action or metamorphism.

The activity of surface agencies is exceedingly complex, and various processes overlap and grade into one another, making clear separation of the specific processes that are responsible for the formation of a mineral deposit difficult or uncertain. Nevertheless, it is possible to group deposits into two major divisions in which certain processes were more instrumental than others in causing their formation. These two divisions, which will be used for further discussion, are shown in the following outline:

Condensed Outline of Deposits Formed by Surface Agencies

I. Deposits formed chiefly by weathering and erosion; mechanical or chemical transportation; and deposition in streams, bodies of standing water, or on the surface of the land.
   A. Placer or alluvial deposits
   B. Sedimentary beds
       1. Formed chiefly by mechanical deposition
       2. Formed chiefly by chemical and biochemical activity
II. Deposits formed chiefly by weathering and the activity of groundwater, either as residues or as deposits from groundwater
   A. Residual deposits
   B. Oxidation and supergene enrichment deposits
   C. Groundwater deposits

Deposits of group I: Placer or alluvial deposits; sedimentary beds

Placer or Alluvial Deposits

A placer deposit is "a mass of gravel, sand, or similar material resulting from the crumbling and erosion of solid rocks and containing particles or nuggets of gold, platinum, tin, or other valuable minerals, that have been derived from rocks or veins." [Chap. 2 (5, p. 517).] Griffith (41) designated all such accumulations *alluvial deposits* and stated that they are known as placer deposits in the United States and Canada. In many European descriptions they are designated alluvial deposits. Some writers designate as *placer deposits* only those that contain gold and use *alluvial deposits* for all others. Here *placer deposit* will be used according to the definition quoted and thus will include all alluvial deposits in which some valuable mineral is associated with sand and gravel. Most placer deposits are concentrated by stream activity, but some are concentrated by wave and current action along shorelines, and some even by wind action. Further, placer deposits, once formed, may be elevated or depressed, may be partly or wholly eroded and reconcentrated, or may be buried beneath sedimentary strata or beneath lava or tuff.

Because placers may be of fluviatile or marine origin and may be elevated or buried, several classifications of placers have been proposed (5, p. 115; 20, p. 29). Lindgren [(10, p. 226) of Chap. 2] proposed a classification that related various types of placer deposits to present and past topographic cycles. This is a very useful general classification, which may be applied in any region, and is well exemplified in the Klondike and the Seward Peninsula, Alaska, in California, Australia, Malaya, and elsewhere. Hence the major elements of it have been used in compiling Table 5-1. Many of the depressed placer deposits become covered with sediments, volcanic rocks, or both, and hence are commonly referred to as buried placers or, as in Australia, deep leads. Some remain essentially uncemented; others become cemented. Cemented stream placers were classed by Mertie (23, p. 108) as hard rock sedimentary reefs. Some buried gravel-plain placers have

TABLE 5-1.   Classification of Placer Deposits*

Fundamental types of present topographic cycle	Types inherited from past topographic cycles
Gulch and creek placers	High level; result from elevation Deep; result from depression; may be buried
River and bar placers	High level and bench; result from elevation Deep; result from depression; may be buried
Gravel-plain placers	High level; result from elevation Depressed; may be buried
Beach placers	High level; result from elevation Depressed; may be buried

* From Waldemar Lindgren, "Mineral Deposits," 4th ed., p. 226, slightly modified. Copyright 1933. McGraw-Hill Book Company. Used by permission.

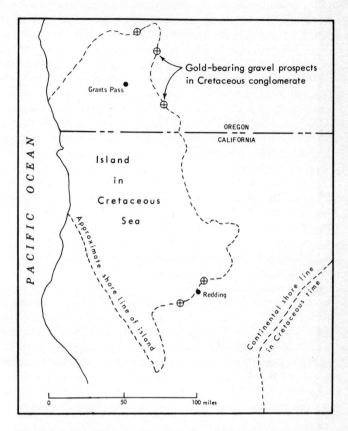

Fig. 5-1   Approximate shore line of an island in the Cretaceous sea, along which gold-bearing gravels accumulated. After J. S. Diller, *U.S. Geol. Survey Bull.* 546, p. 89.

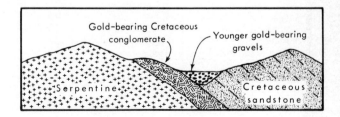

Fig. 5-2   Section showing accumulation of gold-bearing gravels in stream bed as a result of erosion of gold-bearing Cretaceous conglomerate, near Waldo, Oregon. After J. S. Diller, *U.S. Geol. Survey Bull.* 546, p. 94.

been consolidated into conglomerates. Such a conglomerate, formed around the shoreline of a large island in Cretaceous time, has been traced through parts of California and Oregon and in places has been reworked by present streams (Figs. 5-1, 5-2). The suggestion has been made repeatedly that the great and highly productive gold-bearing conglomerate of the Witwatersrand, South Africa, is a buried placer deposit, but opinion is divided and a number of people think the gold was introduced later by hydrothermal solutions.

A type of placer deposit not shown in Table 5-1, but recognized by Lindgren [(10, pp. 219–221) of Chap. 2] is the *eluvial placer*. This type occurs on hillsides or valley slopes and results from decomposition and disintegration of a primary deposit of gold, for example, such as a vein, and from downward movement of material along the slope. In places, as in some of the Malayan cassiterite deposits, such material settled into sinkholes in limestone. Some of the writers who have described the Malayan deposits make a sharp distinction between the eluvial and the alluvial deposits, because the history of their formation and their location in the area differ. The deposits that are designated by some as *residual placers* are, strictly, residual deposits. The gold or other valuable mineral has simply remained as a residue among decomposed rock debris. Such deposits may grade almost imperceptibly into eluvial placers.

The accumulation of a mineral in stream or beach placer deposits depends on the specific gravity of the mineral, its chemical stability under conditions of weathering and transportation by water, and those physical properties that determine whether it will disintegrate easily and form exceedingly small pieces or break into thin flakes. If a mineral is relatively resistant to decomposition during weathering and to solution during transport, its concentration in placer deposits will depend largely on its specific gravity and the size and shape of the particles. Some minerals commonly obtained from placer deposits, their specific gravities, and the specific gravities of some common rock-forming minerals are shown by Table 5-2.

TABLE 5-2.   Specific Gravities of Some Minerals Present in Placer Deposits

Ore minerals		Gem and accessory minerals		Common rock-forming minerals	
Gold	15.6–19.3	Ruby and sapphire (corundum)	3.9–4.1	Pyroxenes	3.1–3.6
Platinum	14.0–19.0*	Spinel	3.5–4.1	Amphiboles	2.8–3.6
Wolframite	7.0– 7.5	Chrysoberyl	3.5–3.8	Micas	2.7–3.1
Cassiterite	6.8– 7.1	Diamond	3.5	Quartz	2.6
Scheelite	5.9– 6.1	Topaz	3.4–3.6	Feldspars	2.5–2.7
Magnetite	5.1	Jade	3.3–3.5		
Monazite	4.9– 5.3	Garnet	3.1–4.3		
Zircon	4.6– 4.7	Tourmaline	2.9–3.2		
Ilmenite	4.5– 5.0	Aquamarine and emerald (beryl)	2.6–2.8		
Chromite	4.1– 4.9				
Rutile	4.1– 4.2				

* 21.0–22.0 if chemically pure.

Among the ore minerals of this table, gold, platinum, and cassiterite are the outstanding placer minerals. Monazite, zircon, ilmenite, and rutile are important in some beach placers. Some ore minerals, galena for example, are heavy enough to accumulate in placers but are chemically unstable and break readily into small pieces because of excellent cleavage and hence are not good placer minerals. Among the gem and accessory minerals, diamond, ruby, and sapphire probably are the most important ones recovered from placers. Among the rock-forming minerals, quartz is the one most common in placer deposits of gold and cassiterite, because gold and cassiterite commonly occur in quartz veins, and cassiterite is associated with granitic rocks. Although the specific gravity of quartz is low, it is not entirely separated from the heavier constituents of placer deposits. Corundum, topaz, garnet, tourmaline, zircon, rutile, and magnetite are commonly present to some extent in many deposits, depending on their presence in the country rocks of the area. The major minerals of this type that accumulate with the placer deposits may impart a distinctive color to the sands, as white sands (quartz), black sands (magnetite and ilmenite), ruby sands (garnet), etc.

The gold that collects in placer deposits varies considerably in size. In general, any large nuggets occur near the source of the primary gold deposit and show little effect of transportation, and gold in the larger stream channels is usually fine to medium fine in size. Lindgren (18, p. 67) stated that very coarse gold is in grains about the size of wheat kernels and that average-size grains are about the size of mustard seed. Gold, by repeated movement along streams, may be reduced to extremely small particles designated *flour gold*. The size of these particles is so small that, according to Lindgren (18, p. 67), 1,000 to 2,000 particles would be required to give the value of one cent (United States money in 1911). Flour gold may be transported a hundred miles or more.

The purity of placer gold generally is greater than that of the primary gold from which it was derived, as has been shown by numerous analyses. Further, it becomes greater with increasing distance from the source, is greater in small grains than in large ones, and greater on the outside of nuggets than on the inside of them. These relations result from the fact that silver, which commonly is alloyed with gold, is dissolved by surface waters. This concept was questioned by Mertie (23, pp. 118–124), who advanced the idea that the apical part of a gold lode contains gold of higher quality than the lower parts and that the apical part, being the first to be attacked by erosion, is carried downstream, is deposited farthest from the source, and thus produces higher-grade material there.

The first step in the process of forming a placer deposit is weathering, which disintegrates and decomposes some of the gangue and rock, and the next step is down-slope movement of material into streams unless material was initially exposed on the bed of the stream. On reaching the bed of the stream the heavier particles tend to settle to the bottom of the mass of sand and gravel. In times of flood, or in places where the gradient is high, material may be moved downstream, but always with a tendency for the heavier material to settle to the bottom and the lighter material to be carried along. Deposition of the heavier material occurs wherever the velocity of the water is checked sufficiently. If material is carried along tributary streams coming from hilly or mountainous country, deposition would likely occur first where those tributaries enter the main valley. Material might then be reworked during periods of high water, be carried downstream, and be redeposited in favorable places. Deposits in such places may accumulate as the floodplain widens and make a rich streak that extends along the valley. Such streaks, generally termed *pay streaks,* or *pay leads,* are somewhat irregular in their distribution, and values may change rapidly (Fig. 5-3).

Gradient, volume, and load of a stream are key factors in the process of transportation and deposition. Irregularities in the bedrock of the stream caused by differences in rock resistance and structure, or, in limestone bedrock, by differential solution, are important sites for deposition of heavier material, which becomes lodged behind these natural riffles and accumulates on the bedrock. In some places clay may collect over gravel some feet above the bedrock of the stream, and gold or other minerals may collect on this clay, which is designated *false bedrock.*

The general process of accumulation of beach placers is similar to that of stream placers, but shore currents and waves are the effective agents. Material may come from weathering and erosion of rocks along the shoreline; it may be brought in by streams and then be worked over by waves and currents; or it may come from previous gravel plains that are reworked by waves.

Sedimentary Beds Formed Chiefly by Mechanical Deposition

The principal sedimentary beds formed by mechanical deposition are those of sand and gravel, sandstone and conglomerate, silt and clay, siltstone, shale, and claystone.

Commercial deposits of sand and gravel are classified, according to origin, into four main groups: (1) fluvial, (2) glacial, (3) marine and lake, and (4) residual [(9, p. 741) of Chap. 2]. A fifth group is eolian deposits, but these are of relatively minor importance. The fluvial, glacial, and marine and lake deposits furnish the major part of the commercial production. All but the residual deposits belong in the group under discussion. Commercial deposits of sand and gravel generally occur either in hilly or mountainous country, around the margins of former glaciers, or in similar places where either high stream gradients or large volumes of water were effective in bringing about their accumulation. Thus certain alluvial fans, deltas, old shorelines, kames, and eskers are sources of commercial deposits. Indurated deposits make sandstone and conglomerate.

The finer detrital materials consist of silt and clay, which may become consolidated into siltstones, shales, and claystones. Clays have been classified as (1) residual clays and (2) transported clays. The residual clays result from prolonged weathering essentially *in situ,* and their origin will be discussed in connection with deposits formed chiefly by weathering and the action of groundwater. Transported clays apparently were formed as residual clays, washed into streams, transported, and finally deposited in the oceans, in estuaries, on floodplains, river terraces, and in lakes. Although classed as transported clays, some deposits may really represent, at least in part, chemical sediments. If the materials were carried as discrete clay minerals and deposited by settling, they would form true detrital deposits; but if they were carried as colloids, precipitated and crystallized in a basin of deposition, they would be chemical precipitates even though they formed lenses surrounded by definitely detrital material, as some do. "It is not known to what extent the clay

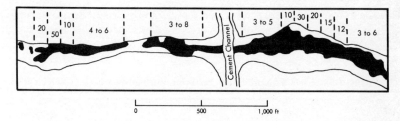

Fig. 5-3 Plan showing areas of pay gravel (solid black) in a buried gold placer, Sierra Nevada region, California. Figures show average value per ton of gravel in various sections of the deposit. The cement channel is a cross-cutting buried channel. After W. Lindgren, *U.S. Geol. Survey Prof. Paper* 73, p. 151.

minerals of pelagic deposits are precipitated from solution in the ocean, or to what proportion they represent precipitates formed shortly after dissolution of the continental igneous source minerals and subsequently transported into the ocean by water or wind."[1]

Sedimentary Beds Formed Chiefly by Chemical and Biochemical Activity

The formation of chemical sedimentary beds depends on (1) material being taken into solution and transported in solution to a suitable site of deposition and (2) the environment at the site of deposition. The first condition determines the separation of chemical materials weathered from the parent rock and which ones may remain *in situ* as a residue; the second condition determines to a large extent where and in what chemical combinations material will be deposited. Some of the conditions under which material may go into solution during weathering will be considered first.

The net result of the decay of rocks is that much potassium, sodium, calcium, and magnesium are carried away in solution; that much aluminum, iron, manganese, and silicon may be concentrated as residues almost at the site of the rocks from which they are derived, but also much iron, manganese, and silicon may be transported in solution and deposited elsewhere; and that varying degrees of separation of aluminum, iron, manganese, and silicon may take place. These results are brought about, at least in part, by a relation known as the *ionic potential,* by the presence or absence of much organic matter, by the presence or absence of electrolytes, by conditions of acidity and alkalinity, conditions of climate, and conditions of time. These factors will be considered briefly.

The relation known as ionic potential is the ratio of the intensity of the charge on the surface of an ion to the size of the radius of the ion, or, in symbols used in some discussions, $Z : r,$ in which Z is the intensity of the charge. If this ratio is low (3 or less), elements tend to go into solution and remain in solution under conditions of weathering and surface or near-surface transportation; if it is intermediate (3 to 12), elements tend to be precipitated as residues near the site of decomposition of the rocks; and if it is high (more than 12), they tend to form

---

[1] From "Researches in Geochemistry," Sedimentation on the Ocean Floor by G. O. S. Arrhenius. Copyright 1959. John Wiley and Sons. Used by permission.

complex ions which are soluble and may be transported long distances. These divisions were suggested by Goldschmidt (40). The expected behavior of some selected elements, based on ionic potential, is shown in Table 5-3.

TABLE 5-3. Ionic Potential and Behavior during Weathering, Selected Elements*

Element	Charge Z	Radius r	Ionic potential Z : r	Ionic potential division	Behavior during weathering and transportation
K	1	1.33	0.75	1. Low: 3 or less (soluble cations)	Remain in solution
Na	1	0.97	1.0		
Ca	2	0.99	2.0		
Mn	2	0.80	2.5		
Fe	2	0.74	2.7		
Mg	2	0.66	3.0		
Fe	3	0.64	4.7	2. Intermediate: 3 to 12 (elements of hydrolysates)	Precipitated by hydrolysis
Al	3	0.51	5.9		
Mn	4	0.60	6.7		
Si	4	0.42	9.5		
B	3	0.23	13	3. High: more than 12† (soluble complex anions)	Form anions containing oxygen; usually again soluble
P	5	0.35	14		
S	6	0.30	20		
C	4	0.16	25		
N	5	0.13	38		

* Compiled chiefly from "Principles of Geochemistry," 2d ed., by Brian Mason. Copyright 1958. John Wiley and Sons. Used by permission.
† The suggestion has been made (12, p. 25) that the grouping be changed to include silicon with the soluble complex anions, making the division at 9.5 instead of 12, because much silica is carried away during weathering.

Ionic potential, as indicated by Table 5-3, would account for much potassium, sodium, calcium, and magnesium being removed during weathering, but it would also indicate that much manganese and iron would follow the same course. This should be true if manganese and iron remained in the bivalent state, but when changed into the tetravalent (Mn) or trivalent (Fe) state, as commonly happens where a ready supply of oxygen is available, the ionic potential is changed; manganese and iron become members of the intermediate group along with aluminum and silicon and tend to accumulate as residues. The high ionic potential of the elements of the third group would indicate that they would form complex anions, as the borates, phosphates, sulfates, etc., and migrate in solution.

The tendencies shown by ionic potential, although they indicate the manner in which elements may behave during weathering, should not be construed to mean that certain things do not migrate in solution if conditions are proper. Vast amounts of iron have been transported and deposited, along with similar amounts of silica, and have formed sedimentary beds that supply most of the iron of the world. Further, bedded cherts are extensive. Similarly, the most important manganese deposits of the world are sedimentary beds. We need to inquire, then, into the problem of the solution and transportation of iron, silica, and manganese, and how these may be separated from alumina and each other.

Moore and Maynard (24), in a study of the solution and transportation of iron and silica, conducted a series of experiments, which indicate that (1) carbonated water and (2) peat solution are important solvents of iron and silica. They concluded (p. 276) that carbonated water could dissolve enough iron and silica from a terrane of basic rocks to form a large sedimentary ore deposit and that organic matter, if present, would assist in dissolving much iron and silica. They also concluded (p. 277) that iron is not carried as a bicarbonate in cold, natural surface waters high in organic matter but that it is probably transported as a ferric oxide hydrosol stabilized by organic colloids,[1] although a relatively small amount might be carried as salts of organic acid or adsorbed by organic colloids. They concluded further that silica in solution in cold, natural water commonly is transported as a colloid. Consequently they concluded that the iron that makes up some of the large sedimentary iron formations was transported chiefly as a ferric oxide hydrosol stabilized by organic matter and that the greater portion of the silica was transported as a silica hydrosol.

The precipitation of iron and silica were also investigated experimentally by Moore and Maynard (24). Their general conclusion was that ferric hydroxide would be precipitated in a few days when colloidal iron and silica solutions, stabilized by organic matter, came in contact with electrolytes of the sea, whereas silica would not be completely coagulated after several months.

Manganese is taken into solution, transported, and deposited in about the same manner as iron. Some differences, however, may cause more or less separation of iron and manganese and, hence, the formation of deposits that are chiefly iron or chiefly manganese; but separation may be only moderate in some cases, leading to the formation of manganiferous iron ore (5 to 10 percent manganese) or ferruginous manganese ore (10 to 35 percent manganese). These relations result, in part, because manganese is dissolved in carbon dioxide–bearing waters more easily than is iron; manganous compounds are more stable in solution than are ferrous compounds; iron is more readily oxidized to the ferric state and later precipitated as ferric hydroxide, whereas manganese, because of a lower affinity for oxygen, remains in solution until most of the iron is precipitated. Colloidal iron and manganese, however, as $Fe(OH)_3$ and $Mn(OH)_4$ are oppositely charged, and if these two hydrosols come in contact, there is mutual deposition and a mixture of iron and manganese hydroxides forms. On the other hand, if a colloidal solution containing ferric and manganic hydroxide comes in contact with an electrolyte, as it would on entering seawater, coagulation of most of the manganic hydroxide would result, while most or all of the iron would remain in solution. Laboratory experiments

[1] Colloidal material consists of very small electrically charged particles, approximately 10^{-3} to 10^{-6} mm across, dispersed throughout some medium. Solid particles dispersed throughout water is the important type in sedimentary processes. Sols, such as ferric oxide hydrosol, resemble liquids in their physical properties. Colloids show great differences in relative stability. Certain organic compounds tend to stabilize inorganic colloids formed during sedimentary activity and thus permit transportation farther than would otherwise be possible. See reference 11, p. 168, of Chap. 2 for discussion.

have shown that the addition of electrolytes to solutions that contain more iron than manganese cause, first, separation of a mixture of hydroxides rich in manganese and, later, precipitation of a large quantity of a mixture of hydroxides rich in iron. Microorganisms apparently play a part in the precipitation of manganese, just as they do in the precipitation of iron, and may bring about separation of the two elements, but information regarding this seems to be lacking [(14, pp. 647–650) of Chap. 2].

Silica, aluminum, and iron behave differently under varying conditions of acidity and alkalinity, that is, different values of pH.[1] Natural waters may range from strongly alkaline (pH more than 9) to strongly acid (pH 5.5 to 2.1), but the great majority of natural waters show pH within one unit either way of 7, that is, they range from 6 to 8. Water of the open sea is slightly alkaline, usually pH 7.5 to 8.4 [(4, p. 764); (13, p. 600) of Chap. 2; (17 p. 3)]. Aluminum will be dissolved if solutions are very acid or very alkaline. Aluminum hydroxide, $Al(OH)_3$, is precipitated from nearly neutral solution. Silica, however, is relatively insoluble in acid solution but becomes increasingly soluble as a solution gets more alkaline. Hence, near neutrality, aluminum hydroxide may be precipitated whereas much silica may remain in solution. Ferric hydroxide, however, is precipitated in very acid solution, so that both ferric hydroxide and a considerable amount of silica should precipitate from such a solution whereas aluminum hydroxide should not. With increasing pH, up to neutrality, less and less silica should precipitate, and aluminum hydroxide should remain in solution, so that the product should be ferric hydroxide with less silica than under very acid conditions; but, near neutrality, aluminum hydroxide should precipitate. If much of the ferric hydroxide had been precipitated, a fair separation of iron and aluminum might result, and the amount of silica might be only moderate.

> *Ferric iron, being able to migrate in true solution only in acid waters, differs from aluminum, which may migrate both in acid and basic solutions, being precipitated only in the neighborhood of the neutral point. . . . When an acid weathering solution is neutralized, ferric hydroxide is first precipitated, but aluminum remains in solution until a higher pH is reached. Consequently, iron and aluminum may become separated, even though the separation is not quantitative.*[2]

We turn next to the second condition that governs the formation of sedimentary beds of chemical or biochemical origin: the environment at the site of deposition. Since chemical sedimentary beds are deposited in water, conditions of (1) salinity, (2) acidity or alkalinity, and (3) oxidation-reduction potential are important environmental factors.

---

[1] pH is the symbol for the logarithm of the reciprocal of the hydrogen ion concentration. In one liter of pure water at 20°C the amount of H^+ is 10^{-7} g, and the amount of OH^- is also 10^{-7} g. The pH of pure water is thus 7, and such pH represents a neutral solution. Hydrogen-ion concentration greater than 10^{-7}, say 10^{-6}, would make the solution acid and be expressed as pH 6, whereas hydrogen-ion concentration less than 10^{-7}, such as 10^{-8}, would make the solution alkaline and be expressed as pH 8.

[2] Reprinted from "Geochemistry," by K. Rankama and T. G. Sahama, by permission of The University of Chicago Press. Copyright 1950.

Salinity relates to the concentration of dissolved solids per kilogram.[1] On the basis of salinity, water may be classed as (1) fresh, salinity $0.1\pm\%_0$; (2) brackish, salinity $10\pm\%_0$; (3) normal seawater, salinity $35\pm\%_0$; (4) hypersaline, salinity $200\pm\%_0$ (17, p. 24). Most chemical deposits apparently were formed in fresh water, brackish water, or normal seawater, so far as salinity is concerned, but one important group, the evaporites, was formed in hypersaline waters.

The pH of a solution determines, in part, whether material will or will not be precipitated and, if precipitated, in what abundance. Thus $CaCO_3$ is not precipitated in acid solution (less than pH 7), is precipitated sparingly in slightly alkaline solution (pH 7.0 to about 7.8), and is precipitated abundantly in alkaline solution (pH more than 7.8). On the other hand, SiO_2 deposits more abundantly at pH 7.0 to 7.8 than it does above pH 7.8, as SiO_2 is less soluble in acid than in alkaline solutions [(17, Fig. 8, p. 26); (11, pp. 157–166) of Chap. 2].

The oxidation-reduction potential, or redox potential,[2] relates to the conditions that favor oxidation or reduction of materials. Determination of the redox potential of more than one thousand samples of bottom sediments gave an Eh range of $+0.35$ to -0.50 volt (29, p. 477). The redox potential is of great importance in determining the kind of compound that will be deposited under given conditions or, conversely, the conditions of deposition indicated by deposits of certain compounds. Thus iron may be deposited abundantly in the ferric state as hematite or limonite at Eh more than 0.0 (oxidizing conditions), in the ferrous states as siderite at Eh between slightly less than 0.0 and about -0.2 (reducing conditions), and as pyrite at Eh less than about -0.2 (strongly reducing conditions) if the pH is about 7.0 to 7.8 (17, Fig. 8, p. 26).

Various investigators have attempted to evaluate the effect of redox potential and hydrogen-ion concentration on the deposition of sediments (4; 10, pp. 91–102; 17; 29). ZoBell (29) stated: "The oxidation-reduction or redox potential is believed to have a pronounced effect upon the composition, chemical activity, diagenesis, color, biological population, and other properties of recent sediments."[3] According to Fleming and Revelle (10), "The distribution of dissolved oxygen is a very important aspect of the conditions in the ocean. It not only reflects certain aspects of the biological conditions, but also has an important bearing on the nature of the sedi-

[1] Salinity can be computed from a knowledge of the amount, in grams, of chloride, bromide, and iodide contained in 1 kg of seawater [(10, p. 61); (11, pp. 184–185) of Chap. 2]. For simplicity, salinity may be considered as a conventional representation of the solids dissolved in seawater, expressed in parts per thousand or grams per kilogram of seawater. It is generally indicated by the symbol $\%_0$.

[2] "The redox potential . . . is a measure of how reducing or how oxidizing the system is with reference to some standard. When referred to hydrogen, the redox potential is commonly expressed as E_h in terms of volts, E being the potential difference between the standard hydrogen electrode and the system of which the redox potential is being measured. . . . E_h is an expression of the tendency of a reversible redox system to be oxidized or reduced." From C. E. ZoBell, *Am. Assoc. Petroleum Geologists Bull.,* vol. 30, p. 479. Copyright 1946. The American Association of Petroleum Geologists. Used by permission.

If $Eh = 0.0$, the material of a system is 50 percent in the reduced form. Positive values of Eh, as $+0.1$, indicate oxidizing conditions, negative values, as -0.1, reducing conditions. This is not strictly so, because the redox potential is affected by the hydrogen-ion concentration.

[3] From C. E. ZoBell, *Am. Assoc. Petroleum Geologists Bull.,* vol. 30, p. 478. Copyright 1946. The American Association of Petroleum Geologists. Used by permission.

ments existing in any area. . . ."[1] Krumbein and Garrels (17) discussed the classification and origin of chemical sediments in terms of pH and oxidation-reduction potentials and presented a diagram to show how some materials might be expected to be precipitated under various conditions.

Conditions of oxidation-reduction and to some extent hydrogen-ion concentration are determined largely by water circulation and depth. Free circulation in the sea results in water that is mildly oxidizing throughout and is alkaline, whereas restricted circulation in basins connected with the sea results in water that ranges from mildly oxidizing to strongly reducing and from neutral to somewhat alkaline. An estimate of Eh and pH of these two types of circulation at the surface and at bottom waters, with humid climate, as shown by Krumbein and Garrels (17), follows:

	pH		*Eh*	
*Type of circulation**	*Surface*	*Bottom*	*Surface*	*Bottom*
Normal marine open	8.2	7.8	0.2–0.4	0.1–0.3
Restricted	8.0	7.0	0.1	−0.3

* Reprinted from *Journal of Geology,* vol. 60, pp. 3–5, Origin and Classification of Chemical Sediments in Terms of pH and Oxidation-reduction Potentials, by W. C. Krumbein and R. M. Garrels, by permission of The University of Chicago Press. Copyright 1952.

The types of materials that might be expected to be deposited in fresh, brackish, and normal marine conditions, as a result of open circulation and varying restrictions on circulation, are shown in Table 5-4. This table is a generalization and an attempt to show some of the most likely materials that would be deposited under the conditions indicated. Natural deposits should not be expected to fit neatly into this scheme, because of fluctuating conditions and later diagenetic changes. Nevertheless, they do fit surprisingly well, judged by field relations and mineral associations.

High salinity of water, restricted circulation, and semiarid to arid climate will bring about deposition of salts containing calcium, sodium, magnesium, potassium, and some other constituents. In this manner gypsum, anhydrite, halite, and various potassium-bearing salts are precipitated. This deposition will be considered more fully in the discussion of evaporites.

Deposition of some specific materials.   The manner in which chemical activity and environment cause deposition of some specific materials as sedimentary beds will be discussed only briefly. Limestone and dolomite, and chert associated with such rocks and with sedimentary iron-bearing rocks, are omitted from discussion. The origins of dolomite and of chert are summarized well by Pettijohn [(13, pp. 421–424, 439–442) of Chap. 2].

[1] From "Recent Marine Sediments," Physical Processes in the Ocean by R. H. Fleming and R. Revelle, p. 91. Copyright 1939. The American Association of Petroleum Geologists. Used by permission.

Diatomite. Diatomite, composed chiefly of diatom tests that consist of opaline silica, originates from the accumulations of these tests that sink after the plant dies. The deposits formed in either marine or fresh water, and the most important ones are of Tertiary and Quaternary age. Many of the large deposits are associated with volcanic formations, and the suggestion has been made that the source of the silica that was extracted from water by the diatoms was volcanic activity [Chap. 2 (9, pp. 305–307)].

Phosphorite and Phosphatic Rocks. Phosphorite (phosphate rock or rock phosphate) is composed chiefly of phosphate minerals, whereas certain other rocks, as limestones and shales, contain only subordinate amounts of phosphate minerals and may be designated *phosphatic rocks,* as phosphatic limestones and phosphatic shales. The phosphorites include (1) the extensive bedded phosphates that are interbedded with other marine sediments, such as the Phosphoria formation of Permian age in the western United States; (2) the residual phosphates, such as the "land pebble" phosphates of Florida; (3) alluvial phosphates, such as the "river pebble" phosphates of Florida; and (4) replacement phosphates, as some of the deposits of Tennessee.

The bedded phosphorites commonly show a close association with organic matter, contain some glauconite and pyrite, contain phosphatic remains of organisms, are associated with black shales, and generally contain very little calcium carbonate. Pettijohn [(13, p. 475) of Chap. 2] stated:

> *The fossils prove the phosphorite to be marine. The black color and commonly associated hydrocarbon materials indicate anaerobic conditions. The absence of fossils of sessile and bottom-living types and the presence of depauperized forms as well as the pyrite and associated black shales further confirm this interpretation. The associated glauconite likewise suggests subnormal aeration. The scarcity of carbonate of lime either in skeletal form or as precipitated carbonate and the presence of much chert suggest a pH slightly less than normal.*[1]

Table 5-4 indicates that phosphorites form most abundantly in an environment of restricted to very restricted circulation and reducing to strongly reducing conditions. Such an environment is characteristic of some types of basins. Fleming and Revelle (10) defined a basin as "an area partially separated by land or submarine barriers from the open ocean and having an entrance, that is, a sill or threshold, shallower than the basin itself."[2] Further, regarding conditions in such a basin, they stated, in part (pp. 95–99), that since free intercommunication of water can take place only above sill depth, if the rate of interchange of water is very slow or effective only at long intervals, all the dissolved oxygen will be used up and stagnant conditions will result. Hydrogen sulfide will gradually dissolve into the water, and the concentration of P_2O_5 and probably organic substances in solution will be extremely high. Finally "The foul bottom waters and the bottom itself will be biologically sterile except for anaerobic bacteria. The bottom will be covered with

[1] From "Sedimentary Rocks," 2d ed., p. 475, by F. J. Pettijohn. Copyright 1957. Harper and Row, Publishers, Incorporated. Used by permission.
[2] From "Recent Marine Sediments," p. 95, Physical Processes in the Ocean by R. H. Fleming and R. Revelle. Copyright 1939. The American Association of Petroleum Geologists. Used by permission.

TABLE 5-4. **Materials Deposited under Various Environmental Conditions (Assuming salinity of water generally not more than about 35⁰/₀₀ to 40⁰/₀₀)***

Environment	Conditions of oxidation-reduction	pH	Materials deposited		
			Most abundant	*Appreciable amount*	*Accessory*
Normal marine open circulation	Strongly oxidizing	7.0–7.8	Hematite, limonite, pyrolusite, psilomelane	Chert, jasper	Calcite, phosphorite, magnetite
		7.8	Calcite, limestone		Hematite, limonite, pyrolusite, psilomelane, chert, phosphorite
Somewhat restricted circulation; less oxygen (aerobic to anaerobic)	Mildly oxidizing to mildly reducing	7.0–7.8	Iron silicates† (chamosite, greenalite, minnesotaite, iron chlorite), glauconite	Magnetite,† chert, jasper	Hematite, siderite, iron-bearing carbonate, phosphorite, calcite
		7.8	Calcite, limestone	Organic matter ?	Organic matter, hematite, limonite, iron silicates, glauconite, chert, phosphorite

Environment	Eh	pH			
Restricted circulation; somewhat stagnant; deficient oxygen (anaerobic)	Reducing	7.0–7.8 7.8	Siderite or iron-bearing carbonate, chert, phosphorite, organic matter Calcite, limestone, organic matter	Iron silicates, glauconite, rhodochrosite	Pyrite, calcite, hematite, primary uranium concentrations Siderite, iron-bearing carbonate, glauconite, rhodochrosite, phosphorite
Very restricted circulation; stagnant; little oxygen (anaerobic)	Strongly reducing	7.0–7.8 7.8	Pyrite, phosphorite, organic matter Calcite, limestone, organic matter	Chert	Siderite, iron-bearing carbonate, rhodochrosite, calcite, primary uranium concentrations, sulfides (chalcocite, chalcopyrite, alabandite ?) Phosphorite, pyrite, other sulfides (?)

* Compiled from three sources and used by permission of each publisher. (1) Reprinted from *Journal of Geology*, vol. 60, pp. 1-33, Origin and Classification of Chemical Sediments in Terms of pH and Oxidation-reduction Potentials by W. C. Krumbein and R. M. Garrels, by permission of The University of Chicago Press. Copyright 1952. (2) Review of Soviet Literature on Petroleum Source-rocks, by G. V. Chilinger (Review of paper by G. I. Teodorovich), *Am. Assoc. Petroleum Geologists Bull.*, vol. 39, pp. 764-767. Copyright 1955. The American Association of Petroleum Geologists. (3) Sedimentary Facies of Iron Formation by H. L. James, *Econ. Geology*, vol. 49, pp. 235-293. Copyright 1954. Economic Geology Publishing Company.
† A number of iron silicates, as well as magnetite, contain both ferric and ferrous iron and thus indicate mildly oxidizing to mildly reducing conditions.

black, stinking soft or flocculated mud. . . . The organic carbon content will be high, owing to decomposition under anaerobic conditions."[1] Woolnough (28) described conditions in the Black Sea, which is a very deep, almost completely landlocked basin, and stated:

> *At a depth of* 200 *meters there is a deficiency of oxygen, aërobic bacteria cease to live, and anaërobic bacteria flourish. While, then, there are more or less abundant plankton and nekton and a not inconsiderable influx of land-derived organic matter, there is a complete inhibition of bottom life. The deeper waters are rendered toxic by the action of sulphate-reducing bacteria which decompose sulphates with production of sulphuretted hydrogen and collateral precipitation of ferrous sulphide. The ferrous sulphide causes highly characteristic blackening of the deposited sediment, and the toxic environment entirely eliminates the benthos, including the scavengers. In these conditions much or all of the organic matter raining down from the superficial sunlit waters eventually reaches the bottom and is "pickled" there.*[2]

W. A. J. M. van Waterschoot van der Gracht suggested to Woolnough that deposits of the type formed in the Black Sea be designated *euxinic* facies, from *Pontus Euxinus,* the Black Sea. We may state, then, that the bedded phosphorites probably were formed under euxinic or near-euxinic conditions.

Evaporites. The evaporites are deposits that have accumulated from waters of high salinity, generally greater than 200‰, in restricted basins in areas of semiarid to arid climate. The essential conditions needed for the accumulation of a large deposit are constant or periodic renewal of supply of material to the basin, lack of many streams that bring fresh water into the basin, and a relatively high rate of evaporation. These conditions are well outlined by Fleming and Revelle (10):

> *The Gulf of Kara-Bugaz on the Caspian Sea is a well known, though perhaps extreme example of the characteristics of basins in semi-arid regions. The Gulf is separated from the Caspian by a bar 60 miles long and has a shallow entrance channel only a few hundred meters wide. No large streams enter the Gulf to offset the continued evaporation from its surface, but there is a constant inflow of water from the Caspian Sea over the threshold. The presence of the sill prevents a compensating outflow, hence the inward transport just balances the amount lost by evaporation. As a result of the evaporation and the continual supply of new material, the salinity of the deeper parts of the Gulf is steadily increasing. The salinity in* 1902 *was* 163.96‰ *as compared with* 12.67‰ *for the Caspian Sea as a whole.*[3]

[1] From "Recent Marine Sediments," pp. 98–99, Physical Processes in the Ocean by R. H. Fleming and R. Revelle. Copyright 1939. The American Association of Petroleum Geologists. Used by permission.
[2] From W. G. Woolnough, *Am. Assoc. Petroleum Geologists Bull.,* vol. 21, p. 1115. Copyright 1937. The American Association of Petroleum Geologists. Used by permission. Plankton are organisms that float freely and inertly in the upper, well aerated, sunlit layers of the sea; nekton are those that can swim more or less freely; benthos are those that live on and in the sea bottom.
[3] From "Recent Marine Sediments," p. 102, Physical Processes in the Ocean by R. H. Fleming and R. Revelle. Copyright 1939. The American Association of Petroleum Geologists. Used by permission.

Precipitation of material from the water of such a basin begins when saturation is attained for the less soluble substances, and, if supply of material and rate of evaporation are proper, precipitation of other materials may take place as saturation is increased. Thus a succession of materials may be deposited. Fluctuation in the amount and salinity of water coming into the basin, and changes in temperature, may bring about repetitions in deposition of some materials. Finally, almost complete evaporation of water, resulting from complete isolation of the basin from the ocean, may bring about precipitation of the most soluble constituents (28, pp. 1125–1127). The process of deposition is not as simple as this may indicate, however, because each substance is affected by the others that are in solution, and deposition is controlled partly by concentration and partly by temperature.

The chemistry of deposition has been studied experimentally. A study by Usiglio (51) in 1849, using seawater, showed that $CaCO_3$ and a minor amount of Fe_2O_3 began to deposit when the volume of solution had been reduced to 0.533 of the original amount; $CaSO_4 \cdot 2H_2O$ and minor $CaCO_3$ when the volume had been reduced to 0.190 of the original amount; NaCl, accompanied by some $CaSO_4 \cdot 2H_2O$ and minor amounts of $MgSO_4$ and $MgCl_2$ at a volume of 0.095 of the original. All four of these constituents, with NaCl far in excess of the others, continued to precipitate until the volume was 0.0302 of the original, when some NaBr also precipitated. Thereafter, until the volume was 0.0162 of the original, $CaSO_4 \cdot 2H_2O$ ceased to precipitate but the other four constituents, NaCl, $MgSO_4$, $MgCl_2$, and NaBr, continued to form, with NaCl still in excess. The final residue, only 0.015 of the original volume, contained NaCl, $MgSO_4$, $MgCl_2$, NaBr, and KCl. [See Chap. 2 (3, Table on p. 220).] A rough calculation shows the following approximate percentage of compounds precipitated and contained in the bittern:[1] NaCl, 77.49; $MgCl_2$, 8.65; $MgSO_4$, 6.46; $CaSO_4 \cdot 2H_2O$, 4.56; NaBr, 1.43; KCl, 1.38; $CaCO_3$, 0.03; Fe_2O_3, 0.001. Most of the $MgSO_4$, $MgCl_2$, NaBr, and all of the KCl, were in the bittern. Consequently the bulk of the material deposited before the volume of the solution became very small consisted of halite (NaCl) and gypsum ($CaSO_4 \cdot 2H_2O$); these two, in fact, are the major constituents of many evaporite deposits.

Other experiments have shown that some dolomite is the next material deposited after calcite and that either gypsum or anhydrite may be precipitated, depending on temperature and salinity. Further, some changes may take place after burial of a deposit and lead to reorganization of constituents (48).

The common evaporite deposits are salt (halite), gypsum, and anhydrite. Less common ones are those generally designated *potash deposits*. These may contain a variety of minerals, some of the most common of which are sylvite, KCl; polyhalite, $2CaSO_4 \cdot MgSO_4 \cdot K_2SO_4 \cdot 2H_2O$; kainite, $MgSO_4 \cdot KCl \cdot 3H_2O$; carnallite, $KMgCl_3 \cdot 6H_2O$; and langbeinite, $K_2SO_4 \cdot 2MgSO_4$.

Evaporates may accumulate in inland basins of semiarid to arid regions as well as in restricted marine basins, but such deposits generally are composed of different materials, because the composition of temporary inland lakes differs not only from that of the ocean but also from one region to another. In general terms, these lakes may be designated *saline lakes* and *alkaline lakes*. The saline lakes range in

---

[1] Bittern is a bitter liquid that is a residue in saltmaking after salt has crystallized from the original liquid. The term is used for the brine left after the extraction of sodium chloride from seawater.

composition from a chloride type, similar to oceanic water, through various gradations that contain more sulfate into a sulfate type. The alkaline lakes, which contain notable amounts of carbonates, range from those in which carbonates are generally in excess of other salts, through those in which carbonates and chlorides predominate, into those containing considerable amounts of carbonates, chlorides, and sulfates, and, less commonly, those in which sulfate is in excess of chloride or carbonate. A few lake waters contain relatively large amounts of B_2O [(3, pp. 156–180) of Chap. 2]. Deposition from some of these lakes furnishes sodium carbonate, sodium sulfate, some potash salts, and borates.

Sulfur. Elemental sulfur occurs in sedimentary beds associated with gypsum and limestone in various parts of the world but especially in the Union of Soviet Socialist Republics and Sicily. The origin of the sulfur in these bedded deposits has been a matter of some controversy, some maintaining that it was formed by replacement of gypsum by circulating waters and hence was of epigenetic origin, others proposing that the sulfur was formed from sulfate by bacterial action at the time of deposition or very shortly thereafter and hence was of syngenetic origin. Murzaiev (46) discussed the origin of some sulfate deposits of the Soviet Union and concluded that the sulfur of those bedded deposits was of syngenetic origin, being deposited in lagoons, with the aid of bacteria, contemporaneously with the enclosing rock, the bacterial reduction of sulfates yielding elemental sulfur. Segui (50) discussed the bedded sulfur deposits of Sicily and postulated that hydrogen sulfide from basaltic eruptions had been dissolved in ground-water, that hydrogen sulfide springs had entered a lagoon or pond into which other springs had carried calcium carbonate, and that both the sulfur and the gypsum of these bedded deposits resulted from reaction of these waters. Others have suggested that the gypsum was a marine deposit formed in the usual way, in which case the sulfur may have formed by bacterial activity.

Iron Deposits. Precambrian iron-bearing beds of world-wide distribution have been studied in some detail. They have been designated *iron formation* in the Lake Superior region of North America, *itabirite* in South America, *banded ironstones* in South Africa, *hematite quartzites* in India, and *quartz-banded ores* in Sweden. James (15, pp. 239–240) proposed a definition for *iron formation* that would be adequate for all these types. He defined *iron formation* as "a chemical sediment, typically thin-bedded or laminated, containing 15 percent or more iron of sedimentary origin, commonly but not necessarily containing layers of chert."[1] He stated, further, that "the definition does not provide a sharp distinction with the non-cherty iron-rich strata of younger age; indeed, insofar as the mineralogy and content of iron is concerned, there is no real difference."[1] The term *iron formation* thus is a convenient designation for all iron-bearing beds that contain 15 percent or more iron of sedimentary origin.

The vast amounts of iron contained in the sedimentary iron formations, the great quantities of silica also present in the iron formations of Precambrian age, commonly as layers interbedded with the iron minerals, and the somewhat general accumulation of iron and, to a less extent, silica as end products of weathering,

---

[1] From H. L. James, *Econ. Geology,* vol. 49, pp. 239–240. Copyright 1954. Economic Geology Publishing Company. Used by permission.

have raised questions regarding the sources, manner of transportation, and deposition of iron and silica.

Two major sources of iron and silica have been postulated: (1) from decomposition of rocks during weathering; (2) from volcanic activity, the iron, or silica, or both being obtained from magmatic emanations or from contact of seawater with hot lava. Van Hise and Leith (52) stated, regarding the sources of iron in the iron formations of the Lake Superior region: "The iron salts may have been transferred from the igneous rocks to the sedimentary iron-formations partly by weathering when the igneous rocks were hot or cold, but the evidence suggests also that they were transferred partly by direct contribution of magmatic waters from the igneous rocks and perhaps in small part by direct reaction of the sea waters upon the hot lavas." Others subsequently expressed somewhat similar ideas. (See review in 15, pp. 276–277.) The volcanic concept apparently was advanced because it was thought that ordinary weathering would not furnish an adequate supply of solutions of the proper type. Work has since shown, however, that weathering under certain tropical or subtropical conditions can supply an adequate amount of iron and silica. Gruner (42) stated: "If an average of three parts of iron per million is assumed for the water (which would be in keeping with the analyses), the Amazon River in 176,000 years could carry 1,940,000 million metric tons of iron to the sea—the amount assumed for the Biwabik formation. . . . The amount of silica carried would be correspondingly large."[1] It seems, therefore, that weathering under proper climatic conditions could well supply enough iron and silica to form a large sedimentary deposit.

Iron in solution is precipitated as various compounds—oxide, silicate, carbonate, sulfide (Table 5-4). James (15) proposed that iron formation could be divided into facies based on mineral composition and that these facies would indicate, approximately, the environment of deposition. He proposed the oxide, silicate, carbonate, and sulfide facies, but stated that the sulfide facies, which in the area studied is represented by a pyritic slate, is really not iron formation but, in places, is typically associated with iron formation. The oxide facies he subdivided into two types, the hematite-banded and the magnetite-banded, and the silicate facies into two types, granular and nongranular. The facies, the more usual minerals of the rocks, and the environments of formation are shown by Table 5-5.

A comparison of this table with Table 5-4 will show the probable environment somewhat more fully, the probable pH and relationships to the possible formation of other minerals, and the accumulation of organic matter.

The magnetite-banded type of iron formation needs some explanation as a primary sedimentary facies. The iron formations of the Lake Superior region, which were used to a large extent by James in formulating the facies of iron formation, have been variously metamorphosed, from a very low degree to a moderately high degree. Many people would, therefore, regard the magnetite in these iron formations as of metamorphic origin. Certainly some of it clearly was formed by metamorphism, and this is recognized by James. A number of workers believe that magnetite forms as a primary sedimentary mineral, but the amount so

[1] From J. W. Gruner, *Econ. Geology,* vol. 17, p. 455. Copyright 1912. Economic Geology Publishing Company. Used by permission.

TABLE 5-5. **Some Features of the Facies of Iron Formation***

Facies	Subdivision	Chief minerals	†Subordinate minerals	Environment
Oxide	Hematite-banded	Hematite and chert or jasper	Magnetite	Strongly oxidizing
	Magnetite-banded	Magnetite	Iron silicates, hematite, chert or jasper, carbonate	Mildly oxidizing to mildly reducing
Silicate	Granular	Iron silicates and chert	Magnetite, hematite, carbonate	Mildly oxidizing to mildly reducing
	Nongranular	Iron silicates	Magnetite, carbonate	Variable, but typically mildly reducing
Carbonate		Iron-rich carbonate and chert	Iron silicates, magnetite, pyrite	Reducing
Sulfide		Pyrite and graphite	Carbonate	Strongly reducing

* Compiled from H. L. James, *Econ. Geology,* vol. 49, pp. 235–293. Copyright 1954. Economic Geology Publishing Company. Used by permission.
† Not all subordinate minerals are included, but those omitted are present in small amounts locally.

formed has been and still remains a matter of doubt, and formation in this manner seems to be exceptional. James based his conclusion partly on the widespread occurrence in the Lake Superior region of magnetite associated with material that was practically unmetamorphosed. He stated (15), regarding the separation of iron formation that contains sedimentary magnetite from iron formation that contains metamorphic magnetite: "Abundance of magnetite in rocks that are essentially unmetamorphosed, as indicated by the fine grain of the chert and the presence of such low-grade minerals as greenalite and minnesotaite, is a valid criterion that should serve to separate the primary magnetite rock from the magnetite-bearing rocks that are products of later metamorphism."[1] James recognized that the formation of magnetite as a sedimentary mineral does not seem to be usual and apparently requires a delicate balance between oxidizing and reducing conditions.

Actual field occurrences of iron-bearing sedimentary rocks are more complex than might be suspected, and many gradations occur between the three facies and, indeed, between the carbonate and the sulfide facies. Further, some bedded hematites and some oolites of hematite contain detrital quartz; some iron-silicate rocks contain considerable amounts of fine detrital material and grade into ferruginous mudstones. Again, some rocks containing iron-bearing carbonate also contain varying amounts of clay and grade into clay ironstone, which contains

[1] From H. L. James, *Econ. Geology,* vol. 49, p. 263. Copyright 1954. Economic Geology Publishing Company. Used by permission.

1 to 30 percent clay. Clay ironstone may form nodules and layers underlying coal seams and may form an iron deposit known as "blackband" because of its dark color. The presence of detrital material in some iron-bearing rocks is not surprising, since oolites, ripple marks, and cross-bedding characterize these rocks in some places, and these features indicate deposition in a somewhat turbulent environment. Further, fluctuating environments of deposition could well account for mixtures of some facies. (See reference 13, p. 449–464, of Chap. 2 for a discussion of iron-bearing sediments.) Changes produced by diagenesis, weathering, and circulation of groundwater add further complexity to the iron-bearing beds that are encountered in the field. Diagenetic changes may bring about some replacements, such as oxides and silicates by carbonate and carbonate by pyrite. Changes produced by weathering and circulation of groundwater have been very great in some places and have brought about enrichment of some deposits. Such changes are discussed in connection with weathering and the circulation of groundwater.

Manganese Deposits. The chemical properties of manganese are similar to those of iron, and hence the origin of sedimentary manganese-bearing beds is similar to that of sedimentary iron formations. Some differences exist, however, since manganese, unlike iron, does not commonly form sedimentary silicates. Hence the manganese deposits are either oxide types or carbonate types. (See Table 5-4.) Teodorovich (Chilinger, 4) makes a distinction among the conditions of oxidation-reduction and alkalinity-acidity that bring about deposition of iron and manganese oxides and iron and manganese carbonates (Table 5-6).

A distinction not shown by Tables 5-4 and 5-6 is that whereas siderite forms nodules and lenses in beds, and also entire layers of considerable extent, rhodo-

TABLE 5-6. Conditions of Oxidation-reduction and Alkalinity-acidity Causing Deposition of Some Iron and Manganese Compounds*

Alkalinity-acidity	*Oxidation-reduction*		
	Oxidizing	*Weakly reducing*	*Reducing*
Weakly alkaline	Oxides of manganese (pyrolusite, psilomelane)	Chiefly carbonates of iron and manganese (siderite, rhodochrosite)	Carbonate and considerable sulfide of iron (siderite, pyrite)
Neutral	Oxides of manganese and iron (pyrolusite, psilomelane, hematite, limonite)	As above	As above
Slightly acid	Oxides of iron (hematite, limonite)		

* From G. V. Chilinger (Review of paper by G. I. Teodorovich), *Am. Assoc. Petroleum Geologists Bull.*, vol. 39, pp. 764–767. Copyright 1955. The American Association of Petroleum Geologists. Used by permission.

chrosite commonly forms only lenses and nodules. Consequently beds that contain siderite are of much greater importance as potential sources of iron deposits than are beds that contain rhodochrosite as potential sources of manganese deposits.

The manganese sulfide, alabandite, although shown in Table 5-4 as a possible mineral that would form in a reducing environment, is rare as a sedimentary mineral. Alabandite may need more strongly reducing conditions for its deposition than does siderite (17, pp. 13–14), and such conditions possibly are not attained in most sedimentary environments.

A very interesting account of investigations of nodules of manganese on the ocean floor was published in 1962 (22). These investigations have shown that concretions of manganese and iron oxides are widely distributed at the surface of the sediments of the three major oceans of the world. They occur as concretions lying on the surface of the soft sea-floor sediments, especially the red muds and the oozes that were formed far from the continental shores but in an oxidizing atmosphere maintained by sea-floor water currents, as indicated by ripple marks in the fine sediments on which the nodules rest. The concretions vary in diameter from 1 to 25 cm, but average about 3 cm. The iron and manganese content of concretions from the Pacific Ocean and the Atlantic Ocean are shown by Table 5-7. Some of the nodules from the Pacific Ocean contain more iron than manganese, and these generally were collected from regions that lie along the continents; but some nodules of this type are known from the central part of the south Pacific.

TABLE 5-7. Percentages of Manganese and Iron in Nodules from the Pacific and Atlantic Oceans*

	Weight percentages, dry weight basis†					
	Pacific Ocean, 54 samples			*Atlantic Ocean, 4 samples*		
Element	*Maximum*	*Minimum*	*Average*	*Maximum*	*Minimum*	*Average*
Mn	50.1	8.2	24.2	21.5	12.0	16.3
Fe	26.6	2.4	14.0	25.9	9.1	17.5

* From J. L. Mero, *Econ. Geology,* vol. 57, p. 761. Copyright 1962. Economic Geology Publishing Company. Used by permission.
† As determined by emission spectrography.

The manner of formation of the nodules is not entirely clear. Apparently any manganese and iron from river waters that escape precipitation near shore, plus additions from volcanic eruptions, sea-floor springs, and the decomposition of submarine igneous outcrops, form hydrated colloidal particles of manganese and iron oxides within the seawater, which settle to the bottom of the ocean and are agglomerated into nodules.

The quantity of these manganese nodules is surprising. Estimates of the abundance of nodules present on the sea floor indicate that about 10^{12} metric tons of nodules should now be at the surface of the sediments of the Pacific Ocean. This amount of material, if concentrated in a much smaller area, would make an ore body of large extent. The time required for the accumulation of these nodules on the floor of the Pacific Ocean is estimated to be 200,000 years. One is forced to wonder why a sedimentary bed of this type has not been noted at some place in the rocks now exposed at the earth's surface.

Deposits of Copper, Lead, and Zinc. The origin of certain deposits of copper, lead, and zinc contained in sedimentary beds has been a matter of debate for many years. The debate has concerned, chiefly, specific deposits and has related to whether the metallic elements were deposited with the sediments as a part of the sedimentary process, that is, were of syngenetic origin, or were introduced later by hydrothermal solutions or perhaps metamorphic activity, and were of epigenetic origin. Ample evidence exists to show that copper can be deposited by sedimentary processes. The evidence regarding lead and zinc is much less definite.

Primary heavy metal sulfides are shown in the diagram of Krumbein and Garrels (17, p. 26) as forming in a strongly reducing environment with pH 7.0 to 7.8. Teodorovich (Chilinger, 4) indicated conditions necessary for the deposition of some sulfides (Table 5-8).

TABLE 5-8. Conditions of Oxidation-reduction and Alkalinity-acidity Causing Deposition of Some Copper-bearing Sulfides*

	Oxidation-reduction	
Alkalinity-acidity	*Strongly reducing*	*Reducing*
Weakly alkaline	Chalcocite, chalcopyrite	
Neutral	Chalcocite, chalcopyrite	
Slightly acid	Chalcocite, chalcopyrite	Chalcocite, chalcopyrite
Acid	Chalcocite	Sulfides

* From G. V. Chilinger (review of paper by G. I. Teodorovich), *Am. Assoc. Petroleum Geologists Bull.*, vol. 39, pp. 764–767. Copyright 1955. The American Association of Petroleum Geologists. Used by permission.

A strongly reducing to reducing environment is one in which organic matter accumulates. Hence copper deposits that formed during the deposition of sediments might be expected to occur in black, bituminous or carbonaceous shales.

Lead and zinc sulfides are not listed specifically by either Krumbein and Garrels or Teodorovich but could be included in the primary heavy-metal sulfides listed by Krumbein and Garrels or in the unspecified sulfides listed by Teodorovich. Both lead and zinc go into solution during weathering, but the length of transportation is uncertain. Zinc should travel much farther than lead, and thus a separation of those two elements would be expected. The amount of zinc in seawater is somewhat more than the amount of manganese, and the amount of lead is only about half of the amount of manganese. The amounts supplied to the sea, however, are vastly different, as shown by Table 5-9. Presumably precipitation of lead and zinc from seawater would take place under somewhat the same conditions as the precipitation of copper. Based on the total amount of material supplied to seawater, the opportunity for the development of sedimentary deposits of copper, zinc, and lead would be much less than that for iron and manganese.

TABLE 5-9. Amounts of Some Elements in Sea Water*

Element	Total amount supplied to seawater, g/ton	Amount present in seawater, g/ton	Transfer percentage
Si	166,320	4	0.002
Mn	600	0.01	0.002
Fe	30,000	0.02	0.00007
Cu	42	0.011	0.03
Zn	79.2	0.014	0.02
Pb	9.6	0.005	0.05

* Reprinted from "Geochemistry," p. 295, by K. Rankama and T. G. Sahama, by permission of The University of Chicago Press. Copyright 1950.

Deposits of Uranium. Primary uranium concentrations, as shown in Table 5-4, may be precipitated under reducing to strongly reducing conditions and thus might be associated with organic matter or phosphorite. Uranium does occur in lignite and coal, in black shale, and in phosphorite, but it also occurs, in greater abundance, in sandstones and conglomerates. The origin of uranium in conglomerates is a matter of much controversy and will be discussed briefly in connection with individual deposits. The uranium in sandstone, however, appears not to have been deposited at the time the sediments were laid down, but to have been deposited somewhat later. Similarly, the uranium in lignite and coal may have been brought in after original deposition. The uranium in the black shales and phosphorites, however, probably was derived from seawater. The black shales and the phosphorites that contain uranium are thus considered by some to be the only truly sedimentary deposits of uranium (21, pp. 514–524).

The uraniferous black shales contain much organic material and iron sulfides and generally are more phosphatic than other shales, containing 0.5 to 3.0 percent P_2O_5. The uranium content is usually less than 0.2 lb per ton of shale and shows little variation in a particular layer over long distances. The uranium is not present as a distinct mineral. The uranium-bearing parts of the uraniferous shales are relatively thin in comparison with other strata formed in the same length of time. The rate of deposition was slow, and thus more time was available for concentration of uranium from seawater than in strata that accumulated more rapidly. Good examples are the Chatanooga shale of Tennessee and the Kölm-bearing alum shale of Sweden.

The uranium-bearing marine phosphorites, such as the Phosphoria formation of the northwestern United States, commonly contain 0.005 to 0.03 percent uranium, considerably more than the amount contained in the black shales. Generally, but not always, the uranium content increases as the phosphate content increases. Distinct uranium minerals commonly are not present in the phosphorites, but the uranium is thought to be in the phosphate mineral, a carbonate-fluorapatite, the uranium substituting for calcium.

A number of problems exist regarding the accumulation of the uraniferous black shales and phosphorites, but the major features are stated by McKelvey, Everhart, and Garrels (21) as follows: "Like the black shales, most of the uranium in the phosphorites and phosphatic nodules probably was derived from seawater at or shortly after the time the rocks were deposited. To a much greater degree than in the shales, however, some of the uranium may have been derived from later solutions or redissolved and redeposited by them."[1]

Deposits of group II: Residual deposits; oxidation and supergene enrichment deposits; groundwater deposits

Residual Deposits

The formation of a residual deposit depends on (1) suitable material from which the deposit may form, (2) favorable topographic situation of this material, (3) favorable climatic conditions, and (4) sufficient length of time.

Material may be suitable for the accumulation of a residual deposit if it contains (1) unstable minerals that may be decomposed under existing climatic conditions, part of the elements removed in solution and part of them converted into a new compound that is stable under the conditions that exist, or (2) stable minerals disseminated throughout the rock in sufficient amount that they can be concentrated into a commercial deposit as a result of decomposition and chemical removal of much of the rock material. Among deposits formed from materials of the first type are some that are mined for aluminum, manganese, iron, iron-nickel-cobalt, and some clays. Among deposits formed from materials of the second type are some that contain platinum, chromite, gold, cassiterite, wolframite, diamond, barite, and quartz. A tabulation of all possible materials that might yield various residual deposits would be difficult because of the variation in types. Some materials that have yielded residual deposits are shown by Table 5-10.

Clay and Bauxite.

Clay and bauxite are discussed together because the manner in which they form is similar and clay may be an intermediate product formed during the conversion of material into bauxite, although that is not always the case.

Some clay minerals come into existence during the chemical weathering of rocks that contain various aluminous silicates. They are not stable end products of weathering in temperate regions but may be converted into stable end products (aluminum hydroxide minerals) if the environment is suitable. The principal clay minerals, the chief aluminum hydroxide minerals, and their chemical compositions are shown in Table 5-11.

Change of clay minerals into aluminum hydroxide minerals, the chief components of bauxite, requires elimination of silicon and also some other elements. Table 5-3 shows that weathering should eventually remove calcium, magnesium, and potassium from montmorillonite and illite and leave aluminum and silicon, thus

---

[1] From V. E. McKelvey, D. L. Everhart, and R. M. Garrels, *Econ. Geology,* Fiftieth Ann. Vol., pt. 1, p. 523. Copyright 1955. Economic Geology Publishing Co. Used by permission.

TABLE 5-10.   Some Materials That Have Yielded Residual Deposits

Original material	Original stable minerals	Newly formed stable minerals	Economic product sought
Syenite, especially feldspathoidal types; some other igneous rocks; clays; some argillaceous limestones; possibly some volcanic tuff and similar material.		Clay minerals, bauxite	Clay, bauxite
Various igneous and metamorphic rocks, but especially schists; some limestones; original low-grade manganese deposits.		Pyrolusite, psilomelane, manganite, wad	Manganese ore
Sideritic and calcareous rocks; gabbroic rocks; pyritic deposits and rocks; low-grade iron deposits	Possibly magnetite, hematite, limonite	Limonite, goethite, hematite	Iron ore
Serpentinized peridotite; dunite	Magnetite, chromite	Limonite, goethite, garnierite, deweylite	Iron ore, nickel ore, chromium ore, cobalt ore
Periodotite; dunite; pyroxenite	Native platinum, chromite, diamond	Limonite, goethite	Platinum, diamond, chromite
Granite; granite pegmatite; greisen	Cassiterite, wolframite, quartz	Clay minerals	Tin and tungsten, possibly kaolinite, quartz
Various rocks and deposits that contain native gold	Native gold	Possibly clay minerals, iron or manganese oxides	Gold
Some limestones	Barite	Clay minerals	Barite, possibly clay

TABLE 5-11.   The Principal Clay Minerals and Aluminum Hydroxide Minerals

Clay mineral	Approximate composition	Aluminum hydroxide mineral	Approximate composition
Kaolinite	$Al_2O_3 \cdot 2SiO_2 \cdot 2H_2O$	Gibbsite*	$Al_2O_3 \cdot 3H_2O$
Montmorillonite	$(Mg, Ca)O \cdot Al_2O_3 \cdot 5SiO_2 \cdot nH_2O$	Boehmite	$Al_2O_3 \cdot H_2O$
Illite	$K_2O \cdot 3Al_2O_3 \cdot 6SiO_2 \cdot 2H_2O$	Diaspore	$Al_2O_3 \cdot H_2O$

* The aluminum hydroxide minerals are also designated *hydrargillite* or *alpha trihydrite* (gibbsite), *alpha monohydrite* (boehmite), and *beta monohydrite* (diaspore).

forming the hydrous aluminum silicate kaolinite. Iron present from the breakdown of any mafic or other iron-bearing minerals might also be precipitated, and a mixture of kaolinite and limonite result. From this mixture bauxite could be formed if both silica and iron were removed, or ferruginous bauxite if some iron remained. Some of the possible ways of achieving separation of aluminum, silicon, and iron were discussed previously, and we should now inquire more specifically into the conditions that would form kaolinite or bauxite.

Much has been written about the formation of kaolinite, bauxite, and laterite. The use of the term *laterite* by various people is not the same. One definition [(5, p. 389) of Chap. 2] is "a name derived from the Latin word for brick earth, and applied many years ago to the red, residual soils, or surface products, that have originated *in situ* from the atmospheric weathering of rocks. They are especially characteristic of the tropics." Sherman (25) stated:

> There is considerable confusion as to the definition of laterite, since there are workers who consider it to be entirely a ferruginous horizon and others who insist that laterite must have a high content of free aluminum oxide. However, in general the former is the accepted definition. The indurated ferruginous clay was given the name "laterite" by Buchanan, not because of its bricklike appearance but rather because bricks were cut from it and used as a building material. Many of the temples of India and Siam are built of this material.[1]

The chemical composition of laterite was not known at the time the name was given to it by Buchanan, but later the original material was found to consist chiefly of clay that approximates kaolinite, and iron oxide (12, p. 218). Further discussion of laterite by Sherman (25, p. 219) indicates that soil scientists think all reference to lithologic aspects should be omitted in defining laterite, and processes of formation should be emphasized, since tropical soils that are ferruginous will develop under one type of environmental condition and those that are rich in alumina under another type. Tropical soils may contain montmorillonite, kaolinite, or free oxide types of clay, depending on the conditions of formation. Sherman suggests the use of *ferruginous laterite* and *aluminous laterite* as adequate terms, and this seems to be a wise usage.

General agreement exists that kaolinite tends to form in an acid environment (37) whereas bauxite tends to form in an alkaline environment. Climate, however, seems to be a critical factor. The conditions necessary for the formation of bauxite (12, pp. 23–24) are (1) a moist, warm climate, one in which rainfall considerably exceeds evaporation and the temperature exceeds 77°F most of the time and in which the microflora can destroy humus faster than the macroflora can produce it; (2) rock located in topographically high and well-drained places, above the level of the permanent water table; (3) a long period of time. Although the rock must be located high enough to permit good drainage, the relief should be such that erosion does not remove residual material that forms. The distinction between the forma-

---

[1] From "Problems of Clay and Laterite Genesis," p. 154, The Genesis and Morphology of Aluminum-rich Laterite Clays by G. D. Sherman. Copyright 1952. The American Institute of Mining and Metallurgical Engineers. Used by permission.

tion of bauxite and kaolinite is summed up by Gordon and Tracey (12, p. 24) as follows:

> *Bauxite is formed . . . when under tropical conditions rain water moves continuously through a porous aluminous rock. Kaolin is formed when the same rock is acted upon by water containing dissolved organic matter, or strong acids. Kaolin is also formed in areas subject to long dry periods, particularly where the annual rainfall is between 25 and 35 in. . . .*[1]

Harder (13, 14) discussed bauxite deposits that illustrate variations in origin. He stated that normal weathering processes of temperate regions form hydrous silicates and hydrous oxides such as those of iron and manganese. The hydrous silicates (the clay minerals) were regarded as intermediate products which, if weathering is carried to completion, as in tropical regions, would be converted into hydrous oxides. He placed bauxite deposits in five groups, as follows (14):

1. *Those derived from rocks rich in alkali-aluminum silicates, particularly the feldspathoids, with alkali feldspars, pyroxenes, amphiboles, and micas, but no quartz;*
2. *Those closely associated with limestones and presumably resulting from the decay of their residual products;*
3. *Those derived from the weathering of hydrous aluminum silicate rocks such as sedimentary clays with little free quartz;*
4. *Those associated with intermediate and basic rocks such as diorite, diabase, dolerite, and basalt, containing abundant calcium-aluminum and calcium-magnesium-iron silicates, but little or no quartz;*
5. *Those resulting from the intensive weathering of various moderately aluminous igneous, metamorphic, and sedimentary rocks such as granite, syenite, gneiss, phyllite, slate, and shale, containing more or less quartz.*[2]

Deposits of the first type are likely to be of high grade, especially if feldspathoids are present, since the rocks from which the deposits formed were initially high in alumina and low in silica. Among such deposits are those of central Arkansas, of Minas Gerais, Brazil, and those in the Los Islands, West Africa.

Deposits of the second type, associated with limestone, are widespread and include important deposits in Europe and elsewhere. These deposits appear to have formed as follows (14):

> *After the widespread deposition of the limy sediments from which the bauxite was eventually formed, extending in some places through several geologic epochs, a period of emergence followed. . . . Then followed a period of profound erosion accompanied by more or less denudation, during which great thicknesses of limestone were worn away, and the resulting residues, mostly*

[1] From "Problems of Clay and Laterite Genesis," p. 24, Origin of Arkansas Bauxite Deposits by M. Gordon, Jr., and J. I. Tracey, Jr. Copyright 1952. The American Institute of Mining and Metallurgical Engineers. Used by permission.
[2] From "Problems of Clay and Laterite Genesis," p. 36, Examples of Bauxite Deposits Illustrating Variations in Origin by E. C. Harder. Copyright 1952. The American Institute of Mining and Metallurgical Engineers. Used by permission.

clayey in character, were in part transported and in part deposited in situ on the unevenly weathered limestone surface. Because of irregularities in solution, this surface presented the usual limestone erosion forms consisting of troughs and pits separated by intervening ridges, knobs, and pinnacles, so that the residual material filling the depressions took the form of elongated or circular funnel-shaped pockets ranging up to 50 or 75 ft., or even more, in depth. Where heavy residual accumulation occurred, such forms often coalesced by the formation of a more or less continuous blanket completely covering the projecting ridges and knobs of limestone and joining together the pockets and trough fillings between them. Thus there developed continuous sheets of residual material with even horizontal, or undulating upper surfaces, but with many pockety irregularities projecting downward from their lower surfaces.[1]

Bauxite deposits occurring with sedimentary clays, the third type given by Harder, are in some places interbedded with sediments, and in others they are close to the surface and are overlain only by alluvium or soil. This type is well represented in the Coastal Plain region of the southeastern United States in which are deposits of kaolinite as well as bauxite. Apparently the bauxite deposits were formed by the weathering of clay deposits.

Bauxite or laterite associated with intermediate and basic igneous rocks, the fourth type, is common in tropical areas. Many of such deposits are ferruginous laterites that contain only a moderate amount of aluminum hydroxides. Some, however, are aluminous laterite or bauxite, and these were formed from rocks that contained larger amounts of aluminous minerals, especially lime-soda feldspars. Even these, however, contain considerable iron and usually are low-grade aluminum deposits, although lenses of medium- to good-grade material may be present. Many of the deposits of India, developed on the Deccan Plateau, are of this type.

Deposits of the fifth type, formed by intensive weathering of a variety of moderately aluminous rocks, include some of the most important ones of the world, such as the deposits of British Guiana and Surinam, and some of the deposits of Brazil and India. These deposits show considerable variation, since they were formed from rocks of different types, but all of them are separated from the original rock by material of considerable thickness that is formed of decomposition products intermediate between bauxite and the parent rock. The bauxite commonly is underlain by residual clay that grades downward into decomposed rock and finally into fresh rock.

Thus bauxite may form from various rock types, but in general the best deposits form from those rocks that are the most feldspathic, that contain little or no quartz and but small amounts of mafic minerals.

Iron and iron-nickel-cobalt-chromium. Iron oxides and hydroxides are stable end products of chemical weathering, and they may be accompanied by compounds of manganese, silicon, and aluminum (Table 5-3). Manganese generally tends to separate from iron, at least to some extent, but a certain amount of admixture of

[1] From "Problems of Clay and Laterite Genesis," p. 42, Examples of Bauxite Deposits Illustrating Variations in Origin by E. C. Harder. Copyright 1952. The American Institute of Mining and Metallurgical Engineers. Used by permission.

iron and manganese is not objectionable in deposits used for some purposes. Many iron ores contain up to 8 or 9 percent silica, and some contain more than that amount. A certain amount of silica, therefore, is not objectionable. Alumina, because of its refractory character, makes recovery of iron from an ore difficult. Alumina is, therefore, objectionable.

The amount of silicon and aluminum that may accumulate in a residual iron deposit will depend in part on the chemical environment and in part on the original rock that undergoes decomposition. Table 5-10 shows some of the most important materials from which residual iron deposits have formed. These will be grouped in four divisions for the purpose of discussion: (1) sideritic and calcareous rocks; (2) low-grade iron deposits; (3) pyritic deposits and rocks; (4) gabbroic rocks and serpentinized peridotites.

Sideritic rocks and calcareous rocks that contain iron generally are low in alumina, although they may contain a considerable amount of silica. The ferrous iron of these rocks would be converted into ferric iron during weathering, and siderite that contains little or no chert is an exceptionally favorable rock for producing a residual iron deposit.

Low-grade iron deposits might be represented by very cherty siderites, by calcareous hematites, siliceous hematites, iron silicate ores, or admixed iron silicate and iron carbonate ores. Regardless of differences in composition, weathering would tend to produce approximately the same end products, siliceous limonite or hematite of varying iron content. Iron formations of many types have been enriched locally at the surface by weathering that has formed a residual deposit over the underlying material. Such deposits properly belong with those formed by secondary enrichment. The manner of formation at the surface is essentially the same as residual concentration, but some secondary enrichment in iron deposits may take place by groundwater circulating to a considerable depth.

Pyritic deposits and rocks, although decomposed readily, are not good sources of residual iron deposits because of the difficulty of removing enough sulfur, which is an undesirable element in iron ores. Further, few of them are large enough to furnish an iron deposit. Many are part of an ore body that contains metals other than iron, and the residual capping formed over such bodies is essentially a part of the work of oxidation and supergene enrichment, although it actually was formed in about the same way as the residual deposits. One of the best examples of the decomposition of huge pyritic lenses to form iron ore is the Huelva area, Spain, where massive hematite was formed, which contains over 50 percent iron and 10 to 15 percent siliceous and argillaceous matter. The average depth of the iron ore is 30 m (35).

Gabbroic rocks and serpentinized peridotites are somewhat similar in composition, as they both contain rather high amounts of iron. The gabbroic rocks, however, are more highly aluminous because of the large percentage of feldspar they contain. The alumina content of gabbros is objectionable, and weathering in tropical climates would tend to produce ferruginous laterites rather than good iron ores. Serpentinized peridotites, in which the original silicates have already undergone some alteration, have furnished iron deposits in some places, especially in Cuba. Deposits developed from serpentine are likely to be important for nickel and cobalt, and also chromium, along with iron, because small amounts of these elements are contained in a number of peridotites. Analyses of iron ore that grades downward into serpentine at Mayari, Cuba, made at 1-ft. intervals, show clearly the changes in amount of various constituents. Selected analyses, enough to show the nature of the changes, are presented in Table 5-12.

TABLE 5-12.   Selected Analyses of Iron Ore That Grades Downward into Serpentine, Mayari District, Cuba*

| Constituent | Depth from surface, ft | | | | | |
	1–2	*6–7*	*20–21*	*24–25*	*25–26*	*28–29*
Fe_2O_3	64.70	72.90	66.55	62.90	50.25	10.14
Fe	45.35	51.00	46.58	40.00	35.12	7.10
Ni and Co	0.33	1.09	1.57	1.43	1.80	0.97
Cr	0.96	2.19	2.27	1.85	1.89	0.20
MgO	—	—	—	0.50	6.49	33.69
SiO_2	2.36	2.20	7.44	17.40	22.54	39.80
Al_2O_3	20.81	8.29	5.13	4.00	4.57	1.39
P	0.022	0.007	0.003	0.003	0.002	0.001
S	0.12	0.19	0.09	0.12	0.06	0.06
H_2O+	10.63	11.35	12.45	11.64	13.65	13.31
Total	99.952	98.217	95.503	99.843	101.252	99.561

* From C. K. Leith and W. J. Mead, *Am. Inst. Mining Engineers Trans.*, vol. 53, p. 76. Copyright 1916. The American Institute of Mining Engineers. Used by permission.

The nickel of deposits of this type tends to concentrate in the lower part of the iron ore (43). (Note Table 5-12.) In places it may migrate downward, become separated to a considerable extent from the iron, and form a nickel deposit. This is well illustrated in New Caledonia, where a relatively low-grade lateritic iron deposit or ferruginous laterite was developed from serpentinized peridotite and is underlain at depths of 20 to 75 ft by a nickel deposit that has been highly productive (38).

Intermediate between the lateritic iron deposits that contain recoverable nickel and the residual nickel deposits, there may be formed low-grade lateritic nickel deposits, such as the one at Nickel Mountain, Riddle, Oregon. The ore there, as it was delivered to the mine, contained an average of 1.5 percent Ni and 8 to 15 percent Fe (33).

Manganese.   The formation of residual manganese deposits is similar to that of iron deposits, but the source rocks generally are different. The most important source rocks throughout the world have been certain schists of India, Africa, and Brazil that contain manganese garnet (spessartite), rhodonite, and some other manganiferous minerals. Less important source rocks have been some limestones and manganiferous sedimentary rocks. The schists, as least in part, were formed by metamorphism of original manganiferous sedimentary rocks.

Prolonged weathering caused the decomposition of the original rocks, removed silica and alumina, and left a residue of pyrolusite, psilomelane, cryptomelane,[1]

---

[1] The term *cryptomelane* is used for a mineral that has the same general appearance as psilomelane but that contains K instead of Ba.

and usually some iron oxides; certain manganese residues may be associated with clay.

Other materials.   Other materials have been concentrated into workable residual deposits at various places. Native gold, native platinum, and cassiterite probably make the most important residual concentrations of metallic elements; scheelite, wolframite, columbite-tantalite, and a few other minerals have accumulated locally. Diamond is the most important residual mineral among the industrial materials. Locally residual deposits of such materials may grade into eluvial placer deposits.

Oxidation and Supergene Enrichment Deposits

The oxidation and supergene enrichment deposits are similar in many ways to residual deposits, since they originate by weathering and the action of groundwater, and in many cases part of the deposit consists of a surface residue. They differ, however, in the amount of activity of groundwater, which carries material downward in solution a few feet to several hundred feet, and from which material may be deposited in part as oxides, carbonates, sulfates, silicates, and similar oxygen-bearing compounds, in part as native metals, or in part as sulfides. Further, the oxidation and supergene enrichment deposits always form from some previously existing mineral deposit, which may range from high grade to low grade and give rise to one or more enriched zones beneath the outcrop or near-surface position of the original deposit. Moreover, many of the minerals of the original deposit are sulfides of various metals admixed with iron sulfides; hence the chemical processes involve to a large extent the decomposition of sulfides and subsequent deposition from sulfate solutions.

The decomposition of sulfides.   The iron sulfides have been shown to be of great importance in promoting decomposition and enrichment of deposits. (See 2, 3, 6, 7, 8, 19.) Experiments by Buehler and Gottschalk (3) indicate that the following equation expresses the reaction that takes place as pyrite decomposes:

$$FeS_2 + 7O + H_2O \rightarrow FeSO_4 + H_2SO_4$$

The ferrous sulfate formed, in the presence of sulfuric acid and additional oxygen, may react as follows to form ferric sulfate, a powerful oxidizing agent.

$$2FeSO_4 + H_2SO_4 + O \rightarrow Fe_2(SO_4)_3 + H_2O$$

Ferric sulfate may also be formed from ferrous sulfate in oxygenated water, thus:

$$6FeSO_4 + 3O + 3H_2O \rightarrow 2Fe_2(SO_4)_3 + Fe_2(OH)_6$$

Some of the ferric sulfate, through hydrolysis, is converted into ferric hydroxide:

$$Fe_2(SO_4)_3 + 6H_2O \rightarrow 2Fe(OH)_3 + 3H_2SO_4$$

Finally, the ferric hydroxide would yield limonite.[1]

[1] Limonite is not a single mineral with a definite composition. It is a very convenient term to use for hydrous iron oxides, the actual composition of which is not known but which probably contain material of the composition of goethite, $HFeO_2$, and lepidochrosite, $FeO(OH)$, with some water, and probably also some hydrous ferric oxide, $Fe_2O_3 \cdot nH_2O$. (See Dana's "System of Mineralogy," 1951, vol. 1, pp. 685–686.)

These equations show that decomposition of pyrite and marcasite yields ferric sulfate and sulfuric acid and thus provides important solvents for other constituents. Pyrrhotite is equally effective. The effectiveness of this decomposition of iron sulfides to form sulfuric acid and then ferric sulfate is illustrated by the use made of it to take chalcocite into solution in the leaching of copper (8, p. 113–114) as follows:

$$Fe_2(SO_4)_3 + Cu_2S \rightarrow CuSO_4 + 2FeSO_4 + CuS$$

The covellite (CuS) formed is then dissolved:

$$Fe_2(SO_4)_3 + CuS + 3O + H_2O \rightarrow CuSO_4 + 2FeSO_4 + H_2SO_4$$

Other sulfides are similarly affected, as shown by the following equations:

$$Fe_2(SO_4)_3 + PbS + 3O + H_2O \rightarrow PbSO_4 + 2FeSO_4 + H_2SO_4$$

$$Fe_2(SO_4)_3 + ZnS + 3O + H_2O \rightarrow ZnSO_4 + 2FeSO_4 + H_2SO_4$$

$$2Fe_2(SO_4)_3 + CuFeS_2 \rightarrow CuSO_4 + 5FeSO_4 + 2S$$

The sulfur is subsequently changed to H_2SO_4 or SO_2 under oxidizing conditions. Sulfuric acid also attacks minerals and converts them into sulfates, as indicated by the decomposition of bornite.

$$Cu_5FeS_4 + 2H_2SO_4 + 18O \rightarrow 5CuSO_4 + FeSO_4 + 2H_2O$$

In addition to the importance of iron sulfides in providing sulfuric acid and ferric sulfate, experiments indicate that they promote an increase in the percentage of lead, zinc, and copper sulfides oxidized, as a result of electrolytic action due to contact with the iron sulfides (3; 8, pp. 109–112).

In the absence of iron sulfides, oxidation takes place slowly as a result of atmospheric air, as indicated by the following equations:

$$CuFeS_2 + 8O \rightarrow FeSO_4 + CuSO_4$$

$$PbS + 4O \rightarrow PbSO_4$$

$$ZnS + 4O \rightarrow ZnSO_4$$

Bacterial activity may also be a factor in promoting oxidation and enrichment of certain copper deposits. Studies by Sutton and Corrick (26) of the U.S. Bureau of Mines have shown that certain bacteria (1) oxidize ferrous iron and thus cause formation of ferric sulfate and (2) oxidize elemental sulfur and cause formation of sulfuric acid. Thus they bring into being the same two constituents, ferric sulfate and sulfuric acid, that were previously shown to attack sulfides and produce sulfate solutions. The bacteria investigated and some results obtained are summarized in Table 5-13.

This table shows that both *F. ferrooxidans* and *T. ferrooxidans* oxidize ferrous iron (Fig. 5-4) but that *F. ferrooxidans* does so in appreciable quantity only in acid solution. The chalcopyrite used in one of the experiments contained traces of siderite, which caused the pH of the environment to remain above 4.6 and inhibited the oxidation of ferrous iron by *F. ferrooxidans*. After one experiment had been in progress fourteen days without appreciable oxidation of ferrous iron in chalcopy-

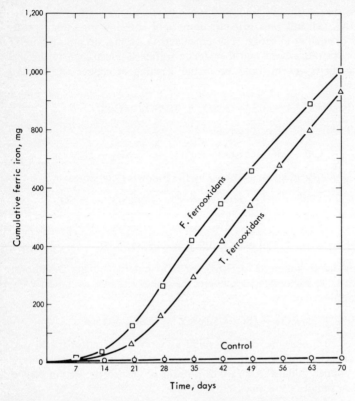

Fig. 5-4 Diagram showing ferric iron produced from pyrite by iron-oxidizing bacteria. After J. A. Sutton and J. D. Corrick, *Mining Engineering,* vol. 15, no. 6, p. 39. Copyright 1963. The American Institute of Mining, Metallurgical, and Petroleum Engineers. Used by permission.

TABLE 5-13.   Some Results of Experimental Microbial Leaching of Copper Minerals*

		Material oxidized		Some results	
Bacteria	pH conditions	Ferrous iron	Elemental sulfur	Greatly accelerate dissolution of iron from pyrite and chalcopyrite	Can bring about dissolution of significant amounts of copper from chalcocite, covellite, and bornite
Ferrobacillus ferrooxidans	Active if pH below 4.5	X		X	X
Thiobacillus ferrooxidans	Active if pH 6.0	X	X	X	
Thiobacillus concretivorous			X		

* Compiled from J. A. Sutton and J. D. Corrick, *Mining Eng.,* vol. 15, no. 6, pp. 37–40. Copyright 1963. The American Institute of Mining, Metallurgical, and Petroleum Engineers. Used by permission.

rite, sulfuric acid was added to the solution a drop at a time until the pH was below 4.6, after which the bacteria became active and oxidation of iron proceeded (Fig. 5-5).

Sutton and Corrick (26) presented five equations to indicate the chemical reactions involved in the microbial dissolution of iron and copper from sulfide minerals. These follow:[1]

$$2FeS_2 + 7O_2 + 2H_2O \rightarrow 2FeSO_4 + 2H_2SO_4 \qquad (5\text{-}1)$$

$$4FeSO_4 + 2H_2SO_4 + O_2 \xrightarrow{\text{bacteria}} 2Fe_2(SO_4)_3 + 2H_2O \qquad (5\text{-}2)$$

$$7Fe_2(SO_4)_3 + FeS_2 + 8H_2O \rightarrow 15FeSO_4 + 8H_2SO_4 \qquad (5\text{-}3)$$

$$Cu_2S + 2Fe_2(SO_4)_3 \rightarrow 2CuSO_4 + 4FeSO_4 + S \qquad (5\text{-}4)$$

$$2S + 3O_2 + 2H_2O \xrightarrow{\text{bacteria}} 2H_2SO_4 \qquad (5\text{-}5)$$

These equations indicate that after the initial oxidation of pyrite (5-1) the bacteria oxidize the ferrous sulfate to ferric sulfate in the presence of oxygen and sulfuric acid (5-2); the ferric sulfate formed can attack pyrite and form more ferrous sulfate and sulfuric acid (5-3), or it can react with a copper sulfide (5-4) and form copper sulfate, ferrous sulfate, and elemental sulfur, after which the ferrous sulfate is reoxidized by the bacteria, bringing about repetition of the cycle. The elemental sulfur, in the presence of oxygen and water, is oxidized to sulfuric acid by *T. concretivorous* (5-5). Experiments indicate that much copper may be extracted from chalocite (Cu_2S) by *F. ferrooxidans* (Fig. 5-6). Sutton and Corrick concluded that *F. ferrooxidans* acts as a catalyst and does not directly attack the iron in the mineral but, instead, oxidizes the ferrous iron released in the solution. Also, *T. concretivorous* does not oxidize the sulfide sulfur, but only any elemental sulfur set free.

Bacterial leaching of copper from mine-dump waste has been under investigation by the Utah Division of the Kennecott Copper Corporation for several years, and it is expected that this process will be used to extract copper from mine waste now in dumps. Their work indicates that best results will be achieved if conditions are carefully controlled—temperature maintained around 35°C (95°F) and pH between 2.0 and 3.5, among other things. Their experiments included not only copper-bearing and iron-bearing materials, but also some that contained sphalerite and molybdenite. These experiments showed that, in addition to extracting much copper from various samples, in 337 days, using a synthetic solution, (1) 19.6 percent of the zinc was extracted from sphalerite; (2) 48.6 percent of the zinc and 100 percent of the iron from sphalerite and pyrite; and (3) in 123 days, 28.2

[1] From J. A. Sutton and J. D. Corrick, *Mining Eng.,* vol. 15, no. 6, pp. 39–40. Copyright 1963. The American Institute of Mining, Metallurgical, and Petroleum Engineers. Used by permission.

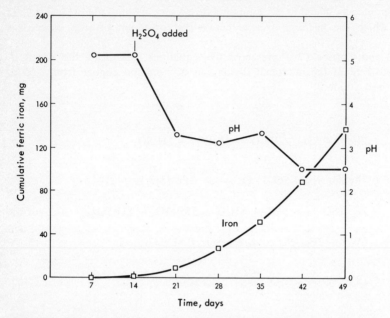

Fig. 5-5 Diagram showing effect of pH upon the ability of *F. ferro-oxidans* to oxidize iron in chalcopyrite. After J. A. Sutton and J. D. Corrick, *Mining Engineering,* vol. 15, no. 6, p. 40. Copyright 1963. The American Institute of Mining, Metallurgical, and Petroleum Engineers. Used by permission.

percent of the copper, 9.8 percent of the iron, and less than 0.1 percent of the molybdenum from copper-bearing molybdenite concentrate.

These experimental results are extremely interesting and strongly suggest that bacterial action may have been an important contributing factor in the oxidation of some mineral deposits.

Migration and precipitation of material.   The sulfates formed by decomposition of sulfides, along with other constituents of weathering, may, if relatively insoluble, as in the case of lead sulfate, remain very near the site of origin, or they may be transported a few tens or a few hundreds of feet and the metals be redeposited either under oxidizing conditions, generally as oxides, carbonates, or silicates, or under reducing conditions as sulfides.

Ferric sulfate may be converted into ferric hydroxide and be deposited as limonite at the site of the decomposition of the original sulfide and within the cavity formerly occupied by the original sulfide (indigenous limonite); or the transformation from ferric sulfate to limonite may be delayed by transportation of the ferric sulfate a short or a long distance from the site of the original mineral, followed by transformation into ferric hydroxide and then by deposition (transported limonite) (19, pp. 37–42, 102–104). The oxidation of ferrous sulfate to ferric sulfate is retarded by the presence of free acid and accelerated by the

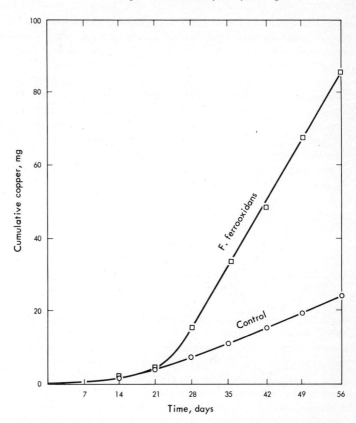

Fig. 5-6 Diagram showing copper extracted from chalcocite by *F. ferrooxidans*. After J. A. Sutton and J. D. Corrick, *Mining Engineering*, vol. 15, no. 6, p. 40. Copyright 1963. The American Institute of Mining, Metallurgical, and Petroleum Engineers. Used by permission.

presence of copper (19, p. 17). The oxidation of pyrite yields free sulfuric acid, and thus the formation of ferric sulfate would be retarded, ferrous sulfate would move downward in solution, and limonite would not be deposited in great amount near the site of decomposition of the pyrite. If some other substance, such as limestone, should be present to neutralize the acid, however, limonite might be deposited near the original pyrite. The oxidation of chalcopyrite furnishes no sulfuric acid, only ferrous sulfate and cupric sulfate, and thus the rate of oxidation of ferrous sulfate to ferric sulfate generally would be more rapid, and limonite would tend to accumulate near the site of decomposition of the chalcopyrite. In the case of massive iron-bearing sulfide deposits, much limonite may accumulate at about the position of the former outcrop of the ore deposit.

If materials are to be deposited as sulfates, oxides, carbonates, or silicates, and not as sulfides, the environment of deposition must be one of oxidation. Experimental work (47) has shown that the action of sulfuric acid and ferric sulfate on many metallic sulfides results in gradual decrease in acidity and a tendency toward neutralization, and this led to the general conclusion that the acidity of descending solutions should decrease with depth and that ultimately the solutions should become neutral or alkaline. Moreover, the type of gangue through which waters move is highly important. The gangue may be inert, as quartz, and cause no change in the solutions; it may be slowly reacting, as sericite; moderately reacting, as coarse-grained feldspar; or rapidly reacting, as carbonates (19, p. 64). If limestone is the country rock through which solutions move, the free acid developed during oxidation reacts with the limestone and the solutions become saturated with CO_2. Consequently much deposition would occur as carbonates of the metals. (See 1, p. 335; 11, p. 154.) If the gangue were inert or only slowly reacting, the solutions might be carried deeper and deeper, gradually changing in character, losing dissolved oxygen, and the environment might change from one of oxidation to one of reduction. Conditions would then be favorable for deposition of materials as sulfides.

The position of the water table at the time the solutions move downward has been cited in many cases as marking the approximate boundary between oxidizing and reducing conditions. In a general way this is true. Lindgren and Ransome (44) noted that, in the Cripple Creek district, Colorado, the depth of oxidation coincided, in a general way, with the depth of the water table but that there were numerous exceptions and qualifications. In some places of impervious and massive rocks, unoxidized minerals occurred high above the water table, and in other places, where oxygen was available, oxidation extended several hundred feet below the water table. Variations of this type have been described at numerous places. Joints, faults, and other larger openings may permit oxygenated water to move considerable distances below the general level of the water table and thus promote oxidation. With exceptions of this type in mind, the water table at the time of enrichment of the deposits may be used as the most likely separation between deposition of oxidized minerals and sulfide minerals. Garrels (11, p. 167) suggested that the boundary between oxidation deposition and sulfide deposition could be established on the basis of oxidation-reduction potential (Eh) and pH, sulfide deposition generally being at Eh lower than 0.2 at pH 2 and changing with Eh and pH so that it would be at Eh lower than about -0.1 at pH 8. Garrels stated (p. 153) that the environment beneath the water table is mildly alkaline and moderately reducing.

The exact control of deposition of material as sulfides has been a matter of considerable investigation, and the concensus now is that precipitation results chiefly from selective replacement of original sulfides of a deposit as sulfate solutions come in contact with them. The early work of Anthon (30) and Schürmann (49) resulted in the establishment, by Schürmann, of a series of metals so arranged that the sulfide of a metal higher in the series would be precipitated at the expense of any sulfide lower in the series. Wells (54) reviewed the series established by Schürmann and noted that Schürmann's series is very nearly the same as the order of solubility of sulfides in pure water, as determined by Weigel (53). This entire matter was discussed by W. H. Emmons (8, pp. 117–119, 137–149), who presented the following comparison of Schürmann's series and Weigel's series of solubilities (Table 5-14). Although the order is not the same throughout, the agreement is relatively close. If the metals of Schürmann's series most com-

TABLE 5-14.   Comparison of Schurmann's Series and Weigel's Series*

Schurmann's series†		Weigel's series		
Order	Metal	Order	Sulfide	Solubility, moles per liter
1	Hg	1	HgS	0.054
2	Ag	2	Bi_2S_3	0.35
3	Cu	3	Ag_2S	0.552
4	Bi	4	CuS	3.51
5	Cd	5	PbS	3.6
6	Pb	6	CdS	9.0
7	Zn	7	NiS	39.81
8	Ni	8	CoS	41.62
9	Co	9	FeS	70.1
10	Fe	10	ZnS	70.6
11	Mn	11	MnS	71.6

* From W. H. Emmons, *U.S. Geol. Survey Bull.* 625, p. 119, 1917.
† Schurmann's series is one in which the sulfide of a metal higher in the series, as Cu, for example, would be precipitated at the expense of the sulfide of a metal lower in the series, as Zn. Weigel's series is the solubility of sulfides in pure water.

monly involved in secondary sulfide enrichment are placed in their relative order and compared with the relative order of solubility in Weigel's series, we have the following:

Schürmann	Weigel
Ag	Ag_2S
Cu	CuS
Pb	PbS
Zn	FeS
Fe	ZnS

The most common secondary sulfides encountered in zones of supergene sulfide enrichment are chalcocite (Cu_2S), less commonly covellite (CuS), and argentite (Ag_2S); complex silver sulfosalts, especially sulfantimonides, are formed also. Garrels (11, pp. 165–166) discussed the general absence of the common sulfides of zinc and lead (sphalerite and galena) as supergene minerals. He noted that if a sulfate is not reduced to a sulfide in the environment that exists beneath the water table, the precipitation as secondary sulfide would depend on the ability of a solution to obtain sulfide ions from preexisting metal sulfides, that is, primary sulfides of the deposit. Hence copper should be precipitated by obtaining sulfide ion from primary zinc or lead sulfides (sphalerite and galena) or perhaps from primary copper-iron sulfide (chalcopyrite). Zinc, on the other hand, being the more soluble (see Weigel's series), would remain in solution. Hence (11) the absence of sphalerite as a supergene sulfide may result from "a deficiency of sulphide ions, caused by the slow rate of reduction of sulphate. Under these circumstances, where the amount of sulphide ions is not sufficient to satisfy all the metals moving in solution, only the least soluble will form."[1] The absence of galena as a supergene sulfide is easily explained by the very low solubility of lead sulfate and lead carbonate. Hence any lead taken into solution precipitates very close to the source as anglesite ($PbSO_4$) or cerussite ($PbCO_3$), and lead sulfate solution does not reach the reducing environment of deeper zones.

Materials other than sulfides may be attacked and decomposed. Iron-bearing carbonates and silicates, and also manganese-bearing carbonates and silicates, are readily decomposed and hydrous or anhydrous oxides formed. Both native copper and native silver are dissolved in dilute sulfuric acid, and the solubility is increased by the presence of an oxidizing agent. Native copper, after being taken into solution, may be deposited as various compounds of copper in the same manner as copper from the decomposition of sulfides. Native silver, however, does not form stable sulfates, carbonates, or oxides. Silver chloride (cerargyrite) and other silver halides are relatively insoluble, however, and thus if the halogens are available, silver may be precipitated in the oxidized zone. Silver may also be precipitated as the native metal, particularly in the absence of ferric sulfate, and as the sulfide (argentite) under reducing conditions. More rarely silver is precipitated as silver sulfantimonide or sulfarsenide (pyrargyrite and proustite). Native gold is insoluble in sulfuric acid and most other solutions that are commonly available, so often it is left behind in the oxidized zone. If both manganese and chlorine are available, however, gold may be taken into solution, as has been shown experimentally (8, pp. 305–307). The gold thus dissolved can be carried in solution in the presence of acid, but if the solution is neutralized, the gold is precipitated as the native metal. The place of deposition will depend to a considerable extent on the character of material encountered, since gold is precipitated by many minerals and also by organic matter. If gold is present as telluride, the gold generally is separated from the tellurium and is left behind in the oxidized zone. (For discussion of decomposition of nonsulfide minerals of manganese, iron, and other constituents, see 8, especially pp. 251–280, 305–324, 437–459.)

Zonal Distribution of Supergene Minerals. Many oxidation and supergene enrichment deposits show characteristic changes from the surface downward and thus a rude zoning. The zones, however, may be very irregular and the continuity may be interrupted here and there by upward or downward extensions of one zone into another; further, zones present in one deposit may be lacking in another, and, indeed, this condition may occur in different parts of the same deposit. This is to be expected, because variations in rock character will either aid precipitation under oxidizing conditions, as in the case of limestone, or permit solutions to move through without precipitation, as in the case of inert gangue. Further, porosity and permeability will make great differences in the ability of solutions and oxygen to move through the rocks.

The zones have been given somewhat different names by various investi-

---

[1] From R. M. Garrels, *Geochim. et Cosmochim. Acta,* vol. 5, p. 166. Copyright 1954. Pergamon Press Ltd. Used by permission.

gators. The terms *gossan, capping, leached zone, oxidized zone, enriched oxidized zone,* and *supergene sulfide zone* may be encountered. Fundamentally, based on processes, two major divisions or zones should exist: a zone of oxidation, or oxidized zone, and a zone of supergene sulfide enrichment, or a supergene sulfide zone. Numerous variations and subdivisions may exist in the oxidized zone, and some of these are indicated by Table 5-15. No attempt is made in the table to show depth of the zones, but merely to show what might be present at one place or another.

TABLE 5-15.   Zones of Oxidation and Supergene Sulfide Enrichment

Major divisions	*All zones developed*	*Some possible variations and subdivisions from the surface downward*				
			Not all zones developed			
Oxidized zone	Gossan or capping	Gossan or capping	Gossan or capping	Leached zone	Gossan or capping
	Leached zone	Somewhat enriched oxidized zone	Leached zone		Much enriched oxidized zone
	Enriched oxidized zone			Somewhat enriched oxidized zone	
Supergene sulfide zone	Supergene sulfide zone	Supergene sulfide zone	Rich supergene sulfide zone	Rich supergene sulfide zone	Practically no supergene sulfide zone
Primary sulfide zone	Primary sulfide zone	Primary sulfide zone	Primary sulfide zone	Primary sulfide zone	Primary sulfide zone

The approximate outcrop position of a former deposit may be marked by a gossan or a capping. Gossan,[1] the older Cornish term, generally relates to a mass of limonitic material that has been referred to as an "iron hat" above the deposit

[1] Gossan has been defined as "a ferruginous deposit filling the upper parts of mineral veins or forming a superficial cover on masses of pyrite. It consists principally of hydrated oxide of iron, and has resulted from the oxidation and removal of the sulphur as well as the copper, etc. . . . Also spelled Gozzan. Iron-hat is also a synonym." (5, p. 311) of Chap. 2. See also reference 7, p. 127, of Chap. 2.

(the *eiserner Hut* of the Germans). *Gossan* was used by Locke (19, p. 167) for the oxidized material at or near the surface of aggregated (massive) sulfide deposits, and *capping* for the oxidized material at or near the surface of disseminated sulfide deposits, and Bateman [(1, p. 251) of Chap. 2] stated that the present tendency is to make this distinction.

Either the outcrop of a former deposit or the material below the capping or gossan may be leached of the valuable metals, although limonite and traces of oxidized minerals may be present to some extent throughout this leached zone. The chief characteristic of the leached zone is that it is not a commercial source of either primary sulfides or minerals formed by secondary enrichment. It may appear at or close to the surface if inert gangue and the formation of free acid were important features, if high permeability permitted downward migration of solutions, and if other similar causes existed.

The oxidized zone is somewhat generally a zone of solution, yet precipitation and enrichment may take place in this zone, especially in its deeper parts. Precipitation is facilitated by limestone wallrock through which solutions pass and by the presence of a carbonate mineral in the gangue, and thus carbonates of copper, zinc, and lead are important in some oxidized zones. Silica in solution also aids in precipitation of copper and zinc silicates. General absence of iron sulfides results in slow oxidation and the accumulation of the oxidized products. The depth of the zone of oxidation may be small or great, depending on permeability and composition of material and the downward migration of the water table. Downward migration of the water table, caused by slow uplift of a region, may result in very deep oxidation. In the Tintic district, Utah, oxidation generally extended to at least 1,600 ft in sedimentary rocks and to 2,300 ft in the Mammoth mine (45). The presence of much calcium carbonate in the waters and the wallrock was a factor that contributed to the deep oxidation there. Change in position of the water table may also cause oxidation of a previously formed supergene sulfide zone. Thus at Chuquicamata, Chile, a former relatively rich supergene sulfide zone containing chalcocite and covellite has undergone oxidation as a result of depression of the water table, change from semihumid to very arid climate, and exposure of the supergene sulfides to oxidation (16, pp. 276–277).

The supergene sulfide zone, as contrasted with the oxidized zone, is generally a zone of deposition, not of solution, and it is here that secondary sulfides, especially those of copper, are deposited by replacement of primary sulfides and perhaps by the production of hydrogen sulfide from some reactions. Although the supergene sulfide zone may begin at approximately the position of the water table, considerable irregularity exists. Lack of oxygen at the site of deposition is most important, and "whether or not there is a relationship between the top of the chalcocite zone and the water table will depend upon the rate of percolation of the ground water and the rate the dissolved oxygen is used in oxidizing sulfide minerals."[1]

Enrichment of deposits of copper, lead, zinc, and silver is the common occurrence, although enrichment of other deposits does occur. Enrichment has been important, also, in some iron and manganese deposits in which the primary

[1] From C. Anderson, *Econ. Geology*, Fiftieth Ann. Vol., pt. 1, p. 329. Copyright 1955. Economic Geology Publishing Company. Used by permission.

mineral was carbonate or silicate, and, indeed, in some iron deposits in which silica has been leached out and originally stable iron oxide left behind. More rarely the gossan of a deposit makes a commercial iron deposit. Native gold generally is not removed, but some enrichment results from removal of other constituents or decomposition of gold-bearing tellurides, the gold being left behind. Much less commonly, if both manganese and chlorine are present, gold is transported downward and enrichment occurs. In copper deposits enrichment has been important at some places in the oxidized zone, but generally it has been more important in the supergene sulfide zone. In lead and zinc deposits the enrichment is in the oxidized zone, and generally the enrichment in lead is above the enrichment in zinc. Lead sulfate (anglesite) is only slightly soluble and commonly is precipitated around the galena from which it comes, thus forming a protective coating that surrounds a core of only partly decomposed galena. Lead carbonate (cerussite) also shows low solubility and generally precipitates near the original mineral. Zinc sulfate, however, is soluble, and very commonly the enrichment in zinc is below the enrichment in lead (Fig. 5-7), especially if limestone is present. Some of the most common relationships among supergene minerals and the zones in which enrichment occurs are shown by Table 5-16.

TABLE 5-16. Relationships among Some Common Supergene Minerals and Zones in Which Enrichment Occurs

Zone	Copper minerals	Silver minerals	Lead minerals	Zinc minerals
Gossan or capping	Perhaps traces of some minerals of the oxidized zone	Cerargyrite	Anglesite Cerussite	
Oxidized zone	Malachite Azurite Chrysocolla Cuprite Tenorite Brochantite† Atacamite† Andorite†	Cerargyrite Native silver	Anglesite Cerussite	Smithsonite* Calamine*
Supergene sulfide zone	Chalcocite Covellite	Argentite Some native silver Pyrargyrite‡ Proustite‡		

* Commonly below lead minerals if those are present.
† Commonly in the more arid regions.
‡ Generally below argentite.

Leached Outcrops. Much study has been given to the characteristics of leached outcrops that might be used as indicators of enrichment of material in depth. These relate especially to the type of limonite (indigenous, contiguous, trans-

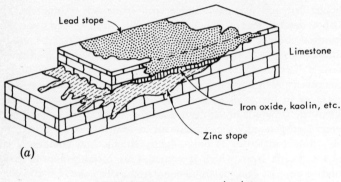

(a)

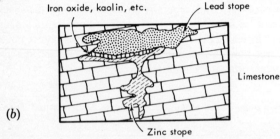

(b)

Fig. 5-7 Block diagram (*a*) and section (*b*) showing accumulation of oxidized zinc ore beneath oxidized lead ore as a result of the greater mobility of zinc in solution. (*a*) shows concentration along bedding, (*b*) shows concentration chiefly along a fractured zone. Ora La Plata mine, Leadville, Colorado. After G. F. Loughlin, *U.S. Geol. Survey Bull.* 681, p. 54.

ported), the color of the limonite, the type of cellular structure (boxwork)[1], and a number of other factors. Blanchard and Boswell (32) made a study of various types of limonite and their significance, and Locke (19) made an extensive study of leached outcrops as guides to copper ore. Anderson (1, pp. 333–338) reviewed the question of leached outcrops and features associated with them. All of these studies are important in the search for ore, but detailed discussion is beyond the scope of the present treatment of mineral deposits.

Groundwater Deposits

The extent to which the circulation of groundwater has given rise to the formation of mineral deposits, except the residual and oxidation-supergene sulfide types, is a matter of conjecture and disagreement. One might expect that in the vicinity of previously formed metallic sulfide deposits there would be considerable lateral migration of solutions in groundwater and that materials might be carried for considerable distances and then reconcentrated at places well removed from the original deposit. Anderson (1, pp. 332–333) reviewed this with respect to copper

---

[1] Boxwork structure, a result of leaching, consists of open boxlike spaces among a network of blades or plates. (See references 6, p. 306, and 7, p. 37, of Chap. 2).

and concluded that, although some cases have been cited in which minor deposits were formed in this way, in general lateral migration of copper is exceptional. Among the metals, however, some copper deposits and some uranium-vanadium deposits have been cited as resulting from deposition from groundwaters that traveled some distance, but disagreement exists about these. Among the non-metals, some deposits of elemental sulfur and associated anhydrite or gypsum originated at least in part by the circulation of groundwater.

Copper ores in "red beds" of Permian to Triassic age show world-wide distribution. Generally the beds in which they occur contain an abundance of coarse material, are cross-bedded, ripple-marked, contain many plant remains, and apparently were formed in shallow water. The copper was deposited as chalcocite around plant remains, some of which were replaced. Later oxidation has changed the chalcocite to malachite and azurite in places, but not uncommonly those minerals surround a chalcocite center. Various ideas have been expressed regarding the origin of these deposits, some favoring a syngenetic origin, in which case the deposits would be classed with the sedimentary beds, others an epigenetic origin by meteoric waters or by hydrothermal solutions. Emmons (34) discussed this problem in connection with the occurrence of deposits of this type in the Colorado Plateau region of the United States and concluded that, for those deposits, all the facts favored leaching of copper by waters from rocks above the deposits and secondary deposition where conditions were favorable. This origin does not necessarily apply to deposits elsewhere, and each deposit should be judged by the evidence that is available there. Generally these deposits are of small economic importance.

Uranium-vanadium deposits, chiefly in sandstone, occur in the Colorado Plateau and elsewhere. These deposits generally contain secondary uranium minerals and, in places, secondary copper minerals, but in depth these are associated with primary minerals. The deposits are irregular in shape, and the boundaries are parallel with the bedding of the sandstone in some places and cut across the bedding in others. The best ore bodies are generally associated with logs and other wood fragments, some of which have been replaced by ore minerals. The bodies are generally elongated in the direction of the long axes of sandstone and conglomerate lenses, and the host rocks have been interpreted as stream-channel deposits and the ore as having been precipitated from groundwater solutions after the sands had accumulated (36). Precipitation from groundwaters has been suggested by various writers, but the origin has remained controversial.

Deposits of elemental sulfur that occur in salt domes of the Gulf Coast region of the United States owe their origin in part, at least, to the circulation of groundwater, although the actual formation of sulfur seems to have been chiefly if not wholly a result of bacterial activity. A typical salt dome consists of a salt plug over which is a cap rock composed of anhydrite, calcite, gypsum, and, in some places, sulfur, along with a number of other minerals that are present in minor amounts. The cap rock may be composed of (1) a lower zone of anhydrite; (2) a transition zone of gypsum, calcite, and sulfur; and (3) an upper zone of calcite that may contain variable amounts of gypsum and sulfur, and also some remnants of anhydrite (27, pp. 49–83). Sulfur may occur with calcite filling fissures, but large amounts of it replace anhydrite and calcite and fill spaces between grains. Taylor (27) concluded that in the original deposition of salt some anhydrite accumulated with it, that circulating groundwaters dissolved the salt but not the less-soluble anhydrite, and that the anhydrite accumulated as a residue from the salt. Some of the anhydrite was later converted into gypsum by circulating groundwater (39), and the calcite and sulfur were formed by alteration of anhydrite and, in part, gypsum.

The origin of Gulf Coast salt-dome sulfur deposits was investigated and reviewed by Feely and Kulp (9), who agreed that the anhydrite of the cap rock was a residue from solution of salt and that some anhydrite was altered to gypsum by circulating groundwater. Feely and Kulp, as a result of isotopic studies of sulfur and carbon, laboratory studies relating to the reduction of sulfate to hydrogen sulfide and the oxidation of hydrogen sulfide to elemental sulfur, and supporting bacteriological investigations by others, concluded (p. 1847) that petroleum reaching the cap rock was oxidized by bacteria with simultaneous reduction of sulfate and formation of hydrogen 'sulfide, that subsequent reoxidation of the hydrogen sulfide caused precipitation of sulfur, and that carbon dioxide produced went into the formation of the calcite of the cap rock.

SELECTED REFERENCES

1. C. Anderson, 1955, Oxidation of copper sulfides and secondary sulfide enrichment, *Econ. Geology,* Fiftieth Ann. Vol., pt. 1, pp. 324–340.
2. H. S. Buehler and V. H. Gottschalk, 1910, Oxidation of sulphides, *Econ. Geology,* vol. 5, pp. 28–35.
3. H. S. Buehler and V. H. Gottschalk, 1912, Oxidation of sulphides (second paper), *Econ. Geology,* vol. 7, pp. 15–34.
4. G. V. Chilinger, 1955, Review of Soviet literature on petroleum source-rocks, *Am. Assoc. Petroleum Geologists Bull.,* vol. 39, pp. 764–767. Review of paper by G. I. Teodorovich, 1954.
5. A. J. Collier, F. L. Hess, P. S. Smith, and A. H. Brooks, 1908, The gold placers of parts of Seward Peninsula, Alaska, *U.S. Geol. Survey Bull.* 328, pp. 114–135.
6. S. F. Emmons, 1900, The secondary enrichment of ore deposits, *Am. Inst. Mining Engineers Trans.,* vol. 30, pp. 177–217.
7. W. H. Emmons, 1913, The enrichment of sulphide ores, *U.S. Geol. Survey Bull.* 529, 260 pp.
8. W. H. Emmons, 1917, The enrichment of ore deposits, *U.S. Geol. Survey Bull.* 625, 530 pp. General features and chemistry of enrichment on pp. 33–152.
9. H. W. Feely and J. L. Kulp, 1957, Origin of Gulf Coast salt-dome sulphur deposits, *Am. Assoc. Petroleum Geologists Bull.,* vol. 41, pp. 1802–1853.
10. R. H. Fleming and R. Revelle, 1939, Physical processes in the ocean, in "Recent Marine Sediments," pp. 48–141, The American Association of Petroleum Geologists, Tulsa, Okla.
11. R. M. Garrels, 1954, Mineral species as functions of pH and oxidation-reduction potentials, with special reference to the zone of oxidation and secondary enrichment of sulphide ore deposits, *Geochim. et Cosmochim. Acta,* vol. 5, pp. 153–168.

12. M. Gordon, Jr., and J. I. Tracey, Jr., 1952, Origin of Arkansas bauxite deposits, in "Problems of Clay and Laterite Genesis," pp. 12–34, The American Institute of Mining and Metallurgical Engineers, New York.

13. E. C. Harder, 1949, Stratigraphy and origin of bauxite deposits, *Geol. Soc. America Bull,.* vol. 60, pp. 887–908.

14. E. C. Harder, 1952, Examples of bauxite deposits illustrating variations in origin, in "Problems of Clay and Laterite Genesis," pp. 35–64, The American Institute of Mining and Metallurgical Engineers, New York.

15. H. L. James, 1954, Sedimentary facies of iron formation, *Econ. Geology,* vol. 49, pp. 235–293.

16. O. W. Jarrell, 1944, Oxidation at Chuquicamata, Chile, *Econ. Geology,* vol. 39, pp. 251–286.

17. W. C. Krumbein and R. M. Garrels, 1952, Origin and classification of chemical sediments in terms of pH and oxidation-reduction potentials, *Jour. Geology,* vol. 60, pp. 1–33.

18. W. Lindgren, 1911, The Tertiary gravels of the Sierra Nevada of California, *U.S. Geol. Survey Prof. Paper* 73, 226 pp.

19. A. Locke, 1926, "Leached Outcrops as Guides to Copper Ore," The Williams and Wilkins Company, Baltimore, 175 pp.

20. R. G. McConnell, 1901, Report on the Klondike gold fields, *Geol. Survey Canada, Ann. Rept.,* pt. B, pp. 5–71.

21. V. E. McKelvey, D. L. Everhart, and R. M. Garrels, 1955, Origin of uranium deposits, *Econ. Geology,* Fiftieth Ann. Vol., pt. 1, pp. 464–533.

22. J. L. Mero, 1962, Ocean-floor manganese nodules, *Econ. Geology,* vol. 57, pp. 747–767.

23. J. B. Mertie, Jr., 1940, Placer gold in Alaska, *Washington Acad. Sci. Jour.,* vol. 30, pp. 93–124.

24. E. S. Moore and J. E. Maynard, 1929, The solution, transportation, and precipitation of iron and silica, *Econ. Geology,* vol. 24, pp. 272–303, 365–402, 506–527.

25. G. D. Sherman, 1952, The genesis and morphology of the alumina-rich laterite clays, in "Problems of Clay and Laterite Genesis," The American Institute of Mining and Metallurgical Engineers, New York, pp. 154–161.

26. J. A. Sutton and J. D. Corrick, 1963, Microbial leaching of copper minerals, *Mining Eng.,* vol. 15, no. 6, pp. 37–40.

27. R. E. Taylor, 1938, Origin of the cap rock of Louisiana salt domes, *Louisiana Geol. Survey Bull.* 11, 191 pp.

28. W. G. Woolnough, 1937, Sedimentation in barred basins, and source rocks of petroleum, *Am. Assoc. Petroleum Geologists Bull.,* vol. 21, pp. 1101–1157.

29. C. E. ZoBell, 1946, Studies on redox potential of marine sediments, *Am. Assoc. Petroleum Geologists Bull.,* vol. 30, pp. 477–513.

MISCELLANEOUS REFERENCES CITED

30. E. F. Anthon, 1837, Über die Anwendung der auf nassem Wege dargestellen Schwefemetalle bie der chemischen Analyse, *Jour. prakt. Chemie,* vol. 10, pp. 353–356.

31. G. O. S. Arrhenius, 1959, Sedimentation on the ocean floor, in "Researches in Geochemistry," pp. 2–4, John Wiley & Sons, Inc., New York, 511 pp.

32. R. Blanchard and P. F. Boswell, various papers published in *Economic Geology,* especially 1925, vol. 20, pp. 613–638; 1927, vol. 22, pp. 419–453; 1929, vol. 24, pp. 791–796; 1930, vol. 25, pp. 557–580; 1934, vol. 29, pp. 671–690; 1935, vol. 30, pp. 313–319. These papers relate to types of limonite associated with leached outcrops.

33. J. R. Bogert, 1960, How ferronickel is produced from low grade laterite by the Ugine process, *Mining World,* vol. 22, pp. 33–37. See also G. F. Kay, 1907, Nickel deposits of Nickel Mountain, Oregon, *U.S. Geol. Survey Bull.* 315, pp. 120–127.

34. S. F. Emmons, 1905, Copper in the red beds of the Colorado Plateau region, *U.S. Geol. Survey Bull.* 260, pp. 221–232.

35. A. M. Finlayson, 1910, The pyritic deposits of Huelva, Spain, *Econ. Geology,* vol. 5, pp. 404–405.

36. R. P. Fischer, 1950, Uranium-bearing sandstone deposits of the Colorado Plateau, *Econ. Geology,* vol. 45, pp. 1–11.

37. A. F. Frederickson, 1952, The genetic significance of mineralogy, in "Problems of Clay and Laterite Genesis," pp. 1–11, The American Institute of Mining and Metallurgical Engineers, New York.

38. E. Glasser, 1904, Rapport sur les richesses minerales de la Nouvelle Caledonie, *Annales des Mines,* ser. 10, vol. 5, pp. 29–154, 503–701.

39. M. I. Goldman, 1952, Deformation, metamorphism, and mineralization in gypsum anyhdrite cap rock [abs.], *Geol. Soc. America Mem.* 50, p. 1.

40. V. M. Goldschmidt, 1937, Principles of distribution of chemical elements in minerals and rocks, *Chem. Soc. (London) Jour.,* pp. 655–673. See B. Mason, 1958, "Principles of Geochemistry," 2d ed., pp. 155–157, for a summary.

41. S. V. Griffith, 1960, "Alluvial Prospecting and Mining," 2d ed., Pergamon Press, New York, 245 pp.

42. J. W. Gruner, 1912, The origin of sedimentary iron formations: The Biwabik formation of the Mesabi Range, *Econ. Geology,* vol. 17, p. 455.

43. C. A. Lamey, 1952, The Cle Elum River nickeliferous iron deposits, Kittitas County, Washington, *U.S. Geol. Survey Bull.* 978-B, pp. 48–55.

44. W. Lindgren and F. L. Ransome, 1906, Geology and ore deposits of the Cripple Creek district, Colorado, *U.S. Geol. Survey Prof. Paper* 54, p. 198.

45. W. Lindgren and G. F. Loughlin, 1919, Geology and ore deposits of the Tintic mining district, Utah, *U.S. Geol. Survey Prof. Paper* 107, p. 160.

46. P. M. Murzaiev, 1937, Genesis of some sulphur deposits of the U.S.S.R., *Econ. Geology,* vol. 32, pp. 69–103.

47. G. S. Nishihara, 1914, The rate of reduction of acidity of descending waters by certain ore and gangue minerals and its bearing upon secondary sulphide enrichment, *Econ. Geology,* vol. 9, pp. 743–757.

48. F. C. Phillips, 1947, Oceanic salt deposits, *Quart. Rev.* 1, pp. 91–111.

49. E. Schürmann, 1888, Über die Verwandschaft der Schwermetalle zum Schwefel, *Liebig's Annelen,* vol. 249, pp. 326–350.

50. C. L. Segui, 1923, The sulphur mines of Sicily, *Econ. Geology,* vol. 18, pp. 278–287.
15. J. Usiglio, 1849, *Ann. Chim. Phys.,* vol. 27, pp. 92–107, 172–191.
52. C. R. Van Hise and C. K. Leith, 1911, Geology of the Lake Superior region, *U.S. Geol. Survey Mon.* 52, p. 516.
53. O. Weigel, 1907, Die Loslishkeit von Schwermetallsulfiden in reinem Wasser, *Zeitschr. phys. Chemie,* vol. 56, pp. 293–300.
54. R. C. Wells, 1915, The fractional precipitation of some ore-forming compounds at moderate temperatures, *U.S. Geol. Survey Bull.* 609, pp. 12–15.

part ***II***

Some Important
Metallic Deposits

chapter 6

Survey of Distribution
and Production of Ores
of Various Metals

Metallic deposits are not distributed uniformly throughout the world; instead, they are concentrated in a relatively few countries. The average production of various ores by a country over a period of years gives a fair index of the distribution of metallic deposits, although not one that is strictly correct, since large undeveloped deposits exist in some countries. The importance of various countries as producers of 20 selected metallic ores during the 15-year period 1945 to 1959 is shown by Tables 6-1 to 6-6. Only those countries that achieved high world rank in production during this period have been included; ranks below six have been omitted. This method of treatment shows clearly the outstanding ores produced by different countries and the rank of a country in world production. Between 64 and 99 percent of the world production of these 20 ores is represented by the figure in the tables.[1] The distribution in various groups follows:

64–71 percent

Zinc	64.4
Lead	66.1
Beryllium	69.4
Tungsten	71.0

75–79 percent

Antimony	75.1
Iron	76.3
Silver	78.2
Manganese	78.7

81–87 percent

Chromium	81.7
Copper	82.2
Tin	84.1

[1] The major sources of information regarding production are shown by the references at the end of this chapter. Statistics for Communist countries are estimates.

Aluminum	84.2
Gold	84.7
Molybdenum	86.4

90–94 percent

Mercury	90.3
Niobium-tantalum	90.4
Cobalt	93.1

96–99 percent

Nickel	96.8
Vanadium	98.3
Platinum	99.8

TABLE 6-1. The Leading World Producers of Ores of Various Metals for the 15-year Period 1945–1959, North America

Country	Metal	World rank and percent of world production	
		Rank	Percent
Canada	Nickel	1	68.0
	Platinum	1	40.0
	Zinc	2	21.0
	Gold	3	11.8
	Silver	3	11.7
	Cobalt	3	8.2
	Lead	4	9.1
	Copper	5	9.0
Mexico	Silver	1	24.1
	Lead	3	11.5
	Antimony	4	12.1
	Zinc	4	8.5
	Mercury	5	7.9
United States	Molybdenum	1	77.9
	Vanadium	1	71.3
	Iron	1	31.4
	Copper	1	29.6
	Zinc	1	21.0
	Lead	1	17.6
	Silver	2	17.6
	Tungsten	2	12.7
	Mercury	3	11.9
	Aluminum	4	11.5
	Cobalt*	4	6.9
	Gold	4	5.4
	Platinum	5	3.0
	Beryllium	6	6.1
	Antimony	6	4.2

* Very little production after 1960.

Uranium is not included in the tables that follow because figures were not released by countries of the free world during a number of years and are not now available for countries of the Communist world. Free-world production of ores of uranium for the five-year period 1958 to 1962, in percent of total free-world production, follows: the United States, 43 percent; Canada, 32; Republic of South Africa, 16; France, 3; Republic of the Congo, 3; Australia, 3.

Among some other ores not listed in the tables, the chief producing countries have been as follows—lithium: Rhodesia (Southern Rhodesia), South-West Africa, the United States, Canada; cesium and rubidium: South-West Africa; thorium and rare earths, titanium and zirconium: Brazil, India, Ceylon, Malagasy Republic (Madagascar), Australia, and to some extent the United States. Bismuth and cadmium generally are not produced from ores of those metals but are obtained as by-products from the treatment of other ores. Further details regarding distribution and production of all metallic ores are contained in the discussions of various deposits that appear in subsequent chapters.

The discussion of the individual deposits has been organized, as nearly as possible, to group together deposits of like origin, although this could not be accomplished strictly because deposits of a single metal may be of diverse origins. The scheme adopted, however, does introduce a fair amount of uniformity into the discussions and tends to bring together certain natural groups. Free use has been made of the vast amount of published material relating to individual deposits, but care has been used to identify the major sources of information. No attempt has been made, however, to present complete bibliographies. Many additional references are listed in the publications cited.

TABLE 6-2.   The Leading World Producers of Ores of Various Metals for the 15-year Period 1945–1959, South America and the Caribbean Area

| Country | Metal | World rank and percent of world production | |
		Rank	Percent
Argentina	Beryllium	2	9.2
Bolivia	Tin	2	19.2
	Antimony	3	17.6
	Tungsten	4	6.6
Brazil	Beryllium	1	28.1
	Niobium-tantalum	4	3.2
	Manganese	5	4.0
British Guiana	Aluminum	3	15.1
Chile	Copper	2	14.9
	Molybdenum	2	4.8
Colombia	Platinum	4	3.9
Cuba	Nickel	4	4.9
Jamaica	Aluminum*	2	17.8
Peru	Vanadium†	3	9.5
	Silver	5	8.8
	Lead	6	5.0
	Zinc	6	4.6
Surinam	Aluminum	1	20.0

* Jamaica ranked first in production of aluminum, 1960 through 1962, and furnished about 25 percent of the world total in 1962.
† Peru has produced practically no vanadium since 1960.

TABLE 6-3. The Leading World Producers of Ores of Various Metals for the 15-year Period 1945–1959, Europe

| Country | Metal | World rank and percent of world production | |
		Rank	Percent
Czechoslovakia	Antimony	5	4.7
Finland	Vanadium	4	2.6
France	Iron	3	12.8
	Aluminum	6	9.5
Germany (West)	Iron	6	4.2
Italy	Mercury	1	29.6
Norway	Niobium-tantalum	3	5.0
Portugal	Tungsten	5	6.6
Spain	Mercury	2	25.6
Sweden	Iron	5	4.8
Turkey	Chromium	2	17.4
United Kingdom	Iron	4	5.0
Soviet Union	Manganese	1	43.6
	Chromium	1	19.4
	Gold	2	24.7
	Nickel	2	18.2
	Iron	2	18.1
	Platinum	3	22.3
	Tungsten	3	12.6
	Molybdenum	3	3.7
	Copper	4	10.0
	Silver	4	10.0
	Mercury	4	8.2
	Aluminum	5	10.2
	Lead	5	9.0
	Zinc	5	8.0
Yugoslavia	Mercury	6	7.1

TABLE 6-4. The Leading World Producers of Ores of Various Metals for the 15-year Period 1945–1959, Africa

| Country | Metal | World rank and percent of world production | |
		Rank	Percent
Congo, Republic of the	Cobalt	1	63.2
(formerly Belgian Congo)	Niobium-tantalum	2	9.3
	Tin	4	8.6
	Copper	6	7.1
Ghana (Gold Coast)	Manganese	3	7.8
Morocco (Southern Zone)	Cobalt	5	5.7
	Manganese	6	3.7

Table 6-4.   Continued

Country	Metal	World rank and percent of world production	
		Rank	Percent
Mozambique	Beryllium	4	8.9
	Niobium-tantalum	6	1.6
Nigeria	Niobium-tantalum	1	68.1
	Tin	6	5.2
Rhodesia (Southern Rhodesia)	Beryllium	3	9.2
	Chromium	4	12.6
South Africa, Republic of	Gold	1	40.0
	Platinum	2	30.6
	Antimony	2	18.0
	Chromium	3	17.1
	Manganese	4	7.3
	Vanadium	5	1.2
South-West Africa	Vanadium	2	13.7
Zambia (Northern Rhodesia)	Cobalt	2	9.1
	Copper	3	11.6

TABLE 6-5.   The Leading World Producers of Ores of Various Metals for the 15-year Period 1945–1959, Asia

Country	Metal	World rank and percent of world production	
		Rank	Percent
China	Tungsten	1	26.4
	Antimony	1	18.5
	Tin	5	7.0
India	Manganese	2	12.3
	Beryllium	5	7.9
Indonesia	Tin	3	15.9
Korea, Republic of	Tungsten	6	6.1
Malaya	Tin	1	28.2
	Niobium-tantalum	5	3.2
Philippines	Chromium	5	12.0

TABLE 6-6.   The Leading World Producers of Ores of Various Metals for the 15-year Period 1945–1959, Australasia

Country	Metal	World rank and percent of world production	
		Rank	Percent
Australia	Lead	2	13.9
	Zinc	3	9.1
	Silver	6	6.0
	Gold	6	2.8
New Caledonia	Nickel	3	5.7

SELECTED REFERENCES

1. *Eng. Mining Jour.*, Annual Survey and Outlook Number, issued annually in February.
2. *Mining Eng.*, Annual Review Number, issued annually in February.
3. *Mining World*, Catalog, Survey, and Directory Number, issued annually in April.
4. *Mining Jour.* (*London*), Annual Review Number, issued annually in May.
5. *U.S. Bur. Mines Minerals Yearbook*, issued annually.

chapter 7

Chromium, Platinum Group Metals, Nickel

Deposits of chromium and the platinum group metals are similar in many respects, and nickel may be associated with such deposits in some places. All three metals characteristically occur in certain ultrabasic rocks, but platinum and nickel, especially nickel, may be associated with rocks of several types. Platinum occurs characteristically as the native metal, less commonly as the arsenide or the sulfide; chromium occurs as the oxide; and nickel occurs chiefly as arsenides and sulfides. Hence both native platinum and chromite may accumulate in placer and residual deposits, whereas nickel does not accumulate in placer deposits but does accumulate, as a secondary nickel silicate, in residual deposits, especially those formed by the decomposition of ultrabasic rocks such as peridotites.

Some magmatic deposits of chromium and platinum occur in layered intrusions.[1] Thayer (48) proposed that two major genetic types of gabbro-peridotite complexes may be distinguished, (1) the stratiform type, in which the layers are regular and some are traceable many miles, and (2) the Alpine type, in which the layers are irregular, lenticular, and are traceable only tens, hundreds, or perhaps a few thousands of feet.

Chromite deposits

The most important chromite production of the world (Fig. 7-1) comes from deposits in the Bushveld igneous complex, Republic of South Africa; the Great Dyke, Rhodesia (Southern Rhodesia); the Ural Mountains, Soviet Union; the Guleman and other areas, Turkey; and the Masinloc and other deposits of the Philippines. Early production differed somewhat from that of 1962. The supply

[1] For a summary of concepts regarding the formation of layered igneous intrusions see W. H. Tauberneck and A. Poldervaart, 1960, Geology of the Elkhorn Mountains, Northern Oregon, pt. 2, Willow Lake intrusion, *Geol. Soc. America Bull.*, vol. 71, pp. 1313–1316.

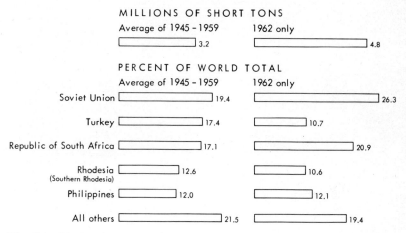

Fig. 7-1 World chromite production. Compiled from *U.S. Bur. Mines Minerals Yearbook,* various years.

about 1901 came chiefly from Turkey; about 1906 it came mostly from New Caledonia; and between 1910 and 1920 the lead was generally held by New Caledonia or Southern Rhodesia. Since 1921 the importance of New Caledonia has declined, so that in 1962 it produced only 0.3 percent of the world total.

Chromite deposits of economic importance are believed by Thayer (47, p. 43) to be of early magmatic origin; he classed them (49, p. 198) as *stratiform* and *podiform*.[1] The stratiform type is exemplified by the deposits of the Bushveld complex, South Africa, and the Stillwater complex, Montana; the podiform type by deposits in Turkey, the Philippines, and the Selukwe deposits, Rhodesia (Southern Rhodesia), among others.

Chromite ores are classed as refractory, metallurgical, and chemical. Specifications for each type are somewhat variable. Refractory ore generally contains less Cr_2O_3 than the other types but more Al_2O_3, a minimum of 57 percent combined Cr_2O_3 and Al_2O_3, and usually iron no more than 10 percent of the total ore and SiO_2 not more than 5 percent. Metallurgical ore generally contains a minimum of 48 percent Cr_2O_3, a minimum ratio Cr/Fe of 3:1, and SiO_2 is not desirable. Chemical ore contains a minimum of 45 percent Cr_2O_3, commonly a ratio Cr/Fe of 1.6:1, and less than 8 percent SiO_2.

Africa

The Bushveld igneous complex (17; 56, p. 290–292) appears in plan as an oval mass about 300 miles long and 100 miles wide (Fig. 7-2) and covers an area of about 12,700 sq. miles in the central Transvaal. It consists essentially of early flows of basic amygdaloidal lava that were followed by flows of acid lava and minor acid intrusions; then by basic sill-like intrusions of gabbro and similar rock; then by a vast lopolithic sill of exceedingly great thickness known as the

[1] The podiform type includes all deposits that occur in podlike masses and includes various types classed by Sampson (43) as evenly scattered, schlieren-banded, sackform, and fissureform.

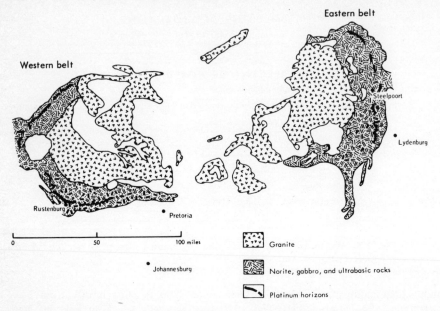

Fig. 7-2 Generalized geologic map showing distribution of the major rock types of the Bushveld igneous complex. Simplified after C. A. Cousins, *Engineering and Mining Journal,* vol. 160, no. 10, p. 114. Copyright 1959. McGraw-Hill Publishing Company. Used by permission.

norite zone but consisting of norite, pyroxenite, dunite, and anorthosite, as well as other rock types; and finally by late sheets, stocks, and dikes of red granite. The great norite zone has the shape of a basin (lopolith) and a thickness estimated by Hall (17, p. 262) to be as much as 17,000 to 18,000 ft. Sekukuniland, eastern Transvaal, was designated (17, p. 270) as the type section of the norite. There it consists of five zones (Table 7-1), a floor of metamorphosed quartzite, shale, and other rocks, and a roof of altered felsites, sedimentary rocks, or granite. A section through the norite is shown by Figure 7-3. The major chromite deposits occur in the critical zone. One of the striking characteristics of the critical zone, in the eastern Transvaal, is layering or pseudostratification (4, p. 1157). This pseudostratification is a general characteristic of the norite zone throughout the Bushveld complex.

Chromite in the Bushveld complex occurs in minute crystals disseminated throughout ultrabasic rocks, as lenticular masses generally not more than 3 ft long in ultrabasic rocks, and in bands, seams, or lenses of chromite rock that follow the pseudostratification of the norite and are closely associated with bronzitite. This last type is the most important and was designated *chromitite* by Wagner (76), a term now somewhat generally applied to material consisting of much chromite and some associated rock-forming minerals. Two major chromite belts are present, the western belt, in the vicinity of Rustenberg, and the eastern belt, in the vicinity of Steelport and Lydenberg (Fig. 7-2). The eastern belt has been the most productive.

TABLE 7-1. **Zones of the Type Section across the Norite of the Bushveld Igneous Complex***

Zone	Dominant rocks	Thickness
5. Upper zone	Bronzite norite; syenitic rocks.	Up to 800 ft (244 m)
4. Main zone	Much bronzite-norite. Near top, magnetite bands rest on anorthosite.	Up to 10,000 ft (3,050 m)
3. Critical zone	Bronzite-norite, diallage-norite, bronzitite, dunite, chromite, anorthosite, etc., serpentine.	Up to 5,000 ft (1,525 m)
2. Transition zone	Bronzite-norite, bronzitite bands.	Up to 2,000 ft (610 m)
1. Basal or chill zone	Fine-grained bronzite-norite.	Up to 400 ft (122 m)
Total maximum thickness		18,200 ft (5,490 m)

* From A. L. Hall, Geol. Survey South Africa, Mem. 28, p. 271, 1932. Extracts from official publications of the Republic of South Africa reproduced under Government Printer's Copyright Authority No. 3390 of 16/9/1964.

Some investigators (15, 17, 25, 51) have placed the chromitite seams in three groups, upper, middle, and lower, which are generally located with reference to depth beneath the platinum-bearing Merensky reef (Fig. 7-3). Individual seams are less than an inch to as much as 6 ft thick. As many as 25 seams have been recognized in the Rustenberg area (15), where they are commonly 18 to 40 in. thick, the lower seams occurring in pyroxenite, the upper ones in anorthosite. Attempts have been made to correlate seams from one part of the complex to another, but the great size of the complex and variations within it make some correlations doubtful (4, pp. 1153, 1166).

Various concepts have been advanced to account for the origin of these layered deposits of the Bushveld complex. Cameron and Emerson (4, p. 1211) concluded that the chromitite seams resulted from magmatic differentiation, and that magmatic currents played an important part in the development of contrasting rock types, but that probably part of the chromite and bronzite crystallized *in situ*. They added, however, that partial resolution of settled crystals seems required as a supplementary process.

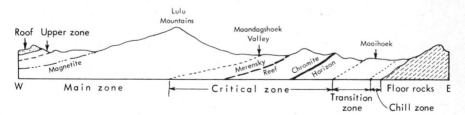

Fig. 7-3 Section showing major zones in the norite of the Bushveld igneous complex north of Steelpoort. Length of section, 19 miles. After E. N. Cameron and M. E. Emerson, *Econ. Geology*, vol. 54, p. 1157. Copyright 1959. Economic Geology Publishing Company. Used by permission.

The remarkable Great Dyke of Rhodesia (Southern Rhodesia) (23, 27, 54, 58), about 330 miles long and 3 to 7 miles wide, extends slightly east of north throughout the central part of Rhodesia (Fig. 16-1). It has the surface form of a dike but is a complex layered intrusion. The bottom part of the dike consists of strongly layered proxenites and peridotites that contain chromitite seams and, near the top, native platinum and sulfides of copper and nickel, a horizon that corresponds to the Merensky reef of the Bushveld complex. Above this ultrabasic part is massive, uniform unlayered norite. Both the norite and the platinum-bearing rock have been eroded from the body for about two-thirds of its length. In cross section the layering gives the Great Dyke the appearance of a shallow syncline, as has been shown diagrammatically by Hess (Fig. 7-4). The seams of chromitite are less than an inch to 30 in. thick, but about 7 in. is the usual thickness.

In addition to the primary deposits, eluvial deposits of chromite occur over a 250-mile stretch along the Great Dyke. Some of these contain as much as 35 percent chromite, compared with 3 percent in the serpentine. Soil cover is less than 2 ft (55).

The approximate reserves of extractable chromite in the Great Dyke were stated by Worst (58, pp. 196–198) as 212,672,000 tons of chemical grade and 113,589,000 tons of metallurgical grade. He also stated that the amount of chromite in seam form could well be ten times that amount. The Great Dyke may produce about 40 percent of the chromite of Rhodesia (Southern Rhodesia), the rest coming from the Selukwe deposit (67).

The Selukwe deposit (29, 43, 61), although near the Great Dyke, apparently is an unrelated ore body. The geology of the area is complex, but the essential elements are a mass of ultrabasic rocks intruded into Precambrian sedi-

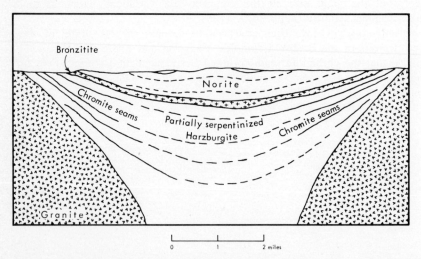

Fig. 7-4 Diagrammatic cross section through the Great Dyke, Rhodesia (Southern Rhodesia), showing general structure. After H. H. Hess, *Geol. Society South Africa, Trans. and Proc.,* vol. 53, p. 163. Copyright 1950. The Geological Society of South Africa. Used by permission.

mentary rocks, both groups older than the Great Dyke. These rocks were sheared and altered, and serpentine and various schists were formed; the group is generally referred to as "talc schists." Chromite occurs chiefly as lenses 150 to 450 ft long, and as many smaller lenses, in talc schist, talc-carbonate schist, and silicified serpentine. Hodges (67) rated the Selukwe deposits as one of the largest single concentrations of high-grade chromite in the world.

Turkey

The chromite deposits of Turkey (3; 10, p. 304–332; 12, 20, 24, 38, 43, 49, 50) occur in (1) the northern Alpine region of Ankara, chiefly around Cankiri; (2) the southern Alpine region, south of Ankara, that extends roughly from Fethiye to Guleman; and (3) the central region, west of Ankara (Fig. 7-5). Tokay (50, p. 83) lists 13 chromite areas in Turkey, 4 of which are in the northern region, 3 in the central region, and 6 in the southern region. The southern Alpine region is the most important one, the deposits at Guleman being by far the most productive and furnishing exceptionally high-grade ore, around 53 percent Cr_2O_3; but deposits in the southwestern part of this southern area (Burdur, Fethiye, and others) have been stated to constitute the best chrome-producing region of the country (65). The central region contains numerous ore bodies, but those in the area around Eskisehir seem to be the most important ones (3, 38).

The deposits of all three regions occur in peridotites, dunites, and their serpentinized equivalents. All deposits are of the podiform type. Ore types have been designated as *sprinkled* (10 to 20 percent Cr_2O_3), *schlieren* and *banded* (20 to 35 percent Cr_2O_3), *spherical* and *spotted* (35 to 40 percent Cr_2O_3), and *massive* (45 to 54 percent Cr_2O_3). Spherical and spotted ore of the Fethiye deposit occurs in lenticular or veinlike bodies that are in sharp contact with ad-

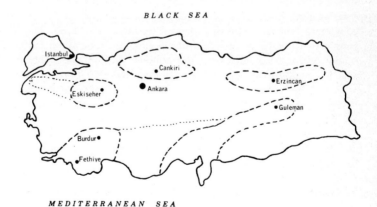

Fig. 7-5 Generalized outline map of Turkey showing major chromite areas (dashed boundaries) and extensions that probably contain chromite (dotted boundaries). After H. F. Kromer, *Eng. Mining Jour.*, vol. 155, no. 4, p. 92. Copyright 1954. McGraw-Hill Publishing Company. Used by permission.

jacent serpentine. Some ore bodies are nearly 1,000 ft long and as much as 15 ft thick. The massive ore is best developed around the Guleman property, where many of the ore lenses are aligned along surfaces of movement, are in sharp contact with serpentine, and some have been described as veinlike occurrences. Internal features indicate that many of these massive deposits were displaced during tectonic movements.

Thayer (47, p. 49; 49, p. 197) included the Turkish deposits in the Alpine type and stated that they probably were carried up as solid inclusions in semi-solid crystal mushes that formed by partial melting and remobilization of rocks differentiated at depth by fractional crystallization. Ergunalp (12, p. 6) suggested that cracks developed in the upper part of the intrusive body and differentiated magma from deeper parts of the chamber came up into those cracks and formed veinlike bodies. Helke (20, p. 961), however, stated that the deposits were originally similar to those of the Bushveld complex and postulated a hypothetical lopolith of basic-ultrabasic rocks, now represented by preserved fragments. These original deposits, according to Helke (20, p. 954), were slightly metamorphosed and somewhat displaced during the Alpine orogeny, and thus passed into the California or Alpine type of Thayer.

Philippines

The most productive chromite deposits of the Philippines are on Luzon, but other deposits are known on Mindanao and Palawan (10, pp. 218–226; 45, 47). The deposits on Luzon have supplied a large part of the refractory chromite of the world and also a considerable amount of metallurgical chromite. The major refractory chromite mine is the Masinloc,[1] the major metallurgical chromite mine, the Acoje. Recent work has disclosed other important ore bodies.

The Zambales Range in western Luzon is, in part, a peridotite-gabbro complex consisting of alternating belts of ultramafic and layered gabbroid rocks classed by Thayer (48, pp. 251, 253) as the Alpine type. The chromite deposits occur on both limbs of a broad northeast-trending syncline, the Acoje deposit on the western limb, the Masinloc deposit on the eastern limb. The Masinloc deposit, along with 10 or 12 smaller deposits, occurs in a zone about 8,000 ft long enclosed in serpentinized saxonite near the contact with olivine gabbro. The major ore body is roughly oval, about 1,800 ft long and 950 ft wide, and has been rated as one of the largest chromite deposits in the world. Most of the ore consists of massive chromite, at the borders of which is disseminated ore. The disseminated ore in places also forms large bodies enclosed in massive ore or lying between massive ore and serpentine. The smaller Acoje deposit, about 10 miles north of the Masinloc, is also in serpentinized saxonite, but the layering is much better developed at Acoje.

Soviet Union

The chromite deposits of the Soviet Union (10, pp. 252–278) occur along a length of about 350 miles in the Ural Mountains, associated with basic and ultra-basic intrusions. The ore bodies are stratiform masses, and both massive and dis-

---

[1] The average percent composition of the ore from the Masinloc mine (45, p. 421) for 1946 to 1952 was Cr_2O_3, 32.33; Al_2O_3, 29.38; MgO, 17.93; Fe, 10.59; SiO_2, 5.06; CaO, 0.54; H_2O, 0.83; total, 96.66.

seminated chromite are present (35, pp. 56–57; 43, p. 113). Chromite occurs in the ultrabasic rocks (chiefly serpentinized peridotite) in the middle and south Urals, the southern terminus of the deposits being the large chromite-bearing mass in the Mugodzhar Mountains at Kempirse to the east of Aktyubinsh. Nalivkin (35, pp. 56–57) rated the deposit at Sarany, in the middle Urals, as the largest chromite deposit in the world.

Other Countries

The deposits of all other countries generally supply less than 20 percent of the total world production of chromite. The most important deposits are those of New Caledonia, Cuba, Canada, and the United States (2, 10, 14, 16, 18, 21, 30, 36, 39, 40, 41, 44, 46, 47). The deposits of New Caledonia are in the Tiebaghi peridotite dome near the northern end of the island, which has supplied most of the production, and also in the southern part of the island. The deposits of Cuba are in serpentinized peridotites and in lateritic residual iron-nickel-cobalt deposits derived therefrom. The deposits of Canada are in the Bird River complex of eastern Manitoba and in the Black ·Lake region, Quebec. The deposits of the United States (75) are in the Stillwater complex, Montana, the serpentine deposits of California, the most productive of which in 1961 was the Butler Estate mine (70), and the reserves of chromite in the black sands in raised beach deposits along the southwest coast of Oregon.

Deposits of the platinum metals

Important production of platinum comes from Canada, the Republic of South Africa, and the Soviet Union (Fig. 7-6). The production of Canada comes from copper-nickel-platinum deposits in which the platinum is contained, at least in

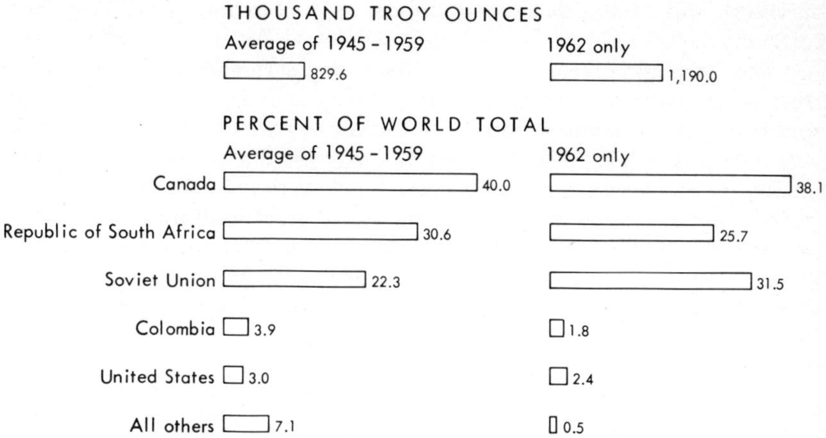

Fig. 7-6 World platinum production. Compiled from *U.S. Bur. Mines Minerals Yearbook,* various years.

part, in sperrylite. The production of the Republic of South Africa and the Soviet Union comes chiefly from the native metal,[1] although both sperrylite and cooperite (PtS), along with some rarer platinum minerals, are present in the South African deposits.

Africa

The principal platinum deposits of Africa are in the Bushveld complex, Republic of South Africa; less productive deposits are in the Great Dyke, Rhodesia (Southern Rhodesia); deposits of still less importance have been reported in Ethiopia (17, pp. 471–486; 22; 34; 37, pp. 209–243; 49). Platinum metals are also recovered as a by-product of gold mining at Witwatersrand.

The deposits of the Bushveld complex are of three major types: (1) pipes in hortonalite[2]-dunite, (2) sulfidic diallage norite deposits of the Merensky reef, and (3) contact-metasomatic deposits. Both of the first two types are in the critical zone of the complex (Fig. 7-3 and Table 7-1). Placer and eluvial deposits occur in some areas of primary ore.

The pipes, about 60 in number, show approximately circular ground plans with diameters as much as 60 by 80 ft, taper downward, and generally cut very steeply through the pseudostratified norite. They occur chiefly in the Lydenburg district. The most famous pipe, the Onverwacht, was opened by mining to a depth of 900 ft. The platinum content varied through wide limits, from as little as 1 dwt (1.55 g) to as much as 1,213 dwt (1,886 g) per short ton; generally values were higher in the central parts of the pipe and in the upper levels of the mine. Below the 250-ft level values decreased steadily; between 550 and 700 ft the pipe contained practically no platinum; good values again occurred at 700 ft but were much lower on the 800-ft level. Earlier work indicated that platinum is present chiefly as native metal, but both sperrylite and cooperite occur also. Work in 1961 (74), making use of the X-ray microanalyzer (electron microscope), indicated that platinum is present not only as the native metal and sperrylite, but also combined with antimony and bismuth as the compounds $PtSb_2$, $PtSb$, and $Pt(Sb, Bi)$, and palladium is present as Pd_3Sb, Pd_8CuSb_3, Pd_2CuSb, and $Pd(Sb,Bi)$.

The Merensky reef (63) lies above the chromitite horizon and below anorthositic norite or anorthosite (Fig. 7-3). The deposits occur in diallage-bronzite-norite that contains patches of sulfides—pyrrhotite, pentlandite, chalcopyrite, and some others. The platinum occurs with the sulfides, in part in solid solution, in part in the minerals cooperite and sperrylite. The principal platinum occurrences are in the Rustenberg and Potgiestersrust districts; less important occurrences are in the Lydenberg district. In the Merensky reef, values vary from about 1 dwt to 219 dwt per short ton, with averages throughout considerable distances of 4.5 to 5.5 dwt in the Rustenburg district, 2.5 dwt in the Potgiestersrust district, and 2 to 3 dwt in the Lydenburg district.

[1] Native platinum is commonly associated with other platinum metals, which consist of two groups: (1) the platinum group (platinum, iridium, osmium, rhenium) and (2) the palladium group (palladium, rhodium, ruthenium, masurium).

[2] Hortonalite is an olivine in which FeO and MgO are present in about equal proportions.

The contact-metasomatic deposits have been noted only in parts of the Potgiestersrust district where the floor of the norite is in part dolomite and banded ironstone. These rocks have been metamorphosed locally and invaded by emanations that brought in pyrrhotite, pentlandite, chalcopyrite, and sperrylite. The contact-metasomatic deposits in the Potgiestersrust district were richer than the deposits of the Merensky reef.

The deposits of the Great Dyke, Rhodesia (Southern Rhodesia), occur near the top of an ultrabasic group of rocks in an horizon that corresponds to the Merensky reef of the Bushveld complex. The general associations are the same as in the Merensky reef, the platinum minerals occurring with sulfides of copper and nickel. The potential reserves of the Great Dyke are large, but metallurgical difficulties have delayed recovery (58, p. 125).

Platinum deposits have been known in Ethiopia since at least 1924 but have not been important producers. The principal deposits occur near Yubdo, western Ethiopia, about 350 miles west of Addis Ababa. The deposits occur in dunite that contains but a small amount of FeO and thus differs from the hortonalite-dunite of South Africa. The chemical composition, however, is almost identical with dunite of the Ural Mountains, Soviet Union, and the general occurrence is similar. Platinum is present in very small quantities in the dunite, values too low for mining, but eluvial deposits have been worked and alluvial (placer) deposits seem to offer future possibilities.

Soviet Union

Placer deposits have furnished most of the production of platinum in the Soviet Union (22, pp. 89–94; 35, p. 57; 37, pp. 163–208). The primary deposits from which the placers have come are in domelike masses of dunite in the Ural Mountains, in the same general area as the chromite deposits.

The primary deposits extend from the northern end of the range to the central part at Nizhni Tagil, where the largest platinum-bearing dunite mass occurs, covering about ten square miles. The domes consist of a dunite core surrounded by pyroxenite which, in turn, is surrounded by gabbro. Platinum values are highest in the dunite and decrease outward through the pyroxenite and gabbro. The greater concentration of platinum is in irregular pockets, bands, and veins of chromite. Values are reported to vary from about 1 to 6.5 dwt and, exceptionally, 183.5 dwt per ton. The primary deposits have been worked to some extent.

The most important placer production has come from the Issovsk district (eastern slope of Urals) and the Nizhni Tagil district (chiefly western slope of Urals). Other placer deposits are present between those of the Issovak district and the northern end of the Ural Mountains and between the Issovsk district and the Nizhni Tagil district, but deposits farther south are of little importance. The placer deposits occur in the beds of some of the larger rivers, in old river beds that lead into the larger rivers, and in terraces. The productive sands in the old river beds, which may be as much as 5 ft thick, are at the base of beds of ancient alluvium and are covered by as much as 35 ft of barren overburden, over which the present streams flow. The deposits are worked by dredges where feasible, but much work is done by hand, especially in the creeks between the rivers and in the higher terraces.

Canada

Canada has, for many years, produced nearly one-third or more of the platinum and two-thirds or more of the nickel of the world. Production of both platinum and nickel has come chiefly from the Sudbury district, Ontario (1). The deposits will be described in the discussion of nickel deposits.

Other Countries

Other platinum deposits that deserve brief mention are in South America (Colombia), Australia (Tasmania and New South Wales), and the United States (Alaska).

Colombia (22; 37, pp. 139–147) was essentially the only producer of platinum before exploitation of the Russian deposits began. Platinum occurs in various places in western Colombia, but the important deposits are placers in a narrow zone west of the Cordillera Occidental in the Intendencia del Choco. The original source of the platinum appears to be ultrabasic rocks, but some platinum may have been reworked from conglomerates and previous placer deposits now represented in terraces.

Tasmania is one of the largest producers of osmiridium in the world (11). The deposits consist of (1) a western group in a belt of serpentine 1 to 5 miles wide, known at various places for about 30 miles from Heaslewood southeasterly to South Dundas, with the Bald Hill district as the most important locality, and (2) a southern group in southwestern Tasmania, in the Adamsfield district, where there is a dikelike mass of ultrabasic rock, chiefly serpentine, about 3 miles long and 1,200 to 2,400 ft wide, that passes into a bulge about a mile wide at its northern end. Both primary and placer (alluvial) deposits have been worked, but production has been chiefly from placers. The placers of the Bald Hill area (both stream and eluvial) furnished the principal production before discovery of the Adamsfield deposits, which, however, have yielded about two-thirds of the total. The primary deposits in the Adamsfield district were traced for a length of more than 900 ft, were worked in places to a depth of 50 ft, and produced some nuggets of osmiridium that weighed as much as 1 oz. Major production there has been from placer and detrital deposits.

The platinum deposits of New South Wales (22, pp. 74–78) are chiefly placers in counties Cunningham and Kennedy, in the vicinity of the village of Fifield, about 335 miles nearly west of Sydney. Both platinum and gold are present in placers known as "deep leads" that mark the sites of old river courses now buried under sand and clay and, in places, basalt. The amount of overburden is as much as 90 ft in some places but thins toward the headwaters of former streams. The Platina lead, which is the most important occurrence, contains from 2 to 5 dwt of platinum and 1 to 3 dwt of gold per ton, but nuggets of platinum that weighed as much as 27 dwt have been recovered. The primary source of the platinum is unknown.

Some platinum has been recovered in the United States as a by-product from placer gold deposits in California and Oregon and from platinum placers in the Goodnews Bay district, Alaska (32, 33, 68). Most of the production has come from the Goodnews Bay district, although platinum has been reported from many places in Alaska. The primary source of the platinum of the Goodnews placer deposits is ultrabasic rock, most of which has been removed by erosion.

Nickel deposits

The most important nickel deposits of the world are in Canada, the older ones being those of the Sudbury district, Ontario, and the more recently discovered ones those of the Lynn Lake district and the Thompson–Moak Lake district, Manitoba. Other deposits, of much less importance, are in the ultrabasic rocks of the Ural Mountains, Soviet Union, the sulfide deposits at Petsamo, Soviet Union (formerly Finland), residual deposits in New Caledonia, residual iron-nickel-cobalt-chromium deposits of Cuba, and low-grade residual nickel deposits of Oregon, United States. Important world production is shown by Figure 7-7.

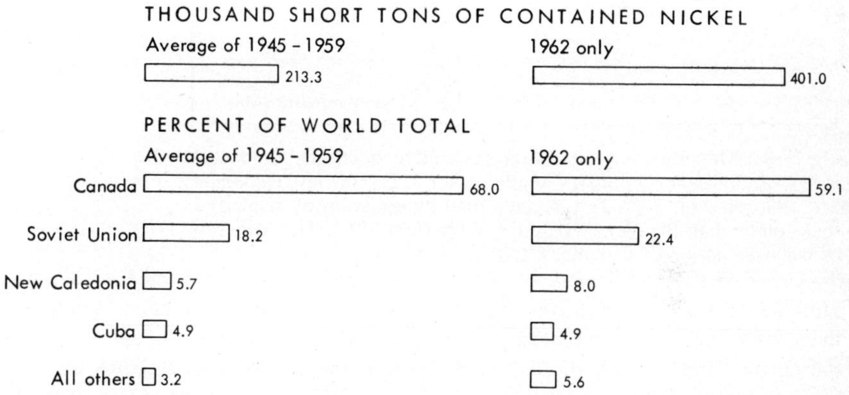

THOUSAND SHORT TONS OF CONTAINED NICKEL

Average of 1945 - 1959

☐ 213.3

1962 only

☐ 401.0

PERCENT OF WORLD TOTAL

Average of 1945 - 1959   1962 only

Canada ☐ 68.0   ☐ 59.1

Soviet Union ☐ 18.2   ☐ 22.4

New Caledonia ☐ 5.7   ☐ 8.0

Cuba ☐ 4.9   ☐ 4.9

All others ☐ 3.2   ☐ 5.6

Fig. 7-7   World nickel production. Compiled from *U.S. Bur. Mines Minerals Year-book,* various years.

Canada

The copper-nickel-platinum deposits of the Sudbury district, Ontario, (19; 22, pp. 60–67; 31; 52) occur around the margins of the Sudbury basin (Fig. 7-8), which is present in a somewhat oval-shaped area about 37 miles long and 16 miles wide underlain by rocks of Precambrian age. The central part of the basin contains rocks of the Whitewater group (volcanic breccias and tuffs overlain chiefly by slate and arkose-graywacke), and the margins consist of norite-micropegmatite, which has been designated the *nickel irruptive* because of its close association with the nickel deposits and is 1 to 4 miles wide. The outer and lower part of the nickel irruptive consists of norite, the inner and upper part of micropegmatite, more siliceous than the norite, into which the norite grades. Associated with the norite and the ore deposits is a quartz diorite, which Yates (60, p. 603–605) stated is the principal host rock for the ore. The quartz diorite belongs to the

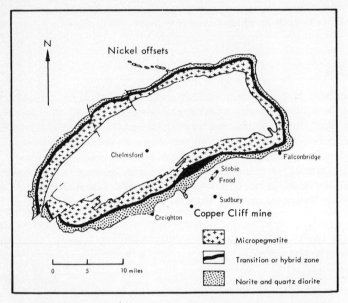

N

Nickel offsets

Chelmsford

Falconbridge

Stobie

Frood

Sudbury

Copper Cliff mine

Creighton

0   5   10 miles

Micropegmatite

Transition or hybrid zone

Norite and quartz diorite

Fig. 7-8 Map showing the major rocks that compose the nickel irruptive, Sudbury, Ontario, Canada. After J. E. Hawley, *Canadian Mineralogist,* vol. 7, pt. 1, p. 4, simplified by omission of detail and rocks other than the nickel irruptive. Copyright 1962. The Mineralogical Association of Canada. Used by permission.

same general period of intrusion as the norite, but indirect evidence indicates that the quartz diorite is later although in places it appears to form apophyses from the norite. Hawley (19, p. 25–26) noted that the quartz diorite seems to be confined to the basal part of the nickel irruptive and the underlying offset dikes, such as Frood-Stobie, and suggested that it may represent a chilled marginal phase of the main intrusive mass.

A striking feature of the basin is the Sudbury breccias, which are extensively developed around the margin and for some miles beyond. These are of great variety and probably of several ages (19, p. 14). The fragments of the breccias are subangular to rounded and extremely variable in size. They attain a length of 200 to 400 ft in large breccia zones (73); an extreme length of 2,600 ft and a width of 900 ft has been reported. The fragments generally consist of the material in the adjacent wallrocks, but some fragments are of rocks that do not crop out in the area and apparently were moved some distance.

The Sudbury deposits are copper-nickel-platinum ores composed chiefly of pyrrhotite, pentlandite, chalcopyrite, and cubanite. In places, especially at the Frood mine, platinum and palladium are present in sperrylite, $PtAs_2$, michenerite, $PdBi_2$, and froodite, $PdBi_2$. Associated gangue minerals depend on the rock in which the ore occurs. The ores have been designated (1) *disseminated,* (2) *massive,* (3) *breccia,* and (4) *vein* or *stringer* ore. Regardless of type of ore, the majority of bodies are lenticular or sheetlike (19, pp. 30–31). Most of the disseminated ore is associated with quartz diorite; the Creighton and Frood deposits apparently are the outstanding examples. The massive ores are composed chiefly of sulfides and may pass into disseminated ores or breccia ores. Breccia ore is that which contains fragments of any rock; quantitatively it is very important. The vein and stringer ores are essentially massive ore confined to fractures and commonly occur below or at the ends of the larger disseminated ore bodies.

Various ideas have been expressed regarding the origin of the ores. Generally there has been agreement that they are genetically related to the nickel irruptive or the magma from which it came. Many investigators have postulated that they are magmatic deposits, either segregated and crystallized *in situ* or injected, but some have suggested that they were formed by hydrothermal solutions and are replacements. Hawley (19), after extensive study, suggested that the ores are dominantly of the magmatic type and derived from the nickel irruptive, which possibly had the composition of a quartz diorite and was almost saturated with sulfides when intruded; that the sulfides were segregated as immiscible liquids, some crystallizing *in situ*, others being injected; and that in places the ores were changed by later emanations and even remobilized locally. Many of the disagreements regarding the origin of the ores stem from disagreements regarding interpretations of the complex geology of the Sudbury area and the formation of the Sudbury basin. Hawley's analysis undoubtedly will not be accepted by everyone who has done detailed work in the area, but he offers strong evidence for a magmatic origin of the principal ores.

A second highly important nickel area of Canada is in Manitoba. The Lynn Lake deposits in the northern part of the province, about 36 miles from the Manitoba-Saskatchewan border, consist of vertical to steeply dipping pipes of more or less massive sulfides in basic rocks. The chief ore minerals are nickeliferous pyrrhotite and chalcopyrite. Production began in 1953, and by 1955 seven ore bodies had been outlined with indicated ore reserves of 14,055,000 short tons containing an average of 1.22 percent nickel (172,000 tons) and 0.62 percent copper (8, pp. 558–560; 31, pp. 16–17). Important annual production has come from these deposits. The Thompson Lake–Moak Lake deposits are chiefly in serpentinized peridotite but in part in gneiss near the contact of the peridotite and the country rock. Much of the ore is of the disseminated type and of low grade (0.3 to 0.7 percent nickel), but some is in seams and stringers of higher grade. The deposits are of enormous size, however, and occur in a zone 270 miles long and 60 miles wide, which might prove to be one of the world's major nickel regions (7; 31, pp. 16–17; 57, p. 820). Production began at the Thompson Lake mine in 1961. A third possible place of future nickel production in Manitoba is deposits of the Maskwa River and Bird River areas in the southeastern part of the province.

Other nickel deposits are known in western Canada (Northwest Territories, British Columbia, Yukon), and in eastern Canada (Quebec), some of which have been productive.

Other Countries

Nickel deposits of other countries furnish but a small part of the total world production, although they are highly important locally. They can be discussed but briefly as sulfide ores, silicate ores, and nickel-iron ores.

The sulfide deposits of Petsamo (formerly Finland, now the Soviet Union) in the Karelo-Finnish area consist of lenses of pyrrhotite containing pentlandite and chalcopyrite. They occur at the contacts of serpentinized peridotite and the country rocks and probably are of large size (66, 72).

Nickel silicate ores of New Caledonia, once the chief source of the world's nickel and still constituting large reserves, are chiefly along the east-central to south-central side of the island. They are lateritic deposits developed from weathering of peridotite. Chetélat (5) classed them as lateritic deposits formed *in situ*, which are relatively unimportant, and as migration deposits formed by downward circulation of groundwater, which are the important type. These are both blanketlike and veinlike deposits, the latter resulting from circulation along joints. Characteristically, enrichment in nickel is close to the original serpentinized peridotite and below the enrichment in iron.

A residual silicate deposit near Riddle, Oregon (62, 64), formed by weathering of peridotite, is reported to contain 6 million tons of ore bearing from 1 to 2 percent nickel. The more nickeliferous material lies close to the original peridotite and beneath more ferruginous material, somewhat the same as it does at New Caledonia. Production began in 1954, the ore being smelted into ferronickel, and expansion of facilities was planned in 1963 (78).

Lateritic residual nickel-iron deposits are the world's largest potential source of nickel, but such deposits generally have not been developed because of their low nickel content and refractory nature (8, 13; 31, p. 37; 59). Large deposits are known in Cuba, the Philippines, Celebes, and parts of Borneo, but production has come only from the Nicaro deposits, Mayari and Moa Bay, Cuba. The upper or surface part of the Nicaro deposits is hematitic ore 4 to 8 ft thick (about 0.8 percent nickel, 49 percent iron) beneath which is limonitic ore 8 to 12 ft thick (about 1.3 percent nickel, a little more than 50 percent iron), followed by disintegrated serpentine 2 to 10 ft thick (about 1.6 percent nickel, 22 percent iron), which passes into more or less unweathered serpentine (9; 31, p. 37). This is a very characteristic distribution of nickel and iron in deposits of this type (69). Important production by United States firms occurred from 1952 to 1960 (77), but in 1960 the Cuban government seized the mines and plants. Since then production appears to have been small, but Cuba hopes to increase production materially (71).

SELECTED REFERENCES

1. C. C. Allen, 1961, The platinum metals, *Canada, Mineral Resources Div., Minerals Rept.* 3, 68 pp.
2. J. E. Allen, 1941, Geological investigations of the chromite deposits of California, *California State Mineralogist's Rept.* 37, 68 pp.
*3. Ö. Barutoglu, 1960, The Eskisehir chrome region with special reference to the Sazak mine (Turkey), in "Symposium on Chrome Ore," Ankara, Turkey, September 1960, Central Treaty Organization, Ankara, Turkey, pp. 137–145.
4. E. N. Cameron and M. E. Emerson, 1959, The origin of certain chromite deposits in the eastern part of the Bushveld complex, *Econ. Geology,* vol. 54, pp. 1151–1213.

5. E. de Chetélat, 1947, La genese et l'evolution des gisements de nickel de la Nouvelle-Caledonie, *Soc. géol. France Bull.*, ser. 5°, vol. 17, pp. 105–160.

6. F. J. Coertze, 1958, Intrusive relationships and ore-deposits in the western part of the Bushveld igneous complex, *Geol. Soc. South Africa Trans.*, vol. 61, pp. 387–400.

7. F. J. Davies, 1960, Geology of the Thompson–Moak Lake district, Manitoba, *Canadian Mining Jour.*, vol. 81, pp. 101–104.

8. H. W. Davis, 1956, Nickel, *U.S. Bur. Mines Bull.* 556, pp. 557–568.

9. D. R. DeVletter, 1955, How Cuban ore was formed—a lesson in laterite genesis, *Eng. Mining Jour.*, vol. 156, pp. 84–87, 178.

10. M. Donath, 1962, "Metallischen Rohstoffe, 14 Band, Chrom," Ferdinand Enke, Stuttgart, 371 pp.

11. J. Elliston, 1953, Platinoids in Tasmania, in *Fifth Empire Min. and Metall. Cong., Australia and New Zealand,* pp. 1250–1254, Australasian Institute of Mining and Metallurgy, Melbourne.

12. F. Ergunalp, 1945, The chromite deposits of Turkey, *Am. Inst. Mining Metall. Engineers, Tech. Pub.* 1746, 11 pp.

13. F. B. Esguerra, 1960, Morphological study of the Surigao nickeliferous laterites, Mindanao, Philippines, *Philippine Geologist,* vol. 14, pp. 29–53.

14. D. E. Flint, F. J. de Albear, and P. W. Gould, 1948, Geology and chromite deposits of the Camagüey district, Camagüey province, Cuba, *U.S. Geol. Survey Bull.* 954-B, pp. 39–63.

15. G. P. Fourie, 1959, The chromite deposits of the Rustenberg area, *South Africa Geol. Survey Bull.* 27, 44 pp.

16. A. B. Griggs, 1945, Chromite-bearing sands of the southern part of the coast of Oregon, *U.S. Geol. Survey Bull.* 945-E, pp. 113–150.

17. A. L. Hall, 1932, The Bushveld igneous complex of the central Transvaal, *Geol. Survey South Africa, Mem.* 28, 560 pp.

18. E. C. Harder, 1910, Some chromite deposits in western and central California, *U.S. Geol. Survey Bull.* 430, pp. 167–183.

19. J. E. Hawley, 1962, The Sudbury ores: their mineralogy and origin, *Canadian Mineralogist,* vol. 7, pt. 1, 207 pp.

20. A. Helke, 1962, The metallogeny of the chromite deposits of the Guleman district, Turkey, *Econ. Geology,* vol. 57, pp. 954–962.

21. A. L. Howland, 1955, Chromite deposits of the central part of the Stillwater complex, Montana, *U.S. Geol. Survey Bull.* 1015-D, pp. 99–121.

22. Imperial Institute, 1936, "Platinum and Allied Metals," 2d ed., Imperial Institute, London, Mineral Resources Department, 137 pp.

23. F. E. Keep, 1930, The geology of the chromite and asbestos deposits of the Umvukve Range, Lomagundi and Mazoe districts, *Southern Rhodesia Geol. Survey Bull.* 16, 105 pp.

24. H. F. Kromer, 1954, Chrome ore mining in Turkey, *Eng. Mining Jour.*, vol. 155, pp. 92–95.

25. W. Kupferburger, B. V. Lombaard, B. Wasserstein, and C. M. Schwellnus, 1937, The chromite deposits of the Bushveld igneous complex, Transvaal Union, *South Africa Geol. Survey Bull.* 10, 48 pp.

26. G. S. J. Kuschke, 1940, The critical zone of the Bushveld igneous complex, Lydenburg district, *Geol. Soc. South Africa Trans.*, vol. 42, pp. 57–81.

27. B. Lightfoot, 1940, The Great Dyke of Southern Rhodesia, *Geol. Soc. South Africa Proc.*, vol. 43, pp. xxvii–xlvi.

28. B. V. Lombaard, 1956, Chromite and dunite of the Bushveld complex, *Geol. Soc. South Africa Trans.*, vol. 59, pp. 59–76.

29. H. P. Maufe, B. Lightfoot, and A. E. V. Zealley, 1919, The geology of the Selukwe mineral belt, *Southern Rhodesia Geol. Survey Bull.* 3.

30. J. C. Maxwell, 1949, Some occurrences of chromite in New Caledonia, *Econ. Geology*, vol. 44, pp. 525–544.

31. W. R. McClelland, 1955, Nickel in Canada with a survey of world conditions, *Canada Dept. Mines and Tech. Surveys, Mines Branch Mem. Ser.* 130, 53 pp.

32. J. B. Mertie, Jr., 1939, Platinum deposits of the Goodnews Bay district, Alaska, *U. S. Geol. Survey Bull.* 910-B, pp. 115–145.

33. J. B. Mertie, Jr., 1941, The Goodnews platinum deposits, Alaska, *U.S. Geol. Survey Bull.* 918, 97 pp.

34. E. W. Molly, 1959, Platinum deposits of Ethiopia, *Econ. Geology*, vol. 54, pp. 467–477.

35. D. V. Nalivkin, 1960, "The Geology of the U.S.S.R., a Short Outline," Pergamon Press, New York, 170 pp. See pp. 15, 35, 42, 53, 57, 121.

36. J. W. Peoples and A. L. Howland, 1940, Chromite deposits of the eastern part of the Stillwater complex, Stillwater County, Montana, *U.S. Geol. Survey Bull.* 922-N, pp. 371–416.

37. H. Quiring, 1962, "Die Metallischen Rohstoffe, 16 Band, Platinmetalle," Ferdinand Enke, Stuttgart, 288 pp.

*38. H. P. Rechenberg, 1960, The chrome ore deposit of Kavak, Eskisehir (Turkey), in "Symposium on Chrome Ore," Ankara, Turkey, September 1960, Central Treaty Organization, Ankara, Turkey, pp. 146–156.

39. S. J. Rice, 1957, Chromite, *California Dept. Nat. Resources, Div. Mines, Bull.* 176, pp. 121–130.

40. G. A. Rynearson and F. G. Wells, 1944, Geology of the Grey Eagle and some nearby chromite deposits in Glenn County, California, *U.S. Geol. Survey Bull.* 945-A, pp. 1–22.

41. G. A. Rynearson, 1946, Chromite deposits of the North Elder Creek area, Tehama County, California, *U.S. Geol. Survey Bull.* 945-G, pp. 191–210.

42. E. Sampson, 1932, Magmatic chromite deposits in southern Africa, *Econ. Geology*, vol. 27, pp. 113–144.

43. E. Sampson, 1942, Chromite deposits, in "Ore Deposits As Related to Structural Features," pp. 110–125, Princeton University Press, Princeton, N.J.

44. C. T. Smith and A. B. Griggs, 1944, Chromite deposits near San Luis Obispo, San Luis Obispo County, California, *U.S. Geol. Survey Bull.* 944-B, pp. 23–44.

45. W. C. Stoll, 1958, Geology and petrology of the Masinloc chromite deposits, Zambales, Luzon, Philippine Islands, *Geol. Soc. America. Bull.*, vol. 69, pp. 419–448.

46. T. P. Thayer, 1942, Chrome resources of Cuba, *U.S. Geol. Survey Bull.* 935-A, pp. 1–74.

47. T. P. Thayer, 1956, Mineralogy and geology of chromium, in *American Chem. Soc. Mon.* 132, pp. 14–52.

48. T. P. Thayer, 1960, Some critical differences between Alpine-type and stratiform peridotite-gabbro complexes, Internat. Geol. Cong., 21st, Copenhagen 1960, pt. 13, sec. 13, Petrographic provinces, igneous and metamorphic rocks, pp. 247–259.

*49. T. P. Thayer, 1960, Application of geology in chromite exploration and mining, in "Symposium on Chrome Ore," Ankara, Turkey, September, 1960, Central Treaty Organization, Ankara, Turkey, pp. 197–223.

*50. M. Tokay, 1960, Chromite in Turkey, in "Symposium on Chrome Ore," Ankara, Turkey, September, 1960, Central Treaty Organization, Ankara, Turkey, pp. 82–91. Contains map showing areas of basic and ultrabasic rocks in Turkey.

51. Union of South Africa, Department of Mines, Geological Survey, 1959, Chromite, in "The Mineral Deposits of the Union of South Africa," 4th ed., pp. 187–201.

52. U.S. Bureau of Mines, 1950, "Materials Survey, Nickel." See especially the following: Chap. 2, History, 27 pp.; Chap. 4, Minerals, ores, and geology, 15 pp.; Chap. 5, Nickel reserves of the world, 39 pp.; Chap. 9, Production, 11 pp.

53. C. F. J. Van der Walt, 1941, Chrome ores of the western Bushveld, *Geol. Soc. South Africa Trans.*, vol. 44, pp. 79–112.

54. P. A. Wagner, 1929, "The Platinum Deposits and Mines of South Africa," Oliver & Boyd Ltd., Edinburgh and London, 326 pp.

*55. W. J. Waylett, 1960, Notes on Rhodesian chrome, in "Symposium on Chrome Ore," Ankara, Turkey, September, 1960, Central Treaty Organization, Ankara, Turkey, pp. 224–231. This is a summary of various features of the Rhodesian deposits.

56. H. D. B. Wilson, 1956, Structure of lopoliths, *Geol. Soc. America Bull.*, vol. 67, pp. 289–300.

57. H. D. B. Wilson and W. C. Brisbin, 1961, Regional structure of the Thompson–Moak Lake mineral belt, *Canadian Mining Metall. Bull*, vol. 54, pp. 815–822.

58. G. B. Worst, 1960, The Great Dyke of Southern Rhodesia, *Southern Rhodesian Geol. Survey Bull.* 47, 239 pp.

59. W. S. Wright, 1958, The Surigao laterite deposits, *Philippine Geologist*, vol. 12, pp. 47–59.

60. A. B. Yates, 1948, Properties of International Nickel Company of Canada, in "Structural Geology of Canadian Ore Deposits," pp. 596–617, Canadian Institute of Mining and Metallurgy, Montreal.

61. A. E. V. Zealley, 1914, The geology of the chromite deposits of Selukwe, Rhodesia, *Geol. Soc. South Africa Trans. and Proc.*, vol. 17, pp. 60–74.

* The "Symposium on Chrome Ore" contains 24 papers relating chiefly to chromite deposits in Iran, Pakistan, and Turkey. Some discuss geology, origin, and structure of the deposits; others discuss mining, beneficiation, reserves, economics, and marketing of the ores.

MISCELLANEOUS REFERENCES CITED

62. J. R. Bogert, 1960, How ferronickel is produced from low grade laterite by the Ugine process, *Mining World,* vol. 22, no. 10, pp. 33–37.
63. C. A. Cousins, 1959, Mining Africa's huge platinum reef, *Eng. Mining Jour.,* vol. 160, no. 10, pp. 114–116.
64. W. A. Foster, 1957, Open pit on Nickel Mountain, *Mining Eng.,* vol. 9, pp. 903–904.
65. R. Gencer, 1960, Memorandum of the Committee of Chrome Producers in Turkey, in "Symposium on Chrome Ore," Ankara, Turkey, September, 1960, Central Treaty Organization, Ankara, Turkey, p. 194.
66. G. I. Gorbunov, 1961, Principles of distribution of sulfide copper-nickel deposits in the region of Pechenga (Kola peninsula), *Econ. Geology,* vol. 56, pp. 221–222.
67. P. A. Hodges, 1954, Chrome mining in Southern Rhodesia shows wide variety of operations, *Mining Eng.,* vol. 6, pp. 791–797.
68. C. Johnston, 1962, Platinum mining in Alaska, *Mining Mag. (London),* vol. 106, no. 4, pp. 204–209.
69. C. A. Lamey, 1951, The Cle Elum River nickeliferous iron deposits, Kittitas County, Washington, *U.S. Geol. Survey Bull.* 978-B, pp. 48–55.
70. R. A. Matthews, 1961, Geology of the Butler Estate chromite mine, southwestern Fresno County, California, *California Div. Mines and Geology, Spec. Rept.* 17, 19 pp.
71. *Mining Jour. (London),* 1962, vol. 259, no. 6626, p. 153; 1963, vol. 260, no. 6659, p. 320. Short notes relating to production of nickel in Cuba.
72. M. Saksela, 1935, Copper ore bodies in Finland, in "Copper Resources of the World," vol. 2, p. 564, Internat. Geol. Cong., 16th, United States, 1933.
73. E. C. Speers, 1957, The age relation and origin of common Sudbury breccia, *Jour. Geology,* vol. 65, pp. 497–514.
74. E. F. Stumpfl, 1961, Some new platinoid-rich minerals, identified with the electron microanalyzer, *Mineralog. Mag.,* vol. 32, pp. 833–841. See also 1962, Some aspects of the genesis of platinum deposits, *Econ. Geology,* vol. 57, pp. 619–623.
75. T. P. Thayer and M. H. Miller, 1962, Chromite in the United States, *U.S. Geol. Survey Mineral Inv. Resource Map* MR-26. Map and pamphlet of 5 pp.
76. P. A. Wagner, 1923, The chromite of the Bushveld complex, *South African Jour. Sci.,* vol. 20, pp. 223–225.
77. Anon., 1959, Freeport Nickel's Moa Bay puts Cuba among ranking Ni-producing nations, *Eng. Mining Jour.,* vol. 160, no. 12, pp. 84–92.
78. Anon., 1963, Hanna Nickel plans expansion of mining-smelting facilities, *Mining World,* vol. 25, no. 11, p. 43.

chapter **8**

Lithium, Beryllium, Niobium, Tantalum, Cesium, Rubidium

Somewhat similar conditions of origin are shown by minerals of lithium, beryllium, niobium, tantalum, cesium, and rubidium. Most of the minerals of these elements occur in pegmatites, and the commercial production of some of them comes almost entirely from pegmatites. Some production comes also from carbonatites and from hydrothermal deposits, and some chiefly from placer and detrital deposits, although the original source of such deposits not uncommonly was pegmatites. The major relationships are shown by Table 8-1.

Lithium

Most of the world's lithium has been obtained from pegmatites, but some has come from brines, especially from Searles Lake, California. The chief producing places have been Rhodesia (Southern Rhodesia), South-West Africa, the United States, and Canada. Before 1930 the United States furnished more than 70 percent of the world production of lithium; in 1930 South-West Africa began to be an important producer and in 1936 supplied 42 percent of the world total, whereas the United States furnished 54 percent; in the period 1937 to 1949 the United States produced 68 percent of the world total, and South-West Africa supplied 22 percent; in the period 1950 to 1954 the United States furnished 42 percent of the total, Rhodesia (Southern Rhodesia) 32 percent (important production began in 1951), and South-West Africa 20 percent. Since 1955, production figures for the United States have not been released; Rhodesia (Southern Rhodesia) has furnished between 85 and 90 percent of the reported world production. Important potential resources are present in Brazil, Argentina, the Republic of the Congo, Australia, and perhaps the Soviet Union.

Four minerals are the chief commercial sources of lithium: spodumene, ambylgonite, petalite, and lepidolite. The first three are used for the extraction of chemical lithium, whereas lepidolite, because of its potash content, is not so used

TABLE 8.1   Occurrence of Minerals of Beryllium, Lithium, Niobium, Tantalum, Cesium, and Rubidium in Various Types of Deposits

		*Importance in various types of deposits**			
Element	*Mineral*	*Pegmatites*	*Hydro-thermal deposits*	*Car-bonatites*	*Placer and detrital deposits*
Lithium	Spodumene	Great in N. America			Minor
	Petalite	Considerable			
	Lepidolite	Considerable			
	Amblygonite	Great abroad			
	Triphylite	Minor			
	Lithiophilite	Minor			
Beryllium	Beryl	Great	Moderate		Minor
	Bertrandite	Minor	Minor		
	Phenakite	Minor	Minor		
	Helvite	Minor	Minor		
Niobium and tantalum	Columbite-tantalite	Considerable	Minor		Considerable Moderate
	Euxenite	Minor			
	Pyrochlore			Great	
Cesium and rubidium	Pollucite	Moderate		Great	

* Mineral is essentially of no importance commercially if nothing is designated.

but is used directly by glass manufacturers to produce flint, opal, and heat-resistant glass. Spodumene is the chief mineral used for chemical lithium in North America, whereas amblygonite is used abroad, where it is more abundant than spodumene. Petalite is known in significant commercial amounts only in South-West Africa near Karibib, in Sweden, and in Australia. Lepidolite is produced especially from deposits in Rhodesia (Southern Rhodesia) (12, 37, 54).

Africa

Lithium-bearing pegmatites are abundant in parts of South-West Africa, the Republic of South Africa, and Rhodesia (Southern Rhodesia) (Fig. 16-1), and these pegmatites also furnish beryllium, niobium, tantalum, tin, and tungsten, as well as feldspar, mica, and quartz. Lithium production has been important since 1937 from South-West Africa but only since about 1951 from Rhodesia (Southern Rhodesia), at present the chief producer. The largest known deposit of lithium is reported to be in the Congo (12), but production from it has been limited because of lack of transportation.

Rhodesia (Southern Rhodesia).   The lithium deposits of Rhodesia (Southern Rhodesia) are chiefly in the southern part of the Bikita tin field. The largest concentration is in a pegmatite about 6,000 ft long that averages about 900 ft wide.

Much lepidolite and some petalite and amblygonite are present in the Salisbury district (1; 60, pp. 28–29). Production by Bikiti Minerals, Ltd., was chiefly lepidolite, with some spodumene and amblygonite, but in 1959 a large reserve of petalite, amblygonite, and eucryptite[1] was obtained by the purchase of an adjacent mine.

South-West Africa. Lepidolite and petalite deposits of South-West Africa are scattered over many miles, but most of the production has come from the Karibib and Omaruru districts, near the west-central part of the country, and the Wambad district, the southern part of South-West Africa north of the Orange River (Fig. 16-1). These deposits have attracted wide attention because some lithium-bearing pegmatites contain cassiterite, beryl, columbite, tantalite, and wolframite (4, 9, 15, 16; 60, p. 29).

The majority of lithium occurrences in South-West Africa are in pegmatites that have intruded ancient sedimentary rocks. The deposits are of two types: (1) nests and books of lepidolite in little-altered pegmatite, the lepidolite always associated with "blows"[1] of quartz, marginally around which the lepidolite is arranged; (2) lepidolite-greisen composed of fine scales and crystals of lepidolite, usually less than 2 mm across, commonly some quartz, cassiterite, topaz, and feldspar altered in part to sericite, muscovite, kaolinite, or lepidolite. The lepidolite-greisen occurs as lenticular bodies, always associated with white pegmatitic quartz that is enclosed by or adjoins pegmatite (9).

The great bulk of the lithium minerals, cassiterite, and other metal-bearing minerals were thought by de Kock (9) and Gevers and Frommurze (16) to have formed by replacement of rock-forming pegmatite minerals and injection of pegmatite quartz after the formation of simple pegmatites. Distribution of lithium-bearing and other minerals was considered to be highly irregular and patchy, and this concept apparently discouraged exploration and exploitation. A zonal concept of origin was postulated by Cameron (4), and this concept should make possible systematic exploration (83) and stimulate production.

South Africa. The chief lithium-bearing pegmatites of South Africa are in Namaqualand, south of the Orange River, in a zone extending generally east-west about latitude 29°S (Fig. 16-1). The pegmatites have been classed in three groups (17), following the scheme of W. H. Emmons (81): (1) interior (core) pegmatites; (2) exterior (hood) pegmatites; (3) exterior (roof) pegmatites. The pegmatites of the first group generally are small simple pegmatites; those of the second group are larger and commonly show extensive mineralization; whereas those of the third group make a zone 9 to 10 miles wide and are highly mineralized. Lithium minerals, beryl, columbite-tantalite, cassiterite, native bismuth, and bismuthinite have been discovered in these pegmatites, in addition to commercial feldspar and mica, but the relatively small quantities reported and the remoteness of the area from ready market cast doubt on the feasibility of economic development.

[1] Eucryptite is an alteration of spodumene.
[1] A blow has been defined as "a large mass of quartz or other gangue, isolated or forming a sudden enlargement on a lode." (5, p. 88, of Chap. 2)

Republic of the Congo (Belgian Congo). Pegmatites of Manono and Kitotolo, Katanga Province, Republic of the Congo, which are mined for cassiterite, contain also columbite-tantalite and spodumene. The deposits are large and spodumene might be recovered as a by-product. See further discussion under niobium and tantalum deposits.

Brazil

Hundreds of pegmatite dikes occur in northeastern Brazil in three states: Rio Grande do Norte, Paraíba, and Ceará. Many of them are zoned and contain lithium minerals, beryl, columbite-tantalite, cassiterite, and muscovite. They are important sources of beryl but produce also spodumene along with some amblygonite and lepidolite. Spodumene crystals 7 to 8 ft long were reported from one pegmatite in Rio Grande do Norte (40, 44, 45; 60, pp. 26–27). They are discussed further with beryl deposits.

United States

Lithium minerals occur in pegmatites in South Dakota, Colorado, New Mexico, Arizona, California, the New England States, the southeastern Piedmont states, and elsewhere, but the major production has been from South Dakota and North Carolina (46; 60, pp. 19–24). The Etta mine, about 1 mile south of Keystone, Pennington County, South Dakota, has been the largest producer in the Black Hills. Before 1883 it was worked for mica, from 1898 to 1960 it yielded spodumene, in the spring of 1960 it was closed, and in 1961 it was sold. It was noted for the huge crystals of spodumene present, one of which, 42 ft long, yielded 90 tons of spodumene; another one was 47 ft long (61; 72, p. 654; 84). The spodumene crystals of other mines were not as large, although some were from 1 to 10 ft long. Throughout the area, however, many are smaller and some are merely grains of microscopic size. No spodumene was produced in South Dakota in 1961, although the total resources of the region are still considerable (43, 46, 47).

The deposits of North Carolina occur in the area known as the tin-spodumene belt, which has a length of 24.5 miles and a maximum width of 1.8 miles, extends southwestward from Lincolnton to Grover (Fig. 8-1), and has been estimated to contain 4,300,000 tons of pegmatite ore averaging 15 percent spodumene, although unusually rich parts contain 30 to 50 percent spodumene. The largest pegmatite body in the area is 3,250 ft long and 395 ft wide (2, 36). The major production comes from Kings Mountain, North Carolina. The deposit, reported to be the largest known spodumene occurrence in the Western Hemisphere, consists of irregular discontinuous dikes with an average width of 50 to 100 ft, a depth of about 900 ft as indicated by core drilling, and may be covered by overburden to a depth of 40 ft. The dikes contain about 20 percent spodumene, 35 percent feldspar, and 2 to 3 percent mica, but material is concentrated to 80 percent spodumene by a specially developed process (2, 19, 24, 36, 74).

The waters of Searles Lake, California, provide a most unusual occurrence of lithium. The lake, on the edge of the Mojave Desert in San Bernardino County, is a playa that contains crystalline salt to a depth of 60 to 90 ft exposed over an

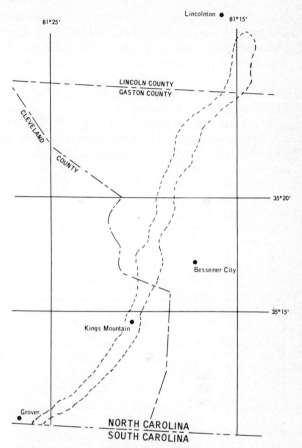

Fig. 8-1 Map showing generalized position of the tin-spodumene belt, North Carolina. After T. L. Kesler, *U.S. Geol. Survey Bull.* 936-J, p. 248.

area of about 12 sq miles. Voids in the salt body, estimated at 25 to 45 percent, are filled with brine that contains 0.048 percent lithium chloride by weight. The total amount of LiO_2 in these brines had been stated as 90,000 tons (60, pp. 17, 20–21).

Canada

Four principal areas of spodumene-bearing pegmatites are at present known in Canada. These are (1) about 25 miles north of Val d'Or, in Lacorne township, Quebec, where the deposit is rated as one of the largest known in the world; (2) south of MacDiarmid, Ontario; (3) near Cat Lake and Bernic Lake, southeastern Manitoba; and (4) about 50 to 90 miles northeast to east of Yellowknife, Northwest Territories. Other deposits have been reported present in Manitoba and Ontario. The spodumene occurs in the intermediate zones of zoned pegmatites, in thickly lenticular pods, and in pegmatite bodies that are essentially spodumene-bearing from wall to wall and that contain the bulk of the present lithium reserves. Diamond drilling and mapping indicate reserves in the order of 36 million tons that contain 1 percent lithia or more (20, 29, 54; 60, pp. 24–26).

Canada did not become an important producer of lithium minerals until 1955, when Quebec Lithium Corporation began operations at its Lacorne mine, Quebec. The lithium-bearing dikes occur in an area extending about 20 miles east-west and 12 miles north-south (Fig. 8-2). The dikes are less than 1 ft but up to 150 ft wide; several are more than 1,000 ft long and an exceptional one is 1.5 miles long. Minerals present, in addition to feldspar, quartz, mica, and spodumene, are columbite-tantalite, microlite, betafite, bismuthinite, bismuthotantalite, hematite, molybdenite, beryl, lepidolite, and garnet. Some of the dikes that contain columbite-tantalite also contain a little uranium and are slightly radioactive. The spodumene is present as fine-grained material, crystals the size of match sticks, and crystals several feet long. A company estimate is that the group of dikes explored to about 1957 contains 15 million tons of ore to a depth of 700 ft, the average grade of which is 1.2 percent lithia. Structural considerations, however, indicate a

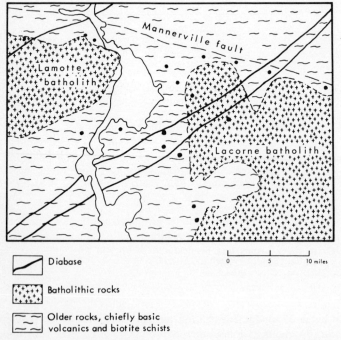

Diabase

Batholithic rocks

Older rocks, chiefly basic volcanics and biotite schists

0   5   10 miles

Fig. 8-2 Generalized geologic map of the Lacorne lithium area, Quebec, Canada. Principal lithium deposits, in dikes, shown by black dots. After W. N. Ingham and M. Latulippe, "The Geology of Canadian Industrial Mineral Deposits," p. 160. Copyright 1957. The Canadian Institute of Mining and Metallurgy. Used by permission.

depth of at least 2,500 ft; if so, the tonnage might be more than three times as great (11, 29).

Siroonian and others (64) made a geochemical study of the pegmatites of the Lacorne region. They concluded that field investigations indicated that the spodumene-bearing pegmatites were formed by fractional crystallization of batholithic magma and that geochemical data substantiate this concept.

Beryllium

Beryllium deposits are of two types, pegmatite and nonpegmatite (3). Commercial deposits at present are almost entirely beryl-bearing pegmatites, although minor amounts of beryllium may be obtained from other minerals contained in pegmatites: bertrandite, chrysoberyl, and phenakite. Nonpegmatite deposits are in part hydrothermal veins or veins closely related to pegmatites, and contact-metasomatic deposits. The veins that contain beryllium are, in order of importance, (1) quartz-tungsten, perhaps also quartz-cassiterite and quartz-molybdenite; (2) quartz-gold; and (3) manganese-lead-zinc. Among these, the first type is by far the most important and is of world-wide distribution (3, pp. 87–95; 26; 69). The beryllium mineral of many of these veins is beryl, although in some it is helvite. Recent discoveries in Utah (65, 67), Colorado (85), and in Mexico (41) show that important deposits, apparently of hydrothermal origin, which contain bertrandite and perhaps some other beryllium minerals, are associated with fluorite deposits. The most important beryllium mineral in contact-metasomatic deposits usually is helvite. Beryllium is present in idocrase (vesuvianite) in some deposits; analyses have shown as much as 4.0 percent BeO in this material. Beryllium is present also in small amounts in some other minerals, including feldspar of pegmatites. The occurrence of beryllium in deposits is shown schematically by Fig. 8-3. Pegmatite beryllium deposits have been especially important in Brazil, Argentina, and Mozambique, although numerous deposits occur in various places throughout Africa, and recently Uganda has been an important source of beryl (Fig. 8-4).

Pegmatite Deposits

Brazil.   Pegmatites occur in two major regions in Brazil: (1) eastern Minas Gerais, which yields most of the mica and gem tourmalines and beryls; (2) in Paraíba, Rio Grande do Norte, and Ceará, chiefly the first two, which are important sources of beryl, columbite-tantalite, lithium-bearing minerals, and also some muscovite and cassiterite.

Estimates in 1963 (88) stated the reserves of beryllium ore in Brazil as 350,000 metric tons, distributed as follows: Minas Gerais, 75 percent; Bahia, 15 percent; Rio Grande do Norte, 4 percent; Ceará, 3 percent; Paraíba, 2 percent; Pernambuco, 1 percent.

A somewhat typical deposit of Minas Gerais is the one at Sabinopolis, where a pegmatite dike contains beryl, fluorite, columbite, monazite, and semiprecious stones (the most important source in Brazil). Some huge crystals of beryl

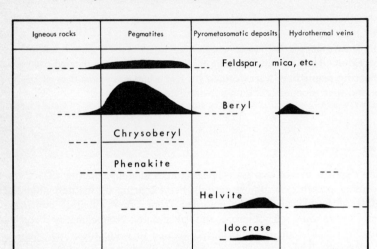

Fig. 8-3 Schematic diagram showing occurrence and distribution of principal beryllium-bearing minerals. After L. A. Warner and others, *U.S. Geol. Survey Prof. Paper* 318, p. 58.

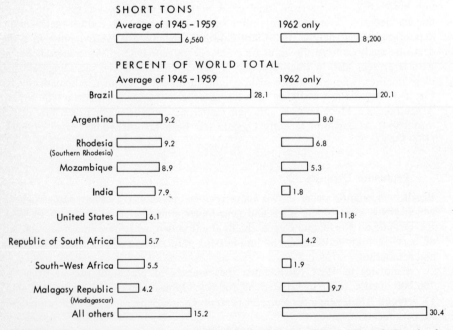

Fig. 8-4 World beryl production. Compiled from *U.S. Bur. Mines Minerals Yearbook*, various years.

occurred, the largest one reported being 70 cm across and 250 cm long (about 2.3 by 8.2 ft) (80).

In Paraíba and Rio Grande do Norte, northeastern Brazil, throughout a large area there are both simple (homogeneous) and complex (heterogeneous, zoned) pegmatites. The complex (zoned) ones yield beryl, columbite-tantalite, and spodumene, as well as some cassiterite, bismutite, and muscovite. Some of them contain very large crystals, weighing tons and measuring in meters (8). These pegmatites yielded approximately 8,000 metric tons of beryl and 600 metric tons of tantalite between 1937 and 1944, and are rated as one of the world's most important sources of beryl and tantalite (34). They are also important potential sources of lithium (40, 44, 45, 60, 80).

Argentina. Argentina is one of the important beryl-producing countries of the world, but little information has been published in English regarding the pegmatites present there, and only a few articles have appeared in Spanish (25, 75, 76). Hence the material presented here is given somewhat in detail and covers more than beryl.

A great many pegmatites are present throughout a zone in the mountainous parts of the provinces of Salta, Catamarca, La Rioja, San Luis, and Córdoba (Fig. 8-5). These pegmatites in general form seams, dikes, and masses in crystalline schists, the thicker bodies in places cutting across the schistosity, the thinner ones more generally following the schistosity; some pegmatites, especially in the province of Salta, cut granite. The pegmatites contain, in addition to quartz, feldspar, and mica, varying amounts of tourmaline, apatite, garnet, beryl, columbite, tantalite, native bismuth, bismuthinite, triplite, lithiophyllite, spodumene, lepidolite, uraninite, gahnite (zinc spinel), and some molybdenite, pyrite, and chalcopyrite; also secondary minerals developed from some of these. Materials mined include feldspar, quartz, mica, beryl, columbite, tantalite, bismuth minerals, and some uranium minerals. Columbite and tantalite are present in many pegmatites, but production has been chiefly from the provinces of Córdoba, San Luis, and especially Salta, where, in the region of El Quemada, the production in 1943 and 1944 was about twice as much as it was in the other two provinces combined. In this same region, El Quemada, the production of bismuth seems to have been greater than elsewhere. Small amounts of uranium have been produced in Córdoba Province. Beryl, although widely distributed, has been produced mostly in the provinces of San Luis and Córdoba, especially the latter. In most cases, mining has been chiefly for feldspar, mica, or quartz, and other materials have been recovered as by-products. A notable exception is Las Tapias mine in Córdoba Province, about 100 km southwest of Córdoba, which has been operated chiefly for beryl and has been rated as probably the largest individual producer of beryl in the world.

The beryl deposits that have been the most productive lie in a zone that starts, in a general way, somewhat west of Córdoba (Córdoba Province) and extends to the vicinity of San Luis (San Luis Province) (Fig. 8-5). The scattered nature of the deposits and the small concentrations in many of them (generally not more than a few tens of tons of beryl) do not, in general, justify installation

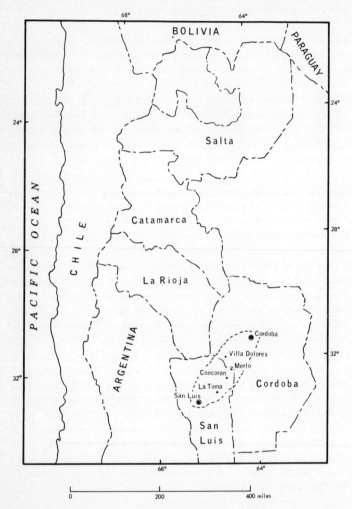

Fig. 8-5 Map showing major pegmatite provinces of Argentina and chief area of beryl production (dashed boundary).

of modern mining equipment. Consequently the work is done chiefly by groups of independent miners, with the assistance of women and children.

The beryl usually mined ranges in diameter from about 2 cm to more than 1 m (about 1 in. to 3 ft) and in length up to 4 m (about 13 ft). Crystals 10 to 20 cm across are not uncommon, and those a meter across have been reported from several mines. Individual crystals have been reported to weigh from 0.5 kg to 2,000 kg (about 1 to 4,400 lb). The beryl occurs as single crystals, as masses in pockets, stringers or shoots, and more or less definite bands. One group of five crystals is reported to have yielded 15 tons of beryl; one pocket, 30 tons; and one zone, 42 tons. Some features of the chief mining areas are shown in Table 8-2.

TABLE 8-2.   Some Features of the Occurrence of Beryl in Argentina

Province	Mine or mines and location	Beryl production	Special features and other materials produced
Córdoba	Las Tapias, 12 km north of Villa Dolores, 100 km southwest of Córdoba.	Most important in Argentina	Zoned pegmatite; bismuth, feldspar, mica recovered; spodumene and columbite present
	El Criollo, near Cosquin, and others around Punilla, vicinity of Córdoba.	Important	Some beryl crystals up to 50 cm diameter; feldspar and quartz recovered; apatite, triplite, columbite-tantalite, bismuthinite present
	La Magdalena and several other mines in the Sierra de Comechingones, southern part of province.	Considerable amount	Pegmatites exploited chiefly for mica; also contain columbite and some uranium minerals
San Luis	Mines in the vicinity of La Toma, about 50 km east-northeast of San Luis. Important mines are Santa Ana, 5 km northwest of La Toma; Las Palomas, 66 km northeast of La Toma; La Esmeralda and La Violeta, 26 km south of La Toma.	Considerable amount	Some large pockets of beryl, but much beryl is by-product of recovery of feldspar, bismuthinite, and quartz; some bismuthinite in nests and pockets, weight 40–50 kg, exceptionally 500 kg
	Los Magos mine, 25 km east-northeast of Santa Rosa, near border of San Luis and Córdoba provinces; also La Berta mine, 26 km east-northeast of Concaran.	Some importance, especially early	Beryl from quartz vein near center of pegmatite (Los Magos) and in crystals 2–5 cm diameter (La Berta); columbite, spodumene, triplite, apatite also present

Africa.   Much of the beryl of Africa comes from the same general areas that furnish lithium minerals (Fig. 16-1). Thus in Rhodesia (Southern Rhodesia) beryl comes from the pegmatites of the Bikita tin fields; in South-West Africa, from the Karibib and Omaruru districts; in the Republic of South Africa, from the districts of Little Namaqualand, Kenhardt, Gordonia, and Cape of Good Hope; in the northern part of Mozambique, from the Alto Molocue area (53) and the Alto Ligonha area (7, 28) along with by-product columbite-tantalite and bismuthinite; and from the Malagasy Republic (Madagascar) (79). The largest production of recent years has been from Mozambique where pegmatites contain, in addition to beryl, columbite, lepidolite, and bismuth. Mining there has been from eluvial and alluvial deposits, kaolinized pegmatites, and in part from relatively unaltered pegmatites. The pegmatites are zoned much in the same manner as those of the United States. Some beryls are 1 ft across and 2 ft long.

United States.   The chief production of beryl in the United States has been from South Dakota, New Mexico, Colorado, and New Hampshire (53); the chief reserves from pegmatites are in South Dakota, New England, Colorado, and, to a lesser extent, Idaho, Nevada, and New Mexico (77, p. 94). Also, potential reserves are high in the North Carolina tin-spodumene belt, the Spruce Pine area of North Carolina, and possibly in Troop County, Georgia (22).

The pegmatites of the Black Hills were investigated in some detail in the 1950s (38, 47, 62). The Beecher No. 3–Black Diamond pegmatite, Custer County, South Dakota, where beryl was discovered in 1952, appears to be one of the most important sources in the United States. The pegmatite contains spodumene also and could yield a large amount of scrap mica as a by-product (63).

Investigations of the pegmatites of Colorado from 1950 to 1960 (18, 23, 66, 68) indicate moderate reserves of beryl, much of which can be recovered only by milling. Small tonnages of scrap mica, feldspar, and columbite-tantalite might be recoverable also.

The Harding pegmatite, Taos County, New Mexico, was noted especially for the production of lithium and tantalum before 1943, when it became an important source of beryl. Some of the deposits are coarse-grained; between 1950 and 1952 hand sorting yielded 350 tons of high-grade concentrates and 150 tons of lower-grade material (33).

Beryl throughout the New England region generally is not abundant enough to justify mining it alone (6). Most beryl in Connecticut is in crystals too small to be recovered by hand sorting and probably could be recovered by milling only in connection with the production of feldspar, scrap mica, and possibly quartz (5). Most of the beryl of New Hampshire is in crystals as least 1 in. in diameter but is so widely scattered throughout the pegmatites that it represents less than 0.10 percent of the rock and is recoverable only as a by-product of the mining of feldspar and mica.

In the North Carolina tin-spodumene belt recovery of beryl probably can be made economically only as a by-product of the mining of spodumene and feldspar (21). If spodumene, feldspar, mica, cassiterite, and columbite, along with beryl, could be recovered efficiently, this area might become the principal source of beryl in the future (77, p. 98).

Nonpegmatite Deposits

Nonpegmatite beryllium deposits in the United States were investigated between 1950 and 1960 (69), but in other countries few similar investigations appear to have been undertaken. In the United States deposits of New Mexico at Iron Mountain and in the Victorio Mountains, deposits of Utah in the Thomas Range, and deposits of Colorado in the Badger Flat area (22) appear to be of potential importance. In Mexico a deposit at Coahuila may be important. In the Soviet Union a deposit in the southern Ural Mountains is known to be an important source of beryllium.

The deposits at Iron Mountain, New Mexico (31), about 60 miles southwest of Socorro, are of the contact-metasomatic type. Paleozoic limestone around intrusions contains massive tactite and "ribbon rock" tactite, both formed by replacement of limestone. The massive tactite consists chiefly of magnetite and andradite, whereas the ribbon rock contains magnetite and fluorite as its chief constituents but also significant amounts of helvite, in places as much as 24 percent. Further, some of the ribbon rock contains beryllium-bearing idocrase. It is a striking rock that is composed of thin layers, which are minutely crenulated and differ in mineralogy from the immediately adjacent layers. A small tonnage of high-grade tactite contains from 0.5 to 3.5 percent BeO,[1] but averages less than 1.0 percent, chiefly in helvite, whereas low-grade tactite contains from 0.1 to 0.85 percent BeO, only a small amount of which is in helvite; this low-grade material would be uneconomical to recover. Although the deposits at Iron Mountain contain several thousand tons of high-grade material, they have not been mined, principally because the material is fine-grained (69, p. 59).

The deposits in the Victorio Mountains, New Mexico (26), in Luna County, exemplify both the contact-metasomatic and the hydrothermal vein types. Helvite forms lenses and zones in tactite, and beryllium-bearing idocrase (0.2 percent BeO) is interlayered with fluorite and garnet; beryl crystals, some as much as 5 cm long and 1 cm wide, form a selvage of a quartz-tungsten vein.

The deposits in the Thomas Range, Utah (65, 67), in the area commonly called the Topaz district, are in altered rhyolite tuff of Tertiary age along the eastern and western flanks of Spor Mountain, an important fluorspar district about 120 miles southwest of Salt Lake City. The fluorite-rich deposits are in the center of the district and the beryllium-rich deposits are around the margins in disseminated fluorspar deposits of low grade. The highest amounts of beryllium are in nodules composed of fluorite, silica minerals (quartz, chalcedony, opal), and, less commonly, calcite. The only beryllium mineral positively identified is bertrandite. Apparently the deposits are of hydrothermal origin and were deposited from relatively low-temperature, nearly spent solutions that moved through persistent fracture channels and deposited most of their fluorite in the central part of the area. The deposit may be an important source of beryllium if economic methods of recovery can be devised. Fluorspar and uranium, which occurs in carnotite in the fluorspar deposits but was not noted in the beryllium deposits, are potential by-products.

Deposits recently exploited in the Badger Flat area, Park County, Colorado (86), consist of hydrothermal veins that contain beryl, bertrandite, and euclase. A considerable amount of production was reported from 1956 through 1960.

Fluorite deposits that contain beryllium occur in the Aguachile district, Coahuila, Mexico (41), about 80 air miles southeast of Marathon, Brewster County, Texas. Deposits in the central part of Aguachile Mountain, along a ring dike of rhyolite porphyry that encircles a subsidence basin, contain relatively high amounts of beryllium and bertrandite, and these are surrounded by deposits that contain insignificant amounts of beryllium and no determined beryllium mineral. Apparently the beryllium deposits were formed by relatively low-temperature hydrothermal solutions (82). Descriptions of the deposits do not indicate their importance as a potential source of beryllium.

In the southern Ural Mountains of the Soviet Union, at the Izumrudnie emerald mines, beryl occurs chiefly in schists and slates that surround beryl-

---

[1] Beryl contains 10.0 to 14.0 percent BeO; helvite contains 10.5 to 15.0 percent. The zones of beryl pegmatites that are mined contain from 0.2 to 0.01 percent BeO.

bearing pegmatites. Less important deposits are in the Sherlovi Mountains, Zabaikal, where beryl occurs in quartz veins and in greisen zones.

Niobium and tantalum

Niobium and tantalum occur in pegmatites, granites, basic alkaline rocks, carbonatites, and placers. Carbonatites probably are the most important future source of niobium (10), but actual production so far has come chiefly from alluvial deposits, decomposed granites, and pegmatites of Nigeria (30, 70). Mining of columbite-bearing eluvial gravels derived from the weathering of pegmatites in Portuguese East Africa may also be economically feasible (28). The following estimate of the reserves and output of niobium and tantalum in 1960 (Table 8-3), classed according to types of deposits, was made by de Kun (10, p. 321).

TABLE 8-3.   Reserves and Output of Niobium and Tantalum in 1960, by Types of Deposits*

	Reserves, percent		*1960 output, percent*	
Type of deposit	*Nb*	*Ta*	*Nb*	*Ta*
Carbonatites	98.5	—	7	—
Placers	1.2	61	64	54
Pegmatites	0.1	39	9	46
Granites	0.2	—	20	—

* From Nicholas de Kun, *Econ. Geology,* vol. 57, p. 381. Copyright 1962. Economic Geology Publishing Company. Used by permission.

World production of niobium and tantalum has come chiefly from Africa, mostly from Nigeria (Fig. 8-6), but Canada became an important producer of niobium in 1962.

Mention has been made of niobium and tantalum in pegmatites in discussions of deposits of lithium and beryllium and of the fact that niobium and tantalum may be recovered as by-products or co-products of the mining of such pegmatites. Hence only two outstanding niobium-tantalum deposits will be presented here, along with a brief statement regarding deposits in the United States and Canada.

Deposits in or Associated with Pegmatites and Granites

Nigeria.   The Jos Plateau of north-central Nigeria, according to de Kun (10, p. 400), contains more columbite than all other deposits combined. Tin-niobium mineralization of the north-central part of Nigeria is associated with the "older" granite and the "younger" or Plateau granite, both provisionally classified as of Precambrian age (30). The granites were later partly covered by the fluvio-

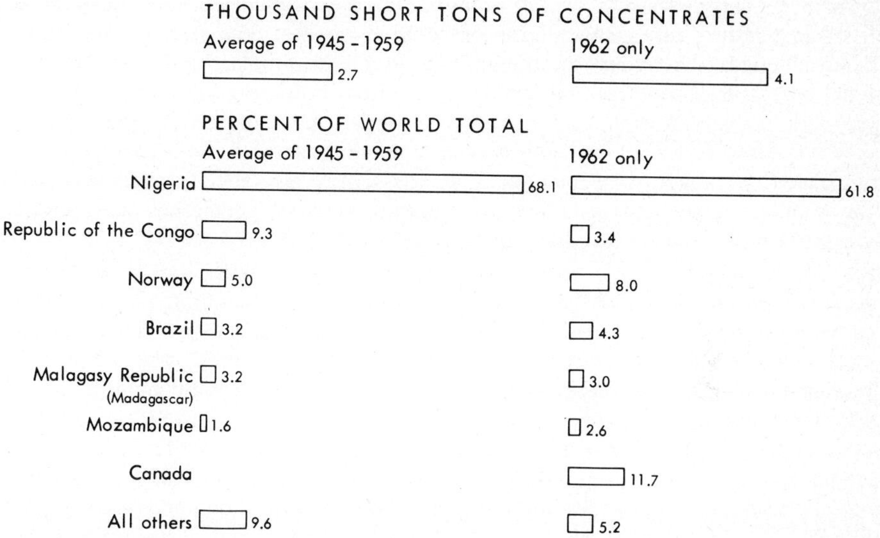

THOUSAND SHORT TONS OF CONCENTRATES

Average of 1945 - 1959   1962 only

[_____] 2.7   [_____] 4.1

PERCENT OF WORLD TOTAL

Average of 1945 - 1959   1962 only

	Average of 1945-1959	1962 only
Nigeria	68.1	61.8
Republic of the Congo	9.3	3.4
Norway	5.0	8.0
Brazil	3.2	4.3
Malagasy Republic (Madagascar)	3.2	3.0
Mozambique	1.6	2.6
Canada		11.7
All others	9.6	5.2

Fig. 8-6 World niobium-tantalum production. Compiled from *U.S. Bur. Mines Minerals Yearbook,* various years.

volcanic series and all these rocks buried in places by newer basalts. Mineralization occurs in veins, stockworks, greisens, and in pegmatite dikes, but most production has been obtained from alluvial deposits associated with the "younger" granite of the Jos Plateau. Production comes also from the Wamba-Jemaa area of the Egbe district. Early alluvial mining indicated that columbite was derived from the same lodes and greisens as cassiterite, since columbite was always found associated with cassiterite. Later work has shown, however, that the Jos-Bukuro complex comprises granitic bodies of several types, some of which contain columbite (70). Investigations disclosed that columbite could be economically concentrated from decomposed granites of this type, some of which were shown by drilling to be decomposed to a depth of more than 100 ft. Some of the granite is soft enough to be mechanically excavated and subsequently concentrated.

The alluvial deposits of Nigeria (10; 13; 35, p. 9–11; 39) contain columbite in the vicinity of niobium-bearing granites. The alluvial deposits have been placed in three groups (39, p. 7): (1) those in the present drainage system; (2) deep leads beneath the newer basalts; (3) deep leads beneath the fluvio-volcanic series. Some leads reach a depth of 100 ft or more; some extend for more than 5 miles across country and occupy old valleys 100 to 400 ft wide. Overburden is 15 to 25 ft thick. These deposits have furnished more than 80 percent of the total columbite production of the world.

The most important known economic deposits of the pegmatite-tin field are in a broad zone extending from the Wamba-Jemaa region in Plateau Province, the most important tin area, to the Egbe district in Kabba Province, where the ratio of columbite-tantalite to cassiterite is much higher than it is in the Wamba-Jemaa area, although cassiterite is still the predominant mineral. The pegmatites have been classified (30) as (1) interior or core pegmatites, (2) marginal or hood pegmatites, and (3) exterior or roof pegmatites. The interior pegmatites occur chiefly in the "older" granite and are essentially simple pegmatites; the marginal pegmatites occur both within the granite and bordering gneisses and schists, are intermediate in composition between simple and complex pegmatites, and contain conspicuous albite with which columbite-tantalite and cassiterite are associated; the exterior pegmatites lie in the schists and gneisses relatively near the granite contact and generally are complex albitized pegmatites that contain the great bulk of the columbite-tantalite and cassiterite.

Republic of the Congo (Belgian Congo).   Niobium and tantalum are present in cassiterite-bearing pegmatites of the Manono-Kitotolo (Katanga) region of the Congo, about 150 miles southwest of Albertville, and a considerable amount of production comes from there as a by-product or co-product of tin mining (35, p. 12). Early production in this area, after discovery of cassiterite in 1912, came from alluvial deposits, then from kaolinized, near-surface "soft pegmatites," and finally from less-altered "hard pegmatites" (73). Two pegmatites, the Kitotolo (westernmost) and the Manono (easternmost), each about 3.5 miles long and 150 to 2,500 ft wide, are separated by about 2 miles of schist and granite. Cassiterite is disseminated almost homogeneously throughout the pegmatite in an amount equal to 4.5 to 5.5 lb of tin per cubic yard, and grade is fairly uniform to a depth of at least 350 ft. Thoreaulite ($SnTa_2O_7$) has been estimated to be about 0.05 as abundant as cassiterite. Some thoreaulite contains as much as 17 percent Nb_2O_5. Important amounts of spodumene are present also (73). The Manono-Kitotolo pegmatite possibly contains more tin, niobium, and tantalum than all the other pegmatites combined, according to de Kun (10, p. 390).

United States and Canada.   Little niobium and tantalum have been produced from pegmatites in the United States (49) and Canada (35). Some columbite-tantalite is present in pegmatites of Colorado and a small amount may be recoverable as a by-product of mining other constituents. A considerable amount of columbite-tantalite has been recovered from pegmatites in the Black Hills, South Dakota, and more than 500 lb of coarse tantalite-columbite averaging 45 to 55 percent Ta_2O_5 was recovered from small placer deposits associated with the Harding beryllium-tantalum-lithium pegmatites of Taos County, New Mexico, but the chief tantalum mineral in that pegmatite is microlite, $(Na,Ca)_2Ta_2O_6(O,OH,F)$ (32). Recent recovery of niobium-tantalum in the United States was from placer deposits of euxenite in Idaho. Columbite-tantalite is known from some pegmatites of Canada, but only a very small amount of production has come from the Yellowknife area of the Northwest Territories (35). Future production in both the United States and Canada may come from pyrochlore in carbonatites; the production from Canada was large in 1962.

Deposits in Carbonatites

Among more than a hundred carbonate-alkaline rock complexes discovered, somewhat less than half contain niobium and some traces of tantalum. About 20 can be considered as economic deposits, most of which are in southeastern Africa, South America, and North America (10, p. 382). Pyrochlore, essentially

$NaCaNb_2O_6F$, is the chief niobium mineral present in carbonatites. Little detail can be given about these complex deposits in an introductory treatment, but a few examples will be cited.

In Africa a recently discovered deposit at Luesche in northern Kivu, Republic of the Congo, is reported to contain 30 million tons of ore that assays 1.34 percent Nb_2O_5. This deposit is rated by de Kun (10, p. 400) as one of two deposits in the world (the other is Araxá, Brazil) that, together, contain more pyrochlore than all other deposits combined. The Luesche deposit (42) lies in the western scarp of the Lake Edward Rift, about 25 miles southwest of the southern shore of Lake Edward and about 43 miles north of Lake Kivu. The Luesche carbonatite is an elliptical plug composed of a central core of cancrinite syenite (busorite) surrounded by a nearly continuous ring of massive calcitic carbonatite (aegyrine-sövite); toward the southeast is massive magnesian carbonatite (rauhaugite) containing 30 to 90 percent $MgCO_3$. The sövite is exceptionally coarse-grained and is composed of calcite crystals up to 1 cm across along with varying amounts of aegerine (the most abundant mineral except calcite), apatite, alkali feldspar, pyrrhotite, and pyrochlore. Although final estimates of reserves were not available in 1960, Meyer and Béthune (42, p. 308) regarded this deposit as one of the largest pyrochlore-apatite deposits in the world and probably the most important one in Africa. Other important African deposits of pyrochlore are known in Tanganyika (Mbeya, at Panda Hill), in Kenya (Mrima Hill), in Uganda (Sukulu), and elsewhere (10, pp. 400–402; 14; 50, p. 1540, Table 1; 56, p. 282). The African carbonatites generally occur close to rift valleys (14, p. 572).

The carbonatite belt of South America starts in Goías and Minas Geraís and extends for about 900 miles southwesterly (10, p. 401). The outstanding carbonatite deposit in this belt is at Araxá, Brazil. One estimate of the reserves is in excess of 200 million tons of 3.5 percent ore; another is 4.6 million tons of Nb_2O_5 from 4 percent ore. Other large deposits have been discovered at Tapira and Serra Negra.

In North America the most noted pyrochlore deposit is the Oka (35, pp. 24–25; 55, 59, 78), 2 miles east of Oka at La Trappe, Quebec, about 20 miles from Montreal. The deposits occur in a complex of carbonate and alkaline rocks that has a maximum width of about 1.5 miles and a length of about 4 miles. Pyrochlore occurs chiefly in the carbonate rocks and in minor amounts in the carbonate matrix of brecciated alkaline rocks. The reserves have not been completely stated, as at least four companies have holdings in the area, but they are known to be large, and estimates of grade range from 0.25 to 0.6 precent Nb_2O_5. Mill concentration of material by St. Lawrence Columbium and Metals Corporation of Canada began in 1961 and was increased from 500 to 1,000 short tons of ore per day in 1962. Production in 1962 was 890 long tons of concentrate containing 50 percent Nb_2O_5, making the company the first producer of niobium concentrate in North America and also, at that time, the world's largest single producer.

Cesium and rubidium

Cesium has become important recently in connection with a new thermo-electrical process for converting heat into electricity (87). Rubidium has important electronic uses (58).

The chief source of cesium is the mineral pollucite, $CsSi_2AlO_6$, which occurs in pegmatites with lithium minerals. The Jooste lithium mines, near Karibib, South-West Africa, have furnished much of the production of the world. Pollucite has also been obtained as a by-product of beryl mining in the Republic of South Africa, and as a by-product of mining in the Bikita tin fields, Rhodesia (Southern Rhodesia) (57). It occurs in great masses associated with lithium minerals in the Varuträsk pegmatite, about 10 km from Skelleftea, Sweden (51).

A new major deposit, rated as the largest in the world, is the Montgary pegmatite at Bernic Lake, southeastern Manitoba, about 100 miles northeast of Winnipeg (27, 71). The pegmatite does not crop out but has been throughly explored by diamond drilling (77 holes from the surface and 33 from underground workings). The known length of the pegmatite is about 1,600 ft (the actual length may be 200 ft more than this), and the greatest known thickness is 237 ft. The pegmatite consists of six zones: border, wall, three intermediate, and core. Further, at least two definite replacement bodies have been recognized. Lithium minerals (spodumene, amblygonite, lepidolite, and lithiophylite-triphylite) occur in some of the intermediate zones and, except lepidolite, in the quartz core. Lepidolite also occurs in a replacement unit that contains about 80 percent lepidolite. Pollucite occurs in three sheetlike units that appear to be the result of replacement. Small amounts of beryl, cassiterite, and columbite-tantalite are present along with some other accessory minerals. The complete reserves of pollucite have not been stated, but published reserves for one unit indicate 125,000 to 150,000 tons of ore averaging 25 to 30 percent Cs_2O and 1.0 percent Rb_2O, extremely high-grade material, as pure pollucite from the deposit contains 36.2 percent Cs_2O.

Rubidium is associated with pollucite and lithium minerals. Lepidolite may contain from a trace to more than 3 percent rubidium and averages about 1.5 percent. Pollucite is stated to contain up to 1 percent Rb_2O, but some African pollucite contains as much as 3.73 percent. The chief known reserves of rubidium are the lepidolite deposits of South-West Africa, which contain about 2.2 percent Rb_2O (58). As pollucite deposits are investigated more thoroughly the most important reserves of rubidium might well be found to exist in those deposits.

SELECTED REFERENCES

1. J. C. Arundale and F. B. Mentch, 1952, Lithium, *U. S. Bur. Mines Minerals Yearbook,* vol. 1, pp. 650–659.
2. M. K. Banks, W. T. McDaniel, and P. N. Sales, 1953, A method for concentration of North Carolina spodumene ores, *Mining Eng.,* vol. 5, pp. 181–186.
3. A. A. Beus, 1962, "Beryllium, Evaluation of Deposits during Prospecting and Exploratory Work," W. H. Freeman and Company, San Francisco, 161 pp.
4. E. N. Cameron, 1955, Occurrence of mineral deposits in the pegmatites of the Karibib-Omaruru and Orange River areas of South West Africa, *Mining Eng.,* vol. 7, pp. 867–874.
5. E. N. Cameron and V. E. Shainin, 1947, The beryl resources of Connecticut, *Econ. Geology,* vol. 42, pp. 353–367.

6. E. N. Cameron and others, 1954, Pegmatite investigations, 1942–1945, New England, *U.S. Geol. Survey Prof. Paper* 255, 352 pp.

7. J. M. Cotalo Neiva and J. M. Correia Neves, 1960, Pegmatites of Alto-Ligonha (Mozambique–Portuguese East Africa), Internat. Geol. Cong., 21st, Copenhagen, 1960, pt. 17, sec. 17, Minerals and genesis of pegmatites, pp. 53–60.

8. S. C. DeAlmeida, W. D. Johnston, Jr., O. H. Leonardos, and E. P. Scorza, 1944, The beryl-tantalite-cassiterite pegmatites of Paraiba and Rio Grande do Norte, northern Brazil, *Econ. Geology,* vol. 39, pp. 206–223.

9. W. P. deKock, 1932, The lepidolite deposits of South-West Africa, *Geol. Soc. South Africa, Trans. and Proc.,* vol. 35, pp. 97–113.

10. Nicholas deKun, 1962, The economic geology of columbium (niobium) and tantalum, *Econ. Geology,* vol. 57, pp. 377–404.

11. D. R. Derry, 1950, Lithium-bearing pegmatites in northern Quebec, *Econ. Geology,* vol. 45, pp. 95–104.

12. D. P. Eigo, J. W. Franklin, and G. H. Cleaver, 1955, Lithium, *Eng. Mining Jour.,* vol. 155, pp. 75–89.

13. J. D. Falconer, 1921, The geology of the Plateau tin-fields, *Geol. Survey Nigeria, Bull.* 1, 55 pp.

14. A. P. Fawley and T. C. James, 1955, A pyrochlore (columbium) carbonatite, southern Tanganyika, *Econ. Geology,* vol. 50, pp. 571–585.

15. T. W. Gevers, 1937, Phases of mineralisation in Namaqualand pegmatites (1936), *Geol. Soc. South Africa, Trans. and Proc.,* vol. 39, pp. 331–378.

16. T. W. Gevers and H. F. Frommurze, 1929, The tin-bearing pegmatites of the Erongo area, South-West Africa, *Geol. Soc. South Africa, Trans. and Proc.,* vol. 32, pp. 111–149.

17. T. W. Gevers, F. C. Partridge, and G. K. Joubert, 1937, The pegmatite area south of the Orange River in Namaqualand, *South Africa Geol. Survey Mem.* 31, 164 pp.

18. M. M. Gilkey, 1960, Hyatt Ranch pegmatite, Larimer County, Colorado, *U.S. Bur. Mines Rept. Inv.* 5643, 18 pp.

19. E. R. Goter, W. R. Hudspeth, and D. L. Rainey, 1963, Mining and milling of lithium pegmatite at King Mountain, N.C., *Mining Eng.,* vol. 5, pp. 890–893.

20. M. F. Goudge and others, 1957, Lithium in Canada, in "The Geology of Canadian Industrial Mineral Deposits," pp. 159–163, Canadian Institute of Mining and Metallurgy, Montreal.

21. W. R. Griffitts, 1954, Beryllium resources of the tin-spodumene belt, North Carolina, *U.S. Geol. Survey Circ.* 309, 12 pp.

22. W. R. Griffitts, D. M. Larrabee, and J. J. Norton, 1962, Beryllium in the United States, *U.S. Geol. Survey Mineral Inv. Resource Map* MR-35.

23. J. B. Hanley, E. W. Heinrich, and L. R. Page, 1950, Pegmatite investigations in Colorado, Wyoming, and Utah, 1942–1944, *U.S. Geol. Survey Prof. Paper* 227, 125 pp.

24. F. L. Hess, 1940, Spodumene pegmatites of North Carolina, *Econ. Geology,* vol. 35, pp. 942–966.

25. G. Hileman, 1936, El berilo, *Petrol. y Minas,* Buenos Aires, vol. 16, no. 181, pp. 7–12.

26. W. T. Holser, 1953, Beryllium minerals in the Victorio Mountains, Luna County, New Mexico, *Am. Mineralogist,* vol. 38, pp. 599–611.

27. R. W. Hutchinson, 1959, Geology of the Montgary pegmatite, *Econ. Geology,* vol. 54, pp. 1525–1542.

28. R. W. Hutchinson and R. J. Claus, 1956, Pegmatite deposits, Alto Ligonha, Portuguese East Africa, *Econ. Geology,* vol. 51, pp. 757–780.

29. W. N. Ingham and M. Latulippe, 1957, Lithium deposits of the Lacorne area, Quebec, in "The Geology of Canadian Industrial Mineral Deposits," pp. 159–163, Canadian Institute of Mining and Metallurgy, Montreal.

30. R. Jacobson and J. S. Webb, 1946, The pegmatites of central Nigeria, *Geol. Survey Nigeria Bull.* 17, 61 pp.

31. R. H. Jahns, 1944, "Ribbon rock," an unusual beryllium-bearing tactite, *Econ. Geology,* vol. 39, pp. 173–205.

32. R. H. Jahns and L. A. Wright, 1944, The Harding beryllium-tantalum-lithium pegmatites, Taos County, New Mexico [abs.], *Econ. Geology,* vol. 39, pp. 96–97.

33. R. H. Jahns and J. W. Adams, 1953, Beryl deposits in the Harding pegmatites, Taos County, New Mexico [abs.], *Econ. Geology,* vol. 48, pp. 328–329.

34. W. D. Johnston, Jr., 1945, Beryl-tantalite pegmatites of northeastern Brazil, *Geol. Soc. America Bull.,* vol. 56, pp. 1015–1070.

35. R. J. Jones, 1957, Columbium (niobium) and tantalum, *Canada Dept. Mines and Tech. Surveys, Mines Branch, Mem. Ser.* 135, 56 pp.

36. T. L. Kesler, 1942, The tin-spodumene belt of the Carolinas, a preliminary report, *U.S. Geol. Survey Bull.* 936-J, pp. 245–269.

37. P. E. Landolt, 1951, Lithium minerals provide unique industrial raw material, *Mining Eng.,* vol. 3, pp. 1045–1048.

38. A. J. Lang, Jr., and J. A. Redden, 1953, Geology and pegmatites of part of the Fourmile area, Custer County, South Dakota, *U.S. Geol. Survey Circ.* 245, 20 pp.

39. R. A. Mackay, R. Greenwood, and J. E. Rockingham, 1949, The geology of the Plateau tinfields—Resurvey, 1945–1948, *Geol. Survey Nigeria Bull.* 19, 80 pp.

40. W. B. Mather, 1954, Lithium: northeast Brazil is potential source, *Mining Eng.,* vol. 6, pp. 897–903.

41. W. N. McAnulty, C. R. Sewell, D. R. Atkinson, and J. M. Rasberry, 1963, Aguachile beryllium-bearing fluorspar district, Coachuila, Mexico, *Geol. Soc. America Bull.,* vol. 74, pp. 735–743.

42. A. Meyer and P. de Béthune, 1960, The Luesche carbonatite (Kivu, Belgian Congo), Internat. Geol. Cong., 21st, Copenhagen, 1960, pt. 13, sec. 13, Petrographic provinces, igneous and metamorphic rocks, pp. 304–309.

43. G. A. Munson and F. F. Clarke, 1955, Mining and concentrating spodumene in the Black Hills, South Dakota, *Mining Eng.,* vol. 7, pp. 1041–1047.

44. J. Murdoch, 1955, Phosphate minerals of the Borborema pegmatites: I—Patrimonio, *Am. Mineralogist,* vol. 40, pp. 50–63.

45. J. Murdoch, 1958, Phosphate minerals of the Borborema pegmatites: II—Boqueirão, *Am. Mineralogist,* vol. 43, pp. 1148–1156.

46. J. J. Norton and D. M. Schlegel, 1955, Lithium resources of North America, *U.S. Geol. Survey Bull.* 1027-G, pp. 325–350.

47. L. R. Page and others, 1953, Pegmatite investigations, 1942–1945, Black Hills, South Dakota, *U.S. Geol. Survey Prof. Paper* 247, 228 pp.

48. J. J. Page and D. M. Larrabee, 1962, Beryl resources of New Hampshire, *U.S. Geol. Survey Prof. Paper* 353, 49 pp.

49. R. L. Parker, 1963, Niobium and tantalum in the United States, *U.S. Geol. Survey Mineral Inv. Resource Map* MR-36. Map and pamphlet of 8 pp.

50. W. T. Pecora, 1956, Carbonatites, a review, *Geol. Soc. America Bull.*, vol. 67, pp. 1537–1555.

51. P. Quensel, 1952, The parageneses of the Varuträsk pegmatite: a preliminary report, *Geol. Mag.* vol. 89, pp. 49–60.

52. J. A. Redden, 1959, Beryl deposits of the Beecher No. 3-Black Diamond pegmatite, Custer County, South Dakota, *U.S. Geol. Survey Bull.* 1072-I, pp. 537–559.

53. H. T. Reno, 1956, Beryllium, *U.S. Bur. Mines Bull.* 556, pp. 95–102.

54. R. B. Rowe, 1954, Pegmatitic lithium deposits in Canada, *Econ. Geology,* vol. 49, pp. 501–515.

55. R. B. Rowe, 1958, Niobium (columbium) deposits of Canada, *Geol. Survey Canada, Econ. Geol. Ser.* 18, 108 pp.

56. J. Sandor, 1961, Columbite and pyrochlore, *Mining Jour. (London),* vol. 257, pp. 282–283.

57. J. D. Sargent, 1956, Cesium, *U.S. Bur. Mines Bull.* 556, pp. 169–172.

58. J. D. Sargent, 1956, Rubidium, *U.S. Bur. Mines Bull.* 556, pp. 751–754.

59. V. B. Schneider, 1961, Niobium (columbium) and tantalum, *Canadian Mineral Industry 1961, Canada Dept. Mines and Tech. Surveys, Min. Res. Div.,* 6 pp.

60. A. E. Schreck, 1961, Lithium, a materials survey, *U.S. Bur. Mines Inf. Circ.* 8053, 80 pp.

61. G. M. Schwartz, 1925, Geology of the Etta spodumene mine, *Econ. Geology,* vol. 20, pp. 646–659.

62. D. M. Sheridan, 1955, Geology of the High Climb pegmatite, Custer County, South Dakota, *U.S. Geol. Survey Bull.* 1015-C, pp. 59–98.

63. D. M. Sheridan and others, 1957, Geology and beryl deposits of the Peerless pegmatite, Pennington County, South Dakota, *U.S. Geol. Survey Prof. Paper* 297-A, pp. 1–47.

64. H. A. Siroonian and others, 1959, Lithium geochemistry and the source of the spodumene pegmatites of the Preissac-Lamotte-Lacorne region of western Quebec, *Canadian Mineralogist,* vol. 6, pt. 3, pp. 320–338.

65. M. H. Staatz, 1963, Geology of the beryllium deposits in the Thomas Range, Juab County, Utah, *U.S. Geol. Survey Bull.* 1142-M, 36 pp.

66. M. H. Staatz and A. F. Trites, 1955, Geology of the Quartz Creek pegmatite district, Gunnison County, Colorado, *U.S. Geol. Survey Prof. Paper* 265, 111 pp.

67. M. H. Staatz and W. R. Griffitts, 1961, Beryllium-bearing tuff in the Thomas Range, Juab County, Utah, *Econ. Geology,* vol. 56, pp. 941–950.

68. W. R. Thurston, 1955, Pegmatites of the Crystal Mountain district, Larimer County, Colorado, *U.S. Geol. Survey Bull.* 1011, 185 pp.

69. L. A. Warner and others, 1959, The occurrence of nonpegmatite beryllium in the United States, *U.S. Geol. Survey Prof. Paper* 318, 198 pp.

70. F. A. Williams and others, 1956, Economic geology of the decomposed columbite-bearing granites, Jos Plateau, Nigeria, *Econ. Geology,* vol. 51, pp. 303–332.

71. C. M. Wright, 1963, Geology and origin of the pollucite-bearing Montgary pegmatite, Manitoba, *Geol. Soc. America Bull.,* vol. 74, pp. 919–945.

72. V. Ziegler, 1914, Mineral resources of the Harney Peak pegmatites, *Mining and Sci. Press,* vol. 108, pp. 604–608, 654–656.

73. Anon., 1952, How Geomines will treat 24,000 tons of "hard" tin ore per day, *Mining World,* vol. 14, pp. 32–37.

74. Anon., 1952, Demand for lithium boosts mining at Kings Mountain, *Eng. Mining Jour.,* vol. 153, pp. 90–91.

75. Anon., 1954, Pegmatitas portadoras de berilo, *Industria Minera (Buenos Aires),* vol. 12, no. 149, pp. 35–41.

76. Anon., 1954, Pegmatitas portadoras de columbitas, tantalitas y minerales de uranio y otros yacimientos uraníferos, *Industria Minera (Buenos Aires),* vol. 12, no. 150, pp. 42–48.

77. Anon., 1960, Beryllium today and tomorrow, *Eng. Mining Jour.,* vol. 161, pp. 93–102, 192–196.

78. Anon., 1960, Quebec's columbium will be tapped, *Eng. Mining Jour.,* vol. 161, pp. 92–95.

79. Anon., 1962, Beryl resources of the Malagasy Republic, *Mining Jour. (London),* vol. 258, no. 6602, pp. 213–215.

80. Anon., 1963, Brazilian beryl, *Mining Jour. (London),* vol. 261, no. 6692, pp. 493–494.

MISCELLANEOUS REFERENCES CITED

81. W. H. Emmons, 1933, On the mechanism of the disposition of certain metalliferous lode systems associated with granitic batholiths, in "Ore Deposits of the Western States" (Lindgren volume), pp. 327–331, The American Institute of Mining and Metallurgical Engineers, New York.

82. A. A. Levinson, 1962, Beryllium-fluorine mineralization at Aguachile Mountain, Coahuila, Mexico, *Am. Mineralogist,* vol. 47, pp. 67–74.

83. J. J. Norton and L. R. Page, 1956, Methods used to determine grade and reserves of pegmatites, *Mining Eng.,* vol. 8, pp. 401–404.

84. W. T. Schaller, 1916, Gigantic crystals of spodumene, *U.S. Geol. Survey Bull.* 610, p. 138.

85. W. N. Sharp and C. C. Hawley, 1960, Betrandite-bearing greisen, a new beryllium ore, in the Lake George district, Colorado, in "Short Papers in the Geological Sciences," *U.S. Geol. Survey Prof. Paper* 400-B, art. 33, pp. B 73–74.

86. J. Simons, 1961, New beryllium discoveries on Colorado's Badger Flats, *Mining World,* vol. 23, no. 3, pp. 28–30.

87. Anon., 1959, Look for cesium minerals in lithia-rich pegmatites for growing use in new thermoelectric generators, *Mining World,* vol. 21, no. 6, p. 28.

88. Anon., 1963, Beryl in Brazil, *Mining Jour. (London),* vol. 261, no. 6681, p. 217.

chapter *9*

Tin, Tungsten, Molybdenum, and Bismuth

Tin, tungsten, molybdenum, and, to a lesser extent, bismuth rather commonly occur together, but usually one metal is more abundant than any of the others, either in a particular deposit or in a zone of a deposit. Further, minerals of some of the metals are deposited throughout a longer temperature range than others and, consequently, may show different associations and occurrences. Some relationships are shown by Table 9-1.

Some similarities not shown by Table 9-1 are rather general association of deposits with silicic igneous rocks, especially granites; quartz as a very prominent

TABLE 9-1. Some Relationships among Deposits of Tin, Tungsten, Molybdenum, and Bismuth

		Commercial importance in various types of deposits			
Metal	*Mineral*	*Hydrothermal*	*Pegmatitic*	*Contact metasomatic*	*Residual, eluvial, alluvial*
Tin	Cassiterite	Very great	Moderate	Minor	Very great
	Stannite, franckeite, teallite	Moderate	None	Little or none	None
Tungsten	Wolframite group	Very great	Moderate to minor	Minor	Moderate to minor
	Scheelite	Very great	Moderate to minor	Great	Moderate to minor
Molybdenum	Molybdenite	Very great	Moderate to minor	Moderate to minor	None
Bismuth	Native bismuth, bismuthinite	Moderate	Moderate	Minor	None

constituent of vein deposits; and silicification, greisenization, or tourmalinization as rather common features of higher-temperature hydrothermal deposits. Bismuth differs from the other metals in commercial production, since most of it is recovered from the metallurgical treatment of slimes of lead and copper refining, although some is recovered as a by-product of the mining of tin and tungsten ores, and some as a by-product from pegmatite mining.

The most important deposits of tin, tungsten, molybdenum, and bismuth will be discussed separately, but complete separation is not possible because of the usual occurrence of two or more of these metals in the same deposit.

Tin

Primary tin deposits, as indicated by Table 9-1, originate chiefly from hydrothermal and pegmatitic activity, and such deposits, if the tin mineral is cassiterite, are the source of important residual, eluvial, and alluvial (placer) deposits. Primary deposits of cassiterite are widespread in pegmatites, but the primary deposits of greatest importance are those classed by some as pneumatolytic, by others as a combination of pneumatolytic and hydrothermal, and by still others as hydrothermal deposits, generally hypothermal (high temperature, deep zone). Cassiterite in such deposits shows a fairly common association with boron and fluorine minerals (tourmaline, topaz, fluorite, and fluorine-bearing white mica) and, less commonly, with lithium minerals that may also contain fluorine (lepidolite, zinnwaldite), and it is this association, at least in part, that has led to the concept of pneumatolytic origin. Another type of tin deposit, best developed in Bolivia, contains, in addition to cassiterite, the complex tin sulfide minerals stannite, Cu_2FeSnS_4, franckeite, $Pb_5Sn_3Sb_2S_{14}$, and teallite, $PbSnS_2$, and a number of complex silver minerals. Such deposits have somewhat generally been classed as xenothermal (high temperature, shallow zone).

The general characteristics of tin deposits were summarized by Ferguson and Bateman (15). The deposits generally occur in veins and stockworks. Many of the veins form lodes in which narrow, closely spaced fissures include tabular plates of country rock, much of which was replaced; consequently almost a solid mass of ore was formed. The stockworks generally consist of a mass of rock cut by a network of small veins, the ore occurring not only in the veins but also in the intervening rock. Veins and stockworks generally lie within granite or in country rock close to a granite intrusion. Greisen, topaz or fluorite, and tourmaline are characteristic associates, and arsenopyrite is a rather common one.

Itsikson (19) studied the world-wide distribution of tin deposits and concluded that the period of transformation of a geosyncline into a folded mountain range was a highly productive time of tin emplacement. Among tin deposits of the world, 63.1 percent are associated with Kimmeridgian (Middle Upper Jurassic) folding, 18.1 percent with Variscan (late Paleozoic) folding, 6.6 percent with Caledonian (approximately Ordovician through Devonian) folding, and 3.3 percent with Precambrian folding. He placed tin deposits in two natural groups, (1) hypabyssal and (2) subvolcanic. The hypabyssal group includes the most important deposits of the world (Indonesia, Malaya, Thailand, China); the subvolcanic group is typified by the important deposits of Bolivia. World production is shown by Fig. 9-1.

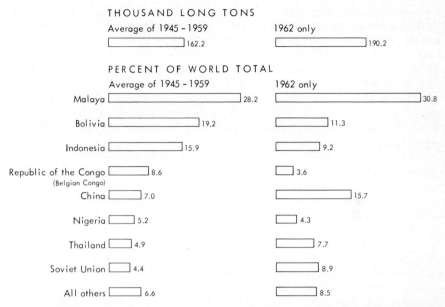

THOUSAND LONG TONS

Average of 1945 - 1959   1962 only

162.2   190.2

PERCENT OF WORLD TOTAL

Average of 1945 - 1959   1962 only

Malaya   28.2   30.8

Bolivia   19.2   11.3

Indonesia   15.9   9.2

Republic of the Congo   8.6   3.6
(Belgian Congo)

China   7.0   15.7

Nigeria   5.2   4.3

Thailand   4.9   7.7

Soviet Union   4.4   8.9

All others   6.6   8.5

Fig. 9-1 World tin production. Compiled from *U.S. Bur. Mines Minerals Yearbook*, various years.

Deposits of the Malay Peninsula Area and Vicinity

The largest and most productive tin-tungsten province of the world extends in an arc throughout a length of about a thousand miles from the Indonesian islands of Belitung (Billiton), Bangka (Banka), and Singkep, between Sumatra and Borneo, through the Federated Malay States, the Malay Peninsula, Thailand (Siam), and into China (14, 17, 18, 20, 22, 24, 27, 28, 33, 35, 36, 37). The deposits of Indonesia and Malaya are dominantly tin producers, those of China are exceedingly important tin and tungsten producers, and those of Burma are in general a link between the Indonesian-Malayan deposits and the Chinese deposits, although perhaps they should be classed as tungsten-tin deposits rather than tin-tungsten deposits. The deposits of Thailand (Siam) form, in the south, a link between the deposits of Malaya and those of Burma but are chiefly tin producers, and, in the west and northwest, they form an extension of the Burmese deposits. The entire Malayan region and vicinity, including China, produced about 62 percent of the tin and 34 percent of the tungsten of the world in the three years 1959 to 1961. The Federation of Malaya, Indonesia, and Thailand combined provided about three-fourths of the tin but less than one-tenth of the tungsten of this region, whereas China furnished about one-fourth of the tin and more than nine-tenths of the tungsten.

The major features of the deposits of this large area are in general similar, although differences exist in detail. Granite, probably of late Mesozoic age, has been intruded as large and small masses into more or less folded Paleozoic and Mesozoic sedimentary rocks that show varying degrees of metamorphism, from slight to great. The granite masses make the backbones of mountain ranges, and the primary tin-tungsten deposits generally lie close to the contact between the granite and the metasedimentary rocks, in places entirely with the granite, in places entirely within the metasedimentary rocks. Generally deposits in the granite are richer in tin than in tungsten, whereas those in schist are richer in tungsten than in tin. The more central parts of granite masses are characterized by minor mineralization but no important deposits. Consequently most of the mining is along the flanks of the mountain ranges, in the foothills and out into the valleys. Some primary deposits are mined, but most of the production is from residual, eluvial, and alluvial (placer) deposits.

Some type of vein deposit is the chief producer almost everywhere the primary deposits are worked. Veins vary in thickness from mere stringers to several feet; many are closely spaced and form lodes; many form networks and pass into stockworks. Some deposits are in greisen and some in pipes developed in limestone and in granite. The major ore minerals are cassiterite and wolframite; the major gangue minerals are quartz and, locally, one or more of tourmaline, topaz, fluorite, and white mica, along with arsenopyrite or pyrrhotite. Some other sulfides, such as pyrite, sphalerite, and galena, are present in some deposits, but generally these seem to have come in later than the principal ore minerals. Only a few of the most important primary deposits will be described briefly as examples of mining districts throughout the region.

Indonesia. The three Indonesian islands of Belitung (Billiton), Bangka (Banka), and Singkep (33, 36) have been producers since the eighteenth century; the existence of tin veins was certainly known as early as 1824. Granite similar to that of the Main Range of Malaya contains, in the marginal parts, quartz veins that carry cassiterite, some wolframite, black tourmaline, and topaz (locally abundant) and that are bordered by greisen. The greisen may contain cassiterite but rarely enough to be mined, although it does furnish some minable residual material. Deposits occur in metasedimentary rocks also and show relationships similar to those that are present in contact-metasomatic deposits: garnet and pyroxene gangue, pyrrhotite, arsenopyrite, and some pyrite, chalcopyrite, sphalerite, galena—minerals practically absent from the veins in granite. Most of the production is from residual, eluvial, and alluvial material.

Malaya. Throughout Malaya (14, 18, 20, 22, 27, 28, 35, 37, 47) granite forms the backbone of the peninsula. The Main Range is paralleled in many places by smaller ranges, one of which, the Kledang Range, forms the western wall of the Kinta Valley, the most productive part of the area. Most of the deposits occur along the western side of the Main Range and around subordinate ranges that lie west of the Main Range. The major mineralized zone extends through the western states of Malaya from Perlis on the north through part of Keyday, Perak, Selangor, and into the area of North Johore. The most productive parts of the area have been Perak and Selangor (Fig. 9-2). The primary ore occurs in quartz veins within the granite and metasedimentary rocks (rarely limestone) and in pipes in limestone; pipes may also be present in the granite, which is intrusive into but beneath the limestone.

The Kinta Valley of Malaya has, since 1890, held the unchallenged position as the largest producer of tin ore in the world (18). From 1876 to 1950 the de-

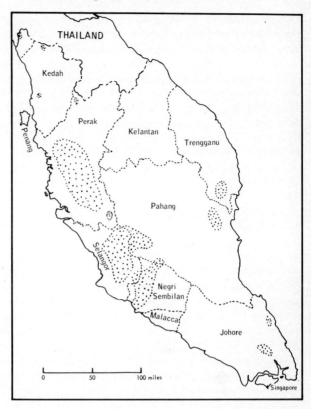

Fig. 9-2   Map showing major tin-mining areas of Malaya.
After J. D. Ridge, Materials Survey, Tin Chap. 4, p. 55.
U.S. Government Printing Office.

posits furnished about 1 million tons of metallic tin. Much of this was from alluvial, eluvial, and residual deposits, but some was from lode deposits. Numerous lode deposits occur, and Willbourne (35) attempted to map and describe briefly all such deposits. The most outstanding types have been summarized by Ingham and Bradford (18).

The floor of the Kinta Valley is chiefly crystalline limestone, along with some schist, shale, and quartzite, known as the "calcareous series." Intrusive granite encircles the sedimentary rocks on three sides. Lode deposits occur in the granite and in the surrounding rocks, in which some deposits are in limestone and others are in schist. Veins and vein systems in granite are exemplified by the deposits of the Menglembu Lode Mining Company, Ltd. The veins and lodes occur *en echelon* along a zone about 1,400 ft long and 100 ft wide, throughout which are some pipelike bodies where the veins are particularly numerous or merge, some of which were worked to a depth of 560 ft. The lodes are composed either of (1) single veins of quartz or two or three parallel veins separated by a few inches or few feet of granite; or (2) a great number of very thin parallel veinlets (known locally as "streaky bacon"), the entire system exceeding 12 ft in width in many places.

Among deposits surrounding the granite, pipes in limestone are the most important. The majority of the deposits are of irregular, chimneylike shape and are either fissure fillings or replacements of limestone, some probably combinations of these.

Fewer workable deposits are known along the east side of the Main Range, but at the Sungei Lembing deposit there is a mine that is rated as one of the largest lode tin mines in the world (27, 47). The mine area covers about 6 sq miles and, since 1888, has yielded more than 60,000 long tons of tin in concentrates from more than 5 million long tons of ore, the recovery being slightly more than 1.2 percent tin. This deposit produced about 1.5 percent of the tin of the world since the First World War. Over 40 veins have been worked, but the major production has come from 8 or 9 of them. Most veins are in shale, are from 50 to 2,000 ft apart, a few inches to 10 ft wide; some have been followed along the strike for over half a mile and have been worked down the pitch for over half a mile.

The most important deposits of the Kinta Valley are rich concentrations of alluvial cassiterite that have accumulated in the wide plain occupied by the Kinta River and its tributaries. The conditions for alluvial concentration have been especially favorable, since the source of the cassiterite is granite which has been likened to a "giant horseshoe" (18, p. xx) that surrounds three sides of the Kinta Valley, which is floored by sedimentary rocks. Further, limestone bedrock, with "trough-and-pinnacle" topography and deep solution channels, has formed a series of natural riffles that retain and concentrate heavy grains of cassiterite washed by erosion from the granite hillsides. Residual materials, derived from a weathered zone 6 to 200 ft thick, follow the edge of the granite in a general way and are washed down and mingle with true alluvial deposits. Consequently eluvial deposits generally cannot be placed in a separate group but are placed in part with the alluvial deposits and in part with the residual deposits. The alluvial deposits have been well summarized by Ingham and Bradford (18, p. 140) who stated, in part:

> *The eroded stanniferous material was carried by streams and rivers from the harder granite hills toward the softer or more soluble sediments which ultimately formed the plain of the Kinta Valley, and, because of its relative heaviness, was deposited wherever the velocity of those streams was checked and their carrying power reduced. The first large scale check of that kind occurred where the streams entered flatter ground: there coarse alluvial ore often combined with eluvial material to form the rich deposits of "contact" mines. . . . When the rivers entered the valley plain their velocity was much reduced, and at that stage even comparatively small checks caused further deposition of tinore. As the valley floor became gradually lowered the irregular surface of the limestone bedrock with its pinnacles and potholes acted as natural riffles which trapped the ore, so that rich alluvial deposits were often formed in deep pockets in the limestone. The smaller the particle size of the ore, the further it was carried by the river; and ultimately a rough sorting was achieved, with the coarser material occurring upstream near the edges of the plain, and the finer material further toward the centre and downstream.*[1]

[1] From The Geology and Mineral Resources of the Kinta Valley, Perak, by F. T. Ingham and E. F. Bradford, *Federation of Malaya, Geological Survey, District Memoir No. 9*, p. 140. Copyright 1960. Used by permission.

Early recovery of cassiterite was by hand labor, both in the weathered granite, which was easily worked with the local digging hoe (*changkol*), and in the alluvial deposits, which were worked by washing material in stream beds with a flat wooden pan (*dulang*) and by other methods. Now much recovery is by dredging, gravel pumping, and hydraulic mining. In 1952 there were 39 dredges operating, digging to a depth of 45 to 135 ft, with a monthly capacity of 149,000 to 396,000 cu yd. During the period from January, 1950, to October, 1952, there were 499 hydraulic and gravel pump mines working.

A highly important new tin deposit in Malaya was announced in 1962. The location is at Kota Tinggi, Johore State. It is stated to be one of the biggest and richest deposits of Malaya with a value of 1,130 million dollars.

Deposits of Bolivia

Bolivia ranks next to the Malayan area in production of tin and also supplies a considerable amount of tungsten. Unlike most tin deposits, however, those of Bolivia furnish important amounts of silver. Silver was mined in parts of Bolivia long before the mining of tin began, and one deposit, at Potosí, was long famous as the greatest silver camp in the world. Although silver mining has declined, and chief attention is now given to tin and tungsten, many of the Bolivian occurrences are tin-silver deposits; they represent the most important occurrence of this peculiar type.

The tin-silver deposits of Bolivia extend from the border of Argentina north-northeast to the vicinity of Lake Titicaca, a distance of about 500 miles, and lie within an area that is about 150 miles wide at the south and about 50 miles wide at the north (Fig. 9-3). Much has been written about these deposits since 1900, and only certain references are cited (12, 13, 29, 30, 31, 32).

Within the region granitic rocks, which make the core of the Andes Mountains, are intrusive into sedimentary rocks, which make the flanks of the mountains. Some of the deposits are closely associated with intrusives, others do not show as clear relationships; the deposits occur both in the igneous and in the sedimentary rocks. Northeast of Oruro (Fig. 9-3) tin is the principal metal obtained, whereas southward both tin and silver are important. Hence some of the deposits may be designated as chiefly tin deposits, others as tin-silver deposits. The 10 most important producing areas, from north to south (25, p. 116) are Araca, Oruro, Machacamarca, Huanuni, Llallagua, Unica, Potosí, Sala Sala, Chocaya, and Chorolque. A comparative study by Turneaure (31) relates to four of these, Potosí and Oruro (tin-silver deposits), Llallagua and Huanuni (tin deposits), and also one not included in the group, Morococala, a tin deposit, which is east of the town of Machacamarca. Multiple veins characterize all the deposits, most of the ore coming from narrow veins and stringer lodes but some also from sheeted zones, stockworks, and breccia lenses. The ore-bearing zones in each of the five deposits reviewed cover but a small area, as shown by Table 9-2. The veins within the ore-bearing zones are lodelike and composed of several mineralized fractures. The length of the main veins generally is 300 to 700 m, the width 30 to 80 cm. Important ore shoots occur, and much ore has come from veins but 3 to 20 cm wide. Intensely sheeted zones occur at Llallagua and Potosí; the one at Potosí attains a width of 175 m, a maximum length of 350 m, and consists of hundreds of parallel

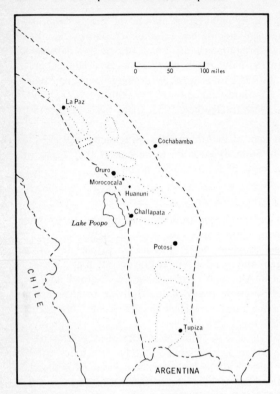

Fig 9-3 Map showing generalized outline of the tin-silver belt of Bolivia and generalized distribution of antimony deposits (dotted boundaries). After J. D. Ridge, Materials Survey, Tin, U.S. Government Printing Office, and D. E. White, Materials Survey, Antimony, U.S. Bur. Mines.

veinlets 1 to 5 cm wide. Stockworks are present at Morococala, and irregular, elongated bodies of breccia at Huanuni and Llallagua.

The wallrocks throughout the districts considered have been much altered, the igneous rocks to such a degree that determination of their original nature is difficult, the sedimentary ones moderately except in the productive zones at Morococala and Huanuni, where they have been intensely altered. Sericite, tourmaline, quartz, and pyrite appear to be the most widespread alteration minerals.

The hypogene ore minerals of these deposits present an apparently confusing array of minerals that generally form at high, moderate, and low temperatures. Thus such minerals as cassiterite and wolframite, for example, generally regarded as forming at high temperature, are present in the same deposit with minerals such as andorite, $PbAgSb_3S_6$, stibnite, Sb_2S_3, and others generally regarded as forming at low or, at most, moderate temperature. This led to much discussion, and many ideas were advanced to account for apparently anomalous mineral associations. (See references 13 and 29 for a review of some concepts.) Turneaure (30) cited Llallagua, a typical tin deposit, as an excellent example of a high-temperature deposit formed in a near-surface, volcanic environment; and Turneaure and Gibson (32) concluded that the deposits at Carguaicolla, in which the principal tin-bearing mineral is teallite, $PbSnS_2$, although some cassiterite is present, showed evidence of some telescoping and represent the xenothermal type of deposit (high temperature, shallow zone). Chace (12), after a thorough study of the tin-silver veins

**TABLE 9-2.   Some Features of the Ore-bearing Zones
of Five Bolivian Tin and Tin-Silver Areas***

District	Type of deposit	Ore-bearing zone		
		Length, m	Width, m	Area, sq km
Potosí	Tin-silver	1,200	1,200 max.	1.44
Oruro	Tin-silver	1,400	900 max.	1.26†
Llallagua	Tin	1,400	1,000 avg.	1.40
Huanuni	Tin	1,400	1,200 max.	1.68
Morococala	Tin	1,000	600 max.	0.60‡

* From F. S. Turneaure, *Econ. Geology,* vol. 55, p. 31. Copyright 1960. Economic Geology Publishing Company. Used by permission.
† Excludes Tetilla vein group.
‡ Includes Reserva group.

of Oruro, in which 28 hypogene ore minerals are present, concluded that hypogene mineralization took place in two stages separated by a period of minor fracturing: (1) formation of veins that contain chiefly quartz, pyrite, and cassiterite; (2) formation of veins that contain stannite, Cu_2FeSnS_4, franckeite, $Pb_5Sn_3Sb_2S_{14}$, various silver-bearing sulfosalts, lead sulfosalts, and others. He also concluded that the ore solutions were probably injected into preheated channelways with a relatively steep thermal gradient, which might explain the presence, in a near-surface environment, of minerals and textures characteristic of a greater depth zone. He placed the deposits in the xenothermal class.

The origin and characteristics of the Bolivian deposits were summarized by Turneaure (31). He concluded that all five of the deposits included in his comparative study show characteristics of the xenothermal class and various degrees of telescoping and stated, in part (p. 591):

> Ore deposition in central Bolivia took place within a relatively shallow zone, probably within a few thousand feet of the surface. Supporting evidence is found in the geological setting of the deposits, vein structure, mineral assemblage, and ore texture. Many of the unusual features exhibited by the ores probably owe their origin to the combination of high temperature and low pressure.[1]

Further, he stated (p. 595):

> The assemblage of vein minerals leaves little doubt that during the early stage of vein growth, particularly in the tin deposits, hypothermal intensities prevailed, with temperatures in the order of 400–500° C. Deposition continued with decreasing temperature, interrupted perhaps by minor reversals, and the last minerals to form were deposited at distinctly low temperature, perhaps in the range of 100–150° C.[1]

[1] From F. S. Turneaure, *Econ. Geology,* vol. 55, pp. 591 and 595. Copyright 1960. Economic Geology Publishing Company. Used by permission.

Deposits of Cornwall and Devonshire, England

Tin mining in the Cornwall area dates back to prehistoric times, and these deposits seem likely to have been the source of tin used in the making of bronze during the Bronze Age (20, p. 98). In 1865 the district produced 41 percent of the tin of the world; in 1875, but 23 percent; in 1913, only 4 percent; and in 1939, but 1 percent (88). Investigators from 1959 through 1963 (91, 100) indicate that 14 areas in the district offer good prospects for further mining development. Regardless of its current economic importance, this district provides an instructive example of a type of tin deposit showing an association with copper. Only the outstanding features will be presented.

Granite, probably of Permo-Carboniferous age, was intruded into sedimentary rocks that are now chiefly slates, phyllites, and schists, locally termed *killas*. An abundance of dikes of quartz-porphyry, known as *elvan*, characterizes the richly mineralized areas. Tourmaline, a common accessory mineral, in some places partly or wholly replaced feldspar and mica and formed a tourmalinized granite known as *luxullianite*. Locally the alteration, instead of tourmalinization, was greisenization, silicification, and pyritization. The chief tin mineral is cassiterite; locally some wolframite or molybdenite occur.

Veins occur in the granite and in the killas but always close to the granite contact. Generally the deposits are zoned both horizontally and vertically, with tin deposits in the granite, tin-copper deposits in the marginal areas, and copper deposits in the killas. Vertical zoning was especially well shown in the Dolcoath mine. Copper ore, with only a small amount of cassiterite, occurred from the surface to a depth of 900 ft; copper-tin ore somewhat poor in both copper and tin, from 900 to 1,440 ft; rich tin ore from 1,440 to 1,800 ft. Cassiterite continued to a depth of 3,300 ft but apparently became less abundant.

Other Deposits

Important tin production between 1950 and 1960, in addition to that from the general Malayan area and vicinity and Bolivia, has come from the Republic of the Congo (Belgian Congo), Nigeria, and the Soviet Union. Also, although Australia has not been an outstanding tin producer, deposits are rather widespread there; the 1961 production was greater than at any time since 1942, and an active exploration program was being conducted (98). Tin in the United States is essentially limited to that present in some pegmatites. The Carolina tin belt is the most extensive area of low-grade mineralization (16).

The production of Nigeria comes chiefly from residual, eluvial, and alluvial deposits of the Jos Plateau and, to some extent, from vein and pegmatite deposits (see niobium). The production of the Republic of the Congo comes in part from the Manono-Kitotolo pegmatite field (see niobium) and in part from quartz veins that traverse granite and the contact zone between granite and adjacent metasedimentary rocks. Wolframite, columbite-tantalite, and beryl are present in some of these veins (27, pp. 88–95; 37, pp. 53–56, 99–102).

Specific information about the Soviet Union's deposits, published in English, is difficult to obtain, although estimates indicate that the Soviet Union ranked fourth in world tin production in 1961. Nalivkin (45, pp. 97, 148) mentioned tin deposits in Transbaikal and elsewhere; Shimkin (48, pp. 133–136) stated that the reserves are all in deposits in Transbaikal (Eastern Siberia) and include peg-matitic, hydrothermal, and placer deposits. Good recent summaries are those by Ahlfeld (37, pp. 108–111), in German, and by Ridge (27, pp. 24–31), in English. Both lode and placer deposits have been worked in the Transbaikal area particularly in the Onan River Valley. Other lode and placer deposits are in northeast Yakutia and in the Maritime Territory of far eastern Siberia. Lode deposits of the general Onan Valley area include cassiterite-bearing pegmatites and cassiterite-wolframite quartz veins, the latter the most important. The lode deposits of northeast Yakutia are present over a large area. They contain both cassiterite and sulfides, and some seem to resemble the Bolivian type. The deposits of the Maritime Territory, al-though classed as the sulfide-cassiterite type, apparently contain predominantly lead, zinc, and copper. Both Ahlfeld (37) and Ridge (27) rated the Yakutia area as potentially, if not actually, the most important. Some of these deposits of the Soviet Union are in regions of permafrost, which would cause difficulty in placer mining.

The tin deposits of Australia are chiefly in Queensland, New South Wales, and Tasmania. Much of the production has been from residual, eluvial, and alluvial deposits, and large new deposits have been discovered recently in Queens-land, but some production has come from underground mining (11, 23), especially in Tasmania. An interesting feature of the Queensland primary deposits is the presence of cassiterite-bearing pipes. These are abundant in granite (more than 100) and in roof pendants of metamorphic rock (more than 60) (38, p. 283). Cylindrical pipes rarely exceed 15 to 25 ft in diameter and 30 to 50 ft in depth; elongated pipes, 20 to 40 ft in width, 6 to 10 ft in thickness, and 30 to 50 ft in depth, in places occur as a succession of pipes with slight, flat offsets, that may persist through a vertical depth of several hundred feet. Some of the vein deposits of Tasmania (27, pp. 72–77; 40, 43, 49) contain, in addition to cassiterite and wolframite, various sulfides that carry about 25 oz of silver to the ton. Such deposits suggest relationships to the Bolivian tin-silver veins. In Tasmania, also, is the famous Mount Bischoff mine in the northeast corner of the island which, for about 30 years, was the most important tin producer not only of Tasmania but of all Australia and for a longer time caused much controversy about the origin of the deposits (14, pp. 126–143; 20, p. 363; 21; 34; 46, p. 82; 49, p. 290). Dikes and sills of quartz porphyry radiate from a center near the summit of Mount Bischoff. These were greisenized and highly altered in places, cut by fractures, and so decomposed that some workers thought that the deposits were of alluvial origin. Certain rock types and structural conditions, in addition to alteration, further complicated the history of the deposits. The many interesting features of these deposits cannot be discussed here.

Tungsten

Tungsten, although commonly associated with tin and niobium in pegmatites and high-temperature vein deposits, also occurs in lower-temperature vein deposits and is highly important in contact-metasomatic deposits. Further, although present in residual and placer deposits, it is generally much less important than tin because the chief tungsten minerals, wolframite and scheelite, disintegrate rather easily, make fine material, and are lost, whereas the chief tin mineral, cassiterite, is highly resistant and readily accumulates in placer deposits. Thus the important tungsten deposits of the world, although in part in the same areas as the tin deposits, show important differences in distribution (Fig. 9-4). China and the Soviet Union, if estimates are correct, produce important amounts of both tin and tungsten; Bolivia generally produces considerably more tin than tungsten; and the Federation of Malaya, Indonesia, and Thailand produce much tin but relatively little tungsten. The United States, the Republic of Korea, and North Korea produce a considerable amount of tungsten but practically no tin.

Asia

The chief tungsten deposits of Asia are in China (about one-third of the world production), Burma, Republic of Korea, North Korea, and Thailand. The tungsten deposits of Thailand, the Federation of Malaya, and Indonesia, occur along with the tin deposits that have been described. Although a few deposits produce tungsten chiefly, on the whole tungsten is a by-product of tin mining, and no additional discussion seems necessary.

China also produces tungsten as a by-product of tin mining, but there is, throughout the country, a gradation from deposits noted chiefly for tin into those noted chiefly for tungsten. Consequently the deposits will be summarized (14, pp. 61–79; 20, pp. 239–245; 27, Chap. 4, pp. 38–42; 37, pp. 176–190; 44, pp. 436–439, 442–444; 51; 71, pp. 20–35; 79, pp. 233–235; 81).

Practically all the tin and tungsten deposits of China are in the Nanling region, which includes southern Yunnan, northern Kwangsi, northern Kwangtung,

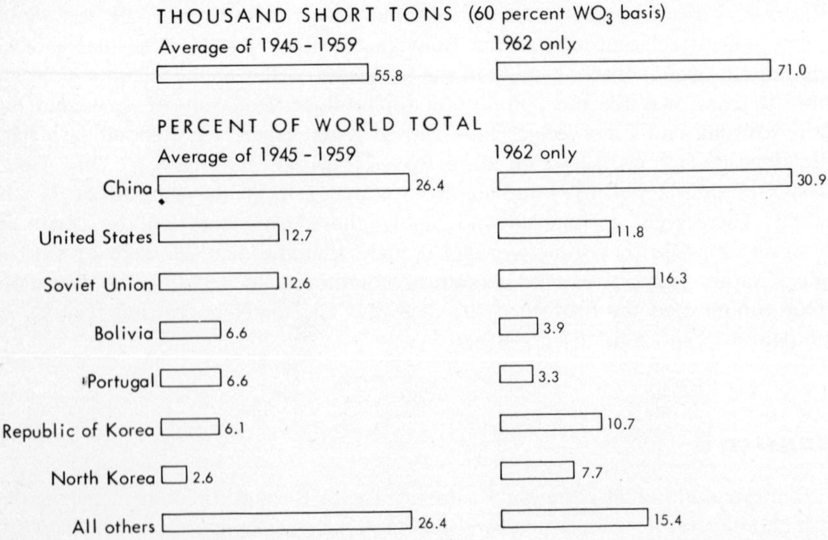

THOUSAND SHORT TONS (60 percent WO_3 basis)

Average of 1945 – 1959	1962 only
55.8	71.0

PERCENT OF WORLD TOTAL

	Average of 1945 – 1959	1962 only
China	26.4	30.9
United States	12.7	11.8
Soviet Union	12.6	16.3
Bolivia	6.6	3.9
Portugal	6.6	3.3
Republic of Korea	6.1	10.7
North Korea	2.6	7.7
All others	26.4	15.4

Fig. 9-4 World tungsten production. Compiled from *U.S. Bur. Mines Minerals Yearbook*, various years.

Fig. 9-5 Generalized outline map showing chief mineral-producing provinces of China and approximate boundary of the tin-tungsten region, the latter generalized after J. D. Ridge, Materials Survey, Tin, U.S. Government Printing Office.

southern Hunan, and southern Kiangsi (Fig. 9-5). Some of the deposits contain, also, bismuth and molybdenum. In general the amount of tungsten in the deposits declines sharply from east to west; the Kiangsi-Hunan section is mainly a tungsten area, the Kwangsi-Yunnan section mainly a tin area (71, p. 23). The region is characterized by zones of intensely and complexly folded mountains and by widespread late Mesozoic granite intrusions.

The principal ore bodies throughout the entire region are regular fissure-filled veins in which the chief minerals are quartz, wolframite, and light-colored micas; among other ore minerals present locally are cassiterite, scheelite, native bismuth, bismuthinite, and molybdenite. Less important deposits occur in (1) pegmatites;

(2) irregular replacements in granite and aplite, in which cassiterite is the principal ore mineral but is accompanied locally by wolframite, scheelite, and bismuthinite; (3) great lens-shaped wolframite-quartz bodies in sedimentary rocks and in granite; and (4) wolframite-quartz stringers in sedimentary rocks.

The fissure-filled veins, the really important type of deposit, range in thickness from paper thin to 2 or 3 m, but the common range is from 0.1 to 0.6 m. The common length of the veins is a hundred to several hundred m, but some of the wider ones attain a length of 1,000 m. In places many veins are closely spaced within a distance of a meter or two, so that all are mined as a unit.

In Burma (39, 41, 42, 58; 62, pp. 996–1001; 71, 76) the mineralized area extends southward from Yengan and Mawmang States of the Southern Shan States through the state of Karenni to the districts of Thaton, Amherst, Tavoy, and Mergui on the south. Production has come chiefly from Mergui (mainly tin), Tavoy (mainly tungsten), and Karenni (tin-tungsten). The main elements of the geology are similar to those in the adjacent regions—sedimentary rocks intruded by granite, and deposits both within the granite and the surrounding country rock. Throughout the Tavoy area the lodes have been classed as (1) wolframite-quartz, (2) cassiterite-quartz, and (3) wolframite-greisen. The wolframite-quartz veins are the most important, but separation of these veins from the cassiterite-quartz veins is based only on predominance of wolframite, since both wolframite and cassiterite occur in veins of each type. In the Karenni area, the Mawchi mine, about 160 miles north-northeast of Rangoon, is one of the largest, if not the largest, tin-tungsten mine in Burma. About 60 lodes are known. In general they consist of parallel quartz veins 3.5 to 5 ft thick (average less than 4 ft) that persist over a length of several hundred feet. In addition to quartz they contain cassiterite, wolframite, some scheelite, tourmaline, muscovite, fluorite, beryl, and small amounts of sulfides.

Korea produced about 16 percent of the tungsten of the world in the three-year period 1959 to 1961, slightly more than half of which came from deposits in the Republic of Korea, the rest from North Korea.

Most of the production of the Republic of Korea came from the Sangdong deposit (37, pp. 184–185; 67), about one hundred miles east and slightly south of Seoul, which is rated as one of the principal tungsten deposits of the world. Scheelite is present in six lenticular beds in the Cambrian Myobong formation, but production has come from only one of these, the Main bed, 40 m below the top of the formation. This bed, 4 to 5 m thick, dipping from 15 to 30°, is known over a stike length of 1,500 m. In a general way the central part of the bed consists of quartz (50 to 95 percent), biotite (up to 40 percent), scheelite (up to 25 percent), fluorite (up to 10 percent), and minor amounts of apatite and sulfides (chiefly pyrrhotite, but also molybdenite, bismuthinite, chalcopyrite, pyrite). Typically scheelite constitutes 1 to 5 percent of this part of the bed, in grains commonly 0.5 to 1.5 mm across. Generally this scheelite-bearing rock is flanked by a fine-grained hornfelsic rock composed of quartz, hornblende or actinolite, diopside, and biotite, with little or no scheelite. A very small amount of garnetiferous rock also is present locally. Klepper (67, p. 474) classed the deposits as high-temperature metasomatic rather than contact metasomatic because the closest known igneous rocks are 3.5 to 4 miles away. He stated that the rocks were reconstituted by recrystallization or by metasomatic replacement and that at the same time scheelite was deposited along a few favorable beds.

Little information is available regarding the deposits of North Korea. Possibly four mines are present, three of them in the southern part of the area. The deposits are chiefly quartz veins that contain wolframite, some cassiterite, molybdenite and some other sulfides (72, p. 210).

North America

Tungsten production in North America is chiefly from deposits in the United States, especially the western part, but some is from Canada and Mexico. Some deposits of Mexico might yield important future production.

Eastern United States. The Hamme district (60, 61, 75), chiefly in Vance County, North Carolina, but extending into Mecklenburg County, Virginia, contains tungsten-bearing veins in a shear zone about 8 miles long and 1 mile wide localized in granodiorite. The shear zone is approximately parallel to the contact between the granodiorite and schist, and most of the veins occur within 1,500 ft of the contact. The richer veins are in the central part of the district in a zone about 2 miles long. The chief tungsten mineral is huebnerite, but some scheelite is present along with various sulfides. The gangue is chiefly quartz, with some fluorite, rhodochrosite, and sericite.

Western United States. Tungsten deposits are present in much of the western United States (Fig. 9-6), but production has come chiefly from California, Colorado, Idaho, Montana, and Nevada.

California. Tungsten deposits are widespread in California (53, 54, 55, 56; 63, pp. 260–277; 66, pp. 135–169; 69, 70, 78), but they are especially abundant along the borders of the Sierra Nevada batholith in Kern and San Bernardino Counties northward to Madera and Mono Counties. The two most important areas have been the Atolia district in the south and the Bishop district in the north (Fig. 9-6). The Atolia district was the chief producer until 1938; the Bishop district has been the chief producer since then.

The Atolia district (56, p. 35; 66, pp. 148–150; 69; 78, pp. 657–659) has produced tungsten from scheelite veins and scheelite placers. The vein deposits occur in a series of roughly parallel, locally branching fractures that cut quartz monzonite. The major production has been from a few quartz-carbonate veins in a narrow belt about two miles long, within which scheelite is localized in nearly vertical ore shoots that average about 100 ft long and vary from a few inches to 17 ft wide. One large mass weighing a ton or more was recovered from the Million Dollar stope of the Union mine. These deposits were rated as the largest high-grade scheelite ore bodies discovered in the United States and possibly in the world. Placers associated with them, which actually led to their discovery in 1904, are extensive but are close to the deposits, since scheelite will not survive during much transportation. Most production has come from an area east of the deposits known as the Spud Patch,[1] so termed because of the presence of chunks of scheelite the

[1] *Spud* is a dialectic or colloquial term for potato, possibly originating from the use of a sharp, narrow spade (a spud) in digging.

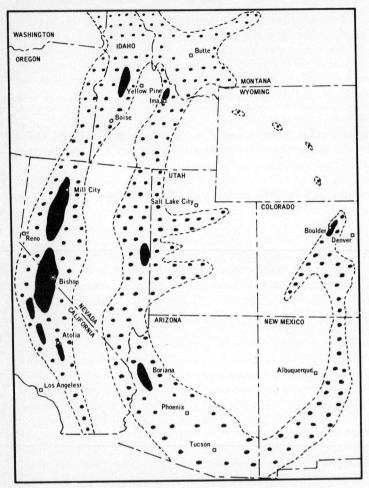

Fig. 9-6   Map showing distribution of tungsten in the western United States, with more productive areas in solid black ovals. After P. F. Kerr, *Geol. Soc. America Mem.* 15, p. 4. Copyright 1946. The Geological Society of America. Used by permission.

size and shape of potatoes; fragments actually ranged from those the size of fine sand to exceptional ones that weighed several hundred pounds. The richest concentrations of scheelite are in channels that have been worked by underground methods and also by stripping and large open-pit operations.

The production of tungsten in the Bishop district (53, 54; 63, pp. 268–276; 66, pp. 139–147) is chiefly from the Pine Creek mine, which contains the largest known reserve of tungsten ore in the United States. Molybdenum, copper, and gold are recovered also. The tungsten deposits of the district are chiefly of the contact-metasomatic type and occur in masses of marble adjacent to intrusive granitic rocks. The Pine Creek deposits occur in a roof pendant, known as the Pine Creek pendant, which is a lens of metamorphic rock nearly 7 miles long and as much as 1 mile wide. Within this lens there is a mass of marble about 3.5 miles long and a maximum of 1,800 ft wide, which gradually pinches northward to about 200 ft. Five ore bodies lie along the contact between the marble and quartz monzonite. Scheelite, the only valuable tungsten mineral present, occurs locally in dark-colored lime-silicate rock (tactite or skarn). Within the ore bodies are well-defined shoots, all of which furnish tungsten and some of which furnish molybdenum, which is present both as molybdenite, MoS_2, and as powellite, $CaMoO_4$.[1] A moderate amount of the powellite molecule is present also in some of the scheelite, as indicated by white to yellow fluorescence.

A potential source of tungsten in California, not yet exploited, is at Searles Lake where brine contains about 70 ppm WO_3. The amount contained in the brine of the lake has been estimated at 170 million lb of WO_3, equal to all other reserves in the United States (86). The tungsten is thought to have accumulated from a combination of leaching during a former stage of the Owens drainage system and waters from hot springs. Investigations regarding a possible method of recovering the tungsten economically are being conducted.

Montana. Deposits opened in Beaverhead County, Montana, north and slightly west of Dillon, appear to be large but relatively low grade (82, 84). Production of considerable extent started about 1954 and later was suspended but resumed in 1959 with planned monthly shipments of about 5,000 tons beginning in January 1960. Production was continuing and increasing in 1961. The deposits are of contact-metasomatic origin and are grouped around quartz-monzonite of Tertiary age and localized in metamorphosed Paleozoic limestone. The ore consists of scheelite-powellite, which is fine grained and disseminated throughout tactite composed to a large extent of garnet (grossularite). Ore is known in a north-south zone 16 to 18 miles long and may be present in a zone 40 miles long.

Colorado. Tungsten deposits are especially abundant in Boulder County, Colorado, in a narrow zone about 10 miles long, chiefly between Boulder and Nederland (66, pp. 82–83; 73, 74). Other deposits occur in the San Juan area, Ouray, San Juan, and San Miguel Counties (57; 66, pp. 85–86). The Boulder district was one of the foremost producers in the United States until 1945, after which production declined considerably. Production in Colorado since about 1960 has been chiefly by-product material from the Climax molybdenite mine. The deposits of the Boulder district consist chiefly of quartz-ferberite veins that cut Precambrian quartz monzonite and apparently were formed during one of the later stages of the Laramide orogeny early in the Tertiary period. Characteristically the ferberite, commonly well crystallized, forms the matrix of a breccia composed of country rock and early vein quartz. Some of the veins have been traced for more than a mile, others for only a hundred yards or less; most are shallow, and production has come chiefly from veins that did not persist below 200 ft. Most of the tungsten occurs in irregular ore shoots not over 100 ft long, which contain 1 to 20 percent WO_3.

---

[1] Pure powellite is $CaMoO_4$, but much material is $Ca(Mo,WO_4)$ with W substituting for Mo up to about Mo/W equals 9:1.

Other States. Tungsten deposits of Nevada, Idaho, and Arizona have yielded appreciable production. The deposits at Mill City, Nevada (66, pp. 182–188), are typical contact-metasomatic deposits in Triassic hornfels that contains some limestone beds and has been intruded by granodiorite. Some steeply inclined limestone members in the hornfels have been completely replaced. The ore is fine-grained scheelite disseminated throughout garnet (andradite and grossularite), epidote, quartz, calcite, and a few other minerals. The Yellow Pine mine, Valley County, central Idaho (59; 66, pp. 124–127), has been the most important producer of that state. The deposit was entirely within quartz monzonite, possibly localized in a shear zone, and had the general shape of a flat upright funnel, with the highest-grade tungsten ore concentrated toward the center of the mass and surrounded by antimony ore containing a little tungsten, with gold deposits clustered around the antimony deposits. The tungsten mineral was scheelite. The ore body was exhausted in 1945. Earlier production in Idaho was from the Ima mine, Lemhi County (66, pp. 122–124). The deposits consist of quartz veins that contain huebnerite and various sulfides. The Boriana mine, Mohave County, Arizona (64; 66, pp. 102–105), 20 miles southeast of Kingman, was sixth in production of tungsten in the United States in the period 1900 to 1945 (59, p. 176). Composite lodes that follow the foliation of phyllite consist of narrow quartz veins containing scheelite and wolframite in about equal amounts, along with some sulfides.

Canada.   Canada has not been an important producer of tungsten, but deposits in British Columbia, in the Northwest Territories, and in Ontario have furnished more ore than others (72). About 90 percent of the total production of Canada to the end of 1952 came from British Columbia.

The Emerald, Feeney, and Dodger ore bodies (52; 72, pp. 105–115), near Salmo, British Columbia, are localized chiefly in Lower Cambrian limestone that overlies argillite, both of which were intruded by Cretaceous (?) granite. The ore bodies dip eastward into the granite. Although the deposits are localized chiefly in limestone, four types of host rock have been recognized: skarn, limestone, quartz, and "greisenized" granite. Scheelite, mostly very finely disseminated, is the chief ore mineral, but some powellite and molybdenite are present also. Quartz is closely associated with the scheelite, quartz-tourmaline veins are abundant in the granite near the contact with other rock, biotite occurs in some quartz veins, and some fluorite is present. Pyrrhotite and biotite generally are present in the skarn, limestone, and quartz ore bodies. The quartz host rock is chiefly silicified limestone intersected by numerous quartz veins, some of which contain crystals of scheelite more than half an inch across. The "greisen" type of ore occurs within altered granite, as much as 40 ft from the contact with limestone; the alteration of the granite appears to have been chiefly kaolinization, sericitization, and silicification, rather than greisenization.

The Red Rose mine, near Hazelton, northwestern British Columbia (72, pp. 51–54; 77) was the largest producer of tungsten in British Columbia during World War II. A quartz-scheelite vein generally 3 to 6 ft wide, but attaining 16 ft at one place, was localized in a sill of diorite where that rock was cut by a shear zone. The shear also cuts other rocks but is much less developed, and veins in such rocks contain no scheelite.

The known deposits of the Northwest Territorities (72, pp. 121–127) are confined to the Yellowknife Mining Division of the district of Mackenzie. More than one thousand tungsten-bearing veins have been reported, but only a few are known to contain enough scheelite to be of economic interest. Production probably would need to be chiefly as a by-product of gold or gold-copper mining. A high-grade ore body, averaging 2.47 percent WO_3 and containing an estimated 1.2 million tons of reserves, yielded initial production early in the 1960s. The ore body is in the upper Flat River Valley of the Mackenzie Mountains, about 200 miles north of Watson Lake near the British Columbia border. A road 200 miles long was constructed from the Alaskan Highway to the newly developed mining town named Tungsten (101).

Tungsten of Ontario (72, pp. 157–178) occurs in quartz veins associated with gold deposits. Generally if appreciable scheelite is present, gold is sparse or absent.

Other Countries

Tungsten production in South America, which generally is less than that of the United States, comes chiefly from Bolivia, with some from Brazil and Argentina. Tungsten is present in various places throughout the Bolivian tin belt and is recovered as a by-product of tin mining, but at Chicote, about 65 km north-northeast of Oruro (Fig. 9-3), within an area of about 4 sq km, 72 veins and veinlets of quartz-wolframite-pyrrhotite make a central zone surrounded by tin-bearing veins (50). This deposit has furnished most of the tungsten of Bolivia, part of which has come from tungsten-bearing talus slides ("Llamperas" in Bolivia). The deposits of Brazil are of contact-metasomatic origin and consist of scheelite localized in tactite or skarn developed in calcareous layers in Precambrian mica schists near granitic intrusions (65). The deposits of Argentina are chiefly wolframite-bearing quartz veins; the largest mine is a few kilometers west of Concoran, San Luis (Fig. 8-5) (72, pp. 215–216).

Information regarding tungsten deposits in the Soviet Union indicates that they are present in a broad zone throughout the Transbaikal region eastward, which is also a region that contains some tin deposits. Both wolframite and huebnerite quartz veins and placers occur, and also contact-metasomatic scheelite deposits (37, pp. 196–198; 45, p. 121; 48, pp. 85–90; 72, p. 211). Some important vein deposits and placers occur in eastern Siberia, around Chita (Tschita), in the vicinity of 51°N, 116°E, and Dzhida, about 51°N, 102°E, and some important contact-metasomatic deposits in the north Caucasus, around Tyrny-auz, 43°22′N, 43°05′E (48, pp. 88–90), where complex tungsten-molybdenum ore is localized in Paleozoic limestones that were intruded by granodiorite (45, p. 121). These may be the most important deposits of the Soviet Union. Numerous other smaller deposits are mentioned.

Australia generally produces somewhat less tungsten than Portugal or Bolivia, but deposits are rather widespread. Production has come chiefly from Queensland (until recent years the leading producer), New South Wales, and Tasmania. The Queensland deposits are chiefly in pipes in granite, which contain wolframite, scheelite, cassiterite, bismuthinite, native bismuth, and molybdenite (see tin deposits). Tasmania has become the leading tungsten producer of Australia since about 1942. Earlier production there was from tin-tungsten deposits, but the major producer since about 1950 has been a contact-metasomatic deposit on King Island, rated as one of the largest tungsten deposits of the world (68; 72, pp. 212–213; 83). Precambrian sedimentary rocks of the Grassy group, originally dolomitic rocks of varying degrees of purity, were intruded by granite, metamorphosed and meta-somatized, and converted into several types of hornfelses. The more calcitic beds were selectively replaced and converted in scheelite-bearing garnet-pyroxene skarn, the economic ore being confined to two beds, an upper ore body 20 to 38 ft thick (average 25 ft), a lower ore body 15 to 110 ft thick (average 90 ft). Most of the scheelite is finely divided in the skarn, but quartz-scheelite veins are numerous in parts of the mineralized zones. These range in width from less than an inch to 6 in. and contain irregularly distributed coarse crystals of scheelite, some of which are 2.5 in. across. Some molybdenum is present also throughout the deposits, in part as minute crystals of molybdenite, in part in powellite. The deposits are known throughout a zone at least 1,400 ft long and 150 to 220 ft thick.

Molybdenum

Molybdenum (3, 4, 7) is commonly associated with tin, tungsten, and bismuth, as has been indicated in discussions of tin and tungsten deposits. Another association of major importance is with copper deposits of the type designated *porphyry copper*. In deposits of this kind, molybdenite is obtained as a by-product of copper recovery, but the molybdenum is so extensive in some deposits that it is a major source of molybdenum. Thus the Utah copper mine at Bingham has become the world's second largest source of molybdenum (99), and since 1940 about 40 percent of the molybdenum produced in the United States has been from by-product sources (94). By-product molybdenum from copper deposits will be discussed in the description of copper deposits.

The most important molybdenum deposit of the world is at Climax, Colorado, and a second highly important deposit is at Questa, New Mexico. The order of world production of molybdenum for a number of years has been (1) the United States, about 78 percent; (2) the Soviet Union, about 10 percent; and (3) Chile, about 5 percent (Fig. 9-7). China has become important since 1958 and produced 4.4 percent of the world total in 1962, and a deposit in Sardinia is reputed to be important (97).

United States

Climax, Colorado. The molybdenum deposit near Climax, Lake County, Colorado, about 1 mile east of Freemont Pass (elevation 11,320 ft) and 600 ft higher, furnished 453,451,153 lb of molybdenum from 1918 to 1953. Total by-product materials from the molybdenum production, 1949 to 1953, were 1,973,086 lb of WO$_3$, 121,420 lb of tin, and 65,000 tons of pyrite (10, p. 41). A conservative estimate of the reserves of ore made in 1961 was 462 million tons, over 90 percent of the world's known reserves.

Within the area of the deposits, dikes and stocks of quartz monzonite of Tertiary age have intruded Silver Plume granite and Idaho Springs formation (schist) of Precambrian age. Hydrothermal alteration was intense around one of

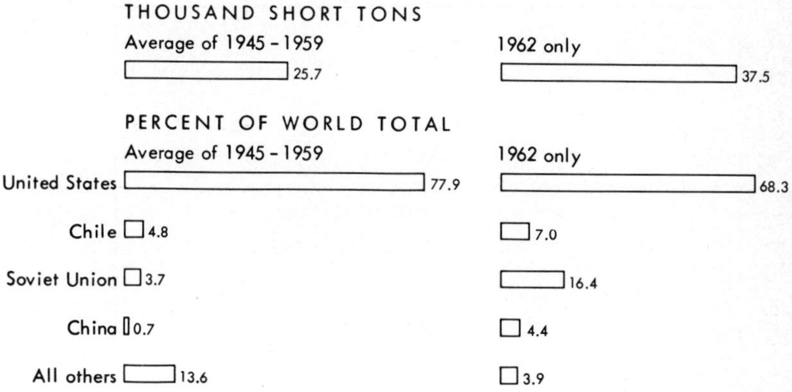

THOUSAND SHORT TONS
Average of 1945 – 1959   1962 only

| | 25.7 | | 37.5

PERCENT OF WORLD TOTAL
Average of 1945 – 1959   1962 only

United States | | 77.9 | | 68.3

Chile ☐ 4.8   ☐ 7.0

Soviet Union ☐ 3.7 | | 16.4

China ▯ 0.7   ☐ 4.4

All others | | 13.6   ☐ 3.9

Fig. 9-7 World molybdenum production. Compiled from *U.S. Bur. Mines Minerals Yearbook*, various years.

the stocks of quartz monzonite, the Climax porphyry stock, and less so within the stock, which was converted into a rock of the mineral composition of granite as a result of elimination of the original plagioclase. Above and around the porphyry stock is a zone of quartz rock (Fig. 9-8) formed by silicification of the Precamrian rocks, and around this quartz rock is the ore zone, in which secondary orthoclase and quartz are the chief alteration minerals, the orthoclase being more abundant than the quartz. Outward from the ore zone the alteration gradually decreases in amount. The rather simple zonation of the upper levels (1) gives way to more complex conditions in depth.

Molybdenite occurs disseminated throughout the rock, in part, but occurs also in numerous small molybdenite-quartz veinlets. These veinlets are commonly less than ¼ in. wide, although exceptionally ½ in. wide, and average 10 to 12 per ft over widths of 300 to 500 ft in the ore zone. Mineralization is present not only in the altered Precambrian rocks but also in the Tertiary dikes and stock. The ore zone is actually designated by assay values 0.4 percent molybdenite and higher. Both inward and outward from the ore zone the amount of molybdenite decreases to less than 0.2 percent.

In addition to the molybdenite-quartz veins the rocks are cut by pyrite-quartz-topaz veins ⅛ to ¼ in. wide, younger and less numerous than the molybdenite-quartz veins. These do not contain molybdenite but do contain huebnerite, small quantities of sulfides, and a few other minerals.

Questa, New Mexico. The molybdenite deposit at Questa, New Mexico (2, 9), is on the western slope of the Sangre de Cristo mountain range, about 70 miles north and slightly east of Santa Fe. The molybdenite there, unlike that in many deposits, is in fissure veins. The veins are localized in fractures in a down-faulted block containing Tertiary volcanics and some older rocks that were intruded by Teritary granitic rocks. Localization is chiefly in a fractured zone parallel to the contact of a granitic mass and the adjacent rocks. Veins are commonly less than 1/16 in. to a foot or more wide and are composed chiefly of quartz and molyb-

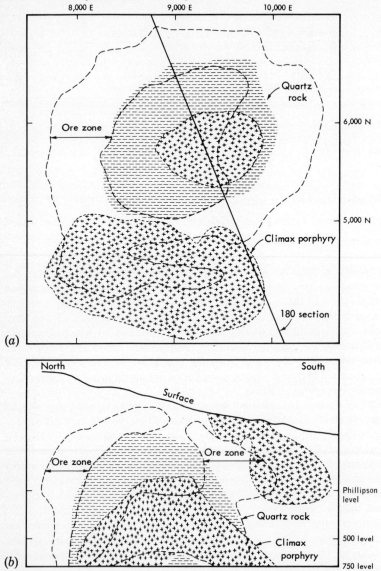

Fig. 9-8 Map (*a*) and 180 section (*b*) on the Phillipson level of the Climax molybdenum deposit, Climax, Colorado, showing porphyry, quartz rock, and ore zone. The country rock is granite and schist. Somewhat simplified after J. W. Vanderwilt and R. U. King, *Mining Engineering,* vol. 7, pp. 42–43. Copyright 1955. The American Institute of Mining, Metallurgical, and Petroleum Engineers. Used by permission.

denite, although the veins were formed in at least three stages and some contain fluorite, carbonates, and pyrite. Hydrothermal alteration, varying from slight to intense, was sericitization followed by kaolinization and that, in turn, by silicification. The molybdenite mineralization started as early as the main stage of sericitization but was especially active during silicification. The deposit is reported to contain 380,000 short tons of molybdenum. This is 760 million lb of molybdenum which, if it could all be recovered, is considerably more than the total molybdenum produced at Climax, Colorado, between 1918 and 1953.

Canada

Canada has never been an important producer of molybdenum, although many occurrences are known (95). Some production has come from the Moss mine, Quyon, Quebec, from pegmatites and aplites, and some as a by-product of copper mining by Keremeos Mines, Ltd., British Columbia, and from the Gaspe copper deposit, Quebec. The most important producing area, which also contains the greatest reserves, is the Lacorne-Preissac district, Quebec (5, 6), near Val D'or, in the same general area as the important lithium deposits (Fig. 8-2). The deposits are in pegmatites and, especially, in pegmatitic quartz veins generally 1 to 7 ft wide. Bismuth is a by-product.

Other Countries

Molybdenum deposits are known to exist in considerable extent in the Soviet Union and in China, and estimates indicate that much production has come from them recently. Production comes also from quartz veins and disseminations in the Knaben mine, Norway, and production has come from the Knaben area since 1885 (4, p. 14; 7, pp. 81–82). Molybdenite occurs in several places in Australia (85, 93): in Queensland, where molybdenum, tungsten, and bismuth occur together; and in New South Wales in pipes, pegmatites, quartz veins, and contact-metasomatic deposits, the most important occurrences being the molybdenum-pipes of the Kingsgate area and the contact-metasomatic deposits of the Yetholme district (89, 90, 92).

Little is known about the deposits of China. Torgasheff (79, p. 278) early reported occurrences in Fukien, Chekiang, and Kwangtung Provinces (Fig. 9-5), the best-known being at Lipikong and Yungtai in Fukien Province. The deposits are pegmatite veins in granite, associated with tungsten and bismuth ores. In Chekiang and Fukien Provinces the deposits have been developed chiefly for molybdenum, whereas in Kwangtung Province tungsten has been the major product and molybdenum has been a by-product (44, p. 440). Three new large deposits were reported in 1962 (96) in the Tsin-Ling Mountains and in Shansi and Kirin Provinces.

Two summaries of the molybdenum deposits of the Soviet Union have been made recently, one (1959) in French (8), the other (1960) in German (7). Many details regarding the relative importance of deposits, and about the deposits themselves, do not appear in the Russian publications from which these summaries were to a large extent prepared. Some information is presented also by Nalivkin (45, pp. 79, 121, 148) and by Shimkin (48, pp. 68–73).

Deposits of the Soviet Union consist of (1) hydrothermal quartz-molybdenite veins, some of which contain scheelite also; (2) contact-metasomatic molybdenite-scheelite deposits; and (3) copper-molybdenite deposits, apparently chiefly of the porphyry copper type.

The quartz-molybdenite type occurs at several places in Siberia, in general between about 50°N to 51°N and 108°E to 133°E. Veins in the Umal'ta area on the upper Bureya River (approximately 51°35'N, 133°15'E) in the Far East contain from 0.60 to 2.42 percent Mo, but reserves are small. Some veins of coarse quartz contain, in places, large flakes of molybdenite, some wolframite and pyrite. A somewhat later stage of mineralization produced fine-grained quartz through which is disseminated molybdenite, fluorite, and sericite. A similar type containing 3 to 4 percent Mo is the Chikoi deposit (approximately 50°15'N, 107°48'E) composed of veins only 4 mm thick and generally not over 50 cm long scattered through highly metamorphosed slates and gneisses. Other deposits occur at Shakhtema and Davenda (approximately 51°25'N, 117°45'E). Quartz-molybdenite deposits occur also in Kazakhstan at Chindigatui (49°50'N, 84°45'E), which differ somewhat from the other types, since they are in close contact with granite.

One of the most important contact-metasomatic deposits is in the Caucasus Mountains at Tyrny-auz (43°22'N, 43°05'E), where granitic material intruded limestone, sandstone, and shale. The deposits are localized in the crest of an anticlinal fold and consist of skarn in which the mineralization is partly scheelite and molybdenite, with a ratio WO_3/MoO_3 about 1:4 to 1:8. This deposit has been stated to contain the principal reserves of tungsten and almost half of the reserves of molybdenum (8, p. 242).

The best known copper-molybdenum deposit is the one at Kounrad (47°00'N., 75°00'E) in Kazakhstan, a porphyry copper deposit, where the average ratio Mo/Cu is 1:146 but reaches as much as 1:100 in some places. Other copper-molybdenum deposits are known in the Transcaucasus region at Zangezur (39°12'N, 46°05'E) and Agarak (39°00'N, 46°10'E).

The relative importance of various deposits is not entirely clear. *Pravda,* in 1957 (7), reported that large molybdenum occurrences had been found in central Kazakhstan, and Nalivkin (45, p. 79) stated that one of the largest deposits of tungsten-molybdenum ore occurs in eastern Kazakhstan. It has been stated also (7) that in the future the copper-molybdenum deposits will furnish the most stable production, that the predominant part of the Russian molybdenum lies in the Caucasus and Transcaucasus, and that the second most important area is central Kazakhstan.

Bismuth

Bismuth generally is not mined for that metal only. An outstanding exception is Argentina, where native bismuth and bismutite (bismuth carbonate) derived therefrom occur with wolframite at the Los Condores mine in San Luco Province. Bismuth is also produced in Argentina along with the mining of beryl. In other countries bismuth is obtained at various places as a by-product of the mining of tin, as in Bolivia; the mining of tungsten, as in Korea; molybdenum, as at the Lacorne mine, Canada; or molybdenum and tungsten, as in Australia. Most of the bismuth produced, however, is obtained from the treatment of refinery slimes of lead ores (e.g., the United States) (87), copper-lead-zinc ores (e.g., Cerro de Pasco Corporation, Peru), and some ores that contain gold. In the United States the most important source of bismuth has been lead-zinc-silver replacement deposits in limestone, such as those at Leadville and Gilman, Colorado, and Tintic, Utah.

SELECTED REFERENCES

MOLYBDENUM

1. B. S. Butler and J. W. Vanderwilt, 1933, Climax molybdenum property of Colorado, *U.S. Geol. Survey Bull.* 846-C, pp. 191–237.
2. R. H. Carpenter, 1960, A resume of hydrothermal alteration and ore deposition at Questa, New Mexico, U.S.A., Internat. Geol. Cong., 21st, Copenhagen 1960, pt. 16, sec. 16, Genetic problems of ores, pp. 79–85.
3. F. L. Hess, 1924, Molybdenum deposits, *U.S. Geol. Survey Bull.* 761, 35 pp.
4. W. McInnis, 1957, Molybdenum: a materials survey, *U.S. Bur. Mines Inf. Cir.* 7784, 77 pp.
5. G. W. H. Norman, 1945, Molybdenite deposits and pegmatites in the Preissac-LaCorne area, Abitibi County, Quebec, *Econ. Geology,* vol. 40, pp. 1–17.
6. G. W. H. Norman, 1948, LaCorne molybdenite deposit, in "Structural Geology of Canadian Ore Deposits," pp. 850–852, Canadian Institute of Mining and Metallurgy, Montreal.
7. H. P. Rechenberg, 1960, Molybdän, "Die metallischen Rohstoffe," vol. 12, Ferdinand Enke, Stuttgart, 128 pp.
8. W. Sakowitsch, 1959, Aperçu sur les gisements de molybdène en U.R.S.S., *Chronique Mines d'Outre-Mer,* vol. 27, no. 278, pp. 241–252.
9. J. H. Schilling, 1956, Geology of the Questa molybdenum (Moly) mine area, Taos County, New Mexico, *New Mexico Bur. Mines and Min. Resources, Bull.* 51, 87 pp.
10. H. W. Vanderwilt and R. U. King, 1955, Hydrothermal alteration at the Climax molybdenite deposit, *Mining Eng.,* vol. 7, pp. 41–53.

TIN

11. E. Broadhurst, 1953, The Heberton tin field, in *Fifth Empire Min. Metall. Cong.,* Australia and New Zealand, pp. 703–717, Australasian Institute of Mining and Metallurgy, Melbourne.
12. F. M. Chace, 1948, Tin-silver veins of Oruro, Bolivia, *Econ. Geology,* vol. 43, pp. 333–383, 435–470.
13. W. M. Davy, 1920, Ore deposition in the Bolivian tin-silver deposits, *Econ. Geology,* vol. 15, pp. 463–496.
14. S. Fawns, undated (1925?), "Tin Deposits of the World," 3d ed., The Mining Journal, London, 306 pp.
15. H. G. Ferguson and A. M. Bateman, 1912, Geologic features of tin deposits, *Econ. Geology,* vol. 7, pp. 209–262.
16. L. C. Graton, 1905, The Carolina tin belt, *U.S. Geol. Survey Bull.* 260, pp. 188–195.

17. F. L. Hess and L. C. Graton, 1905, The occurrence and distribution of tin, *U.S. Geol. Survey Bull.* 260, pp. 161–187.
18. F. T. Ingham and E. F. Bradford, 1960, The geology and mineral resources of the Kinta Valley, Perak, *Federation of Malaya, Geol. Survey, District Mem.* 9, 347 pp.
19. M. I. Itsikson, 1960, The distribution of tin-ore deposits within folded zones, *Internat. Geol. Rev.,* vol. 2, pp. 397–417. (Translated by W. A. Kneller.)
20. W. R. Jones, 1925, "Tinfields of the World," Mining Publications, Ltd., London, 423 pp.
21. C. L. Knight, 1953, Mount Bischoff tin mine, *Fifth Empire Min. Metall. Cong.,* Australia and New Zealand, pp. 1185–1193, Australasian Institute of Mining and Metallurgy, Melbourne.
22. C. L. Mantel, 1949, Tin, *American Chemical Soc. Mon.* 51, pp. 61–88, Reinhold Publishing Corporation, New York, 573 pp.
23. A. A. C. Mason, 1953, The Vulcan tin mine, *Fifth Empire Min. Metall. Cong.,* Australia and New Zealand, pp. 718–721, Australasian Institute of Mining and Metallurgy, Melbourne.
24. C. W. Merrill, 1952, Tin, in W. Van Royan and O. Bowles, "The Mineral Resources of the World," vol. 2, pp. 115–120, Prentice-Hall, Inc., Englewood Cliffs, N.J., 181 pp.
25. B. J. Miller and J. T. Singewald, Jr., 1919, "The Mineral Deposits of South America," McGraw-Hill Book Company, New York, 598 pp. See pp. 94–134 for the Bolivian tin-silver deposits.
26. A. Renick and J. B. Umhau, 1956, Tin, *U.S. Bur. Mines Bull.* 556, pp. 883–903.
27. J. D. Ridge, 1953, Materials survey, tin, Chap. 4, Tin resources, 180 pp., U.S. Government Printing Office, Washington, D.C., 774 pp.
28. J. B. Scrivenor, 1913, The Geology and Mining Industry of the Kinta District, Perak, Federated Malay States, Federated Malay States Government Printing Office, Kuala Lumpur, 90 pp.
29. J. S. Singewald, Jr., 1912, Some genetic relations of tin deposits, *Econ. Geology,* vol. 7, pp. 263–279.
30. F. S. Turneaure, 1935, Tin deposits of Llallagua, Bolivia, *Econ. Geology,* vol. 30, pp. 14–68, 170–190.
31. F. S. Turneaure, 1960, A comparative study of major ore deposits of central Bolivia, *Econ. Geology,* vol. 55, pp. 217–254, 574–606.
32. F. S. Turneaure and R. Gibson, 1945, Tin deposits of Carguaicollo, Bolivia, *Am. Jour. Sci., Daly vol.* 243-A, pp. 523–541.
33. J. Westerveld, 1937, The tin ores of Banca, Billiton, and Singkep, Malay archipelago—a discussion, *Econ. Geology,* vol. 32, pp. 1019–1041.
34. J. G. Weston-Dunn, 1922, The economic geology of the Mount Bischoff tin deposits, Tasmania, *Econ. Geology,* vol. 17, pp. 143–193.
35. E. S. Willbourne, 1924, Notes on the Occurrence of Lode Tin-ore in the Kinta Valley, Federated Malay States Geological Department, Kuala Lumpur, 22 pp.
36. N. Wing-Easton, 1937, The tin ores of Banca, Billiton, and Singkep, Malay Archipelago, *Econ. Geology,* vol. 32, pp. 1–30, 154–182.

TIN-TUNGSTEN

37. F. Ahlfeld, 1958, Zinn und Wolfram, in "Die metallischen Rohstoffe," vol. 11, Ferdinand Enke, Stuttgart, 212 pp.
38. R. Blanchard, 1947, Some pipe deposits of eastern Australia, *Econ. Geology*, vol. 42, pp. 265–304.
39. H. L. Chhibber, 1934, "The Mineral Resources of Burma," pp. 174–210, Macmillan & Co., Ltd., London, 320 pp.
40. H. J. C. Conolly, 1953, The Aberfoyle tin-wolfram mine, in *Fifth Empire Min. Metall. Cong.*, Australia and New Zealand, pp. 1200–1208, Australasian Institute of Mining and Metallurgy, Melbourne.
41. J. A. Dunn, 1938, Tin-tungsten mineralization at Mawchi, Karena States, Burma, *Records Geol. Survey India*, vol. 73, pt. 2, pp. 209–237.
42. J. A. Dunn, 1938, Tin-tungsten mineralization at Hermyingyi, Tavoy district, Burma, *Records Geol. Survey India*, vol. 73, pt. 2, pp. 238–246.
43. A. B. Edwards, 1953, Storey's Creek tin and wolfram mine, in *Fifth Empire Min. Metall. Cong.*, Australia and New Zealand, pp. 1209–1212, Australasian Institute of Mining and Metallurgy, Melbourne.
44. V. C. Juan, 1946, Mineral resources of China, *Econ. Geology*, vol. 41, pp. 399–474.
45. D. V. Nalivkin, 1960, "The Geology of the U.S.S.R., a Short Outline," Pergamon Press, New York, 170 pp., geologic map of the U.S.S.R. See pp. 78–79, 97, 121, 148.
46. P. B. Nye and F. Blake, 1938, The geology and mineral deposits of Tasmania, *Tasmanian Geol. Survey Bull.* 44, pp. 78–87.
47. J. B. Scrivenor, 1928, "The Geology of Malayan Ore Deposits," Macmillan & Co., Ltd., London, 216 pp. Tin, pp. 24–129; tungsten, pp. 130–146.
48. D. B. Shimkin, 1953, "Minerals—a Key to Soviet Power," pp. 88–90, 133–136, Harvard University Press, Cambridge, Mass., 452 pp.
49. A. Spry and M. R. Banks, 1962, The geology of Tasmania, pp. 285–309, *Geol. Soc. Australia Journ.* vol. 9, pt. 2, pp. 107–362.

TUNGSTEN

50. F. Ahlfeld, 1945, The Chicote tungsten deposit, Bolivia, *Econ. Geology*, vol. 40, pp. 394–407.
51. H. F. Bain, 1927, "Ores and Industry in the Far East," pp. 148–152, Council on Foreign Relations, Inc., New York, 229 pp.
52. C. W. Ball, 1954, The Emerald, Feeney and Dodger tungsten ore-bodies, Salmo, British Columbia, Canada, *Econ. Geology*, vol. 49, pp. 625–638.
53. P. C. Bateman, 1945, Pine Creek and Adamston tungsten mines, Inyo County, California, *California Div. Mines Rept.* 41, pp. 231–249.
54. P. C. Bateman, 1956, Economic geology of the Bishop tungsten district, California, *California Dept. Nat. Resources, Div. Mines Spec. Rept.* 47, 87 pp.
55. P. C. Bateman, M. P. Erickson, and P. D. Proctor, 1950, Geology and tungsten deposits of the Tungsten Hills, Inyo County, California, *California Jour. Mines and Geology*, vol. 46, pp. 23–42.
56. P. C. Bateman and W. P. Irwin, 1954, Tungsten in southeastern California,

in Geology of southeastern California, pt. 4, Chap. 8, pp. 31–40, *California Dept. Nat. Resources, Div. Mines Bull.* 170.

57. C. Belser, 1956, Tungsten potential in the San Juan area, Ouray, San Juan, and San Miguel Counties, Colorado, *U.S. Bur. Mines Inf. Circ.* 7731, 18 pp.

58. J. M. Campbell, 1920, Tungsten deposits of Burma and their origin, *Econ. Geology,* vol. 15, pp. 511–534.

59. J. R. Cooper, 1951, Geology of the tungsten, antimony, and gold deposits near Stibnite, Idaho, *U.S. Geol. Survey Bull.* 969-F, pp. 151–197.

60. G. H. Espenshade, 1947, Tungsten deposits of Vance County, North Carolina, *U.S. Geol. Survey Bull.* 948-A, pp. 1–17.

61. G. N. Espenshade, 1950, Occurrence of tungsten minerals in the southeastern states, in "Symposium on Mineral Resources of the Southeastern United States," pp. 56–66, University of Tennessee Press, Knoxville, Tenn., 236 pp.

62. F. L. Hess, 1912, Tungsten, *U.S. Geol. Survey, Min. Resources U.S.,* pt. 1, pp. 987–1001.

63. F. L. Hess and E. S. Larsen, 1922, Contact-metamorphic tungsten deposits of the United States, *U.S. Geol. Survey Bull.* 725, pp. 245–309.

64. S. W. Hobbs, 1944, Tungsten deposits in Boriana district and Aquarius Range, Mohave County, Arizona, *U.S. Geol. Survey Bull.* 940-I, pp. 247–264.

65. W. J. Johnston, Jr., and F. M. de Vasconcellos, 1945, Scheelite in northern Brazil, *Econ. Geology,* vol. 40, pp. 34–50.

66. P. F. Kerr, 1946, Tungsten mineralization in the United States, *Geol. Soc. America Mem.* 15, 241 pp.

67. M. R. Klepper, 1947, The Sangdong tungsten deposit, southern Korea, *Econ. Geology,* vol. 42, pp. 465–477.

68. C. L. Knight and P. B. Nye, 1953, The King Island scheelite mine, in *Fifth Empire Min. Metall. Cong.,* Australia and New Zealand, pp. 1222–1232, Australasian Institute of Mining and Metallurgy, Melbourne.

69. D. M. Lemmon and J. V. N. Dorr, 2nd, 1940, Tungsten deposits in the Atolia district, San Bernardino and Kern Counties, California, *U.S. Geol. Survey Bull.* 922-H, pp. 205–245.

70. D. M. Lemmon and O. L. Tweto, 1962, Tungsten in the United States, *U.S. Geol. Survey Mineral Inv. Resource Map* MR-25. Map and pamphlet of 25 pp.

71. K. C. Li and C. Y. Wang, 1955, The geology of tungsten, in *American Chemical Soc. Mon.* 94, 3d ed., pp. 4–112, Reinhold Publishing Corporation, New York, 506 pp.

72. H. W. Little, 1959, Tungsten deposits of Canada, *Canada Geol. Survey Econ. Geology Ser.,* no. 17, 251 pp. Contains a good summary of many deposits throughout the world.

73. T. S. Lovering, 1941, The origin of the tungsten ores of Boulder County, Colorado, *Econ. Geology,* vol. 36, pp. 229–279.

74. T. S. Lovering and O. L. Tweto, 1953, Geology and ore deposits of the Boulder County tungsten district, Colorado, *U.S. Geol. Survey Prof. Paper* 245, 199 pp.

75. J. M. Parker, III, 1963, Geologic setting of the Hamme tungsten district, North Carolina and Virginia, *U.S. Geol. Survey Bull.* 1122-G, 69 pp.
76. R. H. Rastall, 1918, The genesis of tungsten ores, *Geol. Mag. N. S.,* Decade 6, vol. 5, pp. 193–203, 241–246, 293–296, 367–370.
77. J. S. Stevenson, 1947, Geology of the Red Rose tungsten mine, Hazelton, British Columbia, *Econ. Geology,* vol. 42, pp. 433–464.
78. R. M. Steward, 1957, Tungsten, in mineral commodities of California, pp. 655–667, *California Dept. Nat. Resources, Div. Mines, Bull.* 176, 738 pp.
79. B. P. Torgasheff, 1930, "The Mineral Industry of the Far East," pp. 233–237, Chali Company, Limited, Shanghai, China, 510 pp.
80. H. W. Turner, 1919, Review of recent literature on tungsten deposits of Burma, *Econ. Geol.,* vol. 14, pp. 625–639.
81. P. M. Tyler, 1952, Tungsten, in W. Van Royan and O. Bowles, "The Mineral Resources of the World," vol. 2, pp. 80–83, Prentice-Hall, Inc., Englewood Cliffs, N.J., 181 pp.
82. J. H. Waterhouse, 1952, Montana's Beaverhead tungsten, *Mining World,* vol. 14, pp. 36–39.
83. Anon., 1953, Tungsten from "Down Under," *Mining World,* vol. 15, no. 1, pp. 34–37.
84. Anon., 1955, How Minerals Engineering opens big low-grade tungsten deposit, *Mining World,* vol. 17, no. 1, pp. 38–43.

MISCELLANEOUS REFERENCES CITED

85. E. C. Andrews, 1916, The molybdenum industry in New South Wales, *New South Wales Geol. Survey Mineral Resources, No.* 24, 199 pp.
86. L. G. Carpenter and D. E. Garrett, 1959, Tungsten in Searles Lake, *Mining Eng.,* vol. 11, pp. 301–303.
87. J. R. Cooper, 1962, Bismuth in the United States, *U.S. Geol. Survey Mineral Inv. Resource Map* MR-22. Map and pamphlet of 19 pp.
88. G. H. M. Farley, 1946, Cornwall as a tin producer—today and tomorrow, *Eng. Mining Jour.,* vol. 147, no. 11, pp. 78–79.
89. N. H. Fisher, 1953, The Everton molybdenite mine, in *Fifth Empire Min. Metall. Cong.,* Australia and New Zealand, pp. 1104–1107, Australasian Institute of Mining and Metallurgy, Melbourne.
90. M. D. Garretty, 1953, Bismuth-molybdenite pipes of New England, in *Fifth Empire Min. Metall. Cong.,* Australia and New Zealand, pp. 962–967, Australasian Institute of Mining and Metallurgy, Melbourne.
91. K. F. G. Hosking and J. H. Trounson, 1959, The mineral potential of Cornwall, in Future of non-ferrous mining in Great Britain and Ireland, *Inst. Mining and Metall., London,* pp. 355–369, 383–396.
92. R. L. Jack, 1953, The Whipstick molybdenite and bismuth mines, in *Fifth Empire Min. Metall. Cong.,* Australia and New Zealand, pp. 968–970, Australasian Institute of Mining and Metallurgy, Melbourne.
93. E. J. Kenney, 1924, Antimony. Arsenic. Bismuth. Molybdenum. Tungsten. *New South Wales Dept. Mines, Geol. Survey, Bull.* 5, 47 pp.
94. R. C. Nelson, 1960, A detailed look at present supplies, world molybdenum and tungsten resources, *Eng. Mining Jour.,* vol. 161, no. 5, p. 93.

95. F. M. Vokes, 1963, Molybdenum deposits of Canada, *Canada Geol. Survey Econ. Geology Ser., Rept.* 20, 332 pp.
96. A. W. T. Wei, 1962, Minerals in China in 1961—III, *Mining Jour. (London),* vol. 258, no. 6610, pp. 409–411.
97. Anon., 1960, Huge molybdenite deposits claimed in Sardinia, *Eng. Mining Jour.,* vol. 161, no. 1, p. 110.
98. Anon., 1962, Tin production in Australia, *Mining Jour. (London),* vol. 259, p. 591.
99. Anon., 1962, Molybdenite recovery, *Mining Jour. (London),* vol. 259, p. 539.
100. Anon., 1963, Cornwall, renewed mining activity in ancient mining areas, *Mining Jour. (London),* vol. 260, p. 341.
101. Anon., 1963, New tungsten mine opens in Canada's northwest region, *Eng. Mining Jour.,* vol. 164, no. 1, pp. 75–78.

chapter *10*

Gold and Silver

Gold and silver are commonly associated in many deposits, and native gold may contain 10 to 15 percent silver; if it contains as much as 20 percent it may be designated *electrum*. The gold tellurides krennerite and calaverite commonly contain some silver, with as much as a ratio of 3 Ag/8 Au in krennerite and 1 Ag/4Au in calaverite. A common associate of these two tellurides is the silver-gold telluride sylvanite, in which the ratio Au/Ag is close to 1:1 but the gold content may be considerably more than this. Further, both gold and silver are present in many ores of copper, lead, and zinc. The percentages of gold and silver produced from ores of various types are shown by Table 10-1.

Table 10-1 shows that copper ore is a highly important source of both gold and silver but that lead ore and mixed ores are more important sources of silver than of gold. Placers furnish practically no silver but generally more than 20 percent of the gold production, and as much as 30 percent in some years.

**TABLE 10-1.   Percentages of Gold and Silver Produced
from Ores of Various Types, United States, 1961***

Type of ore	*Gold,* *percent of total production*	*Silver,* *percent of total production*
Dry ore†	47.9	39.1
Copper ore	34.4	29.9
Lead ore	0.2	7.8
Zinc ore	0.1	1.7
Lead-zinc, zinc-copper, lead-copper, and zinc-lead-copper ores	4.3	21.4
Placer ore	13.1	0.1
Total	100.0	100.0

* *U.S. Bur. Mines Minerals Yearbook,* 1961, vol. 1, pp. 603 and 1110.
† Dry ore or dry and siliceous ore is essentially gold ore or silver ore that does not contain enough base metal to meet metallurgical requirements for smelting.

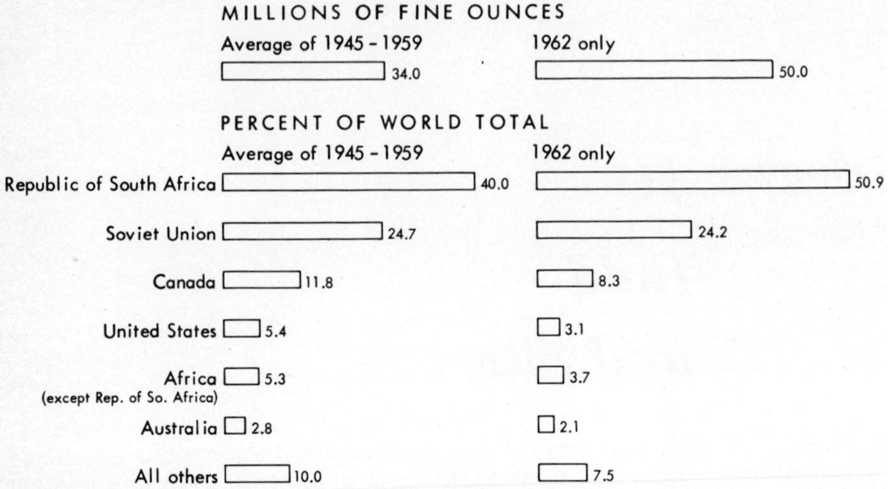

MILLIONS OF FINE OUNCES

Average of 1945 – 1959   1962 only

[_____] 34.0   [_____] 50.0

PERCENT OF WORLD TOTAL

Average of 1945 – 1959   1962 only

Republic of South Africa [_____] 40.0   [_____] 50.9

Soviet Union [_____] 24.7   [_____] 24.2

Canada [_____] 11.8   [____] 8.3

United States [__] 5.4   [_] 3.1

Africa [__] 5.3   [_] 3.7
(except Rep. of So. Africa)

Australia [_] 2.8   [_] 2.1

All others [____] 10.0   [___] 7.5

Fig. 10-1   World gold production. Compiled from *U.S. Bur. Mines Minerals Yearbook,* various years.

Generalization regarding origin of deposits, except the placers, is difficult. In a general way the gold-silver deposits that contain many tellurides and considerable amounts of both gold and silver are apt to be of the low-temperature hydrothermal type, and silver generally shows a strong tendency to accumulate in moderate- to low-temperature hydrothermal deposits. Contact-metasomatic deposits of either gold or silver are rather uncommon, although some important ones do occur.

Although gold and silver may occur together, some areas are noted particularly for the production of one or the other metal, whereas some produce almost equal amounts of each. Thus the Republic of South Africa usually furnishes more than 40 percent of the world's gold but very little silver, whereas Mexico and Peru supply much silver but little gold. In contrast, the Soviet Union, Canada, the United States, and Australia produce appreciable amounts of both gold and silver. The major gold production of the world is shown by Figure 10-1.

Gold

Republic of South Africa

One of the most remarkable and productive gold fields of the world is the Witwatersrand area,[1] commonly referred to as the "Rand," in the general vicinity of Johannesburg, Republic of South Africa (Fig. 10-2).

[1] An interesting history of the discovery and development of this area has been written by Owen Letcher, 1936, "The Gold Mines of Southern Africa," Waterlow and Sons Ltd., London, 580 pp. An interesting and excellent summary of discovery, early work, and development to 1960 is given by R. Borchers (2).

Active production of gold began in 1886 (12), and from then through 1960 this area has furnished about 38 percent of the total gold of the world. At times the production was much greater: in the five-year period 1926 to 1930 it was about 52 percent of the total, and in the five-year period 1956 to 1960 it was about 45 percent. The total production has been in the neighborhood of 650 million oz of gold, the value of which was probably between 1.5 and 2 billion dollars (12, 158). Moreover, iridosmine or iridium and uranium are now recovered as by-products. The amount of gold recovered per ton of ore in the period 1939 to 1955 varied from 3.740 dwt to 4.261 dwt (12, p. 132), equivalent to 0.187 and 0.213 oz per ton, or, at a price of $35 per oz, $6.54 to $7.45 per ton, although

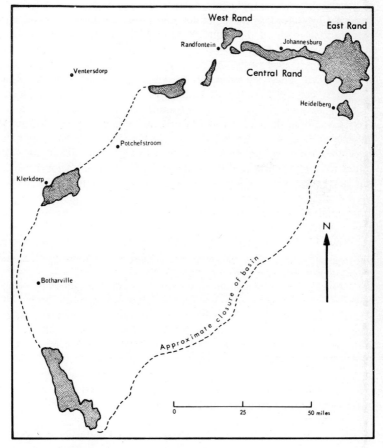

Fig. 10-2   Sketch map showing outline and major areas of the Witwatersrand basin, Republic of South Africa. Somewhat simplified after H. C. M. Whiteside and B. B. Brock, Commonwealth Mining and Metallurgical Congress, 7th, South Africa, Tr. vol. 2, p. 515. Copyright 1961. The South African Institute of Mining and Metallurgy. Used by permission.

the price was not $35 per oz during this entire period. During the year that ended June 30, 1963, the recovery per ton reported for 57 mines varied from 0.4 dwt to 20.4 dwt, but about two-thirds of the mines reported recovery between 3 and 8 dwt, or about $6.25 and $14 per ton (155).

The gold occurs in beds of the Witwatersrand system, which occupy a syncline something more than 100 miles long and about 50 miles wide that trends generally east-west. The Witwatersrand system, about 24,000 ft thick, consists of two divisions, Lower and Upper, the Lower divided into three series (Hospital Hill at bottom, Government Reef, Jeppestown), the Upper into two series (Main-Bird and Kimberley-Elsburg). The Lower Witwatersrand, about 15,000 ft thick, consists of alternating zones of shale or slate, quartzite, and minor conglomerates. The Upper Witwatersrand, about 9,000 ft thick, consists of quartzite, many beds of conglomerate commonly designated *reefs,* and a shale at the base. The beds are displaced by numerous minor faults, and dikes, chiefly basic ones, have been emplaced along many of the faults. Beds around the margins of the basin dip from 60° to nearly 90°. Dipping resistant beds form escarpments along their northern sides, and these gave rise to the name Witwatersrand, which means Ridge of White Waters, as the main watershed of the Transvaal passes along the Rand (12, 32).

Early concepts (1886) were that the reef depth would prove to be shallow, probably between 10 to 15 ft, and surprise was expressed when a shaft 54 ft deep was still on the reef. Small shafts sunk southward began to show that the dip flattened (30 to 40°); by September 1889 a shaft had been sunk to 250 ft; in December 1889 drilling cut reefs at depths of 517 and 581 ft; in 1891 poor gold values were obtained in reefs at a depth of 3,200 ft; in October 1892 a drill hole was started that showed gold values at depth of 2,391 ft; and in 1901 good values were established at a depth of 4,480 ft. In 1960 mining was being conducted at more than 11,000 ft vertically below the surface and operations were planned down to 12,500 ft (12, p. 497). Today the work of geologists has resulted in extension of the field from the Central Rand, where structural conditions are rather simple, westward and southward, and also eastward (Fig. 10-2), where structural conditions are more complex (39).

Gold occurs throughout the entire system but generally not in minable amounts in the Lower Witwatersrand. The conglomerates (reefs) of the Upper Witwatersrand contain the minable gold, and a gold-bearing reef may be designated a *banket,* from a Dutch word for the conglomerate because of its resemblance to an almond sweetmeat. The gold shows a tendency in some places to occur in pay streaks, some of which are as much as 5,000 ft long and 1,000 ft wide.

The gold-bearing conglomerates consist of somewhat oval pebbles generally less than an inch long composed chiefly of vein quartz in a matrix consisting of quartz grains and secondary quartz. Other grains include abundant zircon, fairly common chromite, minor monazite, xenotime, garnet, and even some small diamonds (12, p. 110). Pyrite is a conspicuous component of the conglomerates, being present in amounts ranging from 0.5 to 3.5 percent by weight; where abundant, it is recovered as a by-product for the manufacture of sulfuric acid. Pyrrhotite, small amounts of chalcopyrite, pentlandite, galena, and other sulfides are present in places, as well as various nonmetallic minerals—chloritoid, sericite, muscovite, calcite, and others. The pyrite is in well-defined crystals or aggregates and rounded bodies (buckshot pyrite), in grains of microscopic size to aggregates nearly an inch long.

Native gold, which contains an average of 10 percent silver, is practically confined to pyritic parts of the matrix of the conglomerate. Commonly it is in small particles, up to 0.1 mm across, which generally are jagged grains, although some are more or less rounded. The gold may be accompanied by carbonaceous material, which occurs in granules 2 mm or less in diameter and also in seams up to ⅝ in. thick. The prevalent type is a hydrocarbon-uranium mixture, possibly a variety of thucholite.[1] Uraninite is also present in crystalline form, generally in grains less than 0.08 mm across; also in rounded and oval shapes. Much of the uraninite is associated with hydrocarbon. Thus a sympathetic relationship commonly exists among gold, pyrite, and uranium, so that the content of one increases or decreases with that of the others.

The origin of the Witwatersrand gold deposits has been a matter of argument almost since their discovery. Perhaps as many words have been written about the origin as there have been ounces of gold produced, and, like the mining, there have been times of high and low production. As gold production reached a new high peak between 1960 and 1963, so did the flow of words, and along with the words the generation of much heat which, unfortunately, could not be used as a source of power in mining developments. No attempt will be made to summarize the vast amount of published material. Two major concepts have been advanced in various forms and with various modifications, and these can be stated as follows: (1) the gold, uranium, and much of the pyrite originally accumulated along with the pebbles and matrix of the conglomerate, and subsequent modification took place as a result of low-garde metamorphism, metasomatism, or other rearrangement without much transportation—the deposits are thus slightly modified syngenetic ones; (2) the gold, uranium, much of the pyrite and other sulfides were introduced by hydrothermal solutions of magmatic origin after the conglomerate had been deposited—the deposits are thus epigenetic ones.

Both of the concepts cited have been defended vigorously by various investigators. A summary published in 1959 by the Department of Mines, Geological Survey, Union of South Africa, contains a brief analysis of both concepts and ends with the following statement (12, p. 117):

> *On the whole, the consensus . . . among geologists with experience, in many cases extending over many years, of the gold-fields of the wider Witwatersrand, is that the gold and uraninite and part if not most of the pyrite were deposited with the sediments in which they are found, but that they have experienced a certain amount of metasomatic alteration since their deposition, without, however, having undergone any large-scale transportation during this phase.*[2]

[1] Thucholite is considered to be a hydrocarbon compound that contains large amounts of carbon and small amounts of water, uranium, thorium, rare earth oxides, and silica.

[2] From The mineral resources of South Africa, 4th ed., 1952, p. 117, Dept. Mines, Geological Survey, Union of South Africa. Extracts from official publications of the Republic of South Africa reproduced under Government Printer's Copyright Authority No. 3390 of 16/9/1964.

A strong contrast to this opinion is the conclusion of Davidson (8) who, in 1960, presented a discussion of the present state of the Witwatersrand controversy and cited 96 references dealing with various aspects of the problem. At the beginning of his discussion he stated that five lines of evidence should be considered: (1) age determinations from radioactive material; (2) comparison with modern placers; (3) geochemical evidence; (4) minerographic research; (5) geology. In an earlier paper (7, p. 688) he stated:

> *The evidence that the metals contained in the banket reefs are epigenetic to the reefs themselves is overwhelming; and it therefore becomes of interest to consider whether these ore deposits are epigenetic to the geosyncline as a whole, or whether they arise from mobilization and redistribution of the metals originally present in the geosyncline at the time of deposition of the sediments.[1]*

In the 1960 paper he concluded that the Witwatersrand basin as a whole had been much enriched in uranium and gold and that the uranium must be epigenetic not only to the reefs but also to the geosyncline—that is, it must have been introduced from without. He further stated that age determinations on radioactive material show that the uraninite is contemporaneous with granitization produced at the time of the emplacement of the Bushveld igneous complex and (8, p. 226) that the Witwatersrand mineralization resulted from the Bushveld magmatism.

Thus two opposing schools of thought exist at present. Whiteside and Brock (39, p. 519) summarized the situation well. They stated:

> *There has been no lack of detailed work, down to the most thorough-going chemical and microscopic analyses. . . . Not only has this not provided the answers but the same evidence has been used to support both sides of the conflict between syngeneticists and epigeneticists. The controversy has raged for many years and has in no way eased off with the recent focus on uranium.*
>
> *Yet many of us who have obtained our livelihood from geology on the Witwatersrand feel that if the facts could be assembled there would be little occasion for argument. . . .[2]*

They then pointed out that collection of the actual facts is difficult because of restrictions placed on release of information by many companies, the amount of new data that accumulates daily, and the difficulty of agreement on an acceptable group to bring out an entirely objective report from the data available. Recently there was established the Economic Geology Research Unit of the University of the Witwatersrand, and hope is expressed that progress toward a solution will be made. The first step was the publication of a bibliography (14).

Soviet Union

Estimates indicate that the annual gold production of the Soviet Union since 1955 has been about 25 percent of the world total, yet little information is

---

[1] From C. F. Davidson, *Economic Geology,* vol. 52, p. 688. Copyright 1957. Used by permission.
[2] From The role of geology on the gold mines of the Witwatersrand and the Orange Free State, by H. C. M. Whiteside and B. B. Brock, Commonwealth Mining and Metallurgical Congress, 7th, South Africa, Tr. vol. 2, p. 519. Copyright 1961. The South African Institute of Mining and Metallurgy. Used by permission.

available regarding the deposits. The most important fields appear to be in eastern Siberia. Among these are the Lena-Vitim, the Yenisei, the Aldan, and the Transbaikal (27, p. 35; 33, p. 173). The Lena-Vitim deposits consist of rich placers, apparently derived from gold-bearing pyritic schists (30; 41). These deposits held first place in Russian production from 1884 to about the period 1940 to 1945 (33, p. 168). The Yenesei deposits are gold-bearing quartz veins similar to those of Bendigo, Australia (13). The deposits of Aldan are placers; those of Transbaikal, both placers and lodes. Numerous other deposits, chiefly placers, occur in the Ural Mountains, in Kazakhstan, in central Asia, and the Far East.

A summary of the gold industry of the Soviet Union in 1961 (174) indicates that the most important production comes from the Magadan Province (Kolyma and Chukotka) of eastern Siberia, and the second most important production from the Aldan region (Yakutya), these two regions supplying about 60 percent of the total. Much of the production in these regions is from placer deposits.

Canada

Gold deposits are widespread throughout Canada, generally third in world gold production and furnishing about one-tenth of the total. Some production comes from every province, but by far the most comes from Ontario (about 58 percent in 1961) and Quebec (about 24 percent in 1961). The Northwest Territories furnished about 9 percent. Ontario has, over a period of years, furnished 75 to 85 percent of the Canadian total. About 85 percent of the gold comes from quartz veins, about 13 percent from base-metal deposits, and about 2 percent from placers, chiefly in the Yukon Territory where the entire production is from placers (167). This is distinctly different from the early Canadian production, in the 1860s and 1870s, during the great gold rush to the rich placers in the Cariboo district, British Columbia (150).

The outstanding region of gold deposits at present is the Porcupine–Kirkland Lake–Larder Lake–western Quebec area, extending from Timmins, Ontario, eastward to Louvicourt, Quebec, a distance of about 200 miles and embracing an area of perhaps 5,000 sq miles (11, p. 37). Throughout this area volcanic and sedimentary rocks of Early Precambrian age[1] have been complexly folded, sheared, fractured, faulted, and intruded by granitic and some gabbroic masses and dikes. Some large fault zones are known throughout this region, commonly termed *breaks* in some Canadian literature. Toward the east the Cadillac break is known for a distance of 25 miles or more; in the Larder Lake area the Larder Lake fault is at least 35 miles long. Toward the west is the Porcupine Creek fault zone. Other faults of considerable size are present along this entire zone, and many smaller faults are associated with them. Thomson, in 1948 (37, p. 630), stated that there had been speculation whether the Larder Lake fault may actually be a continuation of the Cadillac–Bouzan Lake fault that extends across northwestern Quebec. Regardless of the actual connection of various faults, it is known that this large area contains perhaps hundreds of faults and that the mines are clustered around the fault

---

[1] *Early Precambrian* is used here in the same manner as it is by the Geological Survey of Canada and includes rocks commonly designated *Keewatin* and *Temiskaming*, although the exact meaning of the latter two terms is not always clear.

zone, extending generally eastward from Timmins, Ontario. Thomson (37, p. 631), in a discussion of this association and the speculation that some of the major faults may be connected, suggested also that the distribution of productive mines throughout 150 miles in Ontario and Quebec near the Larder Lake–Lake Bouzan–Cadillac fault tended to support the hypothesis of some regional structural control.

Only a few of the outstanding districts and mines throughout this great gold zone can be mentioned briefly. In the Porcupine district, around Timmins, Ontario, which is the leading gold-producing district of Canada, 13 mines operated in 1961, among them the famous Hollinger Consolidated Gold Mines, Ltd., from which the production in gold and silver had the value of $363,638,315 from June 1912 through December 1945 (18, p. 464). The deposits in the Porcupine district are quartz veins and lodes that contain native gold in which there is some silver; some veins contain scheelite, which is recovered as a by-product. Some sulfides, chiefly pyrite and pyrrhotite, are present also (9). At the Hollinger mine (18) individual ore bodies are 1 to 70 ft wide but average about 10 ft. In one favorable zone there is a vein system as much as 6,000 ft long and 4,000 ft wide. The deposits of the Kirkland Lake–Larder Lake area are also chiefly gold-bearing quartz veins that were localized in favorable structures. Brittle, competent rocks adjacent to faults and folds fractured readily and produced favorable places for ore deposition (37, p. 632). Fissure veins, formed in the crushed and shattered zones associated with faults, are the most common type and are responsible for the larger ore bodies (38). Other mines farther east, although differing in detail, are also of the gold-quartz type except the Noranda Mines, Ltd., and a few others in the same vicinity in Quebec, near the central part of the zone. Noranda is a base-metal mine, as are a few of the others in the Quebec area, yet about two-thirds of the gold of Quebec was recovered from these base-mental deposits in 1961. The principal base metal recovered is copper. These deposits are, in general, of two types: (1) sulfide replacement deposits, and (2) gold-bearing veins. The first type is by far the more important. In this type the ore consists mainly of chalcopyrite, pyrrhotite, and pyrite, with varying amounts of sphalerite. The deposits are associated with faults and replace fractured rhyolitic and andesitic rocks. Apparently localization was aided by associated dense flow rocks or intrusive rocks that may have served as barriers (40).

Throughout the area some features of the deposits, such as the presence of pyrrhotite, tourmaline, and some arsenopyrite, indicate that they belong to the high-temperature hydrothermal class, yet the presence of some tellurides in places might indicate lower temperature during at least some part of the deposition.

Two other occurrences of gold in Canada should be discussed very briefly, one at the Manitoba-Saskatchewan border, the other in the district of Mackenzie, Northwest Territories. At the first location is the Flin Flon base-metal deposit, which furnishes much of the gold of Manitoba and Saskatchewan. The deposit, a replacement type, is localized in Early Precambrian sheared volcanic rocks. The principal ore bodies are masses of solid sulfides, chiefly pyrite and minor amounts of chalcopyrite and sphalerite. The deposits furnish gold, silver, copper, zinc, and some cadmium, tellurium, and selenium (11, pp. 66–69; 21). In the district of Mackenzie is the Yellowknife district, where the chief producer is the Giant Yellowknife Mines, Ltd. In the Yellowknife district, early Precambrian volcanic and sedimentary rocks have been complexly deformed, sheared, fractured, and faulted, and intruded by rocks of granitic and gabbroic composition. The gold-bearing deposits of most importance occur in quartz-carbonate lenses and quartz veins localized in schist zones in greenstone. Less important deposits occur in metasedimentary rocks. Gold occurs as the native metal and to some extent as gold antimonide, aurostibnite. Quartz is the principal gangue mineral, but moderate amounts of ankerite and calcite are present in places. Among metallic minerals are pyrite and arsenopyrite, which are ubiquitous and present in large amounts; sphalerite and chalcopyrite, widespread but generally not present in large amounts; also various sulfosalts containing antimony along with lead, copper, or iron (3, 6, 15, 16).

Agreement does not exist regarding the origin of the Yellowknife deposits. Some investigators (139, 140, 168) concluded that the ore and gangue elements were present in the country rocks and were concentrated in the deposits by metamorphic agencies and the dilatancy imposed by the great shear-zone systems. Others (6, 134) concluded that the deposits were of hydrothermal origin. It seems well to point out that these differences of interpretation regarding origin come from use of the same data. Further discussion is beyond the scope of the present treatment.

United States

The total production of gold in the United States, excluding Alaska and Hawaii, to the end of 1958 was 267,360,000 oz, and about 77 percent of this came from four states—California, Colorado, South Dakota, and Nevada (Table

**TABLE 10-2. Summary of Most Important
Production of Gold in the United States,
Exclusive of Alaska and Hawaii, through 1958***

Important producing areas	*Ounces*
Entire United States	267,360,000
FOUR LEADING STATES	
California	105,435,000
Colorado	40,425,000
South Dakota	27,120,000
Nevada	26,775,000
Total of above four states	199,755,000
EIGHT LEADING DISTRICTS	
Lead, South Dakota	23,875,000
Cripple Creek, Colorado	19,088,000
West Mountain (Bingham), Utah	10,184,000
Grass Valley–Nevada City, California	9,822,000
Comstock, Nevada	8,620,000
Mother Lode–East Belt, California	7,507,000
Columbia Basin, California	7,500,000
Hammonton–Yuba River, California	4,387,000
Total of above eight districts	90,983,000

* From A. H. Koschmann and M. H. Bergendahl, 1962, *U.S. Geol. Survey Mineral Inv. Resource* Map MR-24.

10-2). Eight districts alone furnished somewhat more than 90 million oz and thirty or more other districts in nine states supplied more than 1 million oz each.

Placer deposits furnished most of the gold of the United States until 1873, but since then lode deposits have supplied more gold than placers. By-product gold from base-metal ores formed but a small part of the total production until 1943, after which production from base-metal deposits increased and, since 1951, has been greater than that from placers (151, 152). The Homestake Mine, Lead, South Dakota, has been the leading gold mine of the country since 1946, and the Utah Copper Mine, Bingham, Utah, has ranked second. Other important mines in 1961 were (3) Knob Hill and Gold Dollar, Republic district, Washington, gold ore; (4) Yuba unit, Yuba River district, California, dredge alluvial ore; and (5) Copper Queen–Lavendar Pit, Warren district, Arizona, copper ore. Important production comes also from deposits in Alaska, Colorado, Nevada, and Montana. Here we will be concerned only with deposits of dry and siliceous ores and placers. Gold-bearing base-metal ores, chiefly copper deposits, will be considered under the appropriate base metals.

South Dakota.   The Homestake mine, Lead, South Dakota, in the northern part of the Black Hills, is in folded Precambrian schists that are surrounded by outward-dipping Paleozoic and Mesozoic rocks. Intrusive igneous rocks of Tertiary age are moderately abundant in the northern part of the Black Hills. The Homestake deposit became known in 1876, apparently as an indirect result of the working of rich gold placer deposits in Deadwood Gulch, which, however, seem to have been derived from outcrops of gold-bearing Cambrian conglomerate (28, p. 222).

The deposits are selective replacements of the Precambrian Homestake formation and are confined to a belt less than a mile wide. The original rock of the Homestake deposit was an iron-magnesium bed perhaps about 60 ft thick (25, p. 564) that possibly contained chert lenses. This was metamorphosed to cummingtonite schist in part of the area then altered hydrothermally to chlorite rock in the ore zones. Five ore bodies are present, each of which is a somewhat steeply inclined spindle-shaped pipe (28, p. 225). These ore bodies are thought to have been localized in zones of cross folding superimposed on earlier isoclinal folds. They consist of veins and masses of quartz in the chloritized rock, accompanied by pyrrhotite, pyrite, and some arsenopyrite, as well as a number of other less-abundant gangue minerals. Gold is present as the native metal, some of which is observable but part of which is known only from assay. Opinion is divided about the age of the gold mineralization. Some think it is Precambrian, others prefer Tertiary, and some think it was in part Precambrian and in part Tertiary. McLaughlin (25, p. 565) concluded that the deposits are of Precambrian age and of hypothermal origin.

California.   The total value of the gold produced in California between 1848 and 1954 was more than 2.3 billion dollars and exceeded that of any other state (5, p. 215). The average silver content of California gold is about 12 percent, so much silver was produced along with the gold. Gold occurs throughout the entire length of California, but deposits are concentrated especially in two regions (Fig. 10-3): (1) the great western Sierra Nevada region, in which are the famous Mother Lode

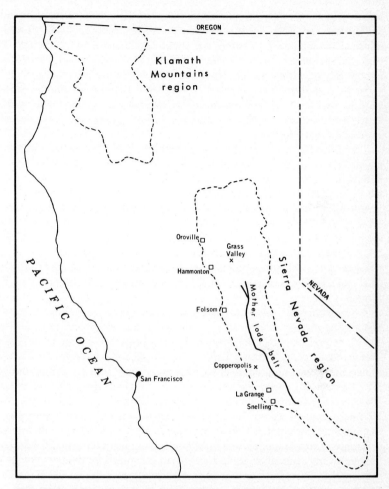

Fig. 10-3 Map of central and northern California showing extent of major gold regions. Chief dredging operations shown by small squares. After W. B. Clark, *Mineral Commodities of California, Bull.* 176, California Dept. Nat. Resources, Div. Mines. California Division of Mines and Geology. Used by permission.

and extensive placer deposits, and (2) the Klamath Mountains, in which are both lode and placer deposits. Only the Mother Lode region will be discussed.

The Mother Lode system of California (5, pp. 215–218; 19, 20) is a great system of *en echelon* to linked steeply dipping veins about 120 miles long and a mile wide on the western foothills of the Sierra Nevada. This vein system traverses steeply dipping slates, schists, and greenstones, and some intrusive masses of serpentine. The slates and schists were originally sedimentary rocks deposited in part during Mississippian time and in part during Upper Jurassic time, and with these are associated some volcanic rocks (green schists and greenstones) and some intrusive igneous rocks varying in age from late Paleozoic to early Jurassic. During late Jurassic or early Cretaceous time the rocks were isoclinally folded and then intruded by magma of various types but chiefly granitic. Although the granitic rocks are prominent in the higher parts of the Sierra Nevada, they occur in but minor amount within the Mother Lode belt, where the principal igneous rocks are serpentines derived from peridotites. The gold deposits of the Mother Lode belt were formed at the end of this period of late igneous activity. They consist of (1) quartz veins and (2) bodies of mineralized country rock. Generally they are classed as mesothermal deposits.

The quartz veins, which vary in thickness from less than a foot to 50 ft and may reach a length of several thousand feet, generally consist chiefly of coarse milk-white quartz in lenticular masses that pinch and swell. At the edges and ends the lenticular veins pass into quartz stringers admixed with country rock. The veins contain minor amounts of ankerite and albite, about 1 to 2 percent of sulfides (chiefly pyrite), native gold, and, rarely, the gold telluride petzite. The ore occurs in shoots, which generally are much shorter than the veins, stope length averaging 200 to 300 ft. In general the gold-quartz ore has averaged only $10 to $20 per ton, although some individual shoots or parts of shoots have yielded $30 to $40 per ton. The more productive quartz veins are characteristic of the northern part of the Mother Lode.

The bodies of mineralized country rock are of somewhat diverse character but may be classed roughly as (1) auriferous greenstone and (2) auriferous schist. They are present either adjacent to quartz veins or in broad zones of fissuring. The mineralized greenstone (gray ore), formed by hydrothermal alteration and replacement, is composed chiefly of ankerite, some sericite, albite, and quartz, and 3 to 4 percent of pyrite and arsenopyrite, with which the gold is closely associated. The gray ore occurs in shoots, many of which are large and average $8 per ton, although some contain as much as $20 per ton. The mineralized schists, originally amphibolite and chlorite schists but altered to pyritic sericite-ankerite schist, generally are of low grade, $2 to $3 per ton, but in places have made very large bodies. Generally they are of a grade too low to be mined profitably, but at one place, Carson Hill, the schist was highly mineralized, rich ore assayed as much as $40 to $50 per ton, and the deposit yielded more than 5 million dollars in gold within a few years, although the ore body did not persist and was soon worked out.

Lode gold deposits occur also east of the Mother Lode in an area designated the *east belt* and west of the Mother Lode in an area designated the *west belt*. In general the veins of the east belt are narrower than are those of the Mother Lode, the ore shoots are smaller, but the gold content is higher—an average of more than $50 per ton in many of them.

At present the two largest and most productive lode gold mines in California are in the Grass Valley–Nevada City district, about 50 miles north to northwest of the Mother Lode. There gold-quartz veins occur in two main systems in granodiorite and metamorphic rocks. Most veins are less than 10 ft thick and contain native gold and some sulfides, chiefly pyrite. One vein, the Empire, has been worked to an inclined depth of more than 11,000 ft (5, p. 217).

The erosion and reconcentration of lode gold deposits in the general Mother Lode region resulted in the formation of important placer or alluvial deposits (5,

p. 218; 23), which may be placed in two categories: (1) older deposits, of early Tertiary age, and (2) younger deposits, of Quaternary age. Eluvial placers are common in some parts of the Mother Lode belt. Collectively placers have furnished more than 40 percent of the gold of California, and some of the Quaternary placers were the cause of the early gold rush (1849).

The older placers were formed as streams cut downward during Cretaceous and early Tertiary times, but toward the end of the Tertiary, rhyolite flows filled many valleys and covered the gold-bearing gravels; then the streams of that time changed their courses, cut downward, and repeatedly crossed old courses and reconcentrated gold in new canyons. Later, enormous volumes of andesitic tuff buried a large number of streams and completely filled the valleys. During late Pliocene and early Quaternary times rapid erosion proceeded in the nonresistant tuffs, and new streams cut canyons to a depth of 1,000 to 4,000 ft below the surface of the volcanic rocks. Some of the Tertiary gravels were uncovered and in part were left stranded in terraces and in part were reconcentrated, but in many places the old gravels remained buried in former Tertiary channels. The gold from some of the buried placers has been recovered by drift mining. At the Mayflower mine a tunnel 4,740 ft long was used to work the main channel for a distance of 3 miles, chiefly from 1888 to 1894. A bed of gravel 2 to 14 ft thick, with an average width of 75 ft, was removed and yielded approximately $1,500,000—about $7 per ton of loose gravel (23, p. 151). At the Hidden Treasure mine an old channel was found beneath a lava cap, and this channel had been filled about 200 ft deep with uncemented gravel, sand, and clay (23, p. 152). At some places the gravels of buried placers had become cemented, and these caused much more difficulty in mining.

The younger (Quaternary) placers were of two general types: (1) very rich deposits of relatively small size and (2) relatively low-grade deposits of great extent. The rich placers generally were soon mined out during the early gold rush. The maximum production of gold in California was in 1852, when the deposits yielded almost 4 million oz of gold, nearly all from placers. The low-grade deposits are now extensively worked by dredging, especially along the Feather River near Oroville, the Yuba River near Hammonton, the American River near Folsom, the Mokelumne River near La Grange, and the Merced River near Snelling (Fig. 10-3). The value of the gold in many of the deposits that are dredged is 25 cents or less a cubic yard, but individual dredges treat vast amounts of gold-bearing material, as much as 5 million cu yd a year (5, p. 218), and this amount, at an average gold value of 25 cents per cu yd, would yield $1,250,000 in gold.

Other Western States. Throughout the western United States, especially in Nevada, Colorado, California, and Idaho, and to a lesser extent in Arizona, New Mexico, Utah, and Washington, are many deposits that are classed as gold-silver or silver-gold occurrences (29). In general in the gold-silver deposits the ratio of gold to silver is more than 1:1, possibly close to 7:3, which is the ratio in Nevada (29, p. 627), whereas in the silver-gold deposits the ratio of gold to silver varies between wide limits, from 1:100 to almost 1:1. These deposits are all in regions of Tertiary volcanism, although the deposits may be in older rocks and are not limited to volcanic rocks. Generally the gangue of the deposits as a whole, including both types, is fine-grained quartz, in places closely banded, and either adularia or alunite may be present in the ore body or the wallrock. The ore minerals include native gold, electrum, various tellurides of gold and silver, some selenides, argentite, numerous sulfosalts, and some base-metal sulfides. Certain differences in mineralogy characterize the two types, but that detail is omitted here. Both types are considered to be of epithermal origin.

In the aggregate, deposits of this type furnish much gold and silver, although at present (1963) but two mining districts are among the leading gold producers of the United States—only the Republic district, Washington, rank third in 1961, ratio of gold to silver about 1:8, and the Cripple Creek district, Colorado, rank eleventh in 1961, ratio of gold to silver about 37:4 for total production to the end of 1946. Among some of the early noted mines are Goldfield, Nevada, ratio of gold to silver about 3:1; Tonopah, Nevada, ratio of gold to silver about 1:100, hence predominantly a silver deposit; and the fabulous Comstock Lode, Nevada, especially famous from 1860 to 1880, ratio of gold to silver about 1:40. Cripple Creek, the Republic district, and Goldfield will be discussed briefly here; Tonopah and Comstock with the silver deposits.

The Cripple Creek district, Teller County, Colorado (22, 24; 29, pp. 631–632), discovered in 1891, had its peak production in 1900 with an output of gold valued at $18,149,645 (878,067 oz at a price then of $20.67 per oz); produced 18,572,959 fine oz of gold and 2,089,367 fine oz of silver to the end of 1946; and was second to the Lead, South Dakota, district for production of a single district through 1958 (151). At one time about 100 mines operated in the district, and 12 mines still were operating in 1961. In this district volcanic eruptions during the Tertiary period broke through Precambrian granite, gneiss, and schist; and one of the sites of these eruptions became the center of localization of the Cripple Creek deposits, which occur in a complex crater or caldera. Although this site has been described as a volcanic neck, Koschmann (22) stated that the basin was formed by intermittent subsidence along steeply dipping faults. Volcanic fragmental rocks are the main ones that occur in the basin. The ore deposits occur chiefly in long, narrow zones of fissures (sheeted zones) rarely more than half a mile long and varying from a few inches to over a hundred feet wide. The persistence of these zones down the dip is roughly proportional to their length. The main gold mineral is the telluride calaverite, but sylvanite and petzite are common. The ore occurs chiefly as a filling of narrow fissures and generally varies in tenor from 1 to 3 or 4 oz per ton, but small amounts mined have contained as much as 2,500 oz per ton.

Goldfield, Esmeralda County, Nevada (29, p. 627; 31), which furnished 4,144,000 oz of gold through 1958 (151), began producing in 1903, and in five years furnished 954,446 oz of gold and 116,188 oz of silver, the total value of which at that time was $19,804,680, an average value of $105.29 per ton. Some of the early ore mined was remarkably rich—$6,000 to $7,000 per ton—but also the ore bodies were extremely irregular and not persistent. Much of the gold was present as the native metal associated with cryptocrystalline quartz. The deposits formed along zones of fissuring that contained fractured and altered country rock, around which the ore minerals occurred in concentric shells.

The Republic district, Ferry County, Washington (17; 29, p. 640), has been the leading gold producer of Washington. Lode gold production in 1961 at the Knob Hill Mines, Inc., of this district, was rated second only to Homestake, South Dakota, although actual production was much less than at Homestake. The ratio of gold to silver was nearly 1:5 in 1961, and most of the silver produced in Washington came from the Republic district. The ore occurs in veins that cut flows, lake deposits, and minor intrusive rocks of Tertiary age.

Alaska. More than 99 percent of the gold produced in Alaska in 1961 came from placer deposits, nearly 40 percent of the total being furnished by the Fairbanks district and about 33 percent by the Nome district. The total production in 1961 was 113,457 oz of gold and 14,788 oz of silver, valued at $3,984,666 (173). This entailed the washing of 11.1 million cu yd of gravel, from which the average recovery was 0.357 oz per cu yd. The value of the gold and accompanying silver furnished by Fairbanks and Nome to the end of 1930 was about 147 million dollars (159). In the Fairbanks region the placers are (1) residual and hillside and (2) stream, the latter being the common type. The stream placers include (a) ancient stream deposits, both bench and buried, and (b) recent stream deposits. Most of the richer placers have been the ancient stream type. The Nome placers of the Seward Peninsula include beach placers and stream placers. The beach placers, present at several levels, extend along the shore of Bering Sea for about 30 miles and were worked extensively in 1899 and 1900. Stream placers, however, appear to be more important.

Lode gold deposits are known throughout a considerable area and are the source of the placer gold. Some rich pockets have been located and mined, but most of the lode deposits are of relatively low grade. Two notable low-grade lode deposits were extensively worked, the Treadwell deposits on Douglas Island and the Alaska Juneau deposits (142, 160, 161). These deposits were somewhat similar and in general consisted of small, closely-spaced gold-bearing veins of quartz and calcite that cut slates and diorite dikes (Treadwell). The tenor at Treadwell averaged about $2 per ton, and at Alaska Juneau as low as $1.12 per ton in 1932 from an output of 11,000 tons of ore per day.[1]

A compilation in 1962 (143) of the occurrences of lode gold in Alaska shows two major areas where mining has occurred: (1) southeastern Alaska, mostly north and northwest of Sitka, and (2) an area around Valdez and Anchorage. Only one mine, the Juneau, produced more than 1 million fine oz of gold. Two mines produced between 250,000 and 1 million fine oz; three produced between 50,000 and 250,000 fine oz; and all others produced less than 50,000 fine oz each.

Australia

Australia supplied only 2.8 percent of the world's gold in the period 1945 to 1959 and but 2.4 percent in 1960, but from 1856 to 1860 Australia produced nearly 41 percent of the world total, the average production being 2,652,125 oz

[1] An interesting historical account of the Treadwell and Alaska Juneau operations was given by T. A. Richard, 1932, "A History of American Mining," pp. 58–62, 65–81, McGraw-Hill Book Company, New York, 419 pp.

per year. The peak production was attained in the period 1901 to 1905, with an average annual supply of 3,600,000 oz, which then was 23 percent of the world total. After 1903, the year of greatest production (3,836,000 oz), the decline was somewhat rapid, so that the average for the 1921 to 1925 period was but 700,371 oz, only about 4 percent of the world total (158). Even though production has declined, Ballarat and Bendigo, of Victoria, and Kalgoorlie and the Golden Mile, of Western Australia, remain names that bring to mind times of fabulous riches. The Australian production of recent years has been chiefly from Western Australia, most of which has come from the Kalgoorlie field from ore containing about 0.30 oz of gold per ton.

Victoria, an intensely mineralized gold province, yielded, to the end of 1950, about 73,201,000 oz of gold, about 45 percent of the Australian production to that time (36). The peak production, 3,055,744 oz, occurred in 1856, but since then the decline has been steady, with production of only 67,825 oz in 1950; and only a few mines operated continuously during 1960. The lode gold of Victoria occurs chiefly as the native metal in quartz reefs that are widespread in Lower Ordovician rocks and probably were formed during pre-Middle Devonian earth movements and occurs subordinately as gold-bearing sulfide ores that may have been formed somewhat later, during pre-Upper Devonian mineralization (36, pp. 979–980). Rich alluvial deposits formed during Tertiary and Recent times, and some of them were buried beneath widespread Tertiary basalt flows and were preserved as "deep leads." The two outstanding districts of Victoria have been Ballarat and Bendigo (Fig. 10-4), which were discussed by Lindgren [Chap. 2 (10, pp. 556–560)] with the mesothermal group of hydrothermal deposits.

The Ballarat field (1), the principal gold-bearing part of which is 5 miles long and 3.5 miles wide, has been noted for (1) exceptionally rich shallow alluvial deposits which passed into buried placers (deep leads) that were followed to a depth of 500 ft beneath basalt flows, were themselves phenomenally rich in places (one famous nugget, the "Welcome," weighed 2,215 oz), and were mined during the first 10 years of the district; (2) the Indicator bed; and (3) the "leather-jacket formations." The Indicator bed was composed of slate, which was extremely thin in places and generally not more than $\frac{1}{8}$ in. thick. Locally it was replaced by a thin film of quartz (1, p. 996). The Indicator was the principal bed in the Ballarat district, along which enrichment of ore occurred, although several less important beds were known. It was worked for a length of 4 miles and to a depth of over 1,000 ft. At its intersection with quartz veins it yielded "slugs" of gold, one of which weighed 444 oz. The leather-jacket formations were formed by a combination of folding and faulting. The district is folded into a series of anticlines and synclines and cut by a series of reverse faults that dip about 45°W. Where these faults intersect steeply dipping beds they form "leather jackets," so called from the leathery nature of the pulverized rock in the fault planes. Along the leather-jacket faults there is a series of more or less lenticular quartz formations, the leather-jacket formations, that extend diagonally upward, commonly for several hundred feet (Fig. 10-5). These, with spurry[1] additions, were up to 100 ft wide and recurred at roughly regular intervals of 250 to 300 ft. They were mineralized and contained small amounts of sulfides in addition to native gold. The leather-jacket formations

---

[1] A spur is "a branch leaving a vein, but not returning to it." [(5, p. 640) of Chap. 2.]

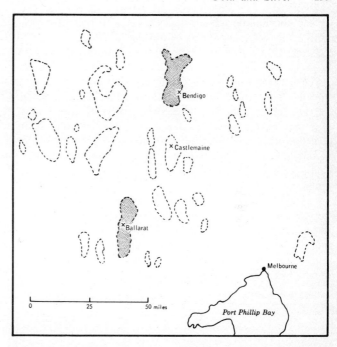

Fig. 10-4 Map showing the general extent of the Ballarat, Bendigo, and other goldfields of Victoria, Australia. Simplified and generalized after D. E. Thomas, 1953, in "Geology of Australian Ore Deposits," p. 976, Fifth Empire Mining and Metallurgical Congress, Australia and New Zealand. Australasian Institute of Mining and Metallurgy.

generally were not barren at any place, but the gold was unevenly distributed. Some rich shoots occurred at intersections with favorable beds (the indicators). In one place throughout a length of 150 ft there occurred as many as 15 nuggets weighing from 25 to 100 oz each. Also, slugs of coarse gold weighing as much as 500 oz were reported (1, p. 996).

The Bendigo field (35) is similar to Ballarat. Much alluvial gold was produced at first, 200,000 oz from the time of discovery in 1851 to the end of that year and more than 475,000 oz in 1859. This alluvial supply gave place in part to production from lodes, and combined production then was between 190,000 and 241,000 oz a year between 1860 and 1870; it reached a peak of 329,500 oz in 1873, oscillated between 148,700 oz (1889) and 249,000 oz (1904), continued to exceed 170,000 oz until 1912, then declined. The total production to the end of 1951 was estimated to be 22,360,000 oz (35). There were several hundred individual mines in the district, which was about 15 miles long and 3 miles wide. In general the maximum depth of mining was 2,000 to 2,700 ft, but a number of mines reached a depth of 3,000 to 3,500 ft, and one, the Victoria Quartz, 4,613 ft. The district is characterized by more than 20 anticlines within a width of 3 miles, but five adjacent ones and one farther east were the most productive. Saddle reefs are the outstanding type of deposit, but with these are associated trough reefs, leg reefs, fault reefs, and spurs. According to Thomas (35, p. 1022) the early miners recognized the saddle-reef type of deposit and used the term *saddle reef* commonly in the 1860s. The reefs generally are not ideal saddle-shaped structures that conform to the bedding and fill an opening at the crest of an anticline, but rather they tend to occur where two strike faults that dip in opposite directions intersect in the crest of an anticline. The cross section of a reef may be roughly triangular or it may be asymmetrical if one of the intersecting faults was more prominent than the other one. Repetition at irregular intervals in depth in the axial planes of anticlines is a characteristic feature of these structures. Twenty-four successive saddle reefs were encountered in a single mine, and they were known to the greatest depth explored, 4,613 ft.

Trough reefs were not as numerous nor as productive as saddle reefs; leg reefs extended down the bedding plane from the saddle or occurred in the bedding plane below the saddle; and fault reefs occupied fissures cutting across the bedding and generally were not of economic importance. Spurs were important at Bendigo just as they were at Ballarat. The quartz veins contain, in addition to gold, small amounts of ankerite, sericite, calcite, rarely albite, and some sulfides, in places as

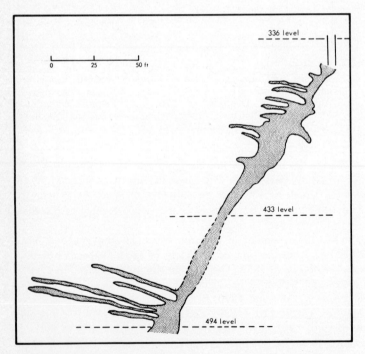

Fig. 10-5 Section through Last Chance Mine, Ballarat East, Victoria, Australia, showing leather-jacket formation and spurs, Pug lode. Simplified after W. Baragwanth, 1953, in "Geology of Australian Ore Deposits," p. 997, Fifth Empire Mining and Metallurgical Congress, Australia and New Zealand. Australasian Institute of Mining and Metallurgy.

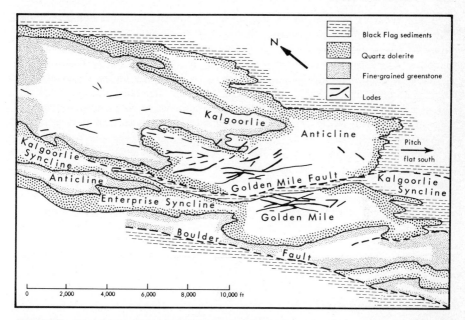

Black Flag sediments

Quartz dolerite

Fine-grained greenstone

Lodes

Fig. 10-6 Geologic map of the Kalgoorlie gold district, Western Australia, showing major gold lodes. Simplified and somewhat generalized after J. D. Campbell, 1953, in "Geology of Australian Ore Deposits," p. 82, Fifth Empire Mining and Metallurgical Congress, Australia and New Zealand. Australasian Institute of Mining and Metallurgy.

much as 1 to 2 percent. Much of the quartz shows lamination or banding caused by remnants of black slate. All of the Bendigo quartz reefs probably contained some gold, but an outstanding feature of the district was irregular distribution of gold in shoots, as much as 3 oz or more per ton, and this caused much unprofitable work prospecting for commercial parts of the reefs.

Many suggestions have been made about the formation of the saddle reefs and other reefs of the Bendigo district. Early concepts regarded filling of cavities formed during folding as the major process; others favored (1) replacement, (2) cavity filling and replacement, (3) fracturing with formation of open spaces and later cavity filling or replacement (162, 163, 164, 165).

The Kalgoorlie gold field (4, 10, 34) of Western Australia (Fig. 10-6) has been noted for the lodes of the Golden Mile, which were responsible for almost all of the gold produced in the field. The Golden Mile really is 6 miles long and about half a mile wide. Rich alluvial gold of small extent was discovered about 3 miles to the north in 1893, but important lode mining did not start until about 1897. Between 1897 and 1950 the amount of ore treated was 55,509,538 tons from which 27,149,234 oz of gold was recovered, an average recovery of 9.8 dwt[1] per ton (4, p. 79). The maximum amount of gold recovered in a single year, 1903, was 1,012,151 oz from ore that averaged 22.8 dwt (1.14 oz) per ton. The recovery in 1950 was 394,696 oz from ore that averaged 4.9 dwt (0.245 oz) per ton.

[1] Equivalent to 0.49 oz per ton, since 1 dwt is equal to 0.05 oz.

The rocks of the Kalgoorlie area, of Precambrian age, consist chiefly of the older and younger greenstone series. A chloritic phase of a quartz dolerite greenstone of the younger greenstone series contains the bulk of the profitable gold deposits (4). The rocks are tightly folded and the structure of the Golden Mile is that of a south-pitching syncline. The lodes, which are pyritic replacements and impregnations, occur in steeply dipping fractures that generally are parallel to the broad trend of the quartz dolerite greenstone. Both native gold and tellurides are present, along with pyrite and some other sulfides, carbonates, and minor quartz. Ten species of tellurides are known, of which coloradoite, HgTe, calaverite, $AuTe_2$, and perhaps krennerite $(Au, Ag)Te_2$, are the most abundant (34). The gold tellurides have produced spectacular bonanzas (4, p. 80). In the oxidized zone the tellurides decomposed and yielded "mustard" gold. Lindgren [(10, pp. 678–682) of Chap. 2] discussed the Kalgoorlie deposits with the hypothermal group, but Stillwell (34, p. 124) stated that the tellurides probably were deposited over a very limited range of moderately low temperatures and that the deposits as a whole apparently formed over a temperature range from 500 to 149°C with deposition taking place throughout a considerable period of time, the tellurides forming late in that period.

Other Countries

Some gold is produced by almost every country, but major amounts have been furnished by only a few in addition to those already discussed. Colombia, from 1601 to 1700, supplied 39 percent of the world's gold (158). During the sixteenth century it was the country to which the Spanish adventurers and explorers traveled in search of El Dorado, the land of gold. Both lode and placer deposits are present, the most important being in the departments of Antiquia, Caldas, and Golima, west central Colombia (26, pp. 354–355). The value of the gold produced to the end of 1886 was estimated at more than 700 million dollars, over two-thirds of which came from placers. Production since about 1895 has generally been less than 1 percent of the world total. The Kolar gold field, Mysore, India, is stated to have produced gold more than a thousand years ago (156). Modern records of production begin with 1882. Between 1891 and 1925 the Kolar field supplied between 2 and 3 percent of the world's gold (158), but since 1940 the amount has been less than 1 percent. The deposits are gold-quartz lodes and gold-bearing sulfide lodes. Mining was being conducted in 1960 over a strike length of 5 miles and to an average depth of 9,000 ft (156). Gold mining in the Philippine Islands antedates written history, but no great progress was made until after 1931. In 1936 the production was greater than that of South Dakota and Alaska, but since then it has declined very much. An interesting and important single mine in Brazil is Morro Velho, at Nova Lima in Minas Gerais. Production began from an open pit, possibly about 1784, and after many vicissitudes underground mining reached a depth of 8,051 ft in 1934, but later work has been at higher levels. Between 1834 and 1955 recovery was 7,319,400 oz of gold from 13,840,000 tons of ore. Reserves at the end of 1955 were estimated at 4,527,108 metric tons of ore containing an average of 13.2 g of gold per ton. The deposits are hydrothermal replacements of bedded quartz-ankerite-dolomite rock and consist of gold associated with sulfides (147).

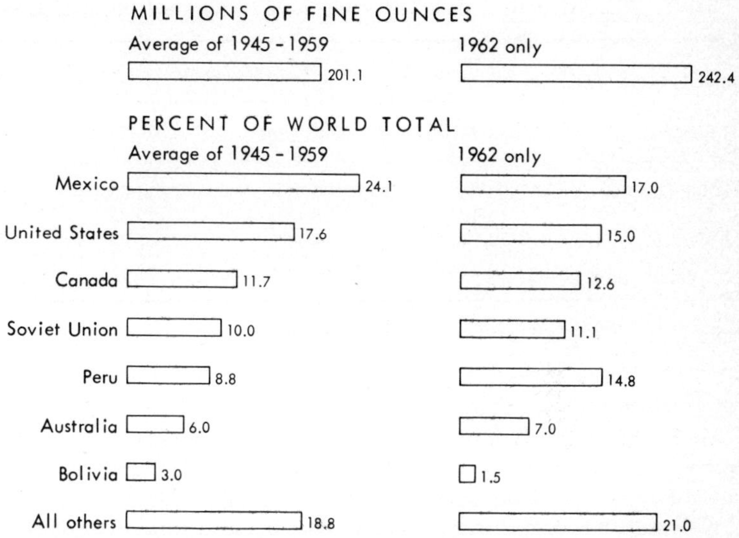

MILLIONS OF FINE OUNCES

Average of 1945 - 1959

201.1

1962 only

242.4

PERCENT OF WORLD TOTAL

Average of 1945 - 1959

1962 only

	Average of 1945-1959	1962 only
Mexico	24.1	17.0
United States	17.6	15.0
Canada	11.7	12.6
Soviet Union	10.0	11.1
Peru	8.8	14.8
Australia	6.0	7.0
Bolivia	3.0	1.5
All others	18.8	21.0

Fig. 10-7 World silver production. Compiled from *U.S. Bur. Mines Minerals Yearbook,* various years.

Silver

The combined production of four countries—Mexico, the United States, Canada, and Peru—furnished 59.4 percent of the world's silver in 1962 and 62.2 percent in the 15-year period 1945 to 1959 (Fig. 10-7). The production of those four countries ranked in the order stated, but in 1961 Peru supplied more silver than Canada and took third rank, which it maintained in 1962 with production almost equal to that of the United States. Mexico still held the lead in 1962 with 17.0 percent. Figure 10-7 indicates that the Soviet Union held fourth rank in production for the period 1945 to 1959 and fifth rank for 1962. The production figures for the Soviet Union are merely estimates, however, and have been listed by the U.S. Bureau of Mines as 25 million oz each year from 1953 through 1961.

During the early history of silver production, between 72 and 88 percent of the world total was supplied by Mexico, Bolivia, and Peru, as shown by Table 10-3. The United States became an important silver producer about 1860, as a result of the exploitation of the Comstock lode, and for the period 1851 to 1900 furnished 33.15 percent of the world total, contrasted with 31.21 percent by Mexico; and for the period 1901 to 1925 the United States supplied 31.31 percent of the total, whereas Mexico produced 32.94 percent. Canada did not become a very large producer until after 1905. The Cobalt district, discovered in 1903, brought about a great increase in production, so that Canada supplied 10.49 percent of the world total in the period 1906 to 1910, and 9.47 percent for the entire period 1901 to 1925, during which time it ranked third in world silver production (103).

Many interesting silver deposits occur throughout the world, but only a few of them can be discussed here.

TABLE 10-3.   Countries Most Important for Silver Production, 1493–1850*

Country	Percent of world total			
	1493–1600	*1601–1700*	*1701–1800*	*1801–1850*
Bolivia	47.58	35.92	11.66	9.52
Peru	12.63	26.14	19.65	14.96
Mexico	12.10	24.11	57.00	56.85
Total	72.31	86.17	88.31	81.37

* From *U.S. Bur. Mines Econ. Paper* 8, pp. 8–10, 1930.

Mexico

About three-fourths of the silver produced in Mexico in 1961 was a co-product or a by-product of lead, zinc, and copper ores, and about one-fourth was from silver ores in the states of Hidalgo, Guanajuato, Guerrero, and Durango. Five states, however, furnished about 84 percent of the silver of Mexico in 1954, as shown by Table 10-4, which also indicates the relative amounts of gold, lead, zinc, and copper furnished by those states. These same five states produced 80 percent of the total silver of Mexico from the beginning of records of production through 1942 (71, p. 99). The state of Chihuahua in 1954 was far ahead of all others and produced not only about 30 percent of the total silver of Mexico but also about 52 percent of the lead and 56 percent of the zinc. The very great importance of Chihuahua as a producer of silver-lead-zinc is more fully understood when one realizes that Mexico normally ranks first in world production of silver and third or fourth in world production of lead and also zinc. The leading state for the production of silver without much lead, zinc, or copper is Hidalgo. Although it furnished about 19 percent of the Mexican total in 1954, contrasted

TABLE 10-4.   Relative Production of Silver, Gold, Lead, Zinc, and Copper from the Five Leading Silver-producing States of Mexico in 1954*

State	Approximate percent of Mexican total				
	Silver	*Gold*	*Lead*	*Zinc*	*Copper*
Chihauhua	30.3	13.6	52.2	56.4	21.8
Hidalgo	19.3	11.0	2.4	2.8	0.2
Zacatecas	16.2	8.9	18.6	19.7	10.9
Durango	11.3	26.2	5.5	0.8	2.7
San Luis Potosi	6.8	3.3	3.7	8.4	2.5
Total	83.9	63.0	82.4	88.1	38.1

* Calculated from information contained in "Riqueza minera y yacimientos minerales de Mexico," 3d ed., 1956, pp. 12–13, by J. Gonzalez Reyna, Internat. Geol. Cong., 20th, Mexico City, 1956, 487 pp.

with the 30 percent supplied by Chihuahua, nearly 60 percent of the total value of the five metals (silver, gold, lead, zinc, copper) produced by Hidalgo was in silver, whereas only 13 percent of the value from Chihuahua was in silver. The major production of Hidalgo comes from the old and world-famous area of Pachuca and Real del Monte.

Since the lead and zinc deposits of Mexico almost invariably contain silver (70, pp. 121–142), and about 62 percent of the silver production of Mexico comes from lead and zinc deposits (115), it is appropriate to consider the general occurrences and relationships among lead, zinc, and silver deposits. Although lead-zinc deposits occur in nearly all states of Mexico, the major lead-zinc-silver mineralization is in the Plateau Province in large areas of Cretaceous limestone but genetically related to various intrusives, commonly of Tertiary age. The most productive deposits are chimneys and mantos that fill preexisting large cavities in limestone and that replaced limestone (70, 154). Other deposits, in places highly important, are veins that cut various rock types (49) and some pyrometasomatic deposits. Generally deposits are in the epithermal to mesothermal group, although some, with silicate gangue minerals, appear to be in part in the hypothermal group. Deposits are not uncommonly zoned, and gradation exists in places from chiefly silver deposits into lead-zinc-silver deposits and lead-zinc deposits that contain little silver.

Actual production of silver from deposits is impossible to state because records are not available. Some deposits may have been worked by the Aztecs before the Spanish conquest and Spanish domination (1519 to 1810). Among these are Pachua (108) and Zimapán (119, 120) in Hidalgo, Santa Eulalia (49) in Chihuahua, and Taxco (66, 109) in Guerrero. Available records show that active mining was conducted in a number of places early in 1500. Much of the early mining was from rich oxidized deposits.

The extent of the mineralized area and the number of mines that are now operating and have operated at one time or another during more than four centuries precludes detailed discussion. Consequently a few areas have been selected for brief treatment in order to show the nature of the deposits.

The deposits of Pachuca and Real del Monte (69, 108) in the state of Hidalgo are 3 miles apart and are, respectively, on the western and eastern slopes of the Pachuca range. The total area of the combined districts is about 80 sq km (about 31 sq miles), in which there is a total extension of about 2,000 km (about 1,200 miles) of underground workings (69). Known exploitation dates from 1526. Total production to 1960 was estimated to be more than 1.2 billion oz of silver and 6.2 million oz of gold, worth roughly 1.3 billion dollars at 1960 prices. This is about 6 percent of the silver produced in the world since the early part of the sixteenth century. In 1959 the district furnished 6 million oz of silver, which was 13.7 percent of the production of Mexico and 3 percent of that of the world. It is rated as the fourth largest silver district of the world (149). The deposits consist of veins that occupy a series of fractures in Tertiary volcanic rocks. Filling of the fractures was accomplished during intermittent movements that produced brecciated structures (148, 166, 169, 170, 171, 172). The hypogene mineralization was comparatively simple and consisted chiefly of quartz, some calcite and rhodonite, argentite (the principal silver mineral), some polybasite and stephanite (sulfosalts of silver), pyrite, and small amounts of galena, sphalerite, and chalcopyrite (44, 69, 108). The maximum length of payable ore bodies is more than 1,000 m; the thickness is generally not more than 7 m and averages about 3 m. Veins tend to become low in silver at depth and pass into sterile galena-sphalerite deposits. Some bonanza bodies started at the surface and continued downward, others did not reach the surface but started 100 to 150 below the surface. One of the largest bonanzas, the San Rafael, occurred at a depth of more than 100 m and was of elliptical form with the greater axis more than 1,000 m long, the lesser axis about 400 m long. Production from this bonanza over a period of 10 years was valued at nearly 14 million dollars (108).

The Zimapán district, Hildago (119, 120), about 170 miles northwest of Pachuca, produced silver valued at something more than 10 million dollars since mining began in 1632. Replacements in Lower Cretaceous limestone formed mantos and chimneys, but the chimneys have yielded most of the ore mined. The largest chimney, in the Lomo de Toro mine, was 65 m long, 35 m wide, and was worked to a depth of 105 m. Mesothermal deposits in which the chief minerals are sphalerite, galena, and pyrite in a gangue of calcite are the predominant type, but some pyrometasomatic deposits occur also. Most if not all the silver in the hypogene ores is contained in galena; some coarse-grained deposits in the Miguel Hidalgo mine contained as much as 2,794 g (90 oz) per ton. The deposits are oxidized to depths of more than 200 m. The oxidized ores probably were the only ones utilized until late in the nineteenth century, and they still contribute a large part of the production. An interesting feature is that the predominant valuable mineral of these oxidized deposits is plumbojarosite, $PbFe_6(SO_4)_4(OH)_{12}$, which has been formed in enormous masses and the total amount of which is many thousands of tons. Ores of this type that appear to be homogeneous megascopically contained 300 to 1,700 g (about 10 to 55 oz) of silver per ton, which may be contained in admixed argentojarosite, $AgFe_3(SO_4)_2(OH)_6$.

Many important deposits exist in the state of Chihuahua. A few of these, located with respect to the city of Chihuahua, starting in the north, are: Namaquipa, about 90 miles to the northwest; Santa Eulalia and San Antonio, about 13 miles to the southeast; Naica, about 55 miles southeast; Hidalgo del Parral, about 120 miles to the south; San Francisco del Oro and Santa Barbara, about 130 miles to the south and only about 10 miles to the southeast of Hidalgo del Parral. These five areas in 1954 produced about 85 percent of the silver of the state of Chihuahua, 87 percent of the lead, 95 percent of the zinc, and 71 percent of the copper. The rank of each area in the production of the four metals in Chihuahua is shown by Table 10-5.

Among the five areas, Hidalgo del Parral, San Francisco del Oro, and Santa Barbara are so near each other that they constitute almost a single area, but generally San Francisco del Oro and Santa Barbara are discussed separately. The Parral area proper (72, 96, 97, 99, 113) began actively producing in 1631, and by 1665 more than 60 mines were in operation around Parral. The early mines produced chiefly from rich oxidized silver ores 30 to 100 m below the surface. In 1750 the exploitation of the richest ores ended, but a second period of mining began about 1890 and, except for disturbances between 1910 and 1925, has continued to the present time. The most important mines now appear to be La Prieta

TABLE 10-5.   Ranks of the Most Important Silver-Lead-Zinc-Copper-producing Areas in the State of Chihuahua, Mexico, in 1954*

| | *Rank* | | | |
Area	*Silver*	*Lead*	*Zinc*	*Copper*
San Francisco del Oro	1	1	1	1
Hidalgo del Parral	2	3	2	2
Santa Barbara	3	4	4	4
†Santa Eulalia–San Antonio	4	2	3	
Naica	5	5	5	5
Namiquipa	6	6	6	6

* Compiled and computed from Memoria geologico-minera del estado de Chihuahua (minerales metalicos), chiefly Table 3, p. 36, and table on p. 204, by J. Gonzalez Reyna, 1956, Internat. Geol. Cong., 20th, Mexico City 1956.
† Practically no copper production.

and Esmeralda. At Esmeralda (96) and La Prieta (99) veins occur chiefly in igneous rocks of Tertiary age. The hypogene ore minerals are galena and sphalerite, both of which carry large amounts of silver, probably as microscopic blebs of argentite (113). Proustite is important at La Prieta, but some of it may be supergene (113). In general there is a marked increase in lead-zinc content between about 130 and 260 m, and there is some copper mineralization (chalcopyrite) in the lowest levels, where the lead-zinc content decreases. Supergene enrichment has been important. Most of the very rich ore, containing 10 kg (330 oz) or more silver per ton, was from the oxidized zone, and rich ore shoots have shown a tendency to diminish greatly in size and richness just below the groundwater surface (113). Secondary silver haloids were abundant.

At San Francisco del Oro (73, 80, 81, 82) and Santa Barbara (74, 79, 114), veins are localized almost entirely along faults in Cretaceous calcareous shale that is overlain by volcanics, probably of Tertiary age. At San Francisco del Oro the veins occur in relatively short fractures, whereas at Santa Barbara they occur in long and persistent fractures. Generally the veins were formed by filling, although some breccia fragments within veins at San Francisco del Oro were replaced. The hypogene ore minerals are chiefly sphalerite, galena, and some chalcopyrite. Silver and gold are closely associated with these sulfides, especially galena. Both pyrite and arsenopyrite are present, the latter more abundant in the lower levels. Gangue minerals present in practically all veins are quartz, calcite, and fluorite, whereas those present in abundance only locally include, among others, pyroxene (apparently in the diopside-hedenbergite series), actinolite, epidote, garnet, and ilvaite, not all present in each area. Koch and Link (83) described zoning in two of the largest veins of the Frisco mine, San Francisco del Oro, below the zones of oxidation and supergene enrichment. In general they noted a decrease in base-metal content with depth and an increase in precious metal content. On the other hand, Scott (114) stated that the only definite relationship at Santa Barbara indicated an increase in copper with depth. Veins of both areas were oxidized near the surface, but the oxidized zones are no longer mined.

The Naica district (126, 133), about half way between Parral and the city of Chihuahua, and the Santa Eulalia district (75, 110) near Chihuahua are characterized by chimneys or pipes (vertical or highly inclined) and mantos (nearly horizontal or only gently inclined) that replace Cretaceous limestone.

In the Naica district the deposits have been classed (126) as silicate-sulfide mantos, crystalline calcite-sulfide mantos ("copper mantos"), sulfide-silicate chimneys, and massive sulfide chimneys, the distinction being based chiefly on the amount of silicate minerals present. The largest of the mantos, the Azules-Navacoyan, shows well the character of the silicate-sulfide type. It consists of a central wollastonite-garnet-idocrase (vesuvianite) zone that grades outward into a hedenbergite-quartz-calcite zone, both of which contain fluorite and some other gangue minerals and various base-metal sulfides, especially galena, sphalerite, chalcopyrite, and pyrite. Silver is present in each zone, apparently associated chiefly with the galena, the ratio of silver to lead being about twice as high in the wollastonite-garnet-idocrase ore as it is in the hedenbergite-quartz-calcite ore. Wilson (133) stated that no definite silver mineral was identified, but Stone (126) listed matildite, $AgBiS_2$, in his paragenetic diagram. The crystalline calcite-sulfide mantos are thin bodies generally associated with the silicate-sulfide mantos. They consist chiefly of coarsely crystalline marble and disseminated sulfides, principally chalcopyrite. The mineral composition of the sulfide-silicate chimneys is somewhat similar to that of the silicate-sulfide mantos but lacks the zoning. The chief silicate minerals are garnet, pyroxene, and tremolite; other characteristic gangue minerals are quartz, fluorite, and calcite. The sulfides include galena, sphalerite, chalcopyrite, pyrite, arsenopyrite, and a small amount of pyrrhotite. The massive sulfide ore of chimneys differs from the silicate-sulfide ore chiefly in the lack of silicates (less than 5 percent), and the two types are gradational. Both types contain silver, the ratio of silver to lead being slightly higher in the silicate type. The localization of the chimneys was controlled by faults. The largest and most important chimney, the Torino-Tehuacan, was emplaced along a fault zone that cut the Azules-Navacoyan manto (Fig. 10-8). Stone (126) classed the deposits as pyrometasomatic and mesothermal and postulated that the silicates and sulfides were introduced by a single fluid, the silicates of the mantos being deposited first, followed by sulfides that crystallized at temperatures of about 500 to 550° C; and that the mantos were cut by silicate-sulfide chimneys, and these were succeeded by chimneys of massive sulfide ore at mesothermal temperatures. The ores are classed as pyrometasomatic rather than contact metasomatic because no major bodies of igneous rock are exposed in the Sierra de Naica.

At Santa Eulalia (75, 110) there are two main types of deposits: (1) an early silver-bearing iron sulfide type and (2) a later silver-lead-zinc type. The silver-bearing iron sulfide type occurs as high-grade tabular bodies that resemble veins, which are economically important, and as large masses of relatively low-grade ore, both characterized by a gangue of lime-iron silicates and thus similar to some of the ore of the Naica district. The silver-lead-zinc type forms chimneys and mantos that have been oxidized to a depth of as much as 1,500 ft, the major products being cerussite, native silver, and cerargyrite. Known primary sulfides are galena, sphalerite, and argentite, but deposits containing these minerals are rare and production is chiefly from oxidized chimneys and mantos, for which the district

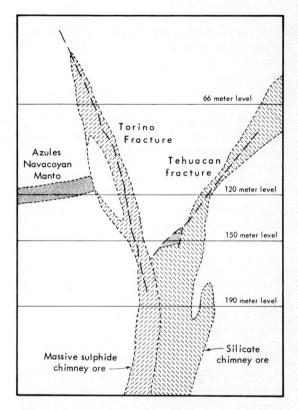

66 meter level

Torino
Fracture

Azules
Navacoyan
Manto

Tehuacan
fracture

120 meter level

150 meter level

190 meter level

Silicate
chimney ore

Massive sulphide
chimney ore

Fig. 10-8 Section through Naica mine, Chihuahua, Mexico, showing manto ore and two types of chimney ore. After J. B. Stone, *Econ. Geology,* vol. 54, p. 1015. Copyright 1959. Economic Geology Publishing Company. Used by permission.

is famous. The magnificent cathedral of the city of Chihuahua, constructed between 1739 and 1750 at a cost of $545,000 (75), was made possible because of the discovery of the Santa Eulalia deposits. The story that seems to be generally accepted (75, 110) is that a band of outlaws from the mission town of Chihuahua took refuge in the mountains and, one night when they had built an especially hot fire, noted molten lead and silver coming from the hearthstones they had used. Whereupon they consulted the padre of the town and, in exchange for complete pardon, offered to show him where he could obtain enough silver to build the most magnificent cathedral in all America. The total production from these deposits, although not accurately known, was enormous, the probable value being between 300 million and 500 million dollars. The height of the bonanza days was during the last quarter of the eighteenth century, but production is still important. It amounted in 1954 to 64,290 kg of silver, 27,753,309 kg of lead, and 30,599,687 kg of zinc. Reserves are great (75).

Many other interesting deposits are present in Chihuahua, some enormously important in the past but now not being worked, others formerly worked, then allowed to become dormant, and reactivated after the establishment of modern methods of treatment. Among the famous abandoned districts is Batopilas, in the Sierra Madre about 170 miles southwest of Chihuahua, noted for native silver (54; 71, pp. 103–116; 84), some of which occurred in remarkably rich bonanzas. The mines yielded about 50 million dollars worth of material between 1880 and 1920 and an unknown amount since 1632, the time of discovery. Large masses of cerargyrite were mined at the outcrops during early operations, but the mines were noted especially for the beauty and variety of specimens of native silver and crystals of ruby silver, a number of which have been put on display in various museums. Krieger (84) concluded that the native silver was hypogene and that the deposits resembled the cobalt-nickel-silver deposits of Ontario, Canada, and other places. A district that was discovered and worked during the time of Spanish occupation, remained idle for a period of time, was operated at intervals from 1916 to about 1946, and then was reactivated, is Namiquipa (116), about 90 miles northwest of the city of Chihuahua. There, veins that cut Tertiary andesite flows contain sphalerite, galena, and chalcopyrite, along with silver, minor gold, quartz, and fluorite. The veins are oxidized to about 325 ft below the surface, with the formation of native silver, cerargyrite, and cerussite. The average grade of combined oxidized and sulfide ore produced from 1948 through August 31, 1955, contained 448 g (14 oz) of silver per metric ton, 3.42 percent lead, and 5.01 percent zinc.

Brief mention should be made of deposits in the state of Zacatecas and the formerly rich deposits of the Guanajuato mining district, state of Guanajuato. The Zacatecas district (98), near the city of Zacatecas, formerly was a great silver camp, but at present few mines are operating. The most important production of Zacatecas now comes from the Fresnillo mining district just south of the city of Fresnillo. Estimated production there since discovery in 1554 to 1956 (125) was: gold, 12,000 kg; silver, 7,750,000 kg; lead, 450,000 metric tons; zinc, 400,000 metric tons; copper, 45,000 metric tons. The total value of this has been estimated to be about 500 million dollars. The deposits (98, 125) consist of a series of narrow, persistent veins, the average width being 1.4 m. Hypogene minerals are abundant galena (generally argentiferous), sphalerite, minor chalcopyrite, pyrargyrite and some other silver sulfantimonides, and possibly argentite. Supergene minerals are native silver in the lower part of the oxidized zone, possibly argentite, and cerargyrite, cerussite, and anglesite, chiefly in the upper part of the oxidized zone. Definite zoning is shown, with few base metals and a high proportion of silver near the surface, abundant base metals and medium silver values at greater depths. The Guanajuato mining district (48, 76, 131) probably produced silver and gold valued at slightly more than 1 billion dollars from 1548 to 1923. Ore occurred in veins formed by filling and replacement along faults and in stockworks formed on the hanging wall sides of faults, apparently caused by extensive shattering as a result of change of dip along the fault. The richest and most persistent vein, Veta Madre (Mother Lode) had stockworks of this type developed, the largest one of which was 200 m long and 60 m wide at its widest part. Rich bonanzas occurred in the Veta Madre stockworks. Values were chiefly in silver and gold, with a ratio of silver to gold about 100:1. Important minerals were chiefly silver sulfides and sulfantimonides and native silver in wire form, in a gangue of quartz and carbonates. Galena, sphalerite, and chalcopyrite occurred in minor amounts. Secondary enrichment apparently was relatively unimportant, and the bonanzas were composed of hypogene ore.

United States

In 1961 five states in the United States produced about 95 percent of the silver. Among them Idaho was far ahead, with about 50 percent of the total, followed by Arizona (15 percent), Utah (14 percent), Montana (10 percent), and Colorado (6 percent). Almost all the silver produced in the United States from the beginning of available records to the end of 1951 was furnished by those same states and three others—Nevada, California, and New Mexico, as shown by Table 10-6. Nevada was a highly important early producer of silver, furnished 47 percent of the United States total during the last years of the "big bonanza" of the Comstock lode, declined to less than 2 percent of the total (1896 to 1900), again increased to 22 percent of the total (1911 to 1915) as a result of production from the Tonopah district, and has since declined so that production in 1961 was only about 1 percent of the total (103, p. 27).

An analysis of the distribution of the 25 leading silver-producing mines in the United States in 1961, along with the type of ore from which the silver was obtained (Table 10-7), shows that only four of the group produced from silver ore whereas seven produced from copper ore.

Discussion of silver production from a number of districts will be given along with discussions of lead, zinc, and copper deposits, because of their close association. Here we will be concerned briefly with the Coeur d'Alene district, Idaho, the Park City and Tintic districts, Utah, the Comstock and Tonopah areas, Nevada,

TABLE 10-6. Silver Production in the United States by the Eight Leading Silver-producing States from Beginning of First Production to the End of 1951*

| | Maximum annual production | | Total production | |
State	Ounces	Year attained	Ounces	Percent of total, United States
Montana	19,038,800	1892	781,718,269	19.2
Utah	21,276,689	1925	756,117,233	18.6
Colorado	25,838,600	1893	745,170,388	18.3
Nevada	16,090,083	1913	597,090,406	14.7
Idaho	19,576,766	1937	583,177,365	14.3
Arizona	9,422,522	1937	317,511,400	7.8
California	3,629,233	1921	113,523,317	2.8
New Mexico	2,343,800	1885	69,970,941	1.7
Total of above states			3,974,289,319	97.4
Total United States			4,067,933,421	100.0

* From *U.S. Bur. Mines Minerals Yearbook,* 1951, vol. 1, p. 624.

TABLE 10-7. Distribution of Mines and Source of Silver among the 25 Leading Silver-producing Mines in the United States in 1961*

State	Number of mines in leading 25	Rank of mines in leading 25	Type of ore and number of mines from which silver was obtained							
			Ag	Pb,Zn Pb-Zn	Pb-Zn, Ag	Pb-Zn, Au-Ag	Au	Cu, Au-Ag	Cu-Pb-Zn	Cu
Idaho	7†	1, 2, 3, 5, 8, 10, 21	4	3						
Arizona	7	9, 12, 16, 18, 20, 23, 25		1				3		3
Montana	4‡	6, 11, 15, 24		1						3
Utah	3	4, 7, 17			1	1				1
Colorado	3	13, 14, 22		1					2	
Washington	1	19					1			
Total distribution			4	6	1	1	1	3	2	7

* From *U.S. Bur. Mines Minerals Yearbook*, 1961, vol. 1, p. 1107.
† All from the Coeur d'Alene district.
‡ Three from the Summit Valley (Butte) district.

and the Leadville area, Colorado. The Comstock, Tonopah, and Leadville areas are no longer productive but formerly were highly important and thus should be included in our discussion.

The Coeur d'Alene district, Idaho (111, 112, 117, 118, 123, 130), extends nearly east and west through Shoshone County for about 25 miles. The maximum width is about 15 miles, but many of the numerous mines lie in a strip about 4 miles wide south of the Osburn fault, which passes through Mullan, Osburn, and slightly south of Kellog (Fig. 10-9). The Precambrian rocks of the Belt series, consisting chiefly of quartzites and argillites, some of which are calcareous, have been complexly folded and faulted and cut by monzonite, diabase, and lamprophyre. The monzonite stocks have been dated by lead-alpha determinations as late Cretaceous (153). Six periods of mineralization have been recognized in the district, two of which are of Precambrian age, one Cretaceous, and three Tertiary. The Cretaceous mineralization, younger than the monzonite stocks, was responsible for the major ore deposits of the district.

The deposits consist of veins that were emplaced along zones of fracturing and shearing that extended to great depths, but individual fractures apparently were faults of small displacement. Most of the veins are in bleached zones that apparently resulted from hydrothermal solutions, and the deposits were formed chiefly by replacement of quartzitic rock. The chief ore minerals are galena, sphalerite, and tetrahedrite; the chief gangue minerals, siderite and quartz. Both the galena and the tetrahedrite commonly are argentiferous. In general the higher silver values seem to be associated with tetrahedrite, some of which, in the Sunshine mine, contained as much as 11 percent silver. During the early history of the district the silver-rich veins were thought not to extend far below the surface, but in 1930 a bonanza ore shoot was opened on the 1,700 ft level of the Sunshine mine. Since then explorations have shown similar deep rich ore shoots in other mines. The largest ore shoot of the district, in the Morning Star vein, is known to have a vertical length of 6,700 ft, a horizontal length of

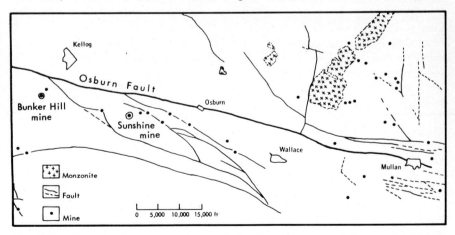

Fig. 10-9 Map showing some features of the Coeur d'Alene mining district, Idaho. Somewhat simplified after R. E. Sorenson, *Eng. Mining Jour.*, vol. 148, no. 10, pp. 70–71. Copyright 1947. McGraw-Hill Publishing Company. Used by permission.

4,000 ft, and a maximum width of 50 ft. Most shoots are smaller than this, but vertical lengths of 1,000 ft are common.

Many great mines characterize the district, among which are the Bunker Hill and the Sunshine. The Bunker Hill mine produced from 1885 through 1960 about 93,082,000 oz of silver, 2,078,000 tons of lead, and 286,000 tons of zinc. In the early 1960s the annual silver production was between 1,750,000 and 2,200,000 oz, along with some 26,000 to 35,000 tons of lead and 9,500 to 12,500 tons of zinc. The mine contains more than 100 miles of underground workings and is about a mile deep. The Sunshine mine between 1904 and 1959 furnished 187,500,311 oz of silver, 123,186,726 lb of lead, and 52,856,877 lb of copper. The production of the entire Coeur d'Alene mining district, 1884 through 1960, was as follows (112):

Gold, fine oz	Silver, fine oz	Lead, short tons	Copper, short tons	Zinc, short tons	Value
439,842	619,803,582	6,546,307	93,340	2,157,485	$1,853,374,797

Much of the silver produced in Utah that does not come from copper deposits of the Bingham area is supplied by the Park City and the Tintic districts, about 25 miles southeast and 90 miles southwest of Salt Lake City.

The Park City district (50, 51), still active in 1961, produced ore valued at $287,627,721 between 1870 and 1930, which included 199,401,578 oz of silver, 1,958,571,369 lb of lead, 322,306,387 lb of zinc, 54,062,149 lb of copper, and gold worth $8,049,187. The principal ore bodies occur along two major northeasterly trending fracture zones and generally lie between walls of quartzite and limestone, where they form veins and lodes and replacement bodies in certain Carboniferous and Triassic limestone beds adjacent to fissures. They are chiefly lead-silver ores in which the principal hypogene ore minerals are argentiferous galena and tetrahedrite, and spahalerite, and the characteristic gangue minerals are quartz and pyrite. The deposits were oxidized to a depth of 1,200 ft in places. Early mining recovered rich lead-silver-copper ores, but present mining is of lower-grade ore that requires concentration.

The Tintic area (47, 58, 61, 62, 87, 88, 91, 94, 95) may be divided into several parts—the Main Tintic district, the East Tintic district, the West Tintic district—but here we will consider the entire area briefly. Unlike some other districts, this one has a history of accelerated and important production many years after discovery. The value of the average annual yield from 1886 through 1905 was about $3,850,000; from 1906 through 1912 it was about $7,700,000; and from 1916 through 1931 it was about $12,800,000 (47). About half of the total value to 1931 ($345,000,000) was from silver, one-fourth from lead, and one-eighth each from copper and gold. An outstanding feature of the district is blind deposits that do not crop out, and some of which are as much as 700 ft below the surface. A world-famous mine of this type, the Tintic Standard, was sunk through rhyolite in 1907 and encountered the rich ore body known as the "Tintic Standard pothole," and the shaft remained in ore for nearly 200 ft. Other discoveries of important blind ore bodies followed, the latest in 1958 with the sinking of the

Burgin shaft. Recently geochemical methods have been used as an aid to supplement rock alteration and other geological work in the location of such ore bodies. The greater part of the district is covered by Tertiary effusive rocks, but monzonite stocks are closely associated with the valuable ore bodies, the best of which have been in Middle Cambrian Ophir limestone and the Ordovician Ajax limestone. The structure of the area is complex, the rocks having been folded, faulted, and crumpled. The larger ore bodies branched upward and outward from a "root" and formed pipes in crumpled limestone, which was largely replaced, as in the Tintic Standard mine (47). The minerals of the replacement deposits are chiefly galena, sphalerite, and pyrite, in a quartz-barite gangue; minor ruby silver is present also. The replacement bodies generally are encased in pyritic jasperoid, which in turn is surrounded by hydrothermal dolomite. Vein deposits of two types occur in quartzite: (1) gold-copper veins characterized by gold, enargite, tetrahedrite, and pyrite; (2) gold telluride veins that contain various gold and silver tellurides, tetrahedrite, and pyrite. Oxidation was almost complete down to the 900-ft level in the Tintic Standard mine, and local pockets of silver-bearing jarosite from the upper levels yielded carload shipments of ore worth $60,000 per car (47, p. 124).

Early silver production in Nevada was largely from the famous Comstock lode, chiefly 1859 to 1882, and then from the Tonopah district. Production was approximately as follows:

Comstock district	Total value	Tonopah district	Total value
1859–1882 1882–1919	$320,000,000 $ 55,000,000	1901–1948	$150,000,000

Much of the early production at Comstock was between about 1873 and 1880 from rich bonanza deposits such as the Consolidated Virginia, which yielded about 16 million dollars a year for the three years 1875 to 1877, and the California, with a yield of almost 14 million dollars a year for the three years 1876 to 1878, a total of more than 90 million dollars from these two bonanzas in this short time (122). Much valuable material was left in the ore tailings from these early operations, and as late as 1950 the fifth place in silver production in Nevada was from tailings recovered by Central Comstock Tailings. The peak production from the Tonopah district was in 1918 with a value of about 9.3 million dollars (59), and for the 18-year period 1907 to 1924 the yearly average was about 6.6 million dollars. Production declined rapidly from 1925 through 1929 but still averaged more than 1 million dollars a year, and from 1930 through 1948 the average yearly production was about $350,000, with practically no production in 1949 and 1950.

Briefly, the Comstock lode (42, 45, 122) was formed in a fault zone that cut Tertiary volcanic rocks and that extended for about 4 miles along the base of the Virginia range, although the main ore bodies were concentrated in the central 2 miles of this zone, and the lode split into diverging branches at each end. The upper section of the lode resembled a V-shaped trough 200 to 1,000 ft wide at the surface and about 100 ft wide at a depth of 500 ft. The upper part of this trough contained many fragments of country rock of all sizes, up to horses thousands of feet long, as well as quartz, clay, and argentiferous material. One of the large bonanzas was composed of crushed quartz, fragments of country rock, and a few narrow seams of rich ore, but almost the whole mass was impregnated to a moderate extent with ore minerals, chiefly hypogene argentite and native gold; minor minerals were chalcopyrite, sphalerite, and galena. The ore averaged only about $80 per ton, including the rich stringers, but was of enormous volume, so that a great amount of good ore was concentrated in a small space and this, rather than extreme richness, accounted for the bonanzas. Most of the ore was hypogene although a small amount of native silver very near the surface was supergene (42). Sixteen large ore bodies were found, most of them within 600 ft of the surface. Hot water in the deeper levels caused much trouble in mining.

The deposits of the Tonopah district (57, 59, 124) consisted of replacement veins along narrow sheeted zones in Tertiary volcanic rocks. Hypogene minerals appear to have been chiefly argentite along with polybasite and perhaps stephanite in a gangue of quartz and adularia. Minor amounts of chalcopyrite, galena, sphalerite, and pyrite were present also. Oxidation extended to a depth of about 700 ft, and in this zone horn silver, chiefly cerargyrite, was plentiful. Some supergene sulfides were present also—pyrargyrite and argentite—and some supergene native silver. Gold was present both in hypogene and in supergene ores, but the ratio of gold to silver generally was about 1:100 by weight.

Silver production in Colorado comes chiefly from ores of lead, zinc, and copper. In 1961 the Eagle mine of the Gilman area, Eagle County, and the Idarado Mining Company, operating in San Miguel and Ouray Counties, were important producers. In the past the Leadville district, Lake County (46, 64, 85), has furnished large silver production. Since its discovery in 1874 it produced, through 1946, material valued as follows (145):

Silver (237,213,930 oz)	$193,690,014
Zinc (1,480,360,690 lb)	101,505,483
Lead (2,093,611,797 lb)	96,659,976
Gold (288,672 oz)	61,721,248
Copper	14,369,845
Manganiferous ore	3,466,070
Total of above	$471,412,636

Water was a major problem in the mines. The Yak tunnel, started in 1895, extended into the eastern part of the district and was over 4 miles long but drained only part of the area. A tunnel to be constructed at government expense was authorized and $1,400,000 appropriated, which was the amount estimated for a total of 17,300 ft of construction, and construction started December 6, 1945. The cost of construction was far more than anticipated, funds were exhausted after completion of 6,600 ft of the tunnel, and the project was abandoned in August 1945 (146). Because of water difficulties all operations except those in the part drained by the Yak tunnel were discontinued.

The hypogene deposits of the Leadville district (64, 85) are of three types: (1) silicate-oxide replacement deposits formed at high temperature; (2) mixed sulfide veins, mainly in siliceous rocks, formed at moderate temperature; (3) mixed sulfide replacement bodies, chiefly in limestone, formed at moderate temperature, partly hot mesothermal, partly cool mesothermal (at the margins) (85). Some more recent work (46) indicates that deposition in the general Leadville area may have taken place at epithermal temperatures and, around the margins, even telethermal temperatures. The first type of deposits, silicate-oxide replacements, was of relatively little commercial importance; the second type predominated in the eastern part of the district; the third type predominated in the western part of the district and furnished most of the production. The limestone replacements formed blanketlike bodies along fractures or sheeted zones beneath impervious covers (Fig. 2-9). The largest replacement bodies, at the top of the Leadville limestone, were more than 2,000 ft long, 800 ft wide, and 200 ft thick. Mineralization consisted chiefly of galena, sphalerite, and pyrite, and the ore generally contained a few ounces of silver and 0.03 to 0.05 oz of gold per ton. The mixed sulfide veins contained pyrite and a little chalcopyrite in a gangue of quartz where they cut siliceous rocks and were valuable chiefly for gold, but where small veins cut limestone separated by porphyry sills or capped by quartzite they expanded into replacement deposits in the limestone which, after a short distance, contained a mixture of galena and sphalerite. Oxidation of deposits produced anglesite and cerussite near the original site of galena, the cerussite being accompanied by a considerable amount of horn silver near the surface. Large bodies of smithsonite were formed where carbon dioxide was available. Some gold and silver enrichment was present beneath the oxidized zone—native gold, silver, and argentite.

The Gilman district (63, 92, 93) is somewhat similar to the Leadville area. The main commercial deposits are bedding replacements or mantos in the Leadville limestone beneath relatively impervious quartz latite porphyry and overlying black shale; other deposits form chimneys, which are thickened roots of the mantos. The major minerals of the mantos are sphalerite, pyrite, and manganosiderite; minor minerals are galena, chalcopyrite, barite, dolomite, and quartz. The chimneys consist of barren pyrite and irregular bodies of pyritic ore containing chalcopyrite, copper-silver sulfosalts, and argentiferous galena. The district was rated in 1961 as the first producer of zinc in Colorado and the second producer of silver, copper, and lead. The zinc comes chiefly from the mantos, the copper and silver from the chimneys. The total value of production to 1958 was estimated at more than 180 million dollars.

The Idorado deposits, of which the Smuggler Union vein (43; 56, p. 648) is characteristic, were, in 1961, first in Colorado in production of silver and second in production of zinc, producing also much copper and lead. The Smuggler Union vein cuts andesite and rhyolite of Tertiary age and contains, as hypogene minerals, quartz, pyrite, sphalerite, galena, argentiferous tennantite, proustite, pearceite, and a little chalcopyrite; also some native gold. Some secondary enrichment produced argentite and native silver, but the rich ores have been the hypogene ones.

Canada

Lead-zinc and silver-lead-zinc ores, chiefly from British Columbia, furnished about 56 percent of the silver production of Canada in 1961; copper, copper-nickel, and copper-zinc ores, about 26 percent; silver-cobalt ores, about 16 percent; and gold ores, about 2 percent (141). Leading producers were the Consolidated Mining and Smelting Company of Canada, Limited (Cominco), British Columbia, and the United Keno Hill Mines Limited, Yukon Territory.

The Cominco production comes from the Sullivan mine (127, 128), the Bluebell mine (78), and the H. B. mine of the Salmo area (67). The Sullivan mine at Kimberley, the largest producer, was discovered in 1892, furnished some ore from 1900 to 1907, and closed because of metallurgical difficulties in treating the ore. It supplied some selected high-grade ore from upper levels of the mine from 1910 to 1915, during which time deep drilling disclosed the presence of an immense body of complex ore, and milling and metallurgical problems were overcome. Finally, about 1923, the mine became a major producer. To the end of 1946 it had furnished about 46,000,000 tons of ore. During 1946 it produced 2,300,000 tons of ore that contained 190,000 tons of lead, 150,000 tons of zinc, and 6,000,000 oz of silver. The reserves in 1946 were estimated at 3,400,000 long tons of lead and 2,700,000 long tons of zinc, with about 2.6 oz of silver per ton of ore (144).

The ore body at the Sullivan mine (Fig. 10-10) is essentially a sulfide replacement of argillites and siltstones of the late Precambrian Aldridge formation and is composed chiefly of galena, sphalerite, pyrrhotite, and pyrite, along with minor chalcopyrite, arsenopyrite, cassiterite, and some other metallic minerals. Many gangue minerals are present, among which tourmaline, chlorite, and albite are prominent, with tourmalinization in the footwall and chloritization and albitization in the hanging wall.

The Bluebell mine, 30 miles east of Nelson, produced ore intermittently from 1895 to 1927, then was reactivated in 1952. The deposits are massive sulfide replacements in the Bluebell limestone, of Proterozoic or early Paleozoic age, which is overlain by quartzite and cut by pegmatite sills. The deposits begin at cross fractures in the limestone at three known centers of mineralization and replace the limestone along the bedding, generally beneath the quartzite or pegmatite sills. The chief sulfides, in order of abundance, are pyrrhotite, sphalerite, and galena; minor but persistent sulfides are pyrite, chalcopyrite, and arsenopyrite. The common gangue mineral is quartz.

In the Salmo area, in addition to the H. B. mine of Cominco, production comes from the Jersey mine of Canadian Exploration Limited and the Reeves MacDonald mine of New Jersey Zinc Exploration Limited; the latter two mines have been the major producers of the Salmo district since 1949 and furnished, from 1949 to 1954, a total of 233,134 oz of silver, 87,659,399 lb of lead, 215,708,856 lb of zinc, and 475,397 lb of cadmium. They have supplied, since the first recorded production of the district in 1906, more than 80 percent of the lead and 95 percent of the zinc of the Salmo area. Estimated reserves at the largest properties are several million tons of ore containing 4 to 7 percent combined lead and zinc (67, p. 105). The deposits in these two mines and in the H. B. mine are sphalerite-galena-pyrite replacements of highly deformed

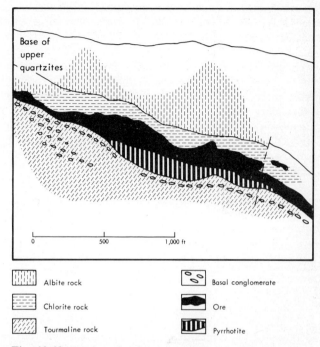

Base of
upper
quartzites

0	500	1,000 ft

Albite rock Basal conglomerate

Chlorite rock Ore

Tourmaline rock Pyrrhotite

Fig. 10-10 Section showing some features of the ore zone, Sullivan mine, British Columbia. Slightly simplified after C. O. Swanson, "Structural Geology of Canadian Ore Deposits," p. 224. Copyright 1948. The Canadian Institute of Mining and Metallurgy. Used by permission.

Laib dolomitized limestone of Cambrian age. Mineralizing solutions apparently moved along a thrust fault, then through numerous small bedding faults close to the main fault, and outward into the limestone.

The Keno Hill–Galena Hill lead-zinc-silver deposits (52, 53) are in central Yukon about 220 miles north of Whitehorse and 35 miles northeast of Mayo. Worked since 1915, these deposits have produced silver, lead, zinc, and cadmium valued at more than 75 million dollars. The ore deposits are veins and lodes that occur chiefly in thick-bedded quartzites and greenstones at the sites of various faults and fractures. Mineralization is complex and apparently involved three periods of hypogene mineralization followed by a long period of supergene alteration and enrichment. The major mineralization consisted of deposition of galena, sphalerite, freibergite (argentiferous tetrahedrite), pyrite, and siderite. Silver and lead were enriched in the oxidized zone but zinc was leached.

Production from silver-cobalt ores in Canada comes chiefly from the Cobalt district (129) and the Gowganda district (105, 106) about 55 miles northwest of Cobalt. Production from the Cobalt area, 1904 to 1954, was 399,087,091 oz of silver, 21,220,979 lb of cobalt, 2,250,536 lb of nickel, and 1,269,952 lb of copper. The total production of the Gowganda district since discovery in 1907 to about 1957 was about 32 million oz of silver and much cobalt (total un-recorded). The annual production from two mines of the Gowganda district that were operating in 1957 was about half a million ounces of silver and con-siderable cobalt. The deposits of the two districts are similar and consist of rela-tively short and narrow veins along faults, and ore bodies commonly are lodes that consist of several veins. Within the two districts the deposits occur in Precambrian rocks consisting of Keewatin greenstones and the Huronian Cobalt sedimentary series cut by the Nipissing diabase sill. In the Cobalt area, early production was from deposits in the lower part of the Cobalt series, whereas recent production has been chiefly from deposits in the Nipissing diabase and the Keewatin rocks. At Gowganda most of the ore has come from the upper 300 to 400 ft of the Nipissing diabase. The ores are very complex, and the mineralogical relationships have been studied in detail (135, 136). The major minerals of the veins, however, are native silver and various cobalt-nickel arsenides, especially cobaltite, skut-terudite (smaltite), and saffrolite, in a gangue of calcite. The native silver may be in small specks and in pieces of various sizes up to large slabs several feet long. The ores are closely associated with the Nipissing diabase, and the opinion has been expressed that both came from the same parent magma.

Peru

Peru, like Bolivia, has produced silver for several centuries. From 1533 to 1600 the annual rate of production was 1,387,647 oz (103, p. 40), and the total production recorded was 94,360,000 oz. From 1601 through 1800 the annual production was more than 3,000,000 oz, and from 1781 through 1810 it was more than 4,000,000 oz. Thereafter it declined but averaged more than 2,000,000 oz per year until the period 1896 to 1900, when the annual rate was more than 5,000,000 oz. Unlike Bolivia, however, Peru has maintained and even strengthened her posi-tion as a world silver producer.

The silver of Peru comes chiefly from lead deposits, copper deposits, or mixed deposits of these metals and zinc. Miller and Singewald (104, p. 446), re-garding deposits of the Peruvian Andes, stated that three types, silver, argentiferous lead, and argentiferous copper, grade into one another in mineral composition and show identical geologic occurrence and geographic distribution. The copper-lead-zinc-silver deposits occur in a large mineralized area of the Andes (Fig. 10-11) that contains many active as well as abandoned mines, but a large part of the production comes from the mines of the Cerro de Pasco Copper Corpora-tion. In general the deposits of the entire area occur in Mesozoic sedimentary rock and Tertiary intrusive rock, are related in orgin to Tertiary rocks of general monzonitic composition, form filled veins and various replacement bodies, and generally belong to a temperature range that covers mesothermal through epithermal types, with a tendency for the deposits that are important chiefly for silver to be in the epithermal group. Both lateral and vertical zoning are shown in some deposits. The nature of the important copper-lead-zinc-silver deposits will be indicated by a brief discussion of the deposits at Cerro de Pasco, Morococha, and Casapalca; the nature of the silver deposits by a discussion of Colquijirca and Quiruvilca.

The Cerro de Pasco deposit (68, 102, 104, 132) is a large crescent-shaped

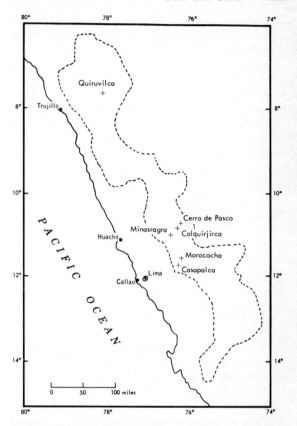

Fig. 10-11 Map showing approximate area in which the major mines of Peru occur. Generalized after B. J. Miller and J. T. Singewald, Jr., "The Mineral Deposits of South America," p. 442. Copyright 1919. McGraw-Hill Book Company. Used by permission.

pyritic replacement body localized along a fault where it is tangential to a volcanic vent about a mile in diameter, which makes the west wall and opposite to which are phyllite and overlying dolomite. The pyritic lens replaced materials of the vent and, to a large extent, the limestone, and the lead-zinc bodies were concentrated especially in the limestone. Galena and sphalerite are the chief ore minerals, but the deposits generally carry some silver. Copper-silver veins cut various rocks and also the pyritic lens, and some copper-silver pipes occur in the western part of the pyritic lens. The chief copper mineral is enargite, with which argentiferous tetrahedrite-tennantite is closely associated. Where the ores are high in silver the deposits contain tetrahedrite-tennantite, argentite, ruby silver, and some other silver minerals, and thus pass into almost straight silver deposits. An oxidized zone, from a few feet to several hundred feet thick, known locally as "pacos," covers much of the central and southern part of the deposit. This is siliceous silver-bearing material that furnished the richer silver ores of the days of early mining and supplied 2,323, 526 lb of silver between 1784 and 1846, excluding the Peruvian civil war years 1821 to 1825 (157). The pacos material that remains contains but 6 to 12 oz of silver per ton and is being slowly used as siliceous flux in smelting operations.

At Morococha (68, 102, 104, 107), nearby, both a central stock of quartz monzonite and the surrounding sedimentary and volcanic rocks are cut by many veins. The deposits occur in veins along fractures, as replacement mantos in certain limestone beds, and as pipes in favorable places. Ore mineralization is known throughout a vertical extent of 3,500 ft. The deposits show rather well-defined zoning. Zone 1, the central zone, is high in copper, low in lead and zinc, with enargite and tennantite the principal copper minerals. Zone 2 is high in copper and zinc, low in silver and lead, with chalcopyrite and sphalerite abundant, tennantite common, and galena rare. Zone 3 is high in silver and zinc, low in copper and lead, with sphalerite abundant, tennantite and galena common, chalcopyrite rare. And zone 4, farthest from the central area, is high in silver and lead, low in copper and zinc, with galena abundant, sphalerite and tennantite common. In a general way, therefore, copper is concentrated in the central area, zinc in a somewhat intermediate area, and lead and silver toward the exterior. One of the mantos, the Ombla, has been especially noteworthy and produced several hundred million pounds of copper (68, pp. 175–177).

At Casapalca (68, 100; 104, pp. 483–485) a zone of discontinuous fractures contains mineralized veins that generally are wide in a brittle conglomerate, very narrow in shale, and make intricately branching stringers in porphyry. The deposits are zoned laterally and vertically. Laterally a central zone contains galena and sphalerite and some silver; an intermediate zone contains galena, sphalerite, tetrahedrite, bournonite, and high silver; and an outer zone contains tetrahedrite, stibnite, realgar, and spotty silver. Vertically the dominant lead-zinc ores, with important silver, pass upward into rich silver ores containing argentite, ruby silver, and other rich sulfosalts, and downward into mineralization marked by an increase in copper and a decrease in silver. The deposits have been worked to a depth of 4,400 ft below the highest outcrops, but the rich silver ores, of epithermal type, were exhausted at a depth of a few hundred feet.

The importance of the three districts—Cerro de Pasco, Morococha, Casapalca —is indicated by the 20-year record shown by Table 10-8.

The mine at Colquijirca (77, 90, 101, 104) is dominantly a silver producer that probably furnished not less than 40 million oz of silver up to 1935. From 1911 through 1919 it supplied nearly 13 million oz of silver, along with some lead and copper, and in 1927 to 1928 it produced 12 million oz of silver. First-class ores averaged 60 oz of silver per ton; second-class ores, 40 oz; poor ores, which contain about 15 percent lead, 20 oz or less. Important mineralization generally follows one horizon, along which mantos are developed in three or four calcareous beds generally separated by sandstone or sandy shale. In places, however, the mantos merge and yield an ore zone as much as 12 m thick. The chief hypogene minerals are galena carrying 8 to 26 oz of silver per ton and argentiferous tennantite; stromeyerite ($AgCuS$) and minor argentite are present. Some question exists regarding the origin of the stromeyerite, whether hypogene or supergene (90, 101). Ores were oxidized down to a depth of 120 m, and important native silver was formed in the oxidized zone. Haapala (77) stated

TABLE 10-8.   Approximate Total Production of Some Metals from Cerro de Pasco, Morococha, and Casapalca, Peru, 1928–1947*

		Approximate percent of total furnished by each district or mine		
Metal	Approximate total production from all three districts or mines	Cerro de Pasco mine†	Morococha district‡	Casapalca mine§
Silver, oz	167,000,000	44	22	34
Gold, oz	473,000	91	9	—
Copper, lb	1,298,000,000	51	44	5
Lead, lb	466,000,000	19	16	65

* Computed from data in article by the Geological Staff, Cerro de Pasco Copper Corporation, Internat. Geol. Cong., 18th, Great Britain 1948, pt. 7, pp. 154–186.
† No lead production, 1931 to 1936.
‡ No lead production, 1932 to 1936.
§ No production of any metals, 1932 to 1933.

that some zoning occurs, with copper mineralization in the south part of the mine and at the lower levels, lead-silver mineralization in the north part and at higher levels.

The Quiruvilca district (55, 86, 104, 121) produces much silver, but within the larger mineralized area of the district the types of deposits differ. In the south there are quartz-silver veins; in the central area, enargite and tetrahedrite veins; and to the northwest, silver veins with barite gangue. Some of the silver veins throughout the district contain much galena and sphalerite. A study by Lewis in 1956 (86) indicated that the deposits of the Quiruvilca mine are filled veins that lie in interbedded andesite flows and flow breccias of post-Cretaceous age. He determined four zones in the deposits, as shown by Table 10-9.

The production from the Quiruvilca district, 1927 to 1931 (55), was 76,860 kg of silver (2,474,892 oz), 661.81 kg of gold (21,310 oz), and 44,065,903 kg of copper (97,147,689 lb). The grade of ore produced in 1951 (86) averaged 4.44 percent copper and 3.73 oz of silver per ton from nine veins.

Silver is produced also in other parts of Peru. Investigations in the early 1950s in the Department of Ancash along the Cordillera Blanco (137) and the Cordillera Negra (138) indicate that mining of deposits that contain lead, zinc, copper, and silver has been done for several hundred years. Base-metal production depends entirely on the silver content of the ore, however, so that in general only argentiferous galena, tetrahedrite-tennantite, and any associated silver sulfides are sought, and sphalerite, which is low in silver, either is not mined or is discarded.

Bolivia

The silver of Bolivia has come from the tin-silver deposits previously discussed. The most important silver producer of the region was Cerro de Potosí (89; 104, pp. 122–126); other famous producers were Colquiri, Oruro, Colquechaca, and Pulacayo. Some idea of the importance of these districts may be obtained from Table 10-10.

TABLE 10-9.   Zones in Deposits at Quiruvilca, Peru*

Zone	Size	Type of deposit	Metallic minerals
Enargite (central)	2,800 m long, 700 m wide	Mesothermal	Enargite; abundant pyrite; some tetrahedrite-tennantite and wurtzite; minor chalcopyrite and galena
Transition	1,200 m wide to west, missing to east	Mesothermal to epithermal	Sphalerite and pyrite most abundant; enargite near central zone, elsewhere tetrahedrite-tennantite; small amounts of chalcopyrite and galena
Lead-zinc	Over 3,000 m wide to west, smaller to north and south, missing to east	Epithermal	Sphalerite; moderate amounts of galena and pyrite; small amounts of chalcopyrite and tetrahedrite-tennantite
Stibnite	Unknown width, may be missing to east	Epithermal	Stibnite associated with sphalerite, argentiferous galena, possibly freibergite (argentiferous tetrahedrite) along inner side of zone, which is important for silver production

* From R. W. Lewis, Jr., *Economic Geology,* vol. 51, pp. 41–63. Copyright 1956. Economic Geology Publishing Company. Used by permission.

The ore from many of the important districts was fabulously rich. At Cerro de Potosí, generally classed as the most productive silver district of the world, the richest ore from the oxidized zone, which reached a depth of about 1,000 ft below the summit of the mountain, contained from 1,500 to 9,000 oz of silver per ton (65). At Oruro (104, pp. 110–114), where there was formerly a rich oxidized zone and also a rich secondary sulfide zone, the sulfide zone contained picked ore that yielded several thousand ounces of silver per ton. In the days of early production from that district ore that contained less than 150 oz of silver per ton was unprofitable and was used to fill the old stopes, but subsequently it was mined. Colquechaca (104, p. 121) was noted especially for an abundance of ruby silver ore. Rich ore contained from 500 to 5,000 oz of silver per ton; second class ore, 100 to 200 oz. Pulacayo, which for many years was the largest silver-producing district in Bolivia (104, pp. 128–130), was noted especially for tetrahedrite that contained 3 to 6 percent silver, and even as much as 10 percent. The tetrahedrite ores contained several hundred ounces of silver per ton; rich shoots of first-grade ore as much as 600 oz, and other shoots of second-grade ore up to 180 oz. In the days of early mining, material thrown on the dumps contained 30 to 45 oz of silver per ton.

In the days of the fabulously rich silver deposits, tin was not mined, but as the very rich silver deposits became depleted, tin was recovered and the region gradually became one in which tin is more important than silver, although Bolivia did produce 3 percent of the world's silver in the 15-year period 1945 to 1959.

TABLE 10-10. Estimated Production of Silver from Some Famous Bolivian Tin-Silver Districts*

Period	District	Production of silver
1553–1910	Cerro de Potosí	At least 1 billion oz, perhaps 2 billion
1595–1944	Oruro	271,000,000 oz
1891	Oruro	1,800,000 oz
1865–1892	Colquechaca (Bartolome tunnel only)	$21,000,000
1884–1892	Colquechaca (Amigos tunnel only)	$6,000,000
1877–1892	Pulacayo	$46,000,000
1891	Pulacayo	6,000,000 oz

* Compiled chiefly from "The Mineral Deposits of South America," by B. J. Miller and J. T. Singewald, Jr. Copyright 1919. McGraw-Hill Book Company. Used by permission. The production for Oruro, 1595–1944, is from F. M. Chace, *Economic Geology*, vol. 43, 1948.

Other Countries

Almost every country produces some silver, especially if deposits of lead, zinc, and copper are present, although some highly important deposits of those metals yield practically no silver. Among the most important silver producers not yet considered are the Soviet Union and Australia. The silver of these countries will be covered with deposits of lead, zinc, and copper.

SELECTED REFERENCES
GOLD

1. W. Baragwanath, 1953, The Ballarat goldfield, in *Fifth Empire Min. Metall. Cong.*, Australia and New Zealand, pp. 986–1002, Australasian Institute of Mining and Metallurgy, Melbourne.
2. R. Borchers, 1961, Exploration of the Witwatersrand system and its extensions, Commonwealth Mining Metall. Cong., 7th, South Africa, *Trans.*, vol. 2, pp. 489–512.
3. R. W. Boyle, 1961, The geology, geochemistry, and origin of the gold deposits of the Yellowknife district, *Canada Geol. Survey Mem.* 310, 193 pp.
4. J. D. Campbell, 1953, The structure of the Kalgoorlie goldfield, in *Fifth Empire Min. Metall. Cong.*, Australia and New Zealand, pp. 79–93, Australasian Institute of Mining and Metallurgy, Melbourne.
5. W. A. Clark, 1957, Gold, in Mineral commodities of California, pp. 215–226, *California Dept. Nat. Resources, Div. Mines, Bull.* 176, 736 pp.
6. L. C. Coleman, 1957, Mineralogy of the Giant Yellowknife gold mine, N. W. T., *Econ. Geology*, vol. 52, pp. 400–425.
7. C. F. Davidson, 1957, On the occurrence of uranium in ancient conglomerates, *Econ. Geology*, vol. 52, pp. 668–693. Lists 73 references.
8. C. F. Davidson, 1960, The present state of the Witwatersrand controversy,

Mining Mag. (London), vol. 102, pp. 84–95, 149–159, 222–229. Lists 96 references.

9. W. R. Dunbar, 1948, Structural relations of the Porcupine ore deposits, in "Structural Geology of Canadian Ore Deposits," pp. 442–456, Canadian Institute of Mining and Metallurgy, Montreal, 948 pp.

10. K. J. Finucane and H. E. Jensen, 1953, Lode structures in the Kalgoorlie goldfield, in *Fifth Empire Min. Metall. Cong.*, Australia and New Zealand, pp. 94–111, Australasian Institute of Mining and Metallurgy, Melbourne.

11. Geological Survey of Canada, 1947, Geology and economic minerals of Canada, 3d ed., *Geol. Survey Canada, Econ. Geology Ser.* 1, 357 pp. Gold, pp. 36–56.

12. Geological Survey, Union of South Africa, 1959, The mineral resources of South Africa, 4th ed., *Dept. Mines, Geol. Survey, Union of South Africa.* Gold, pp. 88–138.

13. N. Gornotajev, 1934, Geochemistry and tectonics of the gold quartz veins of the Soviet mine in the North-Yenisseisky taiga, *Acad. Sci. U.S.S.R., C. R.* 3 (5), pp. 376–382, (Russian and English). Abstract, 1935, *Geol. Zentralbl.*, vol. 54, no. 1, p. 19.

14. R. B. Hargreaves, 1960, A bibliography of the geology of the Witwatersrand, *Univ. Witwatersrand Econ. Geol. Research Unit, Inf. Circ. No* 1.

15. J. F. Henderson and I. C. Brown, 1948, Yellowknife area, Northwest Territories, *Canada Geol. Survey, Paper* 48-17, 10 pp.

16. J. F. Henderson, 1952, The Yellowknife greenstone belt, Northwest Territories, *Canada Geol. Survey, Paper* 52-28, 41 pp.

17. M. T. Huntting, 1955, Gold in Washington, *Washington Div. Mines and Geol., Bull.* 42, 158 pp.

18. W. A. Jones, 1948, Hollinger mine, in "Structural Geology of Canadian Ore Deposits," pp. 464–481, Canadian Institute of Mining and Metallurgy, Montreal, 948 pp.

19. A. Knopf, 1929, The Mother Lode system of California, *U.S. Geol. Survey Prof. Paper* 157, 88 pp.

20. A. Knopf, 1933, The Mother Lode system, Internat. Geol. Cong., 16th, United States 1933, *Guidebook* 17, pp. 45–60.

21. A. A. Koffman, and others, 1948, Flin Flon mine, in "Structural Geology of Canadian Ore Deposits," pp. 295–301, Canadian Institute of Mining and Metallurgy, Montreal, 948 pp.

22. A. H. Koschmann, 1949, Structural control of the gold deposits of the Cripple Creek district, Teller County, Colorado, *U.S. Geol. Survey Bull.* 955-B, pp. 19–60.

23. W. Lindgren, 1911, The Tertiary gravels of the Sierra Nevada of California, *U.S. Geol. Survey Prof. Paper* 73, 226 pp.

24. W. Lindgren and F. L. Ransome, 1906, Geology and gold deposits of the Cripple Creek district, Colorado, *U.S. Geol. Survey Prof. Paper* 54, 516 pp.

25. D. H. McLaughlin, 1933, The Homestake ore bodies, Lead, South Dakota, in "Ore Deposits of the Western States" (Lindgren volume), pp. 563–565, The American Institute of Mining and Metallurgical Engineers, New York, 797 pp.

26. J. B. Miller and J. T. Singewald, Jr., 1910, "The Mineral Deposits of South America," McGraw-Hill Book Company, New York, 598 pp.

27. D. V. Nalivkin, 1960, "The Geology of the U.S.S.R., a Short Outline," pp. 35, 57, 78, 97, 148, Pergamon Press, New York, 170 pp., geologic map of the U.S.S.R.

28. J. A. Noble, 1950, Ore mineralization in the Homestake gold mine, Lead, South Dakota, *Geol. Soc. America Bull.* vol. 61, pp. 221–252.

29. T. B. Nolan, 1933, Epithermal precious-metal deposits, in "Ore Deposits of the Western States" (Lindgren volume), pp. 623–652, The American Institute of Mining and Metallurgical Engineers, New York, 797 pp.

30. V. A. Obruchev, 1935, The problem of gold-bearing pyritic shales of the Lena-Vitim region, *Problems Soviet Geol.* 5 (1), pp. 60–70, (Russian). Abstract, 1935, *Geol. Zentralbl.,* vol. 55, no. 5, p. 216.

31. F. L. Ransome, 1909, The geology and ore deposits of Goldfield, Nevada, *U.S. Geol. Survey Prof. Paper 66,* 258 pp.

32. A. W. Rogers, 1929, Johannesburg, Internat. Geol. Cong., 15th, South Africa, 1929, *Guidebook Exc.* B 8, 16 pp.

33. D. B. Shimkin, 1953, "Minerals—a Key to Soviet Power," Harvard University Press, Cambridge, Mass., 452 pp.

34. F. L. Stillwell, 1953, Tellurides in Western Australia, in *Fifth Empire Min. Metall. Cong.,* Australia and New Zealand, pp. 119–127, Australasian Institute of Mining and Metallurgy, Melbourne.

35. D. E. Thomas, 1953, The Bendigo goldfield, in *Fifth Empire Mining Metall. Cong.,* Australia and New Zealand, pp. 1011–1026, Australasian Institute of Mining and Metallurgy, Melbourne.

36. D. E. Thomas, 1953, Mineralization and its relationship to the geological structure of Victoria, in *Fifth Empire Min. Metall. Cong.,* Australia and New Zealand, pp. 971–985, Australasian Institute of Mining and Metallurgy, Melbourne.

37. J. E. Thomson, 1948, Regional structure of the Kirkland Lake–Larder Lake area, in "Structural Geology of Canadian Ore Deposits," pp. 627–632, Canadian Institute of Mining and Metallurgy, Montreal, 948 pp.

38. W. Ward, and others, 1948, The gold mines of Kirkland Lake, in "Structural Geology of Canadian Ore Deposits," pp. 644–653, Canadian Institute of Mining and Metallurgy, Montreal, 948 pp.

39. H. C. M. Whiteside and B. B. Brock, 1961, The role of geology on the gold mines of the Witwatersrand and the Orange Free State, Commonwealth Mining Metall. Cong., 7th, South Africa, *Trans.,* vol. 2, pp. 513–521. Discussion, pp. 521–523.

40. M. E. Wilson, 1948, Structural features of the Noranda-Rouyn area, in "Structural Geology of Canadian Ore Deposits," pp. 672–683, Canadian Institute of Mining and Metallurgy, Montreal, 948 pp.

41. M. N. Zinberg, 1934, A primary gold deposit in pyritic schists of the Lena-Vitim region (Soviet gold industry), *Sovietskaja Solotopromyschlennost* 5, pp. 32–34 (Russian). Abstract, 1935, *Geol. Zentralbl.,* vol. 54, no. 1, p. 19.

SILVER

42. E. S. Bastin, 1923, Bonanza ores of the Comstock Lode, Virginia City, Nevada, *U.S. Geol. Survey Bull.* 735, pp. 41–63.
43. E. S. Bastin, 1923, Silver enrichment in the San Juan Mountains, Colorado, *U.S. Geol. Survey Bull.* 735, pp. 65–129.
44. E. S. Bastin, 1948, Mineral relationships in the ores of Pachuca and Real del Monte, Hidalgo, Mexico, *Econ. Geology,* vol. 43, pp. 53–65.
45. G. F. Becker, 1882, Geology of the Comstock Lode and the Washoe district, *U.S. Geol. Survey Mon.* 3, 422 pp.
46. C. H. Behre, Jr., 1953, Geology· and ore deposits of the west slope of The Mosquito range, *U.S. Geol. Survey Prof. Paper* 235, 176 pp.
47. P. Billingsley, 1933, Tintic mining district, Internat. Geol. Cong., 16th, United States 1933, *Guidebook* 17, pp. 101–124.
48. W. P. Blake, 1902, Notes on the mines and minerals of Guanajuato, Mexico, *Am. Inst. Mining Metall. Engineers Trans.,* vol. 32, pp. 216–223.
49. A. F. J. Bordeaux, 1908, Silver mines of Mexico, *Am. Inst. Mining Metall. Engineers Trans.,* vol. 39, pp. 357–368.
50. J. M. Boutwell, 1912, Geology and ore deposits of the Park City district, Utah, *U.S. Geol. Survey Prof. Paper* 77, 231 pp.
51. J. M. Boutwell, 1933, Park City mining district, Internat. Geol. Cong., 16th, United States 1933, *Guidebook* 17, pp. 69–82.
52. R. W. Boyle, 1956, The geology and geochemistry of the silver-lead-zinc deposits of Keno Hill and Sourdough Hill, Yukon Territory (preliminary report), *Canada Geol. Survey, Paper* 55-30, 78 pp.
53. R. W. Boyle, 1957, Lead-zinc-silver lodes of the Keno Hill–Galena Hill area, Yukon, in "Structural Geology of Canadian Ore Deposits, Congress Volume," pp. 51–65, Commonwealth Min. Metall. Cong., 6th, Canada, Canadian Institute of Mining and Metallurgy, Montreal.
54. W. M. Brodie, 1907, History of the silver mines of Batopilas, *Mining World,* vol. 30, no. 24, pp. 1105–1110. (Later *Mining and Engineering World,* Chicago.)
55. J. A. Broggi, 1935, El cobre en el Perú, Internat. Geol. Cong., 16th, United States 1933, "Copper Resources of the World," vol. 2, pp. 501–512.
56. W. S. Burbank, 1933, Epithermal base-metal deposits, in "Ore Deposits of the Western States" (Lindgren volume), pp. 641–652, The American Institute of Mining and Metallurgical Engineers, New York, 797 pp.
57. J. A. Burgess, 1909, The geology of the producing part of the Tonopah mining district, *Econ. Geology,* vol. 4, pp. 681–712.
58. J. B. Bush and D. R. Cook, 1960, The Chief Oxide–Burgin area discoveries, East Tintic district, Utah, a case history; part II, Bear Creek Mining Company studies and exploration, *Econ. Geology,* vol. 55, pp. 1507–1540.
59. J. A. Carpenter, R. R. Elliot, and B. F. W. Sawyer, 1953, The history of fifty years of mining at Tonopah, 1900–1950, *Univ. Nevada Bull.,* vol. 47, no. 1, 157 pp. (*Nevada Bur. Mines, Geol. and Min. Ser. No.* 51.)
60. F. M. Chace, 1948, Tin-silver veins of Oruro, Bolivia, *Econ. Geology,* vol. 43, pp. 333–383, 435–470.
61. D. R. Cook (ed.), 1957, Geology of the East Tintic mountains and ore

deposits of the Tintic mining districts, *Guidebook to the geology of Utah, No.* 12, 176 pp., Utah Geological Society, Salt Lake City.

62. G. W. Crane, 1917, Geology and ore deposits of the Tintic mining district, *Am. Inst. Mining Metall. Engineers Trans.,* vol. 54, pp. 342–355.

63. R. D. Crawford and R. Gibson, 1925, Geology and ore deposits of the Red Cliff district, Colorado, *Colorado Geol. Survey, Bull.* 30, 89 pp.

64. S. F. Emmons, J. D. Irving, and G. F. Laughlin, 1927, Geology and ore deposits of the Leadville mining district, Colorado, *U.S. Geol. Survey Prof. Paper* 148, 368 pp.

65. D. L. Evans, 1940, Structure and mineral zoning of the Pailaviri Section, Potosí, Bolivia, *Econ. Geology,* vol. 35, pp. 737–750.

66. G. M. Fowler, R. M. Hernon, and E. A. Stone, 1950, The Taxco mining district, Guerrero, Mexico, Internat. Geol. Cong., 18th, Great Britain 1948, pt. 7, The geology, paragenesis, and reserves of the ores of lead and zinc, pp. 143–153.

67. J. T. Fyles and C. G. Hewlett, 1957, Lead-zinc deposits of the Salmo area, British Columbia, in "Structural Geology of Canadian Ore Deposits, Congress Volume," pp. 104–109, Commonwealth Min. Metall. Cong., 6th, Canada 1957, Canadian Institute of Mining and Metallurgy, Montreal.

68. Geological staff, Cerro de Pasco Copper Corporation, 1950, Lead and zinc deposits of the Cerro de Pasco Copper Corporation in central Peru, Internat. Geol. Cong., 18th, Great Britain 1948, pt. 7, The geology, paragenesis, and reserves of the ores of lead and zinc, pp. 154–186.

69. A. R. Geyne, 1956, Las rocas volcánicas y los yacimientos argentíferos del Distrito Minero de Pachuca-Real del Monte, Estado de Hidalgo, *Internat. Geol. Cong.,* 20th, Mexico 1956, *Guidebook Exc.* A-3 and C-1, pp. 47–57.

70. J. González Reyna, 1950, Geologia, paragenesis y reservas de los yacimientos de plomo y zinc de México, Internat. Geol. Cong., 18th, Great Britain 1948, pt. 7, The geology, paragenesis, and reserves of the ores of lead and zinc, pp. 121–142.

71. J. González Reyna, 1956, Los yacimientos de plata de México, in "Riqueza Minera y Yacimientos Minerales de México," 3d ed., pp. 95–120, Internat. Geol. Cong., 20th, Mexico 1956, 497 pp.

72. J. González Reyna, 1956, Hidalgo del Parral, in "Memoria Geologico-Minera del Estado de Chihuahua," pp. 245–254, Internat. Geol. Cong., 20th, Mexico 1956, 280 pp.

73. J. González Reyna, San Francisco del Oro, in "Memoria Geologico-Minera del Estado de Chihuahua," pp. 242–243, Internat. Geol. Cong., 20th, Mexico 1956, 280 pp.

74. J. González Reyna, 1956, Santa Barbara, in "Memoria Geologico-Minera del Estado de Chihuahua," pp. 234–242, Internat. Geol. Cong., 20th, Mexico 1956, 280 pp.

75. J. González Reyna, 1956, Santa Eulalia, in "Memoria Geologico-Minera del Estado de Chihuahua," pp. 198–204, Internat. Geol. Cong., 20th, Mexico 1956, 280 pp.

76. R. Guiza, Jr., 1956, El distrito minero de Guanajuato, Internat. Geol. Cong., 20th, Mexico 1956, *Guidebook Exc.* A-2 and A-5, pp. 141–148.

77. P. S. Haapala, 1954, Estudio geólogica de la mina Colquijirca, *Mineria,* vol. 2, no. 7, pp. 21–40.

78. W. T. Irvine, 1957, The Bluebell mine, in "Structural Geology of Canadian Ore Deposits, Congress Volume," pp. 95–104, Commonwealth Min. Metall. Cong., 6th, Canada 1957, Canadian Institute of Mining and Metallurgy, Montreal.

79. M. D. Kierans, 1956, Minas de la Compañia Minera Asarco, S. A. en Santa Barbara, Chihuahua, Internat. Geol. Cong., 20th, Mexico 1956, *Guidebook Exc.* A-2 and A-5, pp. 101–108.

80. G. S. Koch, Jr., 1956, Geologia del area vecina a San Francisco del Oro y Santa Barbara, Internat. Geol. Cong., 20th, Mexico 1956, *Guidebook Exc.* A-2 and A-5, pp. 97–100.

81. G. S. Koch, Jr., 1956, Las minas Frisco y Clarines, Internat. Geol. Cong., 20th, Mexico 1956, *Guidebook Exc.* A-2 and A-5, pp. 109–118.

82. G. S. Koch, Jr., 1956, The Frisco mine, Chihuahua, Mexico, *Econ. Geology,* vol. 51, pp. 1–40.

83. G. S. Koch, Jr. and R. F. Link, 1960, Zoning of metals in two veins of the Frisco mine, Chihuahua, Mexico, Internat. Geol. Cong., 21st, Copenhagen 1960, pt. 16, sec. 16, Genetic problems of ores, pp. 192–199.

84. P. Krieger, 1935, Primary native silver ores at Batopilas, Mexico, and Bullard's Peak, New Mexico, *Am. Mineralogist,* vol. 20, pp. 715–723.

85. G. F. Laughlin and C. H. Behre, Jr., 1933, Leadville mining district, Internat. Geol. Cong., 16th, United States 1933, *Guidebook* 19, pp. 77–91.

86. R. W. Lewis, Jr., 1956, The geology and ore deposits of the Quiruvilca district, Peru, *Econ. Geology,* vol. 51, pp. 41–63.

87. W. Lindgren, 1915, Processes of mineralization and enrichment in the Tintic mining district, *Econ. Geology,* vol. 10, pp. 225–240.

88. W. Lindgren and G. F. Laughlin, 1919, Geology and ore deposits of the Tintic mining district, Utah, *U.S. Geol. Survey Prof. Paper* 107, 282 pp.

89. W. Lindgren and J. G. Creveling, 1928, The ores of Potosí, Bolivia, *Econ. Geology,* vol. 23, pp. 233–262, 459.

90. W. Lindgren, 1935, The silver mine of Colquijirca, Peru, *Econ. Geology,* vol. 30, pp. 331–346.

91. T. S. Lovering, 1949, Rock alteration as a guide to ore—East Tintic district, Utah, *Econ. Geology Mon.* 1, 64 pp.

92. Tom G. Lovering, 1958, Temperature and depth of formation of sulfide ore deposits at Gilman, Colorado, *Econ. Geology,* vol. 53, pp. 689–707.

93. T. S. Lovering and C. H. Behre, Jr., 1932, Battle Mountain (Red Cliff, Gilman) mining district, Internat. Geol. Cong., 16th, United States 1933, *Guidebook* 19, *Exc.* C-1, pp. 69–76.

94. T. S. Lovering, V. P. Sokoloff, and H. T. Morris, 1948, Heavy metals in altered rock over blind ore bodies, East Tintic district, Utah, *Econ. Geology,* vol. 43, pp. 384–399.

95. T. S. Lovering and H. T. Morris, 1960, The Chief Oxide—Burgin area discoveries, East Tintic district, Utah, a case history; part I, U.S. Geologi-

cal Survey studies and exploration, *Econ. Geology*, vol. 55, pp. 1116–1147.

96. G. K. Lowther and E. B. Bell, 1956, Geologia de la mina Esmeralda, Parral, Chihuahua, Internat. Geol. Cong., 20th, Mexico 1956, *Guidebook Exc.* A-2 and A-5, pp. 93–96.

97. G. K. Lowther and G. C. Marlow, 1956, Geologia del area de Parral, Internat. Geol. Cong., 20th, Mexico 1956, *Guidebook Exc.* A-2 and A-5, pp. 79–88.

98. E. Mapes Vázquez and J. B. Stone, 1956, Notes sobre la geologia del distrito minero de Zacatecas, Internat. Geol. Cong., 20th, Mexico 1956, *Guidebook Exc.* A-2 and A-5, pp. 133–139.

99. G. C. Marlow and J. M. Smith, 1956, Geologia de la mina la Prieta, Compañia Minera, Asarco, S. A., distrito de Parral, Chihuahua, Internat. Geol. Cong., 20th, Mexico 1956, *Guidebook Exc.* A-2 and A-5, pp. 89–91.

100. H. E. McKinstry and J. A. Noble, 1932, The veins of Casapalca, Peru, *Econ. Geology*, vol. 27, pp. 501–522.

101. H. E. McKinstry, 1936, Geology of the silver deposit at Colquijirca, Peru, *Econ. Geology*, vol. 31, pp. 618–635.

102. D. H. McLaughlin, L. C. Graton, and others, 1935, Copper in the Cerro de Pasco and Morococha districts, Department of Junin, Peru, Internat. Geol. Cong., 16th, United States 1933, "Copper Resources of the World," vol. 2, pp. 513–544.

103. C. W. Merrill, 1930, Summarized data of silver production, *U.S. Bur. Mines Econ. Paper* 8, 58 pp.

104. B. J. Miller and J. T. Singewald, Jr., 1919, "The Mineral Deposits of South America," McGraw-Hill Book Company, New York, 598 pp.

105. E. S. Moore, 1956, Geology of the Miller Lake portion of the Gowganda silver area, *Ontario Dept. Mines, Ann. Rept.*, vol. 64, pt. 5, 41 pp.

106. E. S. Moore, 1957, Gowganda silver area, in "Structural Geology of Canadian Ore Deposits, Congress Volume," pp. 388–392, Commonwealth Min. Metall. Cong., 6th, Canada 1957, Canadian Institute of Mining and Metallurgy, Montreal.

107. R. H. Nagell, 1960, Ore controls in the Morococha district, Peru, *Econ. Geology*, vol. 55, pp. 962–984.

108. E. Ordoñez, 1902, The mining district of Pachuca, Mexico, *Am. Inst. Mining Metall. Engineers Trans.*, vol. 32, pp. 224–241.

109. T. S. Osborne, 1956, Geologia y depositos minerales del distrito minero de Taxco, Internat. Geol. Cong., 20th, Mexico 1956, *Guidebook Exc.* A-4 and C-2, pp. 75–89.

110. B. Prescott, 1915, The main mineral zone of the Santa Eulalia district, Chihuahua, *Am. Inst. Mining Engineers Trans.*, vol. 51, pp. 57–99.

111. F. L. Ransome and F. G. Calkins, 1908, The geology and ore deposits of the Coeur d'Alene district, Idaho, *U.S. Geol. Survey Prof. Paper* 62, 203 pp.

112. R. R. Reid (ed.), 1961, Guidebook to the geology of the Coeur d'Alene mining district, *Idaho Bur. Mines and Geology, Bull.* 16, 37 pp.

113. H. A. Schmitt, 1931, Geology of the Parral area of the Parral district,

Chihuahua, Mexico, *Am. Inst. Mining Engineers Trans.*, vol. 96, pp. 268–289.

114. J. B. Scott, 1958, Structure of the ore deposits at Santa Barbara, Chihuahua, Mexico, *Econ. Geology,* vol. 53, pp. 1004–1037.

115. G. P. Serrano, 1956, The silver-mining industry in Mexico, *Min. Cong. Jour.*, vol. 42, no. 5, pp. 71–74.

116. G. H. Shefelbine, 1957, Silver-lead-zinc mines at Namiquipa, Chihuahua, Mexico, *Mining Eng.,* vol. 9, pp. 1090–1097.

117. P. J. Shenon, 1948, Lead and zinc deposits of the Coeur d'Alene district, Idaho, Internat. Geol. Cong., 18th, Great Britain 1948, pt. 7, The geology, paragenesis, and reserves of the ores of lead and zinc, pp. 78–80.

118. P. J. Shenon and R. H. McConnel, 1939, The silver belt of the Coeur d'Alene district, Idaho, *Idaho Bur. Mines and Geology Pamph.* 50, 9 pp.

119. F. S. Simons and V. E. Mapes, 1956, Geology and ore deposits of the Zimapán mining district, state of Hidalgo, Mexico, *U.S. Geol. Survey Prof. Paper* 284, 128 pp.

120. F. S. Simons and V. E. Mapes, 1956, Geologia y yacimientos minerales del distrito minero de Zimapán, estado de Hidalgo, Internat. Geol. Cong., 20th, Mexico 1956, *Guidebook Exc.* A-3 and C-1, pp. 59–72.

121. J. T. Singewald, Jr., 1935, The Quiruvilca district, Peru, Internat. Geol. Cong., 16th, United States 1933, "Copper Resources of the World," vol. 2, pp. 545–546.

122. G. H. Smith, 1943, History of the Comstock Lode, 1850–1920, *Nevada Univ., Bull. Geol. and Min.* vol. 37, no. 3, 297 pp.

123. R. E. Sorenson, 1947, Deep discoveries intensify Coeur d'Alene activities, *Eng. Mining Jour.,* vol. 148, no. 10, pp. 70–78.

124. J. E. Spurr, 1905, Geology of the Tonopah mining district, Nevada, *U.S. Geol. Survey Prof. Paper* 42, 295 pp.

125. J. B. Stone, 1956, Notes sobre el distrito minero de Fresnillo, Zacatecas, Internat. Geol. Cong., 20th, Mexico 1956, *Guidebook Exc.* A-2 and A-5, pp. 131–132.

126. J. B. Stone, 1959, Ore genesis in the Naica district, Chihuahua, Mexico, *Econ. Geology,* vol. 54, pp. 1002–1034.

127. C. O. Swanson, 1948, The Sullivan mine, Kimberley, B.C., Internat. Geol. Cong., 18th, Great Britain 1948, pt. 7, The geology, paragenesis, and reserves of the ores of lead and zinc, pp. 40–46.

128. C. O. Swanson and H. C. Gunning, 1948, Sullivan mine, in "Structural Geology of Canadian Ore Deposits," pp. 219–230, Canadian Institute of Mining and Metallurgy, Montreal, 948 pp.

129. R. Thompson, 1957, Cobalt camp, in "Structural Geology of Canadian Ore Deposits, Congress Volume," pp. 377–388, Commonwealth Min. Metall. Cong., 6th, Canada 1957, Canadian Institute of Mining and Metallurgy, Montreal.

130. J. B. Umpleby and E. L. Jones, Jr., 1923, Geology and ore deposits of Shoshone County, Idaho, *U.S. Geol. Survey Bull.* 732, 156 pp.

131. A. Wandke and J. Martínez, 1928, The Guanajuato mining district, Guanajuato, Mexico, *Econ. Geology,* vol. 23, pp. 1–44.

132. H. J. Ward, 1961, The pyrite body and copper ore bodies, Cerro de Pasco mine, Peru, *Econ. Geology,* vol. 56, pp. 402–422.

133. I. F. Wilson, 1956, El distrito minero de Naica, Chihuahua, Mexico, Internat. Geol. Cong., 20th, Mexico 1956, *Guidebook Exc.* A-2 and A-5, pp. 63–75.

MISCELLANEOUS REFERENCES CITED

134. R. L. Ames, 1962, The origin of the gold-quartz deposits, Yellowknife, N. W. T., *Econ. Geology,* vol. 57, pp. 1137–1140.

135. E. S. Bastin, 1939, The nickel-cobalt-native silver ore type, *Econ. Geology,* vol. 34, pp. 1–40.

136. E. S. Bastin, 1950, Significant replacement textures at Cobalt and South Lorraine, Ontario, Canada, *Econ. Geology,* vol. 45, pp. 808–817.

137. A. J. Bodenlos and G. E. Ericksen, 1955, Lead-zinc deposits of Cordillera Blanca and northern Cordillera Huayhuash, Peru, *U.S. Geol. Survey Bull.* 1017, 166 pp.

138. A. J. Bodenlos and J. A. Straczek, 1957, Base-metal deposits of the Cordillera Negra, Departmento de Ancash, Peru, *U.S. Geol. Survey Bull.* 1040, 165 pp.

139. R. W. Boyle, 1959, The geochemistry, origin, and role of carbon dioxide, water, sulfur, and boron in the Yellowknife gold deposits Northwest Territories, Canada, *Econ. Geology,* vol. 54, pp. 1506–1524.

140. R. W. Boyle, R. K. Wanless, and J. A. Lowden, 1963, The origin of the gold-quartz deposits, Yellowknife, N. W. T., *Econ. Geology,* vol. 58, pp. 804–807.

141. Canadian Mineral Industry, 1961, chapter on silver, *Canada, Min. Res. Div., Dept. Mines and Tech. Surveys.*

142. R. W. Chaney, 1932, The Alaska Juneau enterprise, *Eng. Mining Jour.,* vol. 133, pp. 460–503.

143. E. H. Cobb, 1962, Lode gold and silver occurrences in Alaska, *U.S. Geol. Survey Mineral Inv. Mineral Resource Map* MR-32.

144. W. H. Dunham, 1950, Table 3, Summarized geology of the principal lead and zinc mines of the world, with statistics on production and reserves, Internat. Geol. Cong., 18th, Great Britain 1948, pt. 7, The geology, paragenesis, and reserves of the ores of lead and zinc, pp. 24–25.

145. N. E. Ebbley, Jr., and J. S. Schumacher, 1949, Examination, mapping, and sampling of mine shafts and under ground workings, Leadville, Lake County, Colorado, *U.S. Bur. Mines Rept. Inv.* 4518, 115 pp.

146. R. A. Elgin and others, 1949, The Leadville drainage tunnel, Lake County, Colorado, *U.S. Bur. Mines Rept. Inv.* 4493, 37 pp.

147. J. E. Gair, 1962, Geology and ore deposits of the Nova Lima and Rio Acima quadrangles, Minas Gerais, Brazil, *U.S. Geol. Survey Prof. Paper* 341-A, 67 pp.

148. A. R. Geyne, 1949, Mineral relationship in the ores of Pachuca and Real del Monte, Hidalgo, Mexico, *Econ. Geology,* vol. 44, pp. 233–234.

149. A. R. Geyne and others, 1961, Pachuca mining district, Hidalgo, Mexico, *U.S. Geol. Survey Prof. Paper* 424-D, pp. 221–222.

150. W. A. Johnston and W. L. Uglow, 1926, Placer and vein deposits of Barkerville, Cariboo district, B.C., *Canada Geol. Survey, Mem.* 149, 246 pp. History of early mining, pp. 4–7.

151. A. H. Koschmann and M. H. Bergendahl, 1961, How about gold? Where mined and future outlook, *Mining World,* vol. 23, no. 1, pp. 26–28.

152. A. H. Koschmann and M. H. Bergendahl, 1962, Gold in the United States, exclusive of Alaska and Hawaii, *U.S. Geol. Survey, Mineral Inv. Resource Map* MR-24.

153. E. S. Larsen, Jr., D. Gotfried, H. W. Jaffe, and C. L. Waring, 1938, Lead-alpha ages of the Mesozoic batholiths of western North America, *U.S. Geol. Survey Bull.* 1070-B, p. 51.

154. P. Sanchez Mejorada, 1959, Horizontes metaliferos del cretacico en una comarca del noreste de la Republica Mexicana, Internat. Geol. Cong., 20th, Mexico 1956, pt. 13, sec. 13, Petrographic provinces, igneous and metamorphic rocks, pp. 357–370.

155. *Mining Jour. (London) Supplement,* 1963, July 26, p. 3. Relates to gold recovery by mines of the Witwatersrand.

156. S. Narayanaswami, M. Ziauddin, and A. V. Ramachandra, 1960, Structural control and localization of gold-bearing lodes, Kolar gold field, India, *Econ. Geology,* vol. 55, p. 1430.

157. C. G. Pfordte, 1894, The Cerro de Pasco mining industry, *Am. Inst. Mining Engineers Trans.,* vol. 24, pp. 107–121.

158. R. H. Ridgway, 1929, Summarized data of gold production, *U.S. Bur. Mines Econ. Paper* 6.

159. P. S. Smith, 1933, Past placer-gold production from Alaska, *U.S. Geol. Survey Bull.* 857-B, p. 96.

160. A. C. Spencer, 1904, The Treadwell ore deposits, Douglas Island, *U.S. Geol. Survey Bull.* 259, pp. 69–87.

161. A. C. Spencer, 1906, The Juneau gold belt, Alaska, *U.S. Geol. Survey Bull.* 287, pp. 93–116.

162. F. L. Stillwell, 1918, Replacement in the Bendigo quartz veins and its relation to gold deposition, *Econ. Geology,* vol. 13, pp. 100–111.

163. F. L. Stillwell, 1921, Vein filling at Bendigo, Victoria, *Econ. Geology,* vol. 16, pp. 153–159.

164. F. L. Stillwell, 1950, Origin of the Bendigo saddle reefs, *Econ. Geology,* vol. 45, pp. 687–701.

165. F. L. Stillwell, 1953, Formation of Bendigo quartz reefs, in *Fifth Empire Min. Metall. Cong.,* Australia and New Zealand, pp. 1028–1033, Australasian Institute of Mining and Metallurgy, Melbourne.

166. C. L. Thornburg, 1945, Some applications of structural geology to mining in the Pachuca-Real del Monte area, Pachuca silver district, Mexico, *Econ. Geology,* vol. 40, pp. 283–297.

167. T. M. Verity, 1961, Gold, in Canadian mineral industry 1961, *Canada, Mineral Resource Div., Dept. Mines and Tech. Surveys.*

168. R. K. Wanless, R. W. Boyle, and J. A. Lowdon, 1960. Sulfur isotope investigations of the gold-quartz deposits of the Yellowknife district, *Econ. Geology,* vol. 55, pp. 1591–1621.

169. E. Wisser, 1937, Formation of the north-south fractures of the Real del Monte area, Pachuca silver district, Mexico, *Am. Inst. Mining Engineers Trans.*, vol. 126, pp. 442–486.

170. E. Wisser, 1942, The Pachuca silver district, Mexico, in "Ore Deposits As Related to Structural Features," pp. 229–235, Princeton University Press, Princeton, N.J., 280 pp.

171. E. Wisser, 1946, Applications of structural geology to mining, Pachuca, Mexico, *Econ. Geology*, vol. 41, pp. 77–86.

172. E. Wisser, 1951, Tectonic analysis of a mining district, Pachuca, Mexico, *Econ. Geology*, vol. 46, pp. 459–477.

173. Anon., 1961, Principal Alaskan dredging operations, *Mining Eng.*, vol. 13, pp. 1335–1336.

174. Anon., 1962, The Soviet gold industry in 1961, *Mining Jour. (London)*, vol. 259, no. 6621, pp. 31–33.

chapter *11*

Lead, Zinc, Copper, Cobalt, and Cadmium

We have seen that deposits of lead, zinc, and copper furnish much silver and gold; further, that not uncommonly lead, zinc, and copper are produced from the same deposit, although in many places one metal is more abundant than either of the others. Lead and zinc are very commonly associated in deposits that lack recoverable copper and that lack also silver to any great extent and, hence, are important for lead or zinc alone. We have also seen that certain silver deposits have cobalt associated with them. The world production of cobalt, however, comes largely from copper deposits of Africa. Cadmium was noted as a by-product from some of the lead-zinc-silver mines of British Columbia, but most of the world's supply of cadmium is recovered from zinc concentrates rather than from the mining of deposits for cadmium alone.

The origin of many deposits of lead, zinc, and copper was discussed with deposits of gold and silver, and it will be remembered that many were hydrothermal deposits, some formed at high temperature but a great many formed at moderate or moderate to low temperature. One world-wide type of lead-zinc deposit that contains essentially no silver, commonly referred to in the United States as the Mississippi Valley type, has been thought by many to have resulted from low-temperature hydrothermal activity, whereas others believe that many such deposits came into existence syngenetically along with the sedimentary beds in which they are inclosed. Similarly, certain copper deposits inclosed in sedimentary beds, such as many of the great deposits of Africa, are thought by some to be low-temperature hydrothermal deposits and by others to be syngenetic sedimentary deposits. Some other types of copper deposits generally differ from deposits of lead or zinc. Here belong the extremely important porphyry copper deposits and the native copper deposits associated with basic lava flows. The porphyry copper deposits are disseminations and replacements in stocklike bodies of porphyry, and some of them furnish important amounts of by-product molybdenum. Generally they appear to have formed under moderate-temperature hydrothermal conditions. The native copper deposits typically occur in amygdaloidal basic lavas and rocks associated with them, both as cavity fillings and replacements. The most important commercial occurence of such deposits has been the Lake Superior region of northern Michigan.

Those deposits of lead, zinc, and copper that were discussed with deposits of gold and silver will be omitted here, except for mention of their importance, and attention will be focused on deposits not previously discussed.

Lead and zinc

Six countries—the United States, Canada, Australia, Mexico, Peru, and the Soviet Union—produced about 66 percent of the lead and 64 percent of the zinc mined in the world in the 15-year period 1945 to 1959, and these same countries furnished nearly 59 percent of the lead and also 58 percent of the zinc mined in the world in 1962 (Figs. 11-1, 11-2). It will be recalled that these same six countries supplied 78 percent of the world's silver in the same 15-year period and nearly 78 percent in 1962. Among these countries Peru and Australia made notable advances in production throughout the 15-year period, both in lead and zinc. Peru furnished about two and a half times as much lead and zinc in the five years 1955 to 1959 as in the corresponding length of time from 1945 to 1949, whereas Australia moved from second place to first place in world production of

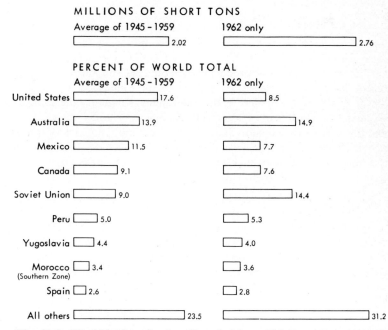

Fig. 11-1 World lead production. Compiled from *U.S. Bur. Mines Minerals Yearbook,* various years.

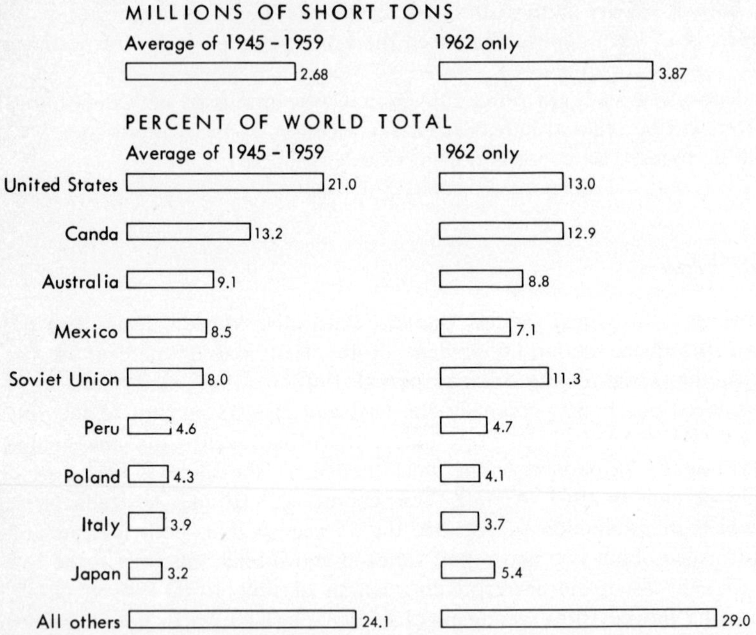

MILLIONS OF SHORT TONS

Average of 1945 – 1959	1962 only
[ ] 2.68	[ ] 3.87

PERCENT OF WORLD TOTAL

	Average of 1945 – 1959	1962 only
United States	[ ] 21.0	[ ] 13.0
Canda	[ ] 13.2	[ ] 12.9
Australia	[ ] 9.1	[ ] 8.8
Mexico	[ ] 8.5	[ ] 7.1
Soviet Union	[ ] 8.0	[ ] 11.3
Peru	[ ] 4.6	[ ] 4.7
Poland	[ ] 4.3	[ ] 4.1
Italy	[ ] 3.9	[ ] 3.7
Japan	[ ] 3.2	[ ] 5.4
All others	[ ] 24.1	[ ] 29.0

Fig. 11-2 World zinc production. Compiled from *U.S. Bur. Mines Minerals Yearbook,* various years.

lead. Estimates indicate that the Soviet Union quadrupled lead production for those same periods, but accurate figures of production are unavailable.

Early distribution of production of both lead and zinc differed radically from recent production. The major differences are indicated by Table 11-1, from which it will be noted that Spain, the United Kingdom, the United States, and Germany supplied nearly 82 percent of the world's lead from 1801 through 1900, and that Germany, Belgium, the United States, and the United Kingdom furnished about 89 percent of the zinc during the same period. The table shows, also, that Spain has produced much lead but not zinc, whereas Germany and Belgium have produced much more zinc than lead.

United States

Distribution of the most important lead and zinc districts or areas of the United States is shown by Table 11-2, which summarizes total production and reserves. The table shows also the major type of deposit of each district.

Four states produced about 87 percent of the lead of the United States in 1961, as follows: Missouri, 37.7; Idaho, 27.3; Utah, 15.6; and Colorado, 6.7. Zinc production was much more scattered, but 11 states supplied about 92 percent of the total: Tennessee, 17.6; Idaho, 12.5; New York, 11.8; Colorado, 9.2; Utah, 8.0; Arizona, 6.4; Virginia, 6.3; Illinois, 5.7; Pennsylvania, 5.0; New Mexico, 4.9; and Washington, 4.3. Among all these states, Missouri produced far more lead

TABLE 11-1.   Major Differences in Distribution of Production of Lead and Zinc during Various Periods

Country	Lead, percent of world production			Zinc, percent of world production		
	1945–1959	*1901–1925*	*1801–1900*	*1945–1959*	*1901–1925*	*1801–1900*
United States	17.6	37.2	19.6	21.0	40.6	12.5
Australia	13.9	9.6	2.9	9.1	1.1	—
Mexico	11.5	7.8	3.3	8.5	—	—
Canada	9.1	2.3	0.4	13.2	0.9	—
Soviet Union	9.0	—	0.3	8.0	0.7	1.9
Peru	5.0	—	—	4.6	—	—
Yugoslavia	4.4	0.2	—	1.8	—	—
Morocco (French, Southern Zone)	3.4	—	—	1.0	—	—
Spain	2.6	15.5	25.4	2.8	1.1	1.5
Germany (West Germany)	2.5	11.4	16.6	2.8	21.8	42.8
United Kingdom	—	1.7	20.3	—	5.7	7.4
Poland	1.6	0.2	—	4.3	1.9	—
Belgium	—	3.3	2.0	—	15.6	26.4
Total	80.6	89.2	90.8	77.1	89.4	92.5

than zinc; Tennessee, New York, and Pennsylvania, chiefly or entirely zinc; Arizona, Virginia, Illinois, Washington, New Mexico, considerably more zinc than lead; and Idaho, Utah, and Colorado, very considerable amounts of both lead and zinc.

The most important lead district in 1961 was Southeastern Missouri, with five mines classed among the leading 25 of the United States (ranks first, fourth, fifth, sixth, and twelfth). The Idaho production of both lead and zinc came chiefly from the Coeur d'Alene district; the Utah production, chiefly from the West Mountain (Bingham) district but in part from the Park City and some other mines; the Colorado production, to a large extent from the Idarado mine of the Upper San Miguel district and the Eagle mine of the Red Cliff (Battle Mountain) district. Much of the production of the western United States comes from ores that yield silver, and consequently some of the outstanding occurrences, such as Coeur d'Alene, Park City, Tintic, and Leadville, have been discussed. Others, such as Bingham, Butte, and Bisbee, will be discussed with copper deposits. Here we will give especial attention to deposits that contain little or no silver, such as those of Southeastern Missouri, the Tri-State district (Missouri, Kansas, Oklahoma), Eastern Tennessee, and St. Lawrence County, New York (Edwards-Balmat area).

The Mississippi Valley Region. The Mississippi Valley region is the outstanding lead-zinc province not only of the United States, but also of the world. Within this region there are six producing areas, four of which have supplied the major amounts mined (6). These are the Tri-State district, the Southeast Missouri district, the Southern Illinois–Kentucky district, and the Upper Mississippi Valley district. Among them the Southeast Missouri and the Tri-State districts are by far the most important.

TABLE 11-2. Most Important Lead and Zinc Districts of the United States*

Area and state	District or mine	Lead production plus reserves		Zinc production plus reserves		Type of deposit
		More than 1,000,000 short tons	50,000–1,000,000 short tons	More than 1,000,000 short tons	50,000–1,000,000 short tons	
WESTERN AREA						
Washington	Metaline	X		X		Replacement bodies in Cambrian dolomite and brecciated dolomitic limestone
Idaho	W. Coeur d'Alene	X		X		Composite replacement veins along shear zones in Precambrian sericitic quartzite
	E. Coeur d'Alene		X	X		Same as above
Montana	Butte (Summit Valley)		X	X		Replacement lodes and veins in quartz monzonite of early Tertiary (?) age
Utah	Bingham (West Mountain)	X		X		Bedded replacement bodies along faults in Pennsylvanian limestone beds between quartzite beds
	Park City	X			X	Bedded replacement and lode deposits along fissures in Pennsylvanian, Permian, and Triassic limestone; lode deposits in Pennsylvanian quartzite and Upper Cretaceous (?) diorite porphyry
	Tintic	X			X	Replacement bodies along fractures in Cambrian, Ordovician, and Mississippian limestone and dolomite; veins in Tertiary igneous rocks
Colorado	Leadville (California)	X		X		Replacement bodies along fissures and chiefly below porphyry sills or shale, in Ordovician and Mississippian limestones
New Mexico	Central (Hanover)		X	X		Replacement veins on faulted contact between porphyry dikes and Carboniferous limestone, or in dikes; contact-metasomatic deposits

District					Remarks
CENTRAL-MIDWESTERN AREA					
Oklahoma — Picher field (Tri-State district)	X		X		Replacement bodies along bedding and within breccias in Mississippian limestone and chert
Missouri — Joplin area (Tri-State district)		X	X		Replacement bodies along bedding and within breccias and collapsed sinks in Mississippian limestone and chert
Viburnum	X			X	Replacement bodies within algal reef facies in Cambrian dolomite along flanks of buried Precambrian knobs
Flat River–Bonne Terre (Southeastern Missouri area)	X				Replacement bodies in Cambrian dolomite near buried Precambrian ridges
Wisconsin — Upper Mississippi Valley		X	X		Replacement bodies and veins along shears and faults in Ordovician dolomite and limestone
EASTERN AREA					
Tennessee — Copper River–Flat Gap			X		Replacement bodies and veinlets in brecciated Ordovician dolomite
Mascot-Jefferson City–New Market			X		Same as above
Virginia — Austinville-Ivanhoe		X	X		Ore along limb of anticline in brecciated Cambrian dolomite
Gossan Lead			X		Replacement lenses along shear zone in Precambrian gneiss and schist
Pennsylvania — Friedensville			X		Fissure filling in highly folded and faulted Ordovician limestone
New Jersey — Franklin–Sterling Hill			X		Replacement bodies in Precambrian marble cut by pegmatite dikes
New York — Balmat-Edwards		X	X		Replacement bodies along channels of microbreccias in Precambrian limestone

* Compiled from *U.S. Geol. Survey Mineral Inv. Resource Map* MR-15 and MR-19, 1962.

The Tri-State district (3, 6, 20, 52), in which the Joplin, Missouri, and the Picher, Oklahoma, areas are the most important, has been productive since 1848 and has supplied ores valued at more than a billion dollars. The deposits, commonly designated the *Mississippi Valley type,* consist chiefly of sphalerite and galena with a gangue of dolomite, calcite, and jasperoid in limestone and chert, chiefly of the Boone formation but to some extent of the overlying Chester group, both of Mississippian age. The Boone formation consists of limestone, chert, limy chert, and some shaly limestone; is 350 to 400 ft thick; and contains ore deposits at places throughout its entire thickness, but more especially in certain favorable units, as some coarse-grained crinoidal and cherty limestones in the Picher area. Below such main ore-bearing units, the Grand Falls chert member, about 35 to 60 ft thick, bears low-grade deposits at some places where it contains, in addition to chert, subordinate, thin, and somewhat lenticular beds of limestone. Locally ore occurs in limestone beneath the Grand Falls chert and above the main ore-bearing units. Although the strata are generally nearly horizontal, shallow folds, some faults, and zones of jointing and shearing are present, and shattering provided space for the introduction of mineralizing solutions.

The deposits may be designated (1) *runs,* which may grade into blanket deposits, (2) *sheet ground deposits,* and (3) *sinkhole deposits.* The runs may follow steep fractures or other zones in broken ground and commonly show a relatively flat floor; in the more favorable beds, the runs may pass locally into rich blanket deposits. The sheet ground deposits, which occur in the Grand Falls chert member, are generally not more than 10 to 12 ft thick. Sinkholes formed during Mississippian time were the sites for the sinkhole deposits, and in these groundwater solution probably aided fracturing and shattering in making space for ore deposition. Fowler (20, 156) regarded structural deformation as a factor of particular importance in the Tri-State district and stated (p. 207) that "the degree and character of structural deformation appear to have controlled the size, shape, and richness of every ore deposit."[1] In general, both ore and gangue minerals replace limestone of favorable beds or fill fractures in brecciated chert; locally open spaces were lined by crystals, many of which are on display in various museums.

In the Tri-State district zinc has been more important than lead, and the ratio of zinc production to lead production has been about 3 or 4 to 1. In 1949 the Tri-State district was the leading zinc producer of the United States, but the district also produced much lead because of the large tonnages of ore mined (34, 35). To the end of 1942 the total production of lead was 2,319,314 tons, and to the end of 1944 the total production of zinc was 20,500,000 tons (26). The grade of the ore is relatively low. In 1946, which was a year of very favorable prices, the average of the whole field was 0.30 percent lead and 1.63 percent zinc. Selective mining increased the grade somewhat, and the ore recovered in 1949 averaged 0.69 percent lead and 1.66 percent zinc (34). The peak production of the district was in 1926, and production has constantly declined since then. In 1961 only one mine of the district was among the leading 25 zinc-producing mines in the United States, and that one ranked 25.

[1] From G. M. Fowler, in "Ore Deposits As Related to Structural Features" (W. H. Newhouse, editor), p. 207. Copyright 1928. Princeton University Press. Used by permission.

Both lead and zinc are recovered in the Upper Mississippi Valley district (1, 6, 24, 25). The deposits are chiefly in the lower part of the Galena dolomite and in the limestone and dolomite underlying the Decorah and Platteville formations, all of Ordovician age. Many of the ore bodies were controlled in part by small folds, are commonly elliptical in outline, and the ore was localized as replacements in inclined shear zones and fractured zones ("pitches") that accompany bedding plane fractures ("flats"), which are also mineralized. Some deposits are replacement veins along vertical joints in the Galena dolomite and are known as "crevice" or gash-vein deposits. The chief ore mineral in the crevice deposits is galena, whereas both galena and sphalerite generally are present in the other deposits. The mineralogy, although similar to that of the Tri-State district, is somewhat more complex, and the galena contains from 0.12 to 2.0 oz of silver per ton. The district produces considerably more zinc than lead, and in 1953 ranked seventh in zinc production and fourteenth in lead production in the United States.

The Illinois-Kentucky district (6) contains veins and tabular bedding replacement deposits in Mississippian limestones. Both types may contain galena and sphalerite, and the bedding deposits have been of considerable commercial importance since 1939. A notable feature of these deposits is the presence of fluorite, which supplied nearly 70 percent of the fluorite mined in the United States in 1961. The deposits will be mentioned again in the discussion of fluorite deposits.

The Southeast Missouri district (6, 13, 27, 44, 50) produced 6,016,742 tons of lead to the end of 1942 (26) and in excess of 8,000,000 tons to 1956, valued at more than a billion dollars. The greater part of this production has come from an area of 50 sq miles around Flat River. Recent exploration has resulted in the development of deposits farther west (150). The chief deposits of the district are disseminations in the Bonneterre dolomite of Upper Cambrian age, which is 450 ft thick. Mineralization exists to some extent throughout the entire formation, but the major deposits are in the lower 270 ft of it. Other deposits, now of little importance but which supplied much of the early production, occur in narrow fractures in the Potosí dolomite, also of Upper Cambrian age, about 400 ft above the Bonneterre formation and west and north of the disseminated deposits. In the major lead belt, production has come chiefly from galena disseminated in the dolomite but also, to some extent, from massive replacements along the bedding and from veinlets and fillings of small cavities. Generally the ore bodies show wide horizontal extent. Localization of most of the ore bodies, according to Snyder and Emery (50, p. 1219), was controlled by "structural centers" that came into existence by deposition over topographic highs. James (27), however, stated that localization was controlled by various faults, fractures, and deformation accompanying folding and by knob and ridge structure related to basement topography at the time of deposition of the sediments.

The mode of origin of the Mississippi Valley lead and zinc deposits has been a matter of debate for many years. The entire question was thoroughly reviewed in 1939 (3) and a critical summary of the origin presented (4). In general two hypotheses have been suggested: (1) solution of materials from rocks of the region by meteoric waters and subsequent deposition in favorable places; (2) bringing in materials by hydrothermal solutions derived from a deeply buried magmatic source and subsequent deposition in favorable places. Igneous rocks are widely scattered throughout the region, but the total area of outcrop is small. They cut various rocks of Paleozoic age, and some igneous rocks may be as young as Middle Cretaceous. They are particularly abundant in the Illinois-Kentucky district, where the fluorite deposits seem to be genetically related to them. Twenty years after the 1939 review, Ohle (43) reviewed new information published since that time. He noted (p. 769) that in 1939 there was the general, but not unanimous, conclusion that the deposits were of hydrothermal origin, but at that time no attempt was made to introduce evidence from other districts of the Mississippi Valley type both in the United States and abroad. He then reviewed briefly the published information regarding deposits in the Mississippi Valley and similar deposits in the United States and throughout the world; summarized the general characteristics of such deposits, which he considered to be sufficiently similar to differentiate them as a group of the Mississippi Valley type; and concluded (p. 787) that all deposits of the Mississippi Valley type apparently had the same mode of origin. At present there seems to be a tendency on the part of one group of geologists to regard deposits of lead, zinc, and copper that are to a large extent confined to certain stratigraphic horizons as syngenetic deposits that accumulated at the same time as the beds in which they occur and were later modified by regional metamorphism or circulating groundwater or both, with some but not great transportation of materials. Another group favors a hydrothermal origin and derivation of material from a magmatic source—an epigenetic concept. Opinion is divided, and the problem is similar to one discussed briefly in connection with the gold deposits of the Witwatersrand.

The Eastern Tennessee–Western Virginia Area.   The Eastern Tennessee–Western Virginia zinc area extends roughly from Knoxville, Tennessee, northeastward to Roanoke, Virginia, a distance of about 300 miles, and is 10 to 40 miles wide. The major producing area is near Knoxville and includes the mining districts of Mascot, Jefferson City, Powell River, and Copper Ridge. A smaller producing area is localized around Austinville, Virginia, about 60 miles southwest of Roanoke.

The Eastern Tennessee district led the entire United States in the production of zinc in 1961. Five mines of the area were among the leading 25 zinc-producing mines of the country and ranked sixth, eighth, eleventh, thirteenth, and nineteenth. The reserves were estimated to be 10 million tons of zinc concentrates.

The Eastern Tennessee district (8, 29, 40, 42) is a mineralized area of Paleozoic rocks of the Appalachian Valley in the vicinity of Knoxville. The deposits occur in folded and faulted limestones of the Knox group (Upper Cambrian and Lower Ordovician), which has been divided into the Upper Cambrian Copper Ridge dolomite and four succeeding Lower Ordovician formations: Chepultepec dolomite, Longview dolomite, Kingsport limestone, and Mascot dolomite.[1] Although mineralization occurs in various formations, major commercial production has come only from the Kingsport limestone, especially the lower half of it; recently some commercial ore bodies have been found in the overlying Mascot dolomite. The ore bodies, in which sphalerite is the chief ore mineral although locally minor galena is present, form (1) replacements roughly along the bedding in coarsely

---

[1] For detailed stratigraphy see Josiah Bridge, 1956, Stratigraphy of the Mascot–Jefferson City zinc district, Tennessee, *U.S. Geol. Survey Prof. Paper 277*, 76 pp.

crystalline dolomitized limestone along with breccia filling in associated fine-grained dolomite and (2) replacements along faults, with breccia and fracture filling near the faults. The large zinc ore bodies are of the first type. Generally they are irregular, with ore shoots forming more or less interconnected branches associated with which are blocks of noncommercial rock known locally as "blank islands" or "limestone islands" (42, p. 8). These ore bodies are generally elongated in the direction of the regional strike of the rocks.

The origin of these deposits, just as the origin of those of the Mississippi Valley, has long been a matter of debate. The two concepts, solution and deposition of zinc by meteoric waters or deposition from hydrothermal solutions, were the ones most usually advanced, but in 1960 Kendall (29) stated that "concentration and deposition from the sea contemporaneous with sedimentation is suggested," and, although a simple syngenetic origin explains many features of the deposits, "it cannot account for the migration of ore into fractured fine dolomite beds above the ore shoots in the limestone beds." His final statement at that time was: "There is some suggestion that the ore is of sedimentary origin, but the presence of fracture-filling ore in overlying beds requires modification of a sedimentary ore bed by later hydrothermal (though not necessarily hypogene derived) activity."[1] C. H. Behre, Jr., in 1962 (151), presented a valuable discussion of the types of evidence that should be considered in connection with deposits such as those of the Eastern Tennessee district.

The Virginia deposits (12, 17, 41), which have been commercially important in the Austinville district, have been considered by some to be a continuation of the Eastern Tennessee deposits. The Austinville deposits are replacements and fracture fillings in the Lower Cambrian Shady limestone. The localization of the ore bodies was effected by a combination of structural control by strike faults and stratigraphic control by incompetent beds, with the result that the ore bodies are pipelike and are enlarged near cross faults. In 1961 Virginia ranked seventh in production of zinc in the United States, and Virginia combined with Tennessee furnished about 24 percent of the total zinc of the country.

The New York–New Jersey Area. The Balmat mine of the Balmat-Edwards district, St. Lawrence County, New York, was rated first among the leading 25 zinc-producing mines in the United States in 1961, and this district furnished the entire output of zinc in New York, which state ranked third in production in the United States. The Edwards mine began operating in 1915, the Balmat mine, 12 miles to the southwest, in 1930. The deposits (9, 10, 11) are replacements of folded Precambrian Grenville limestone that lies between a lower and an upper garnet gneiss. The limestone is highly siliceous, the silica being present as quartzitic zones or silicated zones that contain much tremolite and diopside along with talc and serpentine. Within the limestone are several gneissic members, apparently originally quartzite. Sphalerite and an almost equal amount of pyrite constitute the chief metallic minerals, but some galena is present, along with very minor amounts of chalcopyrite and pyrrhotite. The ore minerals were introduced later than the silicates, except talc and serpentine, apparently by hydrothermal solutions that

---

[1] From D. L. Kendall, *Econ. Geology*, vol. 55, pp. 985 and 1002. Copyright 1960. Economic Geology Publishing Company. Used by permission.

came in as a phase of some of the igneous activity of the area. The alteration of silicates to talc and serpentine seems to have been a late phase of the hydrothermal activity.

The Franklin-Sterling area, Sussex County, New Jersey (30, 46, 47), although of much less commercial importance than many zinc-mining districts, is of great interest because of the number of unusual minerals in the deposits, and thus will be described briefly. Two mines operate in the district, the Franklin, at Franklin, and the Sterling, at Ogdensburg. These mines had been developed to depths of 1,150 and 2,450 ft respectively in 1950. The deposits are in the folded Franklin or "white" limestone of Precambrian age, which is highly metamorphosed and coarsely crystalline and which has been intruded locally by pegmatites of granitic to syenitic composition and by basic dikes. In the Franklin mine pegmatite is close to and is common in the ore, and around the pegmatite are broad zones of skarn, but in the Sterling mine pegmatite is almost lacking in the ore. The ore bodies apparently replace certain beds along folded structures in the limestone. The approximate composition of the ore has been stated to be 40 percent franklinite, 23 percent willemite, slightly less than 1 percent zincite, 11 percent gangue silicates, and 25 percent gangue carbonates (46, p. 82).[1]

Various concepts of origin have been proposed. These were reviewed briefly by Pinger (46), after which he stated that replacement along favorable horizons by a primary oxide ore is the only hypothesis that fits the observed conditions, but no well-defined convictions had been reached. Ridge (47) stated that if replacement was by a primary oxide ore, the deposits must have been formed under conditions of high temperature and pressure and suggested that they are pyrometasomatic deposits, formed at some distance from an unknown igneous source. The deposits had been discussed previously by Lindgren [(10, pp. 737–739) of Chap. 2] with pyrometasomatic types.

Australia

Australia ranked first in world production of lead in 1962, second for the 15-year period 1945 to 1959, fourth in world production of zinc in 1962, and third for the period 1945 to 1959. Moreover, she ranked sixth in world production of silver in 1962 and also in the period 1945 to 1959. Since much of the silver is obtained from the lead-zinc deposits, those are highly important. The principal areas from which the lead-zinc-silver ores are obtained are the Broken Hill district and Captain's Flat, New South Wales; the Mount Isa district, Queensland; and the Read-Rosebury district, Tasmania. Among these Broken Hill has been by far the most important, and Mount Isa has ranked second. The reserves of the Broken Hill district were given in 1950 as 5,750,000 long tons of lead and 5,055,000 long tons of zinc, whereas those of Mount Isa were given as 863,000 long tons of lead and 744,000 long tons of zinc (155).

The Broken Hill district (2; 5, pp. 116–118, 122–123; 22; 31; 32; 51), in the northwestern part of New South Wales, is one of the great lead-zinc-silver districts of the world. Since about 1884 the Broken Hill deposit produced 86 million tons of ore that contained 11.2 million tons of lead, 7.6 million tons of zinc, and 633

---

[1] For detailed mineralogy of these deposits see C. Palache, 1935, The Minerals of Franklin and Sterling Hill, New Jersey, *U.S. Geol. Survey Prof. Paper* 180, 135 p.

million oz of silver (153), as well as some gold and cadmium. Ore is present to a depth of 3,000 ft. Rich oxidized silver ores containing native silver and horn silver, with values up to 300 oz of silver per ton, were the ones chiefly mined between 1885 and 1898. Later the hypogene ores were mined.

The Broken Hill deposits are contained in rocks of the Willyama series of early Precambrian age. These rocks consist of sillimanite-garnet gneisses, schists, and amphibolites, which have been folded and intruded both before and after folding. The deposits are massive lead-zinc sulfide replacements, and the boundaries of the ore bodies are generally parallel to bedding planes in the metamorphosed sedimentary rocks. Two favorable beds have contributed most of the ore, and the ore bodies apparently were formed where these favorable beds were intersected by a zone of intense plastic deformation locally designated the "belt of attenuation." These two favorable beds have given rise to two horizons, the No. 2 lens (upper) and the No. 3 lens (lower). The No. 3 lens has furnished about half of the ore and also the richest ore. The major ore minerals of the No. 3 lens are sphalerite and galena, the sphalerite slightly in excess. These are associated with various gangue minerals, including fluorite, rhodonite, garnet, quartz, and, very rarely, calcite. Among these rhodonite and fluorite are locally abundant. The major ore minerals of the No. 2 lens are also sphalerite and galena but, unlike the No. 3 lens, the galena is slightly in excess. The gangue minerals differ in amount from those of No. 3 lens, because in the No. 2 lens calcite is abundant instead of very rare, rhodonite is present in local small patches, and fluorite is rare. These differences in gangue minerals have caused some to refer to the ores as fluoritic ores (No. 3 lens) and calcitic ores (No. 2 lens) (22, p. 1383), and systematic sampling has established somewhat different metal ratios for the two types. The mineralogy of the No. 2 lens, however, appears to change with depth. Conditions are more complex than this generalization indicates, however, and some other lodes exist that contain much more zinc than lead.

Disagreement exists regarding the origin of the Broken Hill deposits. Some have suggested that the deposits are of hydrothermal origin and belong in the hypothermal class; others wish to assign a metamorphic origin to some gangue minerals and to place the ore minerals partly in the hypothermal class and partly in the mesothermal class; still others suggest that the Willyama sediments once contained simple conformable lead-zinc deposits, the origin of which is entirely conjectural, and that regional metamorphism produced granitization, reorganization of the original deposits, and the present complex mineral composition. Gustafson (23) discussed certain aspects of the deposits and closed with the statement (p. 786): "Broken Hill is still a challenge. It is also a severe testing ground for geological techniques and theories. It is fortunate that geological work there is being vigorously pursued by such capable and independently thinking geologists."[1]

The Mount Isa lead-zinc-silver deposit (5, pp. 118–119; 14; 15), with which is associated a copper deposit, is in western Queensland near the boundary of Northern Territory and about 600 miles west of Townsville, which is on the east coast. The deposits occur in relatively unmetamorphosed Precambrian (Proterozoic) shales, and the ore bodies form lenses in zones of folding and shearing

---

[1] From J. K. Gustafson, *Econ. Geology*, vol. 49, p. 786. Copyright 1954. Economic Geology Publishing Company. Used by permission.

and occupy the middle part of a mineralized zone about 5 miles long from north to south, which reaches a maximum width of about 2,000 ft near its central part and thins toward each end. The ore lenses occur chiefly in thinly bedded shale at certain horizons and are separated by thicker shales that commonly do not contain ore. The major source of the Mount Isa lead-zinc-silver production has been the Black Star group of lenses, the maximum economic length of which is about 2,000 ft, the maximum stratigraphic width, 180 ft; these lenses extend to at least 1,400 ft below the surface (14, pp. 197–198). Other groups of ore lenses occur farther to the south and two of these, the Black Rock and the Rio Grande, although smaller than the Black Star group are of higher grade and have furnished about 20 percent of the production. The average grades of these three bodies are: Black Star, 7.8 percent lead, 8.7 percent zinc, 5.1 oz silver; Black Rock and Rio Grande, 16.8 percent lead, 5.3 percent zinc, 14.2 oz silver. The lead-zinc-silver ores of all groups of lenses show a layered appearance, apparently a result of replacement of certain thin beds of shale along the bedding planes and to some extent on either side of them. The sulfide layers vary in thickness from 0.01 in. to 1 ft, and localization apparently resulted from shearing, crenulation, and some cross fracturing within the thinly bedded shales. The mineral composition is somewhat complex, as deposition apparently took place at three times. The chief primary ore minerals are galena, sphalerite, and argentiferous tetrahedrite; others include minor silver sulfosalts and varying amounts of chalcopyrite. Among gangue minerals pyrite is abundant and widespread, and both pyrrhotite and arsenopyrite are present. Others include coarsely crystalline dolomite and some chlorite. Much of the pyrite is very fine grained (0.005 to 0.025 mm) and some of it is in round or spherical aggregates. Oxidation and secondary enrichment have been important. Silver was largely leached within 80 ft of the surface but was reprecipitated as native silver, cerargyrite, and silver sulfosalts. Native silver was the most prominent silver mineral of the oxidized zone and generally occurred between 90 and 150 ft below the surface; pyrargyrite was the most abundant supergene silver sulfide. There was enrichment in both lead and zinc in the oxidized zone, although zinc was leached down to nearly 150 ft.

Copper production comes from a separate chalcopyrite ore body localized in fractures in a silica-dolomite zone west of the Black Star shales. The crest of the ore body is 800 ft below the surface, its length is about 1,000 ft so far as known, its maximum width, consisting of two lodes, 140 ft. The production from the copper ore body to 1953 (15) was 821,360 tons of ore containing 4.1 percent copper, and the reserves were estimated to be nearly 3 million tons of ore of similar grade. Some lead, zinc, and silver are also present.

The origin of the Mount Isa ore bodies has been considered to be by deposition from hydrothermal solutions, probably of moderate temperature, but this origin has recently been questioned. It has been suggested (33) that at least some of the fine-grained pyrite of the deposits, designated first generation by Carter (14), which is present in spherical forms, may be syngenetic and possibly precipitated by microorganisms. Fisher (18) reviewed evidence regarding origin of the Mount Isa deposits and concluded that, for the lead-zinc-silver ore bodies, it was rather strongly in favor of a syngenetic origin with subsequent recrystallization, but for the copper ore bodies it was fairly strongly in support of a hydrothermal origin.

Canada

The major production of lead and zinc in Canada (5, pp. 137–151; 38, pp. 23–27, 37–40, 54–55, 62) is from the deposits that supply most of the silver of that country—the Sullivan, Bluebell, and H. B. mines of British Columbia, and the Keno Hill–Galena Hill deposits of central Yukon, which were discussed previously. Some lead-zinc production comes also from deposits at Buchans, Newfoundland, and production was expected in 1963 from copper-lead-zinc deposits in the Bathurst area, New Brunswick (152, 158, 183).

The lead-zinc-copper deposits at Buchans (21; 38, pp. 53–54; 39), central Newfoundland, occur chiefly in tuff of the Buchans series, thought to be of Ordovician age, which has been intruded by granite, diabase, and quartz porphyry of pre-Carboniferous age. Two principal groups of ore bodies are known, the Lucky Strike, which is the larger, and the Oriental. In the Lucky Strike area the main ore body, localized at the center of an anticline, consists chiefly of fine-grained sphalerite, galena, chalcopyrite, some gold-silver tellurides, pyrite, barite, and quartz. The deposits are chiefly replacements in tuff. The Oriental ore bodies are localized in weak, incompetent tuff beds between relatively strong beds of arkose on the north limb of the anticline. The ore mined in 1961 from both bodies contained 7.38 percent lead, 12.88 percent zinc, 1.11 percent copper, and 4.59 oz of silver per ton, and production made Newfoundland rank second in amount of lead and third in amount of zinc in Canada. The tonnage of zinc produced in Newfoundland was considerably more than the tonnage of lead.

Much zinc was supplied by the Flin Flon district, Manitoba and Saskatchewan, discussed briefly with gold deposits. Much of the ore mined in the Flin Flon district in 1961 contained 0.2 percent lead, 4.0 percent zinc, 2.45 percent copper, 1.04 oz of silver per ton, and a considerable amount of gold.

Mexico and Peru

The state of Chihuahua, Mexico, furnished 52.2 percent of the lead and 56.4 percent of the zinc of Mexico in 1954, and, in general, production of lead and zinc came from those mines that supplied much silver, which were discussed in Chap. 10. Similarly, much of the lead and zinc of Peru has been furnished by deposits that supply also silver and copper (Chap. 10). Among producing districts of Peru, Cerro de Pasco, Morococha, and Casapalca have been especially important.

Soviet Union

Little specific information can be given about the lead and zinc deposits of the Soviet Union. Although the presence of lead-zinc deposits was known for many years, there was no serious production until about 1925 (34). Three regions apparently contain the most important reserves of lead and zinc (5, pp. 384–401; 48, pp. 126–127, 142), which, in order of importance, are (1) Turkeystan, especially the Altai Mountains with the Ridder, Sokol'noye, and Belousovka deposits, and Central Asia with the Turlan deposits; (2) North Caucasus, the Sadon deposits; and (3) the Far East, the Tetyukhe deposits.

TABLE 11-3.   Composition and Reserves of Ore of Some Important Lead-Zinc Deposits of the Soviet Union*

Region	Deposit	Location, latitude and longitude	Composition, %			Reserves, 1,000 metric tons	
			Pb	Zn	Cu	Pb	Zn
Turkeystan	Ridder	50°18'N, 83°35'E	6.7–2.4	11.9–3.2	0.7–0.3	465.3	740
	Sokol'noye	50°18'N, 83°35'E	5.4–2.2	8.4–4.9	0.4–0.3	336.5	732
	Belousovka	50°07'N, 82°33'E	—	12.6–7.0	—	187.8	842
	Turlan (Achisai)	43°50'N, 68°52'E	17.0–0.6	14.2–2.0	—	343.1†	454
North Caucasus	Sadon-Buron	42°50'N, 44°00'E	7.6–1.8	15.3–4.1	—	377.4	779
Far East	Tetyukhe	44°22'N, 135°50'E	7.8–1.3	10.6–4.3	—	283.6	530
Total reserves of above deposits						1,993.7	4,077
Approximate percent of total reserves of the Soviet Union						68	70

* Reprinted by permission of the publishers from Demitri Shimkin, "Minerals—a Key to Soviet Power." Cambridge, Mass.: Harvard University Press, Copyright 1953, by the President and Fellows of Harvard College.

† Largely exhausted in 1947.

The composition of the ore of the most important deposits is shown in Table 11-3. Apparently most deposits are of hydrothermal origin and deposition was localized in fault zones; the Tetyukhe deposit is of contact-metasomatic origin (48, p. 128).

Other Countries

The major lead and zinc deposits of all countries that furnished 5 percent or more of the world production in the 15-year period 1945 to 1959 have been discussed briefly. A few other general areas deserve mention because of their increasing importance as producers during recent years or because of their world importance in the past. Among countries that have shown marked increase in production recently are Yugoslavia (lead), Morocco (lead), Japan (zinc), and Poland (zinc). Among countries that were formerly of great importance are Spain (lead), the United Kingdom (lead), Germany (zinc and lead), and Belgium (zinc).

Yugoslavia. The most important lead-zinc mining area of Yugoslavia is the Trepca Valley in the southern part of the country, in which are located the Stantrg deposits (5, pp. 307–319; 19). The older rocks of the area are the Stantrg series, a folded complex of schists, phyllites, quartzites, and crystalline limestone, possibly of Ordovician-Silurian age, intruded by small masses and dikes of dolerite and serpentine (Jurassic to Lower Cretaceous). The folded and fractured Stantrg series, along with pre-Tertiary intrusive rocks, are covered by horizontally bedded tuffs of Tertiary age and cut by andesitic intrusive rocks. Sphalerite-galena-pyrite mineralization on a regional scale is attributed to a late stage of Tertiary igneous activity, and the ore bodies formed were (1) replacements in limestone directly in contact with intrusives, as at the Stantrg mine; (2) impregnations, replacements, and irregular veins in limestone; (3) fissure fillings in dolerite, serpentine, or along such contacts; (4) fissure fillings in tuffs or allied sediments; (5) fissure fillings in andesitic intrusive rocks; and (6) lenticular veins in schist. Economic mineralization consists chiefly of replacements and impregnations in limestone and well-developed vein systems in the larger masses of andesitic rock. At the Stantrg mine the ore bodies, which are replacements in limestone around the contact of a breccia pipe composed of a core of dacite surrounded by breccia, are composed chiefly of an intimate mixture of argentiferous galena, sphalerite, pyrite, and pyrrhotite. Gangue minerals, which are minor in amount, arc chicfly quartz, calcite and other rhombohedral carbonates, although locally actinolite, hedenbergite, ilvaite, and garnet are present. Localization of the ore was at contacts between breccia and limestone and between schist and limestone. In 1940 the ore reserves that had been fully proved were approximately 5 million tons that assayed 8.6 percent lead, 3.8 percent zinc, and 4 oz of silver per ton.

Morocco. Lead-zinc-copper mineralization in Morocco extends along a wide orogenic zone in which are the Anti, High, and Middle Atlas Mountains (5, pp. 268–274; 7; 16; 28). Deposits are chiefly replacements in calcareous rocks and veins in rocks of several types. The most important deposit, which is of the replacement type, is the Bou Beker–Touissit (Fig. 11-8), stated to contain four-fifths of the reserves of lead and nine-tenths of the reserves of zinc of Morocco. The Bou Beker–Touissit area of the Middle Atlas range is in the northeastern part of Morocco. The deposits consist of stratiform, discontinuous ore bodies that are replacements of the Lias dolomites at the base of the Jurassic. Generally mineralization is localized in the upper part of the Lias, commonly within the first 15 to 17 m from the top, and follows especially a mottled dolomite. Some ore consists chiefly of sphalerite, and some of galena without sphalerite. In addition to these two, which are both replacement types and constitute most of the deposits, some ore consists of both sphalerite and galena and fills cavities in which it formed not only botryoidal aggregates and similar masses but also perfectly developed crystals that make beautiful coatings (16, p. 489). Considerable amounts of pyrite, and locally chalcopyrite, are present also throughout the deposits. Gangue minerals are practically absent except in some rich zinc ore bodies, which contain a small amount of quartz. According to Claveau, Paulhac, and Pellerin (16, p. 490) the lead-zinc deposits of the Bou Beker–Touissit area compare favorably with the deposits of the Tri-State and Southeastern Missouri area of the United States and belong among the great deposits of the world. Also, they raise the same questions regarding origin and emplacement.

The Silesian Region.   The lead-zinc resources of Germany and Poland are difficult to discuss on a national basis because, over a period of years, accrediting of production has changed from one country to another. The major deposits are in Silesia, so they will be discussed from the viewpoint of the Silesian region. At present Silesia lies partly in northwestern Czechoslovakia and partly in southwestern Poland. We will omit the early history and state that after 1742 and before 1918 the major part of it was in Prussia and a minor part was in Austria. Between 1918 and World War II several changes took place, some territory going to Poland, some to Germany, and some to Czechoslovakia, but eventually much of it was under German control. After World War II much of Silesia again was transferred to Poland and a small part to Czechoslovakia. The important lead-zinc deposits are in Upper Silesia, now belonging entirely to Poland.

The lead-zinc deposits (5, pp. 362–368; 53) occur in the Muschelkalk or Shelly Limestone of Triassic age which, in the ore-bearing areas, has been dolomitized. Ore deposits are present in folded beds that are cut by a great number of faults. There are three ore-bearing horizons which are designated (1) the *chief ore horizon*, (2) the *second ore bed*, and (3) the *third ore bed*. The chief ore horizon, the lowest of the three, is at the base of the dolomitic zone and is underlain by clay, which forms the footwall. The second ore bed is developed locally at varying distances above the chief ore horizon where an upper dolomite has been sheared. The third ore bed is developed only exceptionally. Mineralization consists chiefly of sphalerite, wurzite, galena, pyrite, and marcasite; some cadmium is present in the sphalerite, and rarely greenockite (CdS) occurs; some silver is present in the galena. The minerals fill crevices, joints, and gash veins, some of which were enlarged by solution; in many cases they are disseminated in the dolomite and replace it. Some of the deposits have been oxidized to a considerable depth with the formation of smithsonite, some calamine, and cerussite, which brought about substantial enrichment in some places. Zwierzycki (53, p. 314) stated that the deposits are of telemagmatic (telethermal) type and that the solutions rose along faults and crevices from great depth. Lindgren [(10, pp. 427–428) of Chap. 2] discussed these deposits with the lead and zinc deposits in sedimentary rocks, origin independent of igneous activity, along with the lead-zinc ores of the Mississippi Valley, and called attention to differing concepts of origin. The problems of origin appear to be similar to those of the Mississippi Valley deposits.

The most important mining areas are around Bytom (German *Beuthan*) and Olkusz, about 6 miles to the east, although the ores at Bytom were approaching exhaustion in 1951 (34). Mining began as early as the thirteenth century. Between 1549 and 1668 the mines of Olkusz are said to have produced 17,000 tons of lead, 10,000 tons of lead oxide, and 15,000 kg of silver. Early production was from rich oxidized ores, but as those became exhausted use was made of primary ores. Sphalerite furnished only about 1 percent of the zinc in 1870 but increased to 15 percent in 1880, to 41 percent in 1890, to 63 percent in 1900, to 76 percent in 1910, to 86 percent in 1920, and to 90 percent in 1930. Production from Upper Silesia (53, pp. 320–321) has been chiefly zinc and increased from about 25,000 tons in 1850 to about 554,000 tons in 1929. Nearly 70 percent of the 1929 total came from Polish Upper Silesia, the rest from German Upper Silesia. In some years the lead production was about one-fifth of the zinc production, but in 1929 it was less than one-tenth of the zinc production. Peak production occurred in 1929. Under the conditions of World War II some rich ore shoots were completely worked out, equipment and plants were neglected, and many mines suffered heavily. In 1946 only four mines were operating. Reserves of the region do not appear to be great. The richest part of the Bytom area is nearly exhausted, the rich oxidized ores have been largely depleted, and use of lower and lower grade ores will be necessary. Some of the very rich ore that was mined contained as much as 40 percent zinc, whereas remaining oxidized ores contain 6 to 12 percent zinc. The grade of the Olkusz ores was stated as 10 to 11 percent zinc and 1.5 to 2.0 percent lead in 1951 (34).

Other deposits. Discussion of other interesting deposits of lead and zinc would increase unduly the length of this book. The early important production of zinc in Belgium was from the Moresnet district (5, pp. 91–100), and the production of lead in Spain was from Santander (5, pp. 401–412).

Copper

Six countries of the world—the United States, Chile, Zambia (Northern Rhodesia), the Soviet Union, Canada, and the Congo—produced about 82 percent of the world's copper in the 15-year period 1945 to 1959, and nearly 79 percent in 1962 (Fig. 11-3). Only two of these countries, however, the United States and Chile, were outstanding producers in the century and a quarter from 1801 to 1925 (Table 11-4). Together they furnished 59.0 percent

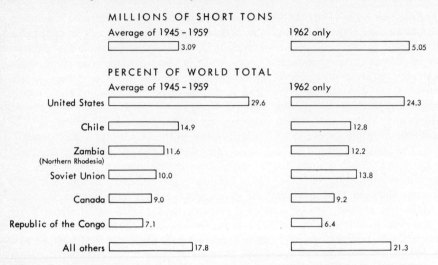

Fig. 11-3 World copper production. Compiled from *U.S. Bur. Mines Minerals Yearbook,* various years.

of the world's copper during that 125 years—the United States 48.1 percent, Chile 10.9 percent (88). The other two most important world producers during the nineteenth century were Spain and England, with 12.9 and 8.4 percent respectively, a total of 21.3 percent compared with 48.0 percent furnished by the United States and Chile. Eighty-five percent of the copper resources of the world are stated to be in 12 districts or mines in six countries (139) (Table 11-5).

TABLE 11-4. Major Differences in Production of Copper during Various Periods*

	Percent of world total	
Country	1801–1900	1901–1925
United States	29.6	56.5
Chile	18.4	7.5
Spain and Portugal	12.9	4.7
England	8.4	0.04
Japan	5.9	6.2
Germany	4.6	2.8
Australia	4.5	3.4
Russia (now the Soviet Union)	4.1	1.4
Mexico	1.3	5.0
Canada	0.6	3.4
Peru	0.3	2.8
Belgian Congo (now the Republic of the Congo)	—	1.9
Northern Rhodesia (now Zambia)	—	0.1

* Compiled from *U.S. Bur. Mines Econ. Paper* 1, 1928.

TABLE 11-5.   Distribution of Major Copper Resources of the World*

Country	Deposits
United States	Butte, Montana; Bingham, Utah; Keweenaw, Michigan; Morenci, Arizona; San Manuel, Arizona
Chile	Chuquicamata; Braden (El Teniente)
Zambia (Northern Rhodesia)	Various mines. The four principal ones are N'Changa, Mufulira, Rhokana, Roan Antelope
Republic of the Congo (Belgian Congo)	Various mines of the Union Miniere du Haut Katanga, the largest of which is the Musonoi open pit
Canada	Sudbury, Ontario
Soviet Union	Various deposits in Kazakhstan and the Ural Mountains

* Compiled from *U.S. Bur. Mines, Materials Survey, Copper,* 1952.

United States

The major production of copper in the United States comes from six states, which in 1962 furnished 98 percent of the copper of the United States: Arizona, 52 percent; Utah, 18; Montana, 8; New Mexico, 7; Nevada, 7; Michigan, 6. These states also contain nearly 98 percent of the copper reserves of the United States (139). They have always furnished most of the copper produced in the United States, but the relative importance of each state has shifted with the passage of time since 1845. Thus Michigan was the outstanding producer early, but Montana surpassed Michigan in production in the period 1886 to 1900 and thereafter, then Arizona surpassed Montana in the period 1906 to 1910 and has maintained the lead in copper production ever since. Utah went ahead of Michigan in the period 1921 to 1925 but remained behind Arizona and Montana, but later surpassed Montana. Some of the history of production is summarized in Table 11-6. Since 1925, production in Arizona, Utah, Nevada, and New Mexico has increased. Peak production in actual tons of copper was attained in Michigan and in Montana in 1916; in Nevada and New Mexico in 1942; in Utah in 1943; and in Arizona in 1962 according to statistics then available, but actually peak production in Arizona probably had not been attained at that time.

The average yield of copper per ton from copper ores in the United States is low and has steadily decreased. During the 37 years 1924 to 1960 the range was from 2.11 percent in 1933, when selective mining was used because of economic conditions, to 0.73 percent in 1960. The yield has been below 1.5 percent since 1936 and below 1.0 percent since 1944. The average for the 37 years was 1.18 percent. During this time, also, more and more mining was by open-pit methods. The recovery of ore by open-pit mining in 1943 was 69 percent whereas in 1960 it was 80 percent. Production would not be economical from some deposits were it not for recovery of silver, gold, or molybdenum also.

TABLE 11-6.   Some Aspects of Copper Production in the United States, 1845–1960*

	Percent of total production in United States				
State	1845–1875	1876–1900	1901–1925	1845–1960	1960 only
Michigan	82.0	36.6	16.5	11.6	5.2
Montana	0.04	40.9	22.0	16.4	8.5
Arizona	0.6	15.6	34.2	37.3	49.8
Utah	0.6	0.9	10.7	17.7	20.1
Nevada	0.1	0.04	4.1	5.7	7.1
New Mexico	0.2	0.4	3.2	4.9	6.0
Total of six states	83.5	94.4	90.7	93.6	96.7

* Compiled from *U.S. Bur. Mines Minerals Yearbook,* 1960, vol. 1, p. 408, and *Econ. Paper* 1, 1928.

A recent summary of the copper resources of the United States (89) placed deposits in four categories based on the amount of copper already produced plus the metal remaining in the deposit that might ,reasonably be expected to be recovered. This means, essentially, past production plus reserves. Deposits of the first magnitude in this grouping are those of more than 1 million short tons of copper; deposits of the second magnitude are those of 50,000 to 1 million short tons of copper. Further, the deposits were placed in five groups based on ore types, as follows: (1) disseminated—disseminations and veinlets of sulfides in minutely fractured rock, commonly designated *porphyry copper* deposits; (2) replacement—includes replacement of beds, manto deposits, pyrometasomatic deposits, replaced shear zones, and irregular replacement deposits; (3) vein—vein-form deposits of both filled and replacement veins and some irregular replacement ore; (4) massive sulfide—massive copper-bearing pyrite or pyrrhotite deposits, commonly along shear zones in metamorphic rocks; (5) native copper—mainly native copper in amygdaloidal mafic lava and in conglomerate. Distribution of deposits of first magnitude, along with types of deposits, are shown by Table 11-7. Only brief description of a few of the many important deposits can be attempted in the following pages.

Arizona and New Mexico.   The greatest copper-producing region of the United States is centered in Arizona but embraces parts of New Mexico and the state of Sonora, Mexico (Fig. 11-4) (89, 125, 127, 135). The importance of Arizona, with eight deposits of the first magnitude, and of New Mexico with two such deposits, is shown by Table 11-7. In addition, in Arizona there are nine districts of second magnitude, and in New Mexico there is one, the Lordsburg district. The second-magnitude districts or mines in Arizona are Eureka (Bagdad area); Superior (Pioneer) district, Magma mine; Globe-Miami district, Castle Dome mine; Globe-Miami district, Copper Cities deposit; Globe-Miami district, Old Dominion and other mines; Silver Bell district; Esperanza deposit; Helvitia district, Copper World and other deposits; Lone Star district.

TABLE 11-7. Distribution of Copper Deposits of the First Magnitude in the United States and Generalized Types of Deposits*

State	Mine or district	Generalized type of deposit
Arizona	United Verde and United Verde Extension	Massive sulfide in metamorphic rock
	Globe-Miami	Enriched disseminated sulfides
	Mineral Creek	Enriched disseminated sulfides
	Old Hat (Mammoth) district, San Manuel mine	Enriched disseminated sulfides
	Ajo district, New Cornelia mine	Disseminated sulfides
	Pima and other mines	Pyrometasomatic replacement of limestone and shale
	Bisbee (Warren) district	Enriched disseminated sulfides and replacement
	Morenci (Copper Mountain) district	Disseminated sulfides and pyrometasomatic replacements and veins
Michigan	Calumet and Hecla and other mines	Native copper, some in amygdaloidal lava, some in conglomerate
	White Pine mine	Sulfides in sandstone
	American Metals Exploration	Sulfides in sandstone
Montana	Butte (Summit Valley) district	Veins and replacements
Nevada	Ely (Robinson) district	Disseminated sulfides
New Mexico	Burro Mountains, Tyrone area	Enriched disseminated sulfides
	Central (Hanover, Santa Rita)	Enriched disseminated sulfides
Tennessee	Ducktown district	Massive sulfides in schist
Utah	Bingham (West Mountain) district	Enriched disseminated sulfides and replacements
Virginia	Gossan Lead and other mines	Massive sulfides in schist

* Compiled from *U.S. Geol. Survey Mineral Inv. Resource Map* MR-13, 1962.

Within Arizona and New Mexico the copper deposits are of several ages but can be grouped in two broad divisions, (1) Precambrian and (2) late Cretaceous or Tertiary (117, p. 13–15; 125, pp. 322–326; 135, pp. 177–179). The only outstanding copper deposit of Precambrian age is near Jerome, Arizona, which was exploited by the famous United Verde and United Verde Extension mines. Hence the major mineralization was in late Cretaceous or Tertiary times.

Copper deposits within this region are of several types. Grouped broadly these are (1) massive sulfides in metamorphic rocks, typified by the Precambrian deposits of the Verde district; (2) contact-metasomatic (pyrometasomatic) replacements; (3) disseminated sulfides or porphyry copper deposits. Oxidation and supergene enrichment have been important in many deposits. The porphyry copper deposits are by far the most important type in the region. Trischka (138) stated in 1953 that there were 13 porphyry copper deposits in Arizona and 2 in New Mexico.

The great importance of the porphyry copper type of deposit, not only in Arizona and New Mexico but elsewhere, makes a short summary of the characteristics of that type desirable. The designations *porphyry copper deposit*

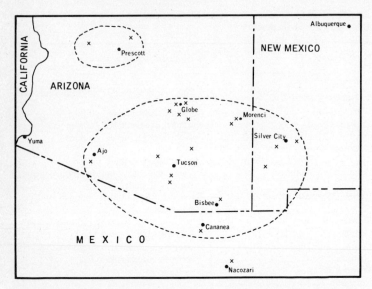

Fig. 11-4 Map showing major copper areas and major mines (X) of the southwestern United States, with extension into Mexico. After H. A. Schmitt, *Mining Engineering,* June 1959, p. 597. Copyright 1959. The American Institute of Mining, Metallurgical, and Petroleum Engineers. Used by permission.

and *porphyry copper ore* are not always used with the same meaning. One definition of porphyry copper ore [(6, p. 310) of Chap. 2] implies that it is disseminated ore and that a porphyry copper deposit is one in which copper-iron sulfides and copper sulfides are disseminated throughout igneous rock that is quartz monzonite porphyry or granite porphyry; and this is, indeed, the character of many porphyry copper deposits. A more general meaning has come into usage, however, regarding which Parsons (108) stated:

> *The essential characteristics of the deposits that are universally, if not quite precisely, designated as Porphyry Coppers are: their huge size, particularly with respect to horizontal dimensions; the relative uniformity with which the copper minerals are disseminated throughout the mass; and the large average per-ton copper content of the exploitable ore.*[1]

In general the masses of rock in which the deposits are localized have been considerably shattered, and many have been hydrothermally altered (128). Minerals have been deposited as disseminations and veinlets in various openings and have also replaced original minerals of the rock. In places where emanations have encountered limestone in some areas, the type of deposit has changed and limestone has been replaced. Consequently, although a major deposit may be classed as a porphyry copper or disseminated type, both veins and limestone replacements may

[1] From "The Porphyry Coppers," pp. 1–2, by A. B. Parsons. Copyright 1933. The American Institute of Mining and Metallurgical Engineers. Used by permission.

be associated with it. This was noted by Locke (95, p. 618), who stated, when discussing disseminated copper deposits:

> *Indeed, the veins, limestone replacements and disseminations seem now to be expressions of the same "mineral fluid" in different structures and rocks; all expand upward, and the peculiarity of dissemination is merely that the fluid that makes it breaks through the confines of the vein walls, and precipitates, not in the free openings of soluble limestone, but in a rock splitting down into blocks and dispersing the sulphide far and wide.*[1]

The mineral composition of the porphyry copper type of deposit is relatively simple. The common hypogene metallic minerals are chalcopyrite and pyrite, but in some deposits there is a considerable amount of molybdenite, so that molybdenite is recovered as a by-product. Minor amounts of bornite, galena, and sphalerite may be present. Quartz is the common gangue mineral.

Two mines of Arizona were far ahead of all others in production of copper in 1961. The Morenci open-pit mine, Greenlee County, which ranked first, extracted 16,286,000 short tons of ore and recovered 111,443 short tons of copper; the San Manuel underground mine produced 12,529,243 short tons of ore that assayed 0.727 percent copper, from which recovery was 82,612 short tons of copper.

Morenci (90, pp. 544–545; 93; 95, p. 618; 108, pp. 97–113; 109, pp. 49–66; 139, Chap. 3, pp. 22–23), early working a high-grade copper type of deposit, changed to working a large porphyry type of deposit about 1907, and at the end of 1931 ranked second only to the Utah Copper mine, Bingham, Utah, among the 12 most important porphyry copper deposits. Published reserves at the end of 1930 were 379,340,000 tons of ore containing 1.02 percent copper, but an unofficial estimate was 450,000,000 tons of ore containing 1.1 percent copper. In 1952 about 1,000,000 tons of copper ore had been mined from the open pit. Within the Morenci areas masses of diorite porphyry, granite porphyry, and quartz-monzonite porphyry of late Cretaceous or early Tertiary age cut Paleozoic sandstone, shale, and limestone. Contact-metasomatic deposits were formed in some of the limestones, veins were formed in both igneous and sedimentary rocks, and the main mass of porphyry was mineralized throughout with pyrite and chalcopyrite and made into a porphyry copper type of deposit. Oxidation occurred to depths of 400 to 600 ft, with formation of azurite, malachite, and cuprite, and, beneath the oxidized products, a secondary sulfide zone of chalcocite. The enriched contact-metasomatic and vein deposits furnished the early-production. Production in 1960 was from the porphyry copper deposits, in which chalcocite was also formed by secondary enrichment.

The San Manuel deposit (109, pp. 244–256; 110; 129; 130), in Pinal County, 35 miles southeast of Tucson, consists of chalcopyrite and pyrite disseminated through quartz-monzonite of Precambrian age, monzonite porphyry doubtfully classed as of Mesozoic or early Tertiary age, and diabase that cuts both the quartz-monzonite and the monzonite porphyry. The known length of the ore body

---

[1] From "Ore Deposits of the Western States," p. 618, Disseminated Copper Deposits by A. Locke. Copyright 1933. The American Institute of Mining and Metallurgical Engineers. Used by permission.

is 6,800 ft, and the greatest width is unknown but is as much as 3,000 ft. The copper content of the rock diminishes both above and below the part that shows maximum grade, so that the ore body is delineated by an arbitrary cutoff grade of 0.5 percent copper. Oxidation and supergene enrichment have been important. The depth of the bottom of the oxidized zone is 285 to 1,630 ft below the surface. The most important mineral of this zone is chrysocolla, and the average copper content of the oxidized zone is about 0.75 percent. An irregular supergene sulfide zone contains chalcocite ore that probably averages about 1 percent copper. The entire deposit may contain (139, Chap. 3, p. 23) around 500 million tons of ore that averages nearly 0.8 percent copper, of which about 100 million tons is oxidized ore.

The Pima district (68, 137), about 20 miles southwest of Tucson, may become one of the major sources of copper in the United States, because of the discovery of several large ore bodies. The Pima ore body, discovered in 1950, was brought into production in 1956 by means of a large open pit. It is a contact-metasomatic type of deposit in which chalcopyrite and pyrite are the chief minerals, and these are concentrated mainly in limestone. Quartz-monzonite of late Cretaceous or early Tertiary age is closely associated with the deposits and is slightly mineralized.

Profitable copper mining has been conducted in the Bisbee district, Arizona (90, pp. 545–546; 108, pp. 301–319; 116; 133; 135, pp. 221–228), since 1880, at first from replacements in limestone and later, since about 1923, from the porphyry copper type of deposit at Sacramento Hill. In the Bisbee district the Precambrian Pinal schist and Paleozoic sedimentary rocks, chiefly limestones, have been intruded by late Cretaceous or early Tertiary granite porphyry that was emplaced as small stocks, dikes, and sills. The largest of the stocks is the one at Sacramento Hill, with which the principal copper deposits are associated. The rocks near the porphyry intrusions were metamorphosed in varying degrees, some converted into marble, some into carbonate-silicate rocks or other types, depending on their original composition, and all rocks, including the porphyry, were hydrothermally altered. Indeed, according to Tenney (133, p. 56), the separation of the effects of contact metamorphism and hydrothermal alteration in the district is largely arbitrary, as they were both caused by hydrothermal solutions, the contact-metamorphic effects resulting from the hottest emanations from the magma chamber. Mineralization was a late phase of the hydrothermal activity, and copper, sulfur, and some silica replaced minerals that were formed previously. The hypogene ore minerals are chiefly pyrite and bornite, but chalcopyrite is present locally. Oxidation was extensive, and enrichment occurred both in the oxidized and the sulfide zones. Great masses of malachite and azurite, which supplied beautiful specimens to many museums, occurred in the mineralized limestones and, just above the secondary sulfides, economic amounts of cuprite and native copper. Chalcocite was the chief mineral of the sulfide zone. The district as a whole has been important not only for copper but also for lead, zinc, silver, and gold, as shown by Table 11–8.

In the general vicinity of Silver City, New Mexico, are, to the northeast, the Central Mining district, which includes the Hanover-Fierro, Bayard, and Copper Flat areas and Santa Rita; to the southwest, the Tyrone area; and con-

TABLE 11-8.   Production from the Bisbee or Warren District, Arizona, 1880–1948*

Metal	Amount	Value
Copper	5,409,052,617 lb	$811,966,127
Silver	78,650,304 oz	53,851,645
Gold	1,848,874 oz	48,335,310
Zinc	245,182,920 lb	28,847,429
Lead	258,602,762 lb	22,653,667
Total value		$965,654,178

* From Arizona Zinc and Lead Deposits, pt. 1, p. 20, Bisbee or Warren District by W. G. Hogue and E. D. Wilson. *Arizona Bur. Mines, Geology ser.* 18, *Bull.* 156, 1950. Used by permission.

siderably farther southwest, the Lordsburg area (91, 92, 106, 107, 126). Geologic formations are somewhat varied, but in general, except in the Lordsburg area, granodiorite stocks of late Cretaceous or early Tertiary age intrude Paleozoic and Cretaceous sedimentary rocks. The granodiorite masses at Tyrone and Santa Rita are mineralized and make important porphyry copper deposits. At Santa Rita much ore was formed in sedimentary rocks adjoining the granodiorite, but at Tyrone the ore is all in porphyry and granite, as all of the bedrock there is igneous. Secondary enrichment was important at both places. At Hanover, a few miles northwest of Santa Rita, are important contact-metasomatic bodies of sphalerite formed by replacement of limestone. At Fierro, a short distance north of Hanover across the Hanover-Fierro stock, are contact-metasomatic deposits of magnetite. In the Lordsburg area grandiorite, presumably of late Cretaceous or early Tertiary age, intruded basaltic andesite and associated volcanic breccias. A distinctly different type of deposit was formed there and consisted of copper-tourmaline veins along fractures. Production came almost entirely from one vein, the Emerald, and from one mine on that vein, which was abandoned in 1932. Before closing, however, ore valued at $18,500,000 was produced.

Nevada and Utah.   Two outstanding copper deposits occur in Nevada and Utah, the important porphyry copper mass at Ely, in west-central Nevada, and the famous Utah Copper open pit and surrounding lead-zinc ores of Bingham, Utah.

At Ely, Nevada (57; 109, pp. 67–80; 111), Paleozoic sedimentary rocks were intruded by a porphyry, designated the *ore porphyry,* which is altered and mineralized, and a younger monzonite porphyry that contains large feldspar phenocrysts which, because they stand out sharply on exposed surfaces, have given rise to the local designation "peanut porphyry." Generally the peanut porphyry is but little altered or mineralized, although locally ore has been formed in it. The ore porphyry was very thoroughly fractured, impregnated by small grains and veinlets of chalcopyrite and pyrite, and thus transformed into an exceedingly productive porphyry copper deposit, which has furnished most of the copper of the district. Some production has come from limestone near the ore porphyry, in part from replacement bodies that were enriched by secondary chalcocite, in part from oxidized bodies that contained copper carbonates, oxides, and native copper.

The West Mountain or Bingham mining district (60; 61; 83; 84; 109, pp. 27–48), 27 miles southwest of Salt Lake City, Utah, is one of the outstanding copper districts of the world. Total production from the district is enormous (Table 11–9). In 1961 the ore mined amounted to 27.8 million tons that contained 16.2 lb of copper per ton and furnished about 211,000 short tons of copper.

TABLE 11-9.   Production from the West Mountain
(Bingham) District, Utah, through 1960*

Copper	15,863,594,000 lb
Lead	3,954,480,000 lb
Zinc	1,556,508,000 lb
Molybdenite	451,086,518 lb
Silver	205,041,444 oz
Gold	10,971,041 oz

* From *Guidebook to the Geology of Utah, No.* 16, 1961, p. 129, History of Mining in the Bingham District by E. D. Hammond. Salt Lake City, Utah, Geological Society. Used by permission.

The central feature of the district is Utah Copper Hill, a stock of monzonite[1] intruded into quartzite and intercalated limestone members of Pennsylvanian age. Disseminated throughout the monzonite stock, which was much fractured, are hypogene chalcopyrite and pyrite, which partially replace minerals of the monzonite and fill small fractures and minute seams. Molybdenite, an important by-product, is present also; and in some small fractures there are veinlets that contain galena and sphalerite along with some copper and iron sulfides. An upper oxidized zone 20 to 150 ft from the surface is succeeded downward by an enriched sulfide zone that extends to as much as 800 ft beneath the base of the oxidized zone, the upper 500 to 600 ft of ore being composed chiefly of secondary sulfides whereas the lower enriched ore consists chiefly of hypogene chalcopyrite and some supergene chalcocite and covellite. This huge mass of disseminated ore is mined from the Utah Copper pit (Fig. 11–5), and in 1961 (86, p. 86) all ore mined was hypogene.

A notable feature of the Bingham district is zoning of the deposits. The central zone, dominated by the Utah Copper stock of monzonite, is the chief copper-molybdenum producer, and outward from this zone is an important lead-zinc zone. Within the copper-molybdenum zone the concentration of both copper and molybdenum increases toward the center, the molybdenum more so than the copper, so that the center is almost a molybdenum zone (86, p. 92). Within these two zones the deposits are of several types: (1) disseminated or porphyry copper; (2) veins, chiefly in limestone; (3) replacements in limestone.

[1] The Utah Copper stock and the Last Chance stock to the south are designated monzonite by several writers, but James, Smith, and Bray (86) refer to the Utah Copper stock as granite porphyry and the Last Chance stock as monzonite porphyry.

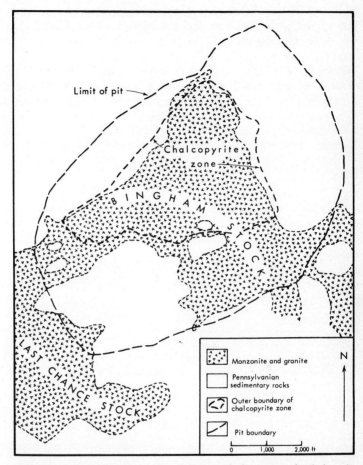

Fig. 11-5 Map showing some features of the copper-lead-zinc deposits at Bingham, Utah. Somewhat generalized after A. James, W. Smith, and E. Bray, *Utah Geological Society Guidebook* 16, 1961, p. 84. Utah Geological and Mineralogical Survey and Utah Geological Society. Used by permisson.

Most of the lead, zinc, and silver have come from replacement bodies of galena and sphalerite in limestone. Some copper also has been produced from such bodies but, although of higher grade than the disseminated ore, the amount furnished by such bodies has been small. Some ore has been obtained, also, from contact-metasomatic deposits. Although the limestone beds in which replacement ore occurs form only about 10 percent of the Pennsylvanian sedimentary rocks, these beds have yielded most of the lead-zinc ore. Replacement of the limestone has been extensive, in places almost completely in stratigraphic thicknesses as much as 100 to 150 ft. Almost all the lead and zinc has been recovered by underground mining from ores that contained about 7 percent lead, 7 percent zinc, less than 1 percent copper, about 7 oz of silver, and 0.1 oz of gold per ton. Production of lead-zinc ores has come from the United States and the Lark mines during recent years.

Montana. The Butte or Summit Valley district (79, 112, 121, 142) certainly deserves a high place among mining districts of the world that have been described in superlative terms. The district has yielded truly impressive amounts of copper and other metals, as indicated by Table 11-10, and Butte has been termed the "richest hill on earth." Most of this production has been from underground mining which, in 1959, had reached the 4,500-ft level, where ore shoots of high-grade copper aggregating more than 2,000 ft in length had been developed (75).

TABLE 11-10.   Production from the Butte District, Montana 1780–1957*

Copper	14,603,722,290 lb
Zinc	4,456,155,423 lb
Manganese	2,629,331,806 lb
Lead	754,799,030 lb
Silver	609,636,341 oz
Gold	2,267,213 oz

* From C. C. Goddard, Jr., *Mining Engineering*, vol. 11, pp. 290–292. Copyright 1959. The American Institute of Mining, Metallurgical, and Petroleum Engineers. Used by permission.

A great expansion program has been under way at Butte since about 1956 (176, 177). The Berkeley open pit began preliminary production in 1955 and by the end of June 1958 had supplied a total of 10,047,390 tons of ore that averaged 0.89 percent copper and 0.12 oz of silver. In 1961 it produced an average of 31,310 tons of ore per day, which required the removal of 3.51 tons of waste for every ton of ore. Further, in addition to the Berkeley pit, new shafts were planned along with the deepening of some old ones in order to reach known ore deposits and to expand underground mining (180). In 1961 one shaft, the Mountain Con, had reached a depth of 5,296 ft, the greatest depth then attained at Butte. It was expected that the shaft development program would be completed early in 1963 and that thereafter production would be materially increased and the life of mining operations greatly extended. The production in 1961, partly because of the material furnished by the Berkeley pit, was 104,000 short tons of copper, the greatest since World War II.

The deposits in the Butte district are veins in quartz monzonite ("Butte granite") that is part of the Boulder batholith in the northern Rocky Mountains. The quartz monzonite has been cut by faults of several ages, and a complex fault pattern has resulted. Altogether there are seven fault systems, the first three of which—Anaconda or east-west, Blue or northwest, and Steward or northeast—were mineralized. Some faults pass into complex branching "horsetail" structures. Among the great amount of detailed information that has accumulated throughout years of active mining and geological work in the district, perhaps two features should be emphasized: (1) pervasive alteration took place along fractures and preceded ore deposition, which was replacement and was related to the degree of alteration; (2) zoning of the deposit is outstanding both horizontally and vertically, and the zoning also is related to the pattern of alteration.

Alteration and vein formation have been intensively studied by the Butte geologists and have been summarized by Sales and Meyer (122, 123). Two zones of alteration were recognized in parts of the mine where alteration could be traced progressively into relatively unaltered quartz monzonite. Next to a fracture is a sericitized zone, representing the highest intensity of alteration, and this passes outward into an argillized zone that consists of two subdivisions, (a) a kaolinite subzone representing a lower intensity of alteration than the sericite zone and (b) a montmorillonite zone representing the lowest intensity of alteration and passing outward into unaltered quartz monzonite. In the central part of the mining area the alteration has been so intense that, in places, all of the quartz monzonite has been sericitized. Outward and upward from this central part alteration is less intense. This is the general picture, although many details are omitted.

Lateral zoning of the district is well shown by Figure 3-2. The central zone is predominantly a copper zone, but some silver is present, in general the ratio of copper to silver being about 1 percent of copper to ½ oz of silver. This also is the zone of most intense alteration. The major ore minerals of this zone are enargite and hypogene chalcocite; the major gangue minerals are quartz and pyrite. The intermediate zone is also predominantly a copper zone but generally contains zinc and lead also and, near the outer margins, some manganese. Alteration is less intense than it is in the central zone. In a general way there is a decrease in copper and an increase in zinc, lead, silver, and manganese from the interior toward the exterior of the intermediate zone. The ratio of copper to silver differs from that of the central zone and is about 1 percent of copper to 1 oz of silver. Among major ore minerals enargite and hypogene chalcocite still persist, but bornite, chalcopyrite, tetrahedrite, and tennantite become important. Sphalerite, galena, rhodonite, and rhodochrosite are present also. The chief gangue minerals are still quartz and pyrite, but pyrite is less abundant than it is in the central zone. The peripheral zone, in which alteration is the least intense, does not contain copper in commercial amounts and is important chiefly for zinc, gold, silver, and some manganese. The most abundant ore minerals are rhodonite, rhodochrosite, and sphalerite with which is associated a considerable amount of galena. Copper minerals are sparingly present. Quartz is the most abundant gangue mineral and pyrite, though common, is less abundant than in other zones. The same type of zoning that exists laterally is present vertically also, and veins worked chiefly for zinc on upper levels may grade into copper producers in depth.

Oxidation and supergene enrichment have gone on to a depth of perhaps 300 to 400 ft, and supergene enrichment of lean ore was the basis for opening the Berkeley pit. Leaching has been intense at the surface, and as much as 250 ft of waste material overlies the enriched zone. In the Berkeley pit area hypogene copper minerals were disseminated through the altered quartz monzonite and also occurred in many small veinlets and some moderate-sized veins in the monzonite.

This material was enriched by sooty chalcocite. In 1956 the reserves in the pit area were estimated to be more than 100 million tons of ore that contains 0.8 percent copper. Oxidation of the hypogene manganese minerals in parts of the mine area has provided the basis for the manganese production (149).

Michigan. The native copper deposits in amygdaloidal lavas and associated conglomerates of the Keweenaw Peninsula, northern Michigan (Fig. 11-6) (63, 67), at one time furnished most of the copper of the United States (Table 11-6) and, although production has declined to such an extent that they supplied only 5.2 percent of the production of the United States in 1960, they still ranked third in total production of the world since recovery began and had yielded almost 5 million tons of copper by 1952 (139, Chap. 3, p. 19). The production from 1957 through 1961 from the native copper mines, from reclamation of previously discarded tailings, and from the newly developed White Pine sulfide ore, is shown by Table 11-11. Distribution of production throughout the major ore horizons to the end of 1925 (67, p. 65) is shown by Table 11-12.

The yield of copper per ton, 1845 to 1925, not including reclamation from mill tailings that were discarded during early processing, was 27.27 lb. The recovery from different types of deposits was quite variable. The highest recorded yield (67, p. 67) was 83.52 lb per ton from the Central fissure deposit, which supplied 51,875,527 lb of copper between 1855 and 1898 from 565,393 tons of rock. The Calumet conglomerate yielded 49.47 lb of copper per ton between 1865 and 1925 from 57,229,052 tons of rock and furnished 2,831,092,153 lb of copper. Yields from amygdaloids ranged from 13.55 to 26.91 lb per ton for 1845 to 1925, and yield from 12,374,823 tons of previously discarded rock (tailings) treated in the Calumet and Hecla reclamation plant, 1915 to 1925, was 10.32 lb per ton. Native silver occurred with the native copper in some mines, but only a few mines appear to have produced appreciable amounts of silver.

New life was injected into the area in 1954 with the beginning of production, in October, at the White Pine mine (144, 161, 163, 171, 174, 178) from a large

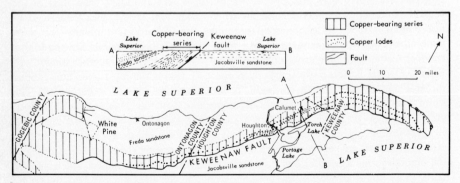

Fig. 11-6 Map and section showing copper-bearing and associated rocks, Keweenaw Peninsula, Michigan. After B. S. Butler and W. S. Burbank, *U.S. Geol. Survey Prof. Paper* 144.

**TABLE 11-11.   Production of Copper from All Sources,
Michigan Copper District, 1957–1961***

	Copper production, lb				
Year	Native copper mines	Sands reclamation	Sulfide ore	Total	Value
1957	36,147,820	8,937,304	68,562,019	113,647,143	$ 33,629,326
1958	35,301,708	5,100,407	81,656,908	122,059,023	31,414,330
1959	34,167,881	9,266,420	69,647,232	113,081,533	35,245,252
1960	31,577,880	8,740,414	75,120,243	115,428,537	37,025,756
1961	31,141,849	7,370,561	103,756,178	142,268,588	42,579,566
Total	168,337,138	39,415,106	398,742,580	606,494,824	$179,894,230

* From Michigan's Mineral Industries, 1961, *Dept. Conservation, Geol. Survey Div.*, 1962, p. 23. Used by permission.

low-grade deposit of chalcocite in sandstone that is stratigraphically several thousand feet above the major native copper deposits and 45 to 70 miles west-southwest of the principal native copper mines (Fig. 11-6). Reserves in this and an adjoining area that has been drilled indicate the presence of 105 million tons of ore that averages 1.3 percent copper. Production must be by underground mining.

Another new development, which had not yet been completely evaluated in 1963, was the discovery by diamond drilling late in 1962 of a new native copper ore body in conglomerate at a Calumet and Hecla property. Cores showed high-grade ore over good widths and apparently throughout a considerable length (182).

The major native copper deposits occur in dark Precambrian amygdaloidal basaltic lavas and interbedded reddish felsite conglomerates of Middle Keweenawan age on the south limb of the Lake Superior syncline and are cut off on the southeast by the great Keweenaw fault. A large number of separate flows and

**TABLE 11-12.   Distribution of Production throughout
Major Ore Horizons, Michigan Copper District, to End of 1925***

Horizon	Percent of total production
Calumet conglomerate	44.9
Kearsarge amygdaloid	15.7
Baltic amygdaloid	11.6
Pewabic amygdaloid	11.6
Other amygdaloids (Osceola, Isle Royale, Atlantic)	10.2
Fissures	2.7
All others	3.3

* From *U.S. Geol. Survey Prof. Paper* 144, p. 65, 1929.

conglomerates are present in the total section, which is many thousands of feet thick, but the major production has come from the Central Mine group. A generalized section showing the most important lodes is given in Table 11-13.

The deposits in lava flows, which constitute the amygdaloid lodes, occur in the vesicular or brecciated tops of certain flows. Most of the important deposits were concentrated in the brecciated or cellular tops of the flows, in which, during solidification and flow, the rock was broken into a jumbled mass of angular fragments and thus became highly permeable. Some important deposits, however, were concentrated in simple scoriaceous tops of flows, and some in tops in which vesicles coalesced and formed long connected passages.

TABLE 11-13. Generalized Section of the Keweenawan Rocks of the Copper District of Northern Michigan, Showing Major Lodes*

Rock unit	*Notes*
Freda sandstone	
Nonesuch shale	Chiefly gray siltstone about 600 ft thick. White Pine chalcocite deposits near bottom.
Copper Harbor group	Conglomerates and flows, thickness 2,300–5,500 ft.
Eagle River group	Conglomerates and flows, thickness 1,400–2,300 ft.
Ashbed group	Chiefly flows, thickness 1,400–2,400 ft. Contains Pewabic and Atlantic amygdaloid lodes.
Central Mine group	Many flows, some conglomerates, thickness about 3,800–25,000 ft. Contains Calumet and Hecla conglomerate, which furnished about 40 percent of the copper and lies below several minor conglomerate and amygdaloid lodes. Successively beneath the Calumet and Hecla conglomerate, with other rocks and minor lodes intervening, are the Osceola amygdaloid lode, the Kearsarge amygdaloid lode, and the Isle Royale amygdaloid lode.
Bohemian Range group	Conglomerates and flows, thickness about 9,500 ft. Includes Mount Houghton felsite and quartz porphyry, which may be intrusive and younger. Includes the Baltic amygdaloid lode and some less productive lodes.

* Compiled from *U.S. Geol. Survey Prof. Paper* 144, p. 16, 1929.

The deposits in the conglomerates, chiefly the Calumet and Hecla, occur as native copper that replaced the finer matrix of the conglomerate.

The fissure deposits or veins were concentrated in various fractures in the rocks, in places in simple fractures that yielded sheets of copper, in places in zones of parallel or interbranching fractures. In some fractures or fracture zones the copper was distributed in small masses throughout the zone, in others it formed masses that weighed a ton or more, and several were estimated to weigh nearly 500 tons each. Some small sulfide and arsenide veins occur, but these have not contributed to production.

Copper beaten from masses lying on the surface was used by the Indians for ornamentation, and such masses led to the discovery of the district by white men. Large masses encountered beneath the surface, however, caused great difficulty in mining.

In 1857 a mass weighing 480 tons was found in the Minnesota mine at a depth of 150 feet. The maximum length was 46 feet, the maximum width was 12½ feet, and the maximum thickness was 8½ feet. The mean thickness was 4 feet. Twenty men labored for 15 months to remove this copper from its rocky encasement; and the work involved the use of both powder and chisels. . . . First the miners excavated the rock next to the mass and exploded six kegs of powder. This proved unavailing, so further blasts with increasing quantities of powder, up to 30 kegs, were exploded. Altogether 110 kegs, or 2750 pounds of powder, were used before the mass was torn by successive explosions.[1]

Gangue minerals of considerable variety are present in the amygdaloidal or vesicular flows. Those most closely associated with the copper of the deposits are quartz, calcite, epidote, pumpellyite, chlorite, prehnite, red feldspars, and, locally, datolite. Among the zeolites laumontite is the most abundant, but generally it occurs where there is little or no copper. Although the deposits are somewhat commonly designated native copper associated with zeolites, actually the zeolites are relatively scarce.

The gangue minerals of the conglomerate lodes are chiefly red feldspar, epidote, pumpellyite, calcite, quartz, and chlorite. Zeolites and allied minerals, such as prehnite, are absent.

Unanimity of opinion does not exist regarding the origin and localization of the Lake Superior native copper deposits. Butler and Burbank (67, pp. 120–146) and Broderick and Hohl (63, p. 280) stated that two contrasting views of origin had been presented: (1) descending meteoric waters leached copper from the rocks and deposited it a considerable depth below the surface; (2) hot ascending waters, that is, hydrothermal waters, brought in the copper, which was deposited in physically and chemically favorable places. Butler and Burbank (67, p. 120) reached the conclusion that

the major mineralization was effected during a single period of somewhat complex activities that followed the completion of all the essential deformation that the rocks reveal. This deformation, it is thought, began during the outpouring of the Keweenawan lavas and the deposition of the associated sediments and was completed shortly after their accumulation was finished; the mineralization probably followed immediately afterward, perhaps overlapping the last structural adjustments.

Regarding the source of the copper-bearing solutions, Butler and Burbank assumed (67, p. 124) "that the copper-bearing solutions were expelled forcibly, at high temperature, from a magma chamber in which they originated." The concept stated by Broderick and Hohl (63, p. 280) is that on crystallizing, underlying intrusives gave off solutions under great pressure, which followed permeable

[1] From "A History of American Mining," pp. 232–233, by T. A. Rickard. Copyright 1932. McGraw-Hill Book Company. Used by permission.

channels upward and entered regions of lower pressure; reaction with highly oxidizing wallrock caused deposition of native copper. This concept was reaffirmed in 1948 (64), 1952 (65), and 1956 (66) after different ideas by others had been published. Cornwall (69), in 1956, discussed the origin of native copper deposits in general, and concluded that the deposits of the Keweenaw Peninsula, Michigan, "were almost certainly formed in permeable channelways by hot ascending solutions, whose composition has not yet been demonstrated," and that "the native state of copper strongly suggests its derivation from the lavas themselves. . . ."[1] Further, he thought that it was unnecessary to postulate a magamatic source for the solutions and that concentration by connate waters might provide a more reasonable explanation. Still later (1959) Stoiber and Davidson (169) stated the opinion that various lines of evidence support an origin of the amygdule minerals, and the copper, by regional hydrothermal metamorphism, with the qualification that hydrothermal was used to mean hot water. They suggested, further (p. 1445) that such alteration occurred during folding and thrusting of the Keweenawan lavas but that evidence is inconclusive regarding a magmatic source of the altering solutions. Neither this concept nor the one suggested by Cornwall is acceptable to a considerable number of geologists.

The White Pine deposits (144) are much different from the deposits just discussed, both statigraphically and geographically. Near the White Pine mine copper mineralization in the Nonesuch shale is almost restricted to a thickness of about 25 ft in the lowermost part of the formation designated for convenience of reference the *cupriferous zone* (144, p. 680). This zone is divided, from the bottom upward, into (1) the lower sandstone, thickness about 5 ft; (2) the parting shale, thickness about 6.5 ft; (3) the upper sandstone, thickness about 4.5 ft; and (4) the upper shale, thickness 20 to 40 ft, only the lower 7 or 8 ft of which contains appreciable copper. About 87 percent of the copper mineral is chalcocite and most of the remaining mineral is native copper, although exceedingly small amounts of bornite, covellite, and chalcopyrite have been observed and a very small amount of native silver is generally present. The deposit is extensive. White and Wright stated (144):

> *The copper in siltstone and shale at White Pine is believed to be derived largely from the waters in which the beds themselves were laid down. It is probably more nearly diagenetic than syngenetic, as the chalcocite and some of the native copper may have been precipitated below the surface of the bottom muds or may have replaced primary sulfide; but the copper is believed to have been introduced before the rock was lithified, deeply buried, and deformed.*[2]

The arguments to support this concept, and to explain occurrences of native copper in sandstone nearby, cannot be discussed here. This syngenetic or diagenetic concept of origin was objected to by Sales (166) and by Joralemon (159) and

[1] From H. R. Cornwall, *Econ. Geology,* vol. 51, p. 621. Copyright 1956. Economic Geology Publishing Company. Used by permission.
[2] From W. S. White and J. C. Wright, *Econ. Geology,* vol. 49, p. 712. Copyright 1954. Economic Geology Publishing Company. Used by permission.

was somewhat modified by White (170). The original ore body ended against the White Pine fault, and drilling on the western downthrown side of that fault disclosed that the ore horizon had been displaced about 2,000 ft vertically but was still an important deposit. A shaft was being sunk in 1961. As mining proceeds new information regarding origin should be available.

Canada

The Sudbury district, Ontario, is the most important producer of copper in Canada and is stated to contain the largest known copper reserves in North America outside of the United States (139, Chap. 3, p. 27). These deposits were described with deposits of nickel. A new district on the Gaspé Peninsula is becoming important and is reported to contain 57 million tons of ore containing an average of 1 percent copper.

Mexico

A considerable amount of copper is produced in Mexico along with silver, lead, and zinc. About half of the total production, however, comes from copper deposits in the state of Sonora (58, pp. 138–141; 76, pp. 182–183; 124, pp. 385–405), in which the Cananea district is the most important. A considerable amount of silver and some gold are also recovered. This part of Sonora, it will be recalled, is really an extension of the great copper province of Arizona and New Mexico (Fig. 11-4).

In the Cananea district (72; 76, pp. 386–387; 113) Paleozoic quartzite and limestone were deformed, covered in part by great amounts of volcanic rocks of rhyolitic to andesitic composition, and later intruded by granitic rocks, which are probably of late Cretaceous age. Mineralization throughout a zone about 2 miles wide and 6 miles long is closely associated with the intrusive rocks. Within this area there are copper deposits of four types: (1) replacements of limestone along bedding; (2) replacements of limestone along and beneath contacts with quartzite and with unconformably overlying volcanic rocks; (3) disseminated deposits in porphyry; and (4) breccia pipes, which are the most important type. The first three types are similar to those described in other deposits. Some of the replacements are in garnetized limestone and have been described as contact-metasomatic deposits. The breccia pipes are remarkable bodies of several types (162). Some are nearly vertical, oval in plan, more than a thousand feet long, contain a bottom part consisting of irregularly mineralized fragments of various rock confined within the pipe and an upper part in which the mineralization has spread out, mushroomlike, into limestone. Others consist of ringlike mineralization surrounding a weakly brecciated interior. Still others are ring-shaped in plan but converge downward in a funnel pattern and then pass into a vertical or inclined pipe. Mineralization differs in the various deposits, but in general the major ore minerals are chalcopyrite and bornite, along with some sphalerite and molybdenite and minor amounts of other sulfides. Secondary enrichment, with the formation of chalcocite, has been important.

The Pilares mine at Los Pilares de Nacozari was in a ringlike brecciated mass of rocks, chiefly volcanic, which apparently formed part of a synclinal structure

(134). Perry (113, p. 416) stated that it resembles some of the pipes of the Cananea district.

Chile and Peru

A vast mineralized province extends through the Andes Mountains of Chile and Peru, in which occur some of the most important copper deposits of the world. The southernmost of these, in Chile, 45 miles southeast of Santiago, is the Braden or El Teniente mine, stated to be the largest underground copper mine in the world (82, p. 865); farther north, and of much less importance, is the Potrerillos mine, which suspended operations in June 1959 after producing 1,762,192 tons of copper; about 20 miles north of the Potrerillos mine is El Salvador, brought into production in 1959 to replace Potrerillos; about 275 miles north of El Salvador is Chuquicamata, regarded as the world's largest economic copper deposit; and northward in Peru are Toquepala and Cerro de Pasco.

The Braden or El Teniente deposit (82, 94; 109, pp. 81–95) yielded almost 300 million tons of ore averaging about 2 percent copper between 1906 and 1959. The central feature of the mine area is the Braden breccia pipe in Tertiary volcanic rocks. The pipe is circular, has a diameter of about 4,000 ft at the surface, tapers downward, is known to extend to a depth of more than 5,200 ft from the surface, and is filled with a variety of rock material designated the Braden formation (82, p. 869 and Fig. 3, pp. 874–875), a semistratified breccia. Two other types of breccia occur around the margins of the pipe: one forms a zone 100 to 200 ft wide entirely outside the pipe and contains abundant fine-grained tourmaline and silicified fragments that contain abundant tourmaline; the other lies just next to this zone but entirely within the pipe along its margin. The ore body surrounds the Braden pipe as a continuous ring with a maximum width of 2,000 ft, and the material within the pipe is not mineralized except for fragments of the ore from the surrounding body. Mineralization consists chiefly of chalcopyrite and bornite with a, quartz-anhydrite gangue, disseminated throughout the material surrounding the pipe, and the deposit is classed as one of the porphyry copper type. The mill head assay in 1960 averaged 1.95 percent copper and 0.08 percent molybdenite. The limit of ore is gradational and must be determined by assay. Supergene enrichment, chiefly as chalcocite, was important and almost doubled the grade of ore in some places. Most of the supergene ore has been mined.

The origin of the breccia pipe is not clear. Lindgren and Bastin (94) regarded it as the result of an explosive volcanic vent, whereas Howell and Molloy (82) thought that it developed after ore deposition upon the site of an older stock and that, at least in part, there was substantial subsidence of material within the pipe.

The older deposit (Andes) at Potrerillos (97; 109, pp. 201–209) and the newly developed one at El Salvador (109, p. 210; 143) are both classed as porphyry coppers. Supergene enrichment by chalcocite is an important feature of each deposit.

The great Chuquicamata ore body is about 160 miles northeast of Antofogasta, Chile, in the Atacama Desert high on the western slope of the Andes Mountains, elevation 9,500 ft. The very early history of the utilization of copper from this extensive deposit is incompletely known, but apparently copper was obtained by the Chuco Indians possibly as much as 600 years ago, then by the Incas before the Spanish conquest, and, between 1560 and 1879, by the Spaniards and the Bolivians. English and Chilean companies conducted the first real mining, 1879 to 1912, from the richer veins. Development of disseminated ore and beginning of open-pit operations started in 1915 and has continued to the present time. Oxidized ores furnished all the production until about 1950, when underlying sulfide ores began to be utilized (71; 132, p. 475). The total amount of copper produced from 1915 to September 30, 1952, was 10,212 million lb (172). In 1961 the ore removed amounted to 24 million tons, which required removal of almost 13 million tons of overburden. The copper recovered from this amount of ore was 275,194 tons (181), a recovery of about 1.15 percent copper. An estimate of reserves has not been published recently, but they are known to be very great. An unofficial estimate in 1935 was about 940 million tons of ore averaging 2.15 percent copper (139, Chap. 3, p. 31). Parsons (108, p. 257) in 1933, stated: "The figure for the aggregate copper content of the reasonably assured ore at Chuquicamata is breath-taking at 50,000,000,000 pounds."[1]

The Chuquicamata ore body (58, pp. 103–109; 96; 109, pp. 157–173; 114; 132) is stated to be localized in granodiorite (96, p. 677; 132, p. 476) or monzonite porphyry and granodiorite (114, p. 1167) probably of late Cretaceous or early Tertiary age. At the surface the ore body is somewhat pear-shaped, decreasing in size southward, and occupies an area about 2 miles long and about 3,600 ft wide at its greatest extent. The west boundary is a fault, to the west of which is relatively unaltered barren granodiorite, whereas to the east, in the ore-bearing zone, is rock so highly altered that its original character is masked. Further, this altered rock is extensively fractured and cut by a multitude of fissures. Taylor (132, pp. 477–479) recognized several degrees of alteration. In a general way, with local exceptions, the rock for some distance to the east of the western-bounding fault has been converted into either (1) a siliceous rock consisting almost entirely of quartz and sericite, with original texture obliterated, in which the quartz occurs as aggregates and interstially and as veinlets of several ages, or (2) a sericitic rock in which almost all the original feldspars have been altered to sericite but in which the quartz is both original and introduced. Both of these altered types have been highly fractured, the siliceous type more so than the sericitic one. Generally the sericitic rock is marginal to the siliceous rock and gives place to less-altered types. The siliceous rock is characterized by high porosity, the sericitic rock by medium porosity, and the major mineralization has been localized in these two types of altered rock. According to Perry (114):

A 300 to 600 ft. width of quartz and sericite, containing the strongest iron and copper mineralization in the orebody, parallels the West fissure and is separated from it by 50 to 200 ft. of highly altered but poorly mineralized rock. Branching from it to form the eastern part of the ore zone, there is a system of strongly mineralized subsidiary fractures with northeast, east, and southeast strikes that intersect and ramify as a network of veins and veinlets,

---

[1] From "The Porphyry Coppers," p. 257, by A. B. Parsons. Copyright 1933. The American Institute of Mining and Metallurgical Engineers. Used by permission.

lacing altered and mineralized rock. Sales states in a private report "The Chuquicamata Orebody," March 1950: "The intensely mineralized belt constituting the orebody contains veins and innumerable criss-cross vein structures with intense mineralization and alteration of intervening porphyry. Although on a much larger scale, the structural pattern here shows a striking similarity to the Leonard vein at Butte."[1]

The ore body has been classed as a porphyry copper but, as indicated by the foregoing quotation, it consists of innumerable veins.

The upper part of the ore body was leached and oxidized, and enriched both in the oxidized zone and in the secondary sulfide zone. Various oxidized minerals were formed, including a number of unusual ones (87, pp. 254–255). The most abundant oxidized mineral was antlerite, $Cu_3SO_4(OH)_4$. The history seems to have been somewhat complex and to have involved changes in climate and level of groundwater. Apparently an earlier secondary chalcocite zone was formed in part of the area, and later this chalcocite ore was oxidized.

Little was known about the primary mineralization during the early history of mining because of the extent of oxidation. The chief primary mineral apparently was enargite, but bornite, chalcopyrite, tennantite, and tetrahedrite were known to be present in small amounts (96). As mining proceeded, however, more and more primary ore was uncovered, and a drainage tunnel was driven approximately 1,200 ft below the original surface. It then appeared (114, p. 1168) that enargite was the important primary mineral but that some chalcocite also was primary. Perry stated that deep holes indicate diminishing secondary enrichment at depth but continuity of enargite with chalcocite and that they were bottomed in commercial grade copper ore. In places chalcocite-enargite-pyrite mineralization passes into chalcopyrite-covellite mineralization and finally into chalcopyrite and specularite. Bornite, molybdenite, and small amounts of galena and sphalerite are present locally. Some of the covellite appears to be primary.

The Cerro de Pasco deposits of Peru were described briefly in the discussion of silver deposits. The Toquepala ore body is farther south, almost at the Chilean border. Production began in 1960 after stripping 125 million tons of waste from the ore body, which is estimated to contain 400 million tons of ore containing 1 percent copper (179). The ore body is classed as a porphyry copper deposit (118). It lies in a somewhat elliptical mineralized zones about 2 miles long, the central part of which is a large breccia pipe around which, within a 2-mile radius, are small satellitic pipes. The ore body is mushroom-shaped and (118) "consists of a flat-lying enriched zone of predominant chalcocite with a stem-like extension of hypogene chalcopyrite ore in depth within and round the pipe."[2] The chalcocite zone is about 500 ft thick in the center but only a few feet near the margins. The rocks of the mineralized zone have been extensively altered, the principal products being sericite and quartz. The rocks of the area are Mesozoic (?) and Tertiary volcanics that have been intruded by diorite apophyses of the Andean batholith. The rocks, along with the zone of alteration about the ore body, are intruded by small stocks and dikes of dacite porphyry.

[1] From V. D. Perry, *Mining Engineering*, vol. 4, p. 1167. Copyright 1952. The American Institute of Mining, Metallurgical, and Petroleum Engineers. Used by permission.
[2] From K. Richard and J. H. Courtright, *Mining Engineering*, vol. 10, p. 262. Copyright 1958. The American Institute of Mining, Metallurgical, and Petroleum Engineers. Used by permission.

Zambia (Northern Rhodesia) and Katanga, Africa

One of the great copper provinces of the world lies partly in Zambia (Northern Rhodesia) and partly in adjoining Katanga of the Congo (Fig. 11-7). In general the deposits are in an arc-shaped zone of deformed rocks that extends for more than 400 miles in Rhodesia and Katanga and is itself part of a much larger base-metal province which extends to the southwest (102, p. 268). The part in Zambia is generally designated the *Copperbelt* and is a zone about 30 miles wide next to the Congo border. The importance of copper production in the Zambia-Katanga area is indicated in Table 11-5 and Figure 11-3, which show that this area, which furnished only 2 percent of the copper of the world from 1901 through 1925, produced 18.7 percent of it in the period 1945 to 1959 and 18.6 percent in 1962. The production from Zambia has been considerably greater than that from Katanga.

The total amount of ore milled from mines in the Copperbelt to June, 1960, was 306,512,800 short tons, which contained 3.22 percent copper and from which was obtained 7,437,962 long tons of copper and 24,242 long tons of cobalt. The reserves as estimated in March and June, 1960, are 837,302,800 short tons of ore containing 3.50 percent copper, which would amount to something more than 29 million short tons of copper. The cobalt reserves from three mines (Baluba, Chibuluma, Rhokana) were estimated to be 400,000 short tons from copper ore containing 0.16 to 0.18 percent cobalt.

Total production from the mines of Union Miniere du Haut Katanga is not available, but in 1955 the company produced 259,000 short tons of copper, 9,440 short tons of cobalt, 126,000 short tons of zinc concentrate, 58 short tons of cadmium, and also important amounts of radium, germanium, silver, and

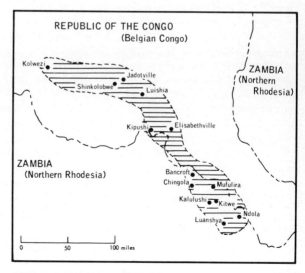

Fig. 11-7 Map showing major copper and cobalt area of Zambia (Northern Rhodesia) and the Republic of the Congo (Belgian Congo). Boundary of the mineralized area generalized and approximate.

concentrated uranium (168). The largest mine of the area, The Musonoi open pit, furnished about half of the total copper output and 85 percent of the cobalt (141, p. 40).

The rock formations in the two areas are much the same. The major units, their general characteristics, and their degrees of mineralization are shown by Table 11-14, which gives the succession as it is known in the Copperbelt (73, 98, 99, 115).

The units shown in Table 11-14 also occur in Katanga (70, 77, 120), although some names and characteristics differ. The Upper Roan of Zambia (Northern Rhodesia) is known as Dolomites of the Série des mines in the vicinity of Elizabethville, Katanga. In each place the copper deposits are in the Mine series although the stratigraphic position in the Mine series is not everywhere the same. Robert (119, p. 535) stated that good results are obtained in Zambia by the use of the Katanga stratigraphic column. The ore in Katanga occurs in the dolomitic and banded siliceous rocks that correspond to the Upper Roan, whereas the ore in Zambia occurs in the Lower Roan.

The structure of the area as a whole is complex. According to Garlick (74) the Copperbelt

> *occupies the southeast end of a 500-mile arc of folded Katanga sediments that have been bowed northeastwards between the Kibaran massif on the west and the Fort Rosebery granite massif on the east. The structure of the Copperbelt is dominated by folds trending and generally plunging to the northwest, and by domes and basins.*[1]

Brock (62) indicates that the majority of copper deposits of the Copperbelt make a straight alignment over a length of more than 200 miles in a northwesterly direction to and beyond the southern border of Katanga, although varied types of structure exist at individual mines—synclines, monoclines, domes, and some others. Numerous small faults are present throughout the area, and perhaps two major ones. In Katanga Province the beds have been tightly folded, and the principal copper mines are associated with the more intense folds (70, p. 55). In the area of the Musonoi open pit, Katanga, beds of the Mine series have been thrust over rocks of the Kundelungu system (141, p. 40).

Mendelsohn (100) stated that the rocks of the Copperbelt have been subjected to moderately low-grade metamorphism of regional type and that generally metamorphism of the Katanga system increases from the Copperbelt westward and southwestward. He stated that the principal effect of metamorphism was recrystallization and that various factors "indicate that metamorphism and growth of new minerals could have been and probably were achieved without the addition of material from an external source."[2]

Outstanding features of the ore deposits of the Copperbelt are their wide

[1] From "The Geology of the Northern Rhodesian Copperbelt," p. 89, Structural Evolution of the Copperbelt by W. G. Garlick. Copyright 1961. Macdonald & Co. (Publishers) Ltd., London, England. Used by permission.
[2] From "The Geology of the Northern Rhodesian Copperbelt," p. 11, Metamorphism by F. Mendelsohn. Copyright 1961. Macdonald & Co. (Publishers) Ltd., London, England. Used by permission.

extent, their even tenor of mineralization, and their confinement to particular stratigraphic units (101, p. 117). Mendelsohn (101) stated that the most important and probably the best-known single type of deposit is that in the Ore shale of the Ore formation (Lower Roan). Other types are the Footwall (quartzite) deposits (see Table 11-14) below the Ore formation, which have been important at a few places in the central part of the Copperbelt; and the quartzite ore bodies of the Mufulira syncline, where the Ore formation consists of five quartzite beds, as follows (157):[1]

	A quartzite
	Lower dolomite and shale
	B quartzite
Ore Formation	Mud seam
	C quartzite
	D quartzite
	E quartzite

The quartzite deposits of the Mufulira syncline are typically thicker and less extensive than are the Ore shale deposits.

The most abundant sulfide minerals in all types of deposits of the Copperbelt are pyrite, chalcopyrite, bornite, and chalcocite, in approximate order of abundance. Some are molded onto and some replace rock-forming minerals, and generally the grain size of the ore minerals varies with the grain size of the rocks in which they occur, being fine in fine-grained rocks, such as argillite, coarser in quartzite, and still coarser in some metamorphosed dolomite. The sulfides are disseminated through the rocks and also concentrated along some bedding planes (101, p. 127). The order of formation of the major sulfides was (1) pyrite, (2) chalcopyrite, (3) bornite, (4) chalcocite (55, 101). Some of the chalcocite appears to be primary, some and possibly much of it is supergene (101, p. 128). Other ore minerals present locally are covellite, possibly digenite, carrollite (Co_2CuS_4), linnaeite (Co_3S_4), and traces of molybdenite and scheelite. Mendelsohn stated (101): "The ore-bearing beds, particularly in the vicinity of the orebodies, are slightly richer in uranium than the barren formations."[2] Zoning of the primary sulfide minerals exists in most of the deposits, but the pattern differs from place to place. Oxidation and leaching have progressed to a depth of about 150 to 200 ft from the surface in most deposits, and locally partial oxidation has penetrated much deeper. In the southern part of the Copperbelt supergene enrichment apparently was relatively unimportant, but elsewhere it contributed much to the value of the deposits. The principal supergene minerals in the oxidized zone were malachite, cuprite, chrysocolla, some tenorite and native copper, and, in the sulfide zone, chalcocite.

In Katanga (77, 105, 120) the ore occurs in dolomitic rocks that correspond to the Upper Roan. The deposits were much oxidized near the surface, but lower

[1] From "The Geology of the Northern Rhodesian Copperbelt," p. 407, Mufulira Area, Luansobe, by H. A. Green. Copyright 1961. Macdonald & Co. (Publishers) Ltd., London, England. Used by permission.

[2] From "The Geology of the Northern Rhodesian Copperbelt," p. 128, Ore Deposits by F. Mendelsohn. Copyright 1961. Macdonald & Co. (Publishers) Ltd., London, England. Used by permission.

TABLE 11-14. Generalized Column of the Precambrian Rocks of the Copperbelt, Zambia (Northern Rhodesia)*

Series	Group or other unit	Formation	Rock types	Notes regarding mineralization
Kundelungu	Upper	—	Gabbro; intrusive, commonly into the Upper Roan dolomites	Generally not mineralized.
	Middle	—	Shale, quartzite Shale Tillite	
	Lower	Kakontwe	Shale Dolomite and shale	
		Tillite	Tillite	
Mine (2,500–6,000 ft thick)	Mwashia	—	Carbonaceous shale, argillite; minor dolomite and quartzite	Generally not mineralized.
	Upper Roan	—	Dolomite and argillite; minor quartzite, breccia Argillite and quartzite	
	Lower Roan	Hanging wall	Quartzite Argillite and feldspathic quartzite; minor dolomite	All ore deposits of Copperbelt are in the Lower Roan series.
		Ore	Argillite, impure dolomite, micaceous quartzite; minor graywacke, arkose	Argillite, together with dolomite in many places, known as the "Ore shale"; appears at same stratigraphic horizon along whole length of the

Katanga system

Footwall	Footwall conglomerate, minor or scarce; argillaceous quartzite, feldspathic quartzite, aeolian quartzite, conglomerates	Copperbelt. Contains about two-thirds of the ore of the Copperbelt, but important deposits present both above and below it. Ore shale extends nearly 80 miles, from Bancroft to Roan Antelope, and ore deposits within one strip about 5 miles wide and about 80 ft thick, with ore bodies about 25–30 ft thick. Ore occurs in argillite, impure dolomite, sericitic quartzite, carbonaceous sericitic quartzite (graywacke), to a limited extent in feldspathic quartzite, and locally in conglomerate. Ore does not occur in pure dolomite or aeolian quartzite.
MAJOR UNCONFORMITY		
Muva system	Schist, quartzite, conglomerate	
Granite	Chiefly gray granite; some red granite and porphyritic gneiss	
Basement complex — Lufubu system	Chiefly mica schist, micaceous quartzite, gneiss; minor carbonate, conglomerate, subgraywacke, calcareous arkose	Disseminated pyrite and chalcopyrite throughout, but more abundant in granitic rocks, some of which contain bornite also. Basement sedimentary and granitoid rocks mineralized where ore bodies in Katanga system rest directly on basement topographic highs. Generally mineralization is greatest in the regolith near top of pre-Katanga rocks, but in many places it extends well below this.

* Compiled from "The Geology of the Northern Rhodesian Copperbelt," pp. 21–54, articles by W. G. Garlick, F. Mendelsohn, and P. J. Pienaar. Copyright 1961. Macdonald & Co. (Publishers) Ltd., London, England. Used by permission.

they contain primary sulfides, chiefly bornite and carrollite, which appear to replace the dolomite. Bornite is partly replaced by chalcocite, and penetration of secondary chalcocite has been so deep that definite recognition of any primary chalcocite that might be present has been obscured.

The genesis of the deposits of the Copperbelt and those of Katanga has been a matter of controversy for many years, and the concepts regarding genesis can be summarized only briefly. In general they are of the same kind as the concepts regarding the gold deposits of the Rand and the lead and zinc deposits of the Mississippi Valley type. One group advocates an epigenetic origin and deposition from hydrothermal solutions derived from a magmatic source. Another group postulates a syngenetic origin with deposition of copper at the same time as the deposition of the sediments and subsequent reorganization, recrystallization, and replacement of rock-forming minerals by metamorphism with essentially no addition of material and only limited migration of the ore-forming constituents; zoning, in this concept, also resulted from metamorphic reorganization. As in the case of the Rand gold deposits, separation of actual facts from opinions and prejudices is difficult, and what one person considers to be an established fact another one will not accept as such. The following is a brief summary of the opinions or conclusions of some of the authors of the recent (1961) book, "The Geology of the Rhodesian Copperbelt." F. Mendelsohn, pp. 130–146, and W. G. Garlick, pp. 146–165, favored a syngenetic theory, but Mendelsohn stated that some features are not adequately explained by such a theory. O. Winfield, pp. 328–342, favored a syngenetic origin for the Chibuluma deposits. R. T. Brandt, C. C. J. Burton, S. C. Maree, and M. E. Woakes, pp. 411–458, concluded (p. 459) for the Mufulira deposits: "The syngenetic theory is the most likely explanation of ore genesis but does not account for all the features of the deposit."[1] D. M. McKinnon and N. J. Smitt, pp. 234–275, concluded (p. 274), for the Nchanga deposits: "The available evidence does not demonstrate beyond doubt either the epigenetic or syngenetic origin of the copper."[1] P. J. Pienaar, pp. 467–484, who discussed the Bwana Mkubwa mine, the oldest one in the Copperbelt, suggested (p. 484), regarding the origin of the primary mineralization at that deposit:

> *The ore beds have been faulted and intruded by a quartz gabbro, which is the geochemical abode of copper. Thus, from geochemical and structural view points, the area is the ideal setting for hydrothermal introduction of copper minerals. To date no relationships have been established between the mineralization and the faults and/or the basic intrusion. However, the information on the deposit is far from complete.*[1]

He suggested, also, two factors that seemed difficult to explain by a syngenetic origin. B. B. Brock, pp. 81–89, discussed the structural setting of the Copperbelt, called attention to the remarkable alignment of the deposits, suggested (p. 87) that, if the alignment resulted from a fundamental fracture that visibly affected only the basement rocks, such a fracture would be available as a channelway for mineralizing solutions, and concluded (p. 89): "Any theory regarding the origin

---

[1] Quotations by various authors cited are from "The Geology of the Northern Rhodesian Copperbelt." Copyright 1961. Macdonald & Co. (Publishers) Ltd., London, England. Used by permission.

of the copper that ignores this remarkable alignment might be laying itself open to a charge of arbitrary selectivity of data."[1]

Many others have participated in the discussions regarding the origin of the copper deposits of Zambia (Northern Rhodesia) and Katanga. References are too numerous to list here. Some discussions may be found in various issues of *Economic Geology*, volumes 49–58.

Other Countries

The estimated copper production of the Soviet Union in 1962 was about 14 percent of the world total. A summary of the deposits was published in English in 1935 by Nekrasoff (104) and in German in 1941 by Berg and Friedensburg (58, pp. 152–161). Since then brief summaries have been published in 1950 by the U.S. Bureau of Mines (139, Chap. 3, pp. 35–36) and in 1953 by Shimkin (48, pp. 118–120). Also, brief mention of the deposits was made in 1960 by Nalivkin (103, pp. 77, 78, 97, 121).

The chief deposits apparently are in the Ural Mountains, in Kazakstan (Kazak Steppe and Altai mining districts), in Central Asia, and in the Caucasus. The principal types of deposits are (1) porphyry copper, (2) copper-bearing pyritic lenses, and (3) copper-bearing sandstones stated to be similar to the deposits of Northern Rhodesia. Less important types are contact-metasomatic deposits and veins. The largest porphyry copper deposits are Kounrad (47°00′N, 75°00′E) and Amalyk (40°49′N, 69°39′E). Kounrad is important for molybdenum also. The pyritic deposits are widespread throughout the Ural Mountains and are less abundant in the Caucasus and Kazakstan. The copper-bearing sandstones are concentrated in the region around Dzhezkazgan (47°51′N, 67°14′E)— also spelled Djeskasgan—where the mineralized area is stated to cover 1,000 sq km of rocks of Carboniferous age. Important vein deposits are present at Kafan (Zangezur) (39°12′N, 46°24′E). Nalivkin (103) rated the order of importance of these four deposits as (1) Dzhezkazgan, (2) Kounrad, (3) Almalyk, (4) Kafan (Zangezur).

The copper-bearing pyritic deposits of the Huelva district in the Sierra Morena, northwest of Sevilla, Spain (58, pp. 166–172; 80; 85), are among the large deposits of the world—indeed, Heim (80, p. 635) regarded them as the largest sulfide ore deposits in the world. The ores have been worked since Phoenician times or earlier. Many mines have furnished production in the Huelva district, the largest of which by far are the Río Tinto mines, which were especially famous in the nineteenth century. The high content of pyrite has made the deposits valuable as a source of material for the manufacture of sulfuric acid, for which pyrite that contains less than 1 percent of copper is used directly. Oxidation has produced gossan that, in parts of the area, has been used as iron ore. Copper and gold were concentrated at the base of the gossan.

Australia produced slightly less than 3 percent of the copper of the world in 1961. The most important copper-producing districts have been Mount Lyell on the west coast of Tasmania, 21 miles south of Rosebury; Mount Morgan, Queens-

[1] From "The Geology of the Northern Rhodesian Copperbelt." Copyright 1961. Macdonald & Co. (Publishers) Ltd., London, England. Used by permission.

land, 23 miles southwest of Rockhampton; and Cobar, north-central New South Wales, about 460 miles northwest of Sydney (54; 58, pp. 78–81; 131; 136; 140). The most important ore bodies at Mount Lyell are massive copper-bearing pyrite lodes along a schist-conglomerate contact. Much of the ore contained about 2 percent copper, but the famous Mount Lyell bonanza contained 850 tons of ore that averaged 1,011 oz of silver per ton and large quantities of copper and gold. Between 1883 and September 30, 1950, the Mount Lyell deposits yielded 516,703 tons of copper, 619,108 oz of gold, and 18,472,139 oz of silver. Mount Morgan is a copper-gold deposit in a roof pendant of flow and sedimentary rocks flanked by granitic rocks. Oxidized ore near the surface passed into pyrite-gold ore at a depth of about 425 ft and that, in turn, into chalcopyrite-pyrite-gold ore at a depth of about 700 ft. The pyrite-gold ore contained 1.4 percent copper and 7.7 dwt gold per ton, whereas the chalcopyrite-pyrite-gold ore contained 3.5 percent copper and 8.0 dwt gold. Some rich ore shoots, however, contained 10 to 12 percent copper; and some contained 300 to 470 oz of gold per ton. From 1886 to 1925, when a disastrous fire caused closing of the original gold mine, the production was 5,345,000 oz of gold and 140,000 tons of copper. A new company was formed in 1929 to reopen the mine and develop it as a large open pit. Between 1932 and 1950 the yield was 6,225,540 oz of gold and 190,735 tons of copper. Reserves estimated in 1951 were something more than 2,000,000 oz of gold and 155,000 tons of copper. Cobar was another gold-copper deposit that yielded, from 1871 to 1950, 1,199,143 oz of gold, 1,916,512 oz of silver, 132,065 tons of copper, and 3,225 tons of lead.

Cobalt

Most of the world's cobalt (147, pp. 24–44; 148, pp. 21–26) is produced in Africa from the Congo, Zambia (Northern Rhodesia), and Morocco. Less important production has come from Canada and the United States, but the United States practically ceased to furnish cobalt in 1960. The early source of cobalt was chiefly New Caledonia (147, p. 17), where production began about 1874 although very active mining began early in the 1880s. From then until the exploitation of the silver-copper deposits of Ontario, Canada, about 1903, New Caledonia furnished most of the cobalt of the world. Ontario became the chief producer from about 1905 to 1925 (147, p. 19); in 1925 by-product cobalt was recovered from the copper ores of Katanga in the Congo, and this soon became the chief world source. Cobalt from Zambia (Northern Rhodesia) became important about 1930, and from French Morocco about 1938 (147, pp. 1–2). The chief reserves of the free world (147, p. 26) are in Africa (1,147,000 tons of cobalt contained in ore), New Caledonia (440,000 tons), Canada (193,000 tons), and the United States (136,000 tons). It should be noted that much of the cobalt in the reserves of the United States will be unavailable without a high price for cobalt or a subsidy.

The copper deposits of Katanga, previously described, are by far the most important source of cobalt (146, pp. 225–227; 147, pp. 43–44; 148, p. 21), especially the Musonoi open pit (139, 173). Cobalt production in Katanga began about 1925 with 126 metric tons of cobalt, which was increased to 1,012 metric

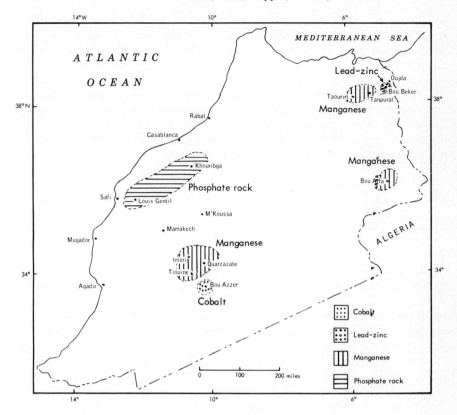

Fig. 11-8 Map showing major areas of cobalt, lead-zinc, manganese, and phosphate rock in Morocco. Manganese locations after Materials Survey, Manganese, U.S. Bur. Mines. Boundaries of all mineralized areas generalized from locations given in various reports.

tons in 1940 and to 5,148 metric tons in 1950 (167), which at that time was 71.5 percent of the total world production.

The production of Zambia (Northern Rhodesia) (146, pp. 240–242; 147, pp. 43–44; 148, pp. 21–22) comes chiefly from deposits at Baluba, Chibuluma, and Rhokana, where the reserves of these three deposits are estimated to be 400,000 short tons from copper ore containing 0.16 to 0.18 percent cobalt. Chibuluma is a relatively new mine from which the first ore was hoisted in October, 1955. Ore reserves in 1956 were estimated to be 7.3 million tons with an average grade of 5.23 percent copper and 0.25 percent cobalt (175), but on July 1, 1959, they were stated (146, p. 22) to be nearly 10 million short tons averaging 4.98 percent copper and 0.19 percent cobalt.

The cobalt of Morocco (Southern Zone) is obtained from deposits at Bou Azzer (146, pp. 258–262; 147, p. 43; 148, p. 22), in the Anti-Atlas Mountains (Fig. 11-8), about 250 km (155 miles) east of the port of Agadir and 170 km (105 miles) southeast of the nearest railroad at Marrakech. Ore is transported to

Marrakech by truck, the road distance being about 186 miles. Precambrian gneisses, mica schists, serpentines, and gabbroic rocks are cut by quartz-carbonate veins that average 1 to 2 m in thickness but which locally enlarge to as much as 12 to 15 m. Ore occurs especially in veins that cut serpentine. The chief ore minerals are smaltite, skutterudite, saffrolite, and rammelsbergite; other metallic minerals include niccolite and other nickel-arsenic minerals, arsenopyrite, and lollingite. The chief gangue mineral is a carbonate. Both gold and silver are present. The average commercial ores contain 11 to 12 percent cobalt, but some are used that contain only 0.75 to 1.25 percent. The gold content generally is 5 to 30 g per ton but exceptionally rises to 300 g. Silver is present in varying amounts but always less than 50 g per ton.

The cobalt of Canada (147, pp. 30–32; 148, pp. 22–23) comes chiefly from the silver-cobalt ores of the Cobalt and the Gowganda districts, the nickel-copper ores of the Sudbury district, all in Ontario, and from the nickel-copper ores of the Thompson and Lynn Lake district, Manitoba, all previously described.

Cobalt production in the United States has been chiefly from the Blackbird mine, Lemhi County, Idaho (145; 147, pp. 25–27; 148, pp. 23–24). The deposits, which are veins and lodes that are irregular and discontinuous, occur in zones of fractured and schistose quartzite of the Belt series of Precambrian age. Four types of lodes have been described: (1) cobalt-tourmaline lodes consisting of tourmalinized quartzite impregnated with cobaltite; (2) cobalt-biotite lodes, in which biotized quartzite was impregnated with cobaltite; (3) cobalt-quartz lodes, similar to the second type but containing, in addition, bodies of quartz that contain cobaltite; (4) gold-copper-cobalt lodes, in which material of the second type contains stringers and lenses of quartz that contain varying amounts of cobaltite, native silver, chalcopyrite, gold, and electrum. Three stages of mineralization have been recognized. Tourmaline was confined to the first stage, biotite chiefly to the first and second stages, and gold and electrum to the third stage. Discovered in 1893, the district was not fully developed until 1949 (154). The deposits are in a difficultly accessible area, but they continued to be worked because of United States government subsidies for the production of cobalt. These were withdrawn at the end of 1958, and the mine was closed in the summer of 1960.

Cobalt was produced to a limited extent from the lead ores of the Southeastern Missouri area (147, p. 27; 148, p. 23), starting about 1954 (160), but the refinery at Fredericktown, Missouri, which processed those ores, was closed early in 1961 because of the depletion of ore bodies that contained cobalt.

Some cobalt has been recovered as a by-product from magnetite iron ores at Cornwall and Morgantown, Pennsylvania (147, p. 27; 148, p. 23).

Probably a large amount of cobalt is contained in residual iron-nickel-cobalt ores of Cuba (147, p. 33; 148, p. 24), which are chiefly in Oriente Province and were mentioned briefly in the discussion of nickel deposits. The recovery of cobalt is problematical. In some ores cobalt is recovered in the concentrate with nickel, in others it has not been recovered because of the small amount present. Also, political conditions in Cuba may retard production.

A small amount of cobalt is obtained from copper-bearing pyrite of Outokumpo, Finland (148, p. 24), some of which contains 0.18 to 0.19 percent cobalt.

Cadmium

Cadmium, unlike most metals, is not recovered from ores of cadmium but is recovered as a by-product of the treatment of zinc ores and other base-metal concentrates. Generally the United States and Canada produce more than half of the cadmium metal of the world. During the period 1956 to 1961 the United States furnished 48 percent of the world total, Canada furnished 10 percent. The production in the United States came not only from domestic zinc ores but also from imported zinc ores and concentrates and other imported base-metal concentrates. Flue dust from Mexico was the source of 18 percent of the total cadmium metal produced in the United States. Mexico actually is an important source of cadmium because, in addition to the material sent to the United States, a considerable amount of cadmium is recovered in Mexico (160). The cadmium produced in Canada comes to a large extent from zinc concentrates of the Sullivan mine, British Columbia, and the Keno Hill mines, Yukon Territory.

SELECTED REFERENCES

LEAD AND ZINC

1. A. A. Agnew, 1955, Application of geology to the discovery of zinc-lead ore in the Wisconsin-Illinois-Iowa district, *Mining Eng.,* vol. 17, pp. 781–795.
2. E. C. Andrews, 1950, Geology of Broken Hill, New South Wales, Internat. Geol. Cong., 18th, Great Britain 1948, pt. 7, The geology, paragenesis, and reserves of the ores of lead and zinc, pp. 187–194.
3. E. S. Bastin (ed.), 1939, Lead and zinc deposits of the Mississippi Valley region, *Geol. Soc. America Spec. Paper* No. 24, 156 pp.
4. E. S. Bastin and C. H. Behre, Jr., 1939, Origin of the Mississippi Valley lead and zinc deposits—a critical summary, in Lead and zinc deposits of the Mississippi Valley region, pp. 121–143, *Geol. Soc. America Spec. Paper No.* 24, 156 pp.
5. G. Berg, F. Friedensburg, and H. Sommerlatte, 1950, Blei und Zink, in "Die Metallischen Rohstoffe," no. 9, Ferdinand Enke, Stuttgart, 468 pp. Major deposits as follows: Australia, pp. 115–132; Belgium, pp. 91–100; Canada, pp. 137–151; Yugoslavia, pp. 307–319; Mexico, pp. 320–334; Morocco, pp. 268–274; Peru, pp. 352–361; Poland, pp. 362–368; Spain, pp. 401–412; the United States, pp. 424–456; the Soviet Union, pp. 384–401.
6. C. H. Behre, Jr., A. V. Heyl, Jr., and E. T. McKnight, 1950, Zinc and lead deposits of the Mississippi Valley, Internat. Geol. Cong. 18th, Great Britain 1948, pt. 7, The geology, paragenesis, and reserves of the ores of lead and zinc, pp. 51–69.
7. J. Bouladon, 1952, Plomb et zinc, in Géologie des gîtes mineraux Marocains, Internat. Geol. Cong., 19th, Algiers 1952, *Mon. Régionales,* ser. 3, Maroc, no. 1, pp. 179–216.
8. A. L. Brokaw, 1950, Geology and mineralogy of the East Tennessee zinc district, Internat. Geol. Cong., 18th, Great Britain 1948, pt. 7, The

geology, paragenesis, and reserves of the ores of lead and zinc, pp. 70–76.

9. J. S. Brown, 1936, Structure and primary mineralization of the zinc mine at Balmat, New York, *Econ. Geology,* vol. 31, pp. 233–258.

10. J. S. Brown, 1942, Edwards-Balmat zinc district, New York, in "Ore Deposits As Related to Structural Features," pp. 171–174, Princeton University Press, Princeton, N.J., 280 pp.

11. J. S. Brown, 1947, Porosity and ore deposition at Edwards and Balmat, New York, *Econ. Geology,* vol. 58, pp. 505–546.

12. W. H. Brown, 1935, A quantitative study of the zoning of ores at the Austinville, mine, Wythe County, Virginia, *Econ. Geology,* vol. 30, pp. 425–433.

13. A. H. Buehler, 1932, The disseminated-lead district of southeastern Missouri, Internat. Geol. Cong., 16th, United States 1933, *Guidebook* 2, pp. 45–55.

14. S. R. Carter, 1950, Mount Isa geology, paragenesis and ore reserves, Internat. Geol. Cong., 18th, Great Britain 1948, pt. 7, The geology, paragenesis, and reserves of the ores of lead and zinc, pp. 195–205.

15. S. R. Carter, 1953, Mount Isa mines, in *Fifth Empire Min. Metall. Cong.,* Australia and New Zealand, pp. 361–377, Australasian Institute of Mining and Metallurgy, Melbourne.

16. J. Claveau, J. Paulhac, and J. Pellerin, 1952, The lead and zinc deposits of the Bou Beker–Touissit area, eastern French Morocco, *Econ. Geology,* vol. 47, pp. 481–493.

17. L. W. Currier, 1935, Structural relations of southern Appalachian zinc deposits, *Econ., Geology,* vol. 30, pp. 260–286.

18. N. H. Fisher, 1960, Review of evidence of genesis of Mt. Isa orebodies, Internat. Geol. Cong., 21st, Copenhagen 1960, pt. 16, sec. 16, Genetic problems of ores, pp. 99–111.

19. C. B. Forgan, 1950, Ore deposits at the Stantrg lead-zinc mine, Internat. Geol. Cong., 18th, Great Britain 1948, pt. 7, The geology, paragenesis, and reserves of the ores of lead and zinc, pp. 290–307.

20. G. M. Fowler, 1942, Ore deposits of the Tri-State zinc and lead district, in "Ore Deposits As Related to Structural Features," pp. 206–211, Princeton University Press, Princeton, N.J., 280 pp.

21. P. W. George, 1937, Geology of lead-zinc-copper deposits at Buchans, Newfoundland, *Am. Inst. Mining Metall. Engineers Tech. Pub.* 816, 23 pp.

22. J. K. Gustafson, H. C. Burrell, and M. D. Garretty, 1950, Geology of the Broken Hill ore deposit, Broken Hill, N.S.W., Australia, *Geol. Soc. America Bull.,* vol. 61, pp. 1369–1438.

23. J. K. Gustafson, 1954, Geology of Australian ore deposits, Broken Hill, *Econ. Geology,* vol. 49, pp. 783–786.

24. A. V. Heyl and others, 1955, Zinc-lead-copper resources and general geology of the Upper Mississippi Valley district, *U.S. Geol. Survey Bull.* 1015-G, pp. 227–245.

25. A. V. Heyl and others, 1959, The geology of the Upper Mississippi Valley zinc-lead district, *U.S. Geol. Survey Prof. Paper* 309, 310 pp.

26. W. R Ingalls, 1946, The great lead and zinc mines, *Mining and Metallurgy,* vol. 27, pp. 469–473.

27. J. A. James, 1952, Structural environments of the lead deposits in the Southeastern Missouri mining district, *Econ. Geology,* vol. 47, pp. 650–660.

28. G. Jouravasky and others, 1950, Deux types de gisements de plomb au Maroc Français, Internat. Geol. Cong., 18th, Great Britain 1948, pt. 7, The geology, paragenesis, and reserves of the ores of lead and zinc, pp. 222–233.

29 D. L. Kendall, 1960, Ore deposits and sedimentary features, Jefferson City mine, Tennessee, *Econ. Geology,* vol. 55, pp. 985–1003.

30. P. F. Kerr, 1932, Zinc deposits near Franklin, New Jersey, Internat. Geol. Cong., 16th, United States 1933, *Guidebook* 8, pp. 2–14.

31. H. F. King and E. S. O'Driscoll, 1953, The Broken Hill lode, in *Fifth Empire Min. Metall. Cong.,* Australia and New Zealand, pp. 578–600, Australasian Institute of Mining and Metallurgy, Melbourne.

32. H. F. King and B. P. Thomson, 1953, The geology of the Broken Hill district, in *Fifth Empire Min. Metall. Cong.,* Australia and New Zealand, pp. 533–577, geologic map, Australasian Institute of Mining and Metallurgy, Melbourne.

33. L. G. Love and D. O. Zimmerman, 1961, Bedded pyrite and microorganisms from the Mount Isa shale, *Econ. Geology,* vol. 56, pp. 873–896; discussion by J. T. Greensmith, 1962, *Econ. Geology,* vol. 57, pp. 118–119; by D. O. Zimmerman, 1962, *Econ. Geology,* vol. 57, pp. 459–460; by L. G. Love, 1962, *Econ. Geology,* vol. 57, pp. 460–462.

34. E. T. McKnight and G. Luttrell, 1951, Lead resources of the world, in "Materials Survey, Lead," pp. 9–53, U.S. Bureau of Mines.

35. E. T. McKnight and G. Luttrell, 1952, Zinc resources of the world, in "Materials Survey, Zinc," pp. 7–49, U.S. Bureau of Mines.

36. E. T. McKnight and others, 1962, Lead in the United States, exclusive of Alaska and Hawaii, *U.S. Geol. Survey Mineral Inv. Resource Map* MR-15. Map and pamphlet of 22 pp.

37. E. T. McKnight and others, 1962, Zinc in the United States, exclusive of Alaska and Hawaii, *U.S. Geol. Survey Mineral Inv. Resource Map* MR-19. Map and pamphlet of 18 pp.

38. R. E. Neelands and D. B. Fraser, 1958, Zinc in Canada with comments on world conditions, *Canada Dept. Mines and Tech. Surveys, Mines Branch, Mem. Ser No.* 137, 87 pp.

39. W. H. Newhouse, 1931, The geology and ore deposits of Buchans, Newfoundland, *Econ. Geology,* vol. 26, pp. 399–414.

40. M. H. Newman, 1932, The Mascot–Jefferson City zinc district of Tennessee, Internat. Geol. Cong., 16th, United States 1933, *Guidebook* 2, pp. 152–161.

41. C. R. L. Oder and J. W. Hook, 1950, Zinc deposits of the southeastern states, in "Symposium on Mineral Resources of the Southeastern United States, " pp. 72–87, University of Tennessee Press, Knoxville, Tenn.

42. C. R. L. Oder and J. E. Ricketts, 1961, Geology of the Mascot–Jefferson City zinc district, Tennessee, *Rept. Inv. No.* 12, *Div. Geology,* Nashville, Tenn., 29 pp.

43. E. L. Ohle, 1959, Some considerations in determining the origin of ore deposits of the Mississippi Valley type, *Econ. Geology,* vol. 54, pp. 769–789.

44. E. L. Ohle and J. S. Brown, 1954, Geologic problems in the Southeast Missouri lead district, *Geol. Soc. America Bull.,* vol. 65, pp. 201–222.

45. E. W. Pehrson, 1929, Summarized date of zinc production, *U.S. Bur. Mines Econ. Paper* 2, 47 pp.

46. A. W. Pinger, 1950, Geology of the Franklin-Sterling area, Sussex County, New Jersey, Internat. Geol. Cong., 18th, Great Britain 1948, pt. 7, The geology, paragenesis, and reserves of the ores of lead and zinc, pp. 78–87.

47. J. D. Ridge, 1952, The geochemistry of the ores of Franklin, New Jersey, *Econ. Geology,* vol. 47, pp. 180–192.

48. D. B. Shimkin, 1953, "Minerals—a Key to Soviet Power," Harvard University Press, Cambridge, Mass., 452 pp.

49. L. A. Smith, 1929, Summarized data of lead production, *U.S. Bur. Mines Econ. Paper* 5, 44 pp.

50. F. G. Snyder and J. A. Emery, 1956, Geology in development and mining, Southeast Missouri lead belt, *Mining Eng.,* vol. 8, pp. 1216–1224.

51. F. L. Stilwell, 1953, Mineralogy of the Broken Hill lode, in *Fifth Empire Min. Metall. Cong.,* Australia and New Zealand, pp. 601–626, Australasian Institute of Mining and Metallurgy, Melbourne.

52. S. Weidman, 1932, The Tri-State zinc-lead region, Internat. Geol. Cong., 16th, United States 1933, *Guidebook* 2, pp. 74–91.

53. J. Zwierzycki, 1950, Lead and zinc ores in Poland, Internat. Geol. Cong., 18th, Great Britain 1948, pt. 7, The geology paragenesis, and reserves of the ores of lead and zinc, pp. 314–324.

COPPER

54. J. M. Alexander, Geology of the Mount Lyell field, in *Fifth Empire Min. Metall. Cong.,* Australia and New Zealand, pp. 1129–1144, Australasian Institute of Mining and Metallurgy, Melbourne.

55. A. M. Bateman, 1930, Ores of the Northern Rhodesian Copperbelt, *Econ. Geology,* vol. 25, pp. 365–418.

56. A. M. Bateman, 1935, The Northern Rhodesia copper belt, in "Copper Resources of the World," vol. 2, pp. 713–740, Internat. Geol. Cong., 16th, United States 1933.

57. A. M. Bateman, 1935, The copper deposits of Ely, Nevada, in "Copper Resources of the World," vol. 1, pp. 307–321, Internat. Geol. Cong., 16th, United States 1933.

58. G. Berg and F. Friedensburg, 1941, Kupfer, "Die Metallischen Rohstoffe," no. 4, Ferdinand Enke, Stuttgart, 195 pp. Contains a description of the major copper deposits of the world up to the time of publication.

59. Y. S. Bonillas and others, 1916, Geology of the Warren mining district, *Trans. Am. Inst. Mining Engineers,* vol. 55, pp. 284–355.

60. J. M. Boutwell, 1905, Bingham mining district, Utah, Part II, Economic geology, *U.S. Geol. Survey Prof. Paper* 38, pp. 71–413.

61. J. M. Boutwell, 1935, Copper deposits at Bingham, Utah, in "Copper Resources of the World," vol. 1, pp. 347–359, Internat. Geol. Cong., 16th, United States 1933.

62. B. B. Brock, 1961, The structural setting of the Copperbelt, in "The Geology of the Northern Rhodesian Copperbelt," pp. 81–89, Macdonald & Co. (Publishers), London, 523 pp.

63. T. M. Broderick and C. D. Hohl, 1935, The Michigan copper district, in "Copper Resources of the World," vol. 1, pp. 271–284, Internat. Geol. Cong., 16th, United States 1933.

64. T. M. Broderick, C. D. Hohl, and H. N. Eidemiller, 1946, Recent contributions to the geology of the Michigan copper district, *Econ. Geology,* vol. 41, pp. 675–725.

65. T. M. Broderick, 1952, The origin of the Michigan copper deposits, *Econ. Geology,* vol. 47, pp. 215–220.

66. T. M. Broderick, 1956, Copper deposits of the Lake Superior region, *Econ. Geology,* vol. 51, pp. 285–287.

67. B. S. Butler and W. S. Burbank, 1929, The copper deposits of Michigan, *U.S. Geol. Survey Prof. Paper* 144, 238 pp.

68. J. R. Cooper, 1960, Some geologic features of the Pima mining district, Pima County, Arizona, *U.S. Geol. Survey Bull.* 1112-C, pp. 63–103.

69. H. R. Cornwall, 1956, A summary on ideas of the origin of native copper deposits, *Econ. Geology,* vol. 51, pp. 615–631.

70. G. Demesmaekar, 1961, The Katanga system in Katanga, in "The Geology of the Northern Rhodesian Copperbelt," pp. 54–56, Macdonald & Co. (Publishers), London, 523 pp.

71. D. M. Dunbar, 1952, History of Chuquicamata copper, *Mining Eng.,* vol. 4, pp. 1164–1165.

72. S. F. Emmons, 1910, Cananea mining district of Sonora, Mexico, *Econ. Geology,* vol. 5, pp. 312–356.

73. W. G. Garlick, 1961, Muva system, in "The Geology of the Northern Rhodesian Copperbelt," pp. 21–30, Macdonald & Co. (Publishers), London, 523 pp.

74. W. G. Garlick, 1961, Structural evolution of the Copperbelt, in "The Geology of the Northern Rhodesian Copperbelt," pp. 89–105, Macdonald & Co. (Publishers), London, 523 pp.

75. C. C. Goddard, Jr., 1959, Anaconda's Berkeley pit, history and geology, *Mining Eng.,* vol. 11, pp. 290–292.

76. J. González Reyna, 1956, Los yacimientos de cobre de México, in "Riqueza Minera y Yacimientos Minerales de México," 3d ed., pp. 165–184, *Internat. Geol. Cong.,* 20th, Mexico 1956, 497 pp.

77. A. Gray, 1930, The correlation of the ore-bearing sediments of the Katanga and Rhodesian copper belt, *Econ. Geology,* vol. 25, pp. 783–804.

78. E. D. Hammond, 1961, History of mining in the Bingham district, Utah, in Geology of the Bingham mining district and Oquirah Mountains, *Guidebook to the geology of Utah, No.* 16, Salt Lake City, Utah Geological Society, pp. 120–129.

79. L. H. Hart, 1935, The Butte district, Montana, in "Copper Resources of

the World," vol. 1, pp. 287–305, Internat. Geol. Cong., 16th, United States 1933.

80. A. Heim, 1935, The cupriferous pyrite ores of Huelva, Spain—a tectonic sketch, in "Copper Resources of the World," vol. 2, p. 635–648, Internat Geol. Cong., 16th, United States 1933.

81. W. G. Hogue and E. D. Wilson, 1950, Bisbee or Warren district, in Arizona zinc and lead deposits, pt. 1, *Arizona Bur. Mines, Geol. Ser.* 18, *Bull.* 156, pp. 17–29.

82. F. H. Howell and J. S. Mollby, 1960, Geology of the Braden orebody, Chile, South America, *Econ. Geology,* vol. 55, pp. 863–905.

83. R. N. Hunt and H. G. Peacock, 1950, Lead and lead-zinc ores of the Bingham district, Utah, Internat. Geol. Cong., 18th, Great Britain 1948, pt. 7, The geology, paragenesis, and reserves of the ores of lead and zinc, pp. 92–96.

84. R. N. Hunt, 1933, Bingham mining district, Internat. Geol. Cong., 16th, United States 1933, *Guidebook* 17, pp. 45–56.

85. Instituto Geológico y Minero de España, 1935, Los yacimientos y la minería de cobre en España, in "Copper Resources of the World," vol. 2, pp. 621–634, Internat. Geol. Cong., 16th, United States 1933.

86. A. James, W. Smith, and E. Bray, 1961, Bingham district, a zoned porphyry ore deposit, in Geology of the Bingham mining district and Oquirah Mountains, *Guidebook to the geology of Utah, No.* 16, Salt Lake City, pp. 81–100, Utah Geological Society.

87. O. W. Jarrell, 1944, Oxidation at Chuquicamata, Chile, *Econ. Geology,* vol. 39, pp. 251–286.

88. C. E. Julihn, 1928, Summarized data of copper production, *U.S. Bur Mines Econ. Paper* 1, 32 pp.

89. A. R. Kinkel, Jr., and N. P. Peterson, 1962, Copper in the United States (exclusive of Alaska and Hawaii), *U.S. Geol. Survey Mineral Inv. Resource Map* MR-13. Map and pamphlet of 15 pp.

90. A. Knopf, 1933, Clifton-Morenci district, Arizona, in "Ore Deposits of the Western States," pp. 544–545; Bisbee, Arizona, pp. 545–546; American Institute of Mining and Metallurgical Engineers, New York, 797 pp.

91. S. G. Lasky, 1935, The Lordsburg district, New Mexico, in "Copper Resources of the World," vol. 1, pp. 337–341, Internat. Geol. Cong., 16th, United States 1933.

92. S. G. Lasky and A. D. Hoagland, 1950, Central mining district, New Mexico, Internat. Geol. Cong., 18th, Great Britain 1948, pt. 7, The geology, paragenesis, and reserves of the ores of lead and zinc, pp. 97–110.

93. W. Lindgren, 1905, The copper deposits of the Clifton-Morenci district, Arizona, *U.S. Geol. Survey Prof. Paper* 43, 375 pp.

94. W. Lindgren and E. S. Bastin, 1935, The Braden copper deposit, Rancagua, Chile, in "Copper Resources of the World," vol. 2, pp. 459–472, Internat. Geol. Cong., 16th, United States 1933.

95. A. Locke, 1933, Disseminated copper deposits, in "Ore Deposits of the Western States," pp. 616–623, American Institute of Mining and Metallurgical Engineers, New York, 797 pp.

96. V. M. Lopez, 1939, The primary mineralization at Chuquicamata, Chile, S.A., *Econ. Geology,* vol. 34, pp. 674–711.

97. W. S. March, 1935, Ore deposits at Potrerillos, Chile, in "Copper Resources of the World," vol. 2, pp. 485–500, Internat. Geol. Cong., 16th, United States 1933.

98. F. Mendelsohn, 1961, Lubufu system; granite, in "The Geology of the Northern Rhodesian Copperbelt," pp. 18–21, Macdonald & Co. (Publishers), London, 523 pp.

99. F. Mendelsohn, 1961, Katanga system, in "The Geology of the Northern Rhodesian Copperbelt," pp. 41–54, Macdonald & Co. (Publishers), London, 523 pp.

100. F. Mendelsohn, 1961, Metamorphism, in "The Geology of the Northern Rhodesian Copperbelt," pp. 106–116, Macdonald & Co. (Publishers), London, 523 pp.

101. F. Mendelsohn, 1961, Ore deposits, in "The Geology of the Northern Rhodesian Copperbelt," pp. 117–129, Macdonald & Co. (Publishers), London, 523 pp.

102. D. M. McKinnon and N. J. Smit, 1961, Nchanga, in "The Geology of the Northern Rhodesian Copperbelt," pp. 234–275, Macdonald & Co. (Publishers), London, 523 pp.

103. D. V. Nalivkin, 1960, "The Geology of the U.S.S.R., a Short Outline," Pergamon Press, New York, 170 pp., geologic map of the U.S.S.R.

104. B. Nekrasoff, 1935, Copper-ore regions of the Union of Soviet Socialist Republics, in "Copper Resources of the World," vol. 2, pp. 649–662, Internat. Geol. Cong., 16th, United States 1933.

105. R. Oosterbosch, 1951, Copper mineralization in the Fungurume region, Katanga, *Econ. Geology,* vol. 46, pp. 121–148.

106. S. Paige, 1932, The region around Santa Rita and Hanover, New Mexico, Internat. Geol. Cong., 16th, United States 1933, *Guidebook* 14, Exc. C-1, pp. 23–40.

107. S. Paige, 1935, Santa Rita and Tyrone, New Mexico, in "Copper Resources of the World," vol. 1, pp. 327–335, Internat. Geol. Cong., 16th, United States 1933.

108. A. B. Parsons, 1933, "The Porphyry Coppers," American Institute of Mining and Metallurgical Engineers, New York, 581 pp.

109. A. B. Parsons, 1957, "The Porphyry Coppers in 1956," American Institute of Mining, Metallurgical, and Petroleum Engineers, New York, 270 pp.

110. J. D. Pelletier, 1957, Geology of the San Manuel mine, *Mining Eng.,* vol. 9, pp. 760–762.

111. E. N. Pennebaker, 1942, The Robinson mining district, Nevada, in "Ore Deposits As Related to Structural Features," pp. 128–131, Princeton University Press, Princeton, N.J., 280 pp.

112. E. S. Perry, 1933, The Butte mining district, Montana, Internat. Geol. Cong. 16th, United States 1933, *Guidebook* 23, 25 pp.

113. V. D. Perry, 1935, Copper deposits of the Cananea district, Sonora, Mexico, in "Copper Resources of the World," vol. 1, pp. 413–418, Internat. Geol. Cong., 16th, United States 1933.

114. V. D. Perry, 1952, Geology of the Chuquicamata orebody, *Mining Eng.*, vol. 4, pp. 1166–1168.

115. P. J. Pienaar, 1961, Mineralization in the Basement Complex, in "The Geology of the Northern Rhodesian Copperbelt," pp. 30–41, Macdonald & Co. (Publishers), London, 523 pp.

116. F. L. Ransome, 1904, The geology and ore deposits of the Bisbee quadrangle, Arizona, *U.S. Geol. Survey Prof. Paper* 21, 168 pp.

117. F. L. Ransome, 1932, General geology and summary of ore deposits, *Internat. Geol. Cong.*, 16th, United States 1933, *Guidebook* 14, *Exc.* C-1, pp. 1–23.

118. K. Richard and J. H. Courtright, 1958, Geology of Toquepala, Peru, *Mining Eng.*, vol. 10, pp. 262–266; discussion, pp. 699–700.

119. M. Robert, 1931, An outline of the geology and ore deposits of Katanga, Belgian Congo, *Econ. Geology,* vol. 26, pp. 531–539.

120. M. Robert and R. du Trieu de Terdonck, 1935, Le bassin cuprifère du Katanga méridional, in "Copper Resources of the World," vol. 2, pp. 703–712, Internat. Geol. Cong., 16th, United States 1933.

121. R. H. Sales, 1913, Ore deposits at Butte, Montana, *Am. Inst. Mining Engineers Trans.*, vol. 46, pp. 3–109.

122. R. H. Sales and C. Meyer, 1948, Wall rock alteration at Butte, Montana, *Am. Inst. Mining Metall. Engineers Trans.*, vol. 178, pp. 9–35.

123. R. H. Sales and C. Meyer, 1949, Results from preliminary studies of vein formation at Butte, Montana, *Econ. Geology,* vol. 44, pp. 465–484.

124. M. Santillán, 1935, El cobre en México, in "Copper Resources of the World," vol. 1, pp. 379–406, Internat. Geol. Cong., 16th, United States 1933.

125. H. Schmitt, 1933, Summary of the geological and metallogenetic history of Arizona and New Mexico, in "Ore Deposits of the Western States," pp. 316–326, American Institute of Mining and Metallurgical Engineers, New York, 797 pp.

126. H. Schmitt, 1942, Central mining district, New Mexico, in "Ore Deposits As Related to Structural Features," pp. 73–77, Princeton University Press, Princeton, N.J., 280 pp.

127. H. Schmitt, 1959, The copper province of the southwest, *Mining Eng.*, vol. 11, pp. 597–600.

128. G. M. Schwartz, 1947, Hydrothermal alteration in the "porphyry copper" deposits, *Econ. Geology,* vol. 42, pp. 319–352.

129. G. M. Schwartz, 1949, Oxidation and enrichment in the San Manuel copper deposit, Arizona, *Econ. Geology,* vol. 44, pp. 253–277.

130. G. M. Schwartz, 1953, Geology of the San Manuel copper deposit, Arizona, *U.S. Geol. Survey Prof. Paper* 256, 65 pp.

131. H. R. E. Staines, Mount Morgan copper and gold mine, in *Fifth Empire Min. Metall. Cong.*, Australia and New Zealand, pp. 732–750, Australasian Institute of Mining and Metallurgy, Melbourne.

132. A. V. Taylor, Jr., 1935, Ore deposits at Chuquicamata, Chile, in "Copper Resources of the World," vol. 2, pp. 473–484, Internat. Geol. Cong., 16th, United States 1933.

133. J. B. Tenney, 1932, The Bisbee mining district, Internat. Geol. Cong., 16th, United States 1933, *Guidebook* 14. *Exc.* C-1, pp. 40–67.

134. J. B. Tenney, 1935, The Pilares mine, Los Pilares de Nacozari, Sonora, Mexico, in "Copper Resources of the World," vol. 1, pp. 419–424, Internat. Geol. Cong., 16th, United States 1933.

135. J. B. Tenney, 1935, The copper deposits of Arizona, in "Copper Resources of the World," vol. 1, pp. 167–235, Internat. Geol. Cong., 16th, United States 1933.

136. B. P. Thomson, 1953, Geology and ore occurrence in the Cobar district, in *Fifth Empire Min. Metall. Cong.,* Australia and New Zealand pp. 863–896, Australasian Institute of Mining and Metallurgy, Melbourne.

137. R. E. Thurmond and others, 1958, Pima: a three-part story, *Mining Eng.,* vol. 10, pp. 453–462.

138. C. Trischka, 1953, The 16 southwest porphyry coppers now in period of greatest activity, *Mining World,* vol. 15, pp. 43–47.

139. U.S. Bureau of Mines, 1952, Resources, Chap. 3, 37 pp., in "Materials Survey, Copper."

140. M. L. Wade and M. Solomon, 1958, Geology of the Mt. Lyell mines, Tasmania, *Econ. Geology,* vol. 53, pp. 367–416.

141. E. Weberg, 1957, How United States equipment is used at Musonoi Cu-Co open pit, *Mining World,* vol. 19, pp. 40–44.

142. W. H. Weed, 1912, Geology and ore deposits of the Butte district, Montana, *U.S. Geo. Survey Prof. Paper* 74, 262 pp.

143. W. H. Weed, V. D. Perry, and others, 1960, El Salvador, *Mining Eng.,* vol. 12, pp. 339–348.

144. W. S. White and J. C. Wright, 1954, The White Pine copper deposit, Ontonagon County, Michigan, *Econ. Geology,* vol. 49, pp. 675–716.

COBALT

145. A. L. Anderson, 1947, Cobalt mineralization in the Blackbird district, Lemhi County, Idaho, *Econ. Geology,* vol. 42, pp. 22–46.

146. G. Berg and K. Horalek, 1944, Nickel und Kobalt, in "Die Metallischen Rohstoffe," no. 6, Ferdinand Enke, Stuttgart, 280, pp. Description of the chief deposits of nickel and cobalt in the world to date of publication.

147. J. H. Bilbrey, Jr., 1962, Cobalt, a materials survey, *U.S. Bur. Mines Inf. Circ.* 8103, 140 pp.

148. R. S. Young (ed.), 1960, Cobalt, its chemistry, metallurgy, and uses, *American Chem. Soc. Mon. Ser. No.* 149, 424 pp. Chapter 2 is a summary of the occurrences of cobalt, including major and minor deposits, with 86 references.

MISCELLANEOUS REFERENCES CITED

149. P. L. Allsman, 1956, Oxidation and enrichment of the manganese deposits of Butte, Mont., *Mining Eng.,* vol. 8, pp. 1110–1112.

150. J. V. Beall, 1963, What's behind the mining boom in Southeast Missouri, *Mining Eng.,* vol. 15, no. 7, pp. 71–76.

151. C. H. Behre, Jr., 1962, Types of evidence for genesis of ore deposits in the East Tennessee and other lead zinc deposits, *Econ. Geology*, vol. 57, pp. 115–118.

152. Canadian mineral industry, 1961, *Canada, Mineral Resource Div., Dept. Mines and Tech. Surveys*.

153. D. S. Carruthers and R. D. Pratten, 1961, The stratigraphic succession and structure in the Zinc Corporation Ltd. and New Broken Hill Consolidated Ltd., Broken Hill, New South Wales, *Econ. Geology*, vol. 56, pp. 1088–1102.

154. E. B. Douglas, 1956, Mining and milling cobalt ores, *Mining Eng.*, vol. 8, pp. 280–281.

155. K. C. Dunham, 1950, Summarized geology of the principal lead and zinc mines of the world, with statistics of production and reserves, Internat. Geol. Cong., 18th, Great Britain 1948, pt. 7, The geology, paragenesis, and reserves of the ores of lead and zinc, pp. 30–31.

156. G. M. Fowler, 1960, Structural deformation and ore deposits, *Eng. Mining Jour.*, vol. 161, no. 6, pp. 183–188.

157. H. A. Green, 1961, Mufulira area, Luansobe, in "The Geology of the Northern Rhodesian Copperbelt," p. 407, Macdonald & Co. (Publishers), London, 523 pp.

158. G. P. Jenney, 1957, New Brunswick develops a major mining camp, *Eng. Mining Jour.*, vol. 158, no. 6, pp. 95–96.

159. I. B. Joralemon, 1959, The White Pine copper deposit, *Econ. Geology*, vol. 54, p. 1127.

160. G. P. Lutjen, 1953, Cobalt at Fredericktown, *Eng. Mining Jour.*, vol. 154, no. 12, pp. 72–76.

161. R. F. Moe, 1954, White Pine mine development, *Mining Eng.*, vol. 6, pp. 381–386.

162. V. D. Perry, 1961, The significance of mineralized breccia pipes, *Mining Eng.*, vol. 13, pp. 367–376.

163. R. H. Ramsey, (ed.), 1953, White Pine copper, *Eng. Mining Jour.*, vol. 154, no. 1, pp. 72–76.

164. J. González Reyna, 1956, El cadmio en México, in "Riqueza Minera y Yacimientos Minerales de México," 3d ed., pp. 333–335, Internat. Geol. Cong., 20th, Mexico, 1956, 497 pp.

165. T. A. Rickard, 1932, "A History of American Mining," pp. 232–233, McGraw-Hill Book Company, New York. 419 pp.

166. R. H. Sales, 1959, The White Pine copper deposit, *Econ. Geology*, vol. 54, pp. 947–951.

167. E. B. Sengier, 1951, Katanga's mineral empire, *Eng. Mining Jour.*, vol. 152, no. 11, pp. 86–89.

168. E. B. Sengier, 1957, Union Miniere's golden jubilee, *Mining World*, vol. 19, no. 2, p. 38.

169. R. E. Stoiber and E. S. Davidson, 1959, Amygdule zoning in the Portage Lake lava series, Michigan copper district, *Econ. Geology*, vol. 54, pp. 1250–1277, 1444–1460.

170. W. S. White, 1960, The White Pine copper deposit, *Econ. Geology,* vol. 56, pp. 402–409.
171. W. S. White and J. C. Wright, 1954, The White Pine copper deposit, Ontonagon County, Michigan, *Science,* vol. 119, p. 354.
172. G. S. Wyman and L. E. Fish, 1952, Open-pit mining operations, *Mining Eng.,* vol. 4, p. 1169.
173. Anon., 1953, From Africa comes cobalt, *Mining World,* vol. 15, no. 11, pp. 48–52.
174. Anon., 1954, White Pine uses new methods and equipment for mine development, *Mining World,* vol. 16, no. 4, pp. 34–38.
175. Anon., 1956, Chibuluma starts moving concentrates, *Eng. Mining Jour.,* vol. 137, no. 6, pp. 102–104.
176. Anon., 1956, Anaconda maps greatest expansion program in history of Butte, *Mining World,* vol. 18, no. 11, pp. 56–61.
177. Anon., 1956, Anaconda pushes underground project, *Mining World,* vol. 18, No. 12, pp. 47–50.
178. Anon., 1960, Copper Range to develop Michigan copper deposit, *Mining World,* vol. 22, no. 9, p. 57.
179. Anon., 1960, Southern Peru Copper Corporation starts production at Toquepala, *Eng. Mining Jour.,* vol. 161, no. 4, unpaged supplement preceding p. 93.
180. Anon., 1961, Mining goes deeper with Anaconda, *Mining World,* vol. 23, no. 2, pp. 23–25.
181. Anon., 1962, Chuqui—50 years old but still a youngster, *Eng. Mining. Jour.,* vol. 163, p. 114.
182. Anon., 1963, Calumet and Hecla evaluates Kingston conglomerate discovery, *Mining World,* vol. 25, no. 6, p. 19.
183. Anon., 1963, Huge New Brunswick metal find sets off exploration boom, *Eng. Mining Jour.,* vol. 154, pp. 101–104, 202, 206, 208.

chapter *12*

Antimony
and Mercury

The principal ore mineral of antimony, stibnite, and the principal ore mineral of mercury, cinnabar, both generally form under low-temperature hydrothermal conditions. Hence they not uncommonly occur together and are associated with the same minerals, especially pyrite, marcasite, chalcedony, quartz, calcite, dolomite, and ankerite. Either one may be formed by hot-spring activity, but such formation appears to be more characteristic of cinnabar than of stibnite. Although stibnite and cinnabar may occur together, the important commercial deposits consist chiefly of one or the other of these minerals, and thus the leading antimony producers of the world and the leading mercury producers commonly differ, China and the Republic of South Africa being noted for production of antimony, Italy and Spain for the production of mercury. Another factor that causes some difference in commercial production is the occurrence of antimony in a considerable number of minerals that are mined for copper, lead, and silver, from which antimony is recovered as a by-product, whereas mercury generally does not so occur and is not recovered as a by-product. Further, stibnite is associated with gold in some deposits.

Antimony

Leaders in world production of antimony for the 15-year period 1945 to 1959, and also for 1962, were China, the Republic of South Africa, and Bolivia (Fig. 12-1). During 1945 to 1959 the production from these three countries was nearly the same and amounted to about 54 percent of the world total, but in 1960 China forged well ahead of the other two, with 31.1 percent of the world total, and Bolivia fell well behind, with only 9.6 percent of the world total, although the three countries combined then furnished nearly 63 percent of the antimony mined.

China

China (5, pp. 444–447; 12, pp. 47–50) was the outstanding world source of antimony for many years and furnished nearly two-thirds of the total world production from 1913 to 1937. During part of this time (1930 to 1932, 1934)

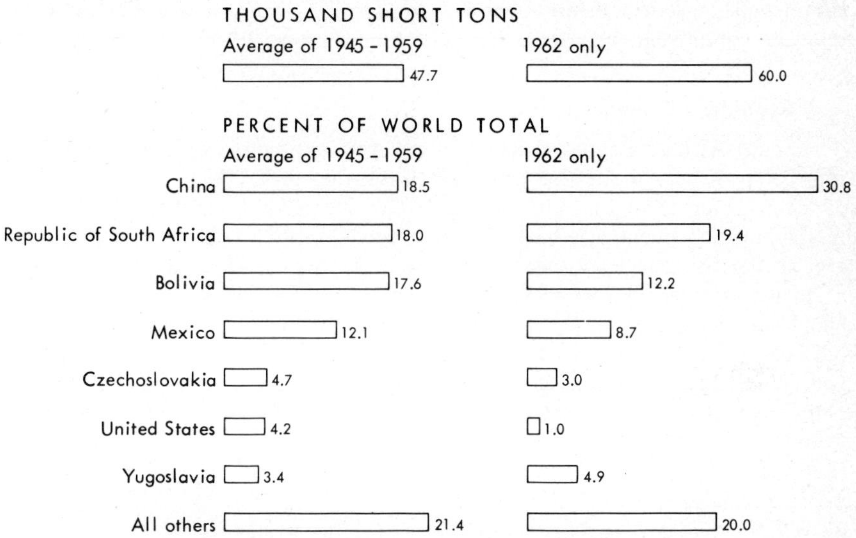

THOUSAND SHORT TONS

Average of 1945 - 1959	1962 only
47.7	60.0

PERCENT OF WORLD TOTAL

	Average of 1945 - 1959	1962 only
China	18.5	30.8
Republic of South Africa	18.0	19.4
Bolivia	17.6	12.2
Mexico	12.1	8.7
Czechoslovakia	4.7	3.0
United States	4.2	1.0
Yugoslavia	3.4	4.9
All others	21.4	20.0

Fig. 12-1   World antimony production. Compiled from *U.S. Bur. Mines Minerals Yearbook,* various years.

the amount was 70 percent or more of the world total, and it reached 76 percent in 1932. Mining was interrupted after 1937 because of Japanese occupation and political change, but began to increase again about 1950. China furnished only about 5 percent of the world total in the period 1945 to 1949, increased its production to about 19 percent in 1950 to 1954, then to nearly 28 percent in 1955 to 1959, and to about 31 percent in 1960. Hunan Province, in which antimony was mined on a small scale as early as the sixteenth or seventeenth century, furnished 99 percent of the total output of China before 1934, but since then developments in Kwangsi and Kweichow Provinces (Fig. 9-5) have been important, and the production from Hunan Province fell to about 80 percent of the Chinese total. The major reserves, however, which are the highest in the world, are in Hunan Province, where there are about 24 mines or deposits, the most important of which are Hsi-k'uang-shan mines about 20 miles northeast of Hsin-hua near the central part of the province.

The deposits of China are fissure fillings and replacements. The fissure fillings, also known as stibnite-cinnabar lodes, are the most important type. They form veins and stockworks in fractures and shear zones in siliceous rock, chiefly quartzitic sandstone and slate of Paleozoic age but in part of Precambrian age, and commonly occur near quartz diorite of late Jurassic or early Tertiary age. The chief hypogene ore mineral is stibnite; other metallic minerals, which are not abundant, are cinnabar and pyrite. Quartz is the chief or only gangue. Individual bodies of nearly pure stibnite in the Hunan deposits have been as large as 50 by 11 by 2 ft. Some veins in Hunan Province are 0.5 to 25 ft wide and as much as 2,000 ft long. Veins in Kweichow Province average about 1 ft wide. The average grade of the Hunan deposits is 6 to 9 percent stibnite. The replacement deposits, which are in limestone, are also known as stibnite-galena lodes. These deposits are less abundant and generally smaller than the stibnite-cinnabar lodes but are of high grade, 20 to 57 stibnite.

Republic of South Africa

South Africa has been a major producer of antimony since 1940. Production was about 9 percent of the world total for the period 1945 to 1949, rose to 19 percent for 1950 to 1954, to more than 23 percent for 1955 to 1959, and was about 22 percent in 1960 and 19 percent in 1962. The deposits (7, 12, 17) are in the Murchison Range of northern Transvaal (Fig. 16-1), which was originally prospected for gold as early as 1870, was the scene of a gold rush about 1886, and furnished gold from a number of mines until 1911, when only one mine was producing. In 1928 recovery of antimony as a by-product of gold ores became feasible because of greater demand and increased price for antimony, and in 1934 the Consolidated Murchison Goldfields and Development Company was formed to recover both gold and antimony. The major mine of the area is the Gravelotte, near the town of that name, but a number of other mines are also worked. Production in 1959 from the Gravelotte mine was 22,185 short tons of cobbed antimony and concentrates and 2,383 oz of gold.

Highly metamorphosed rocks of the Precambrian Swaziland system have been folded into steeply dipping parallel anticlines and synclines that trend east-northeast. Antimony mineralization extends along the southern limb of the Murchison syncline in a shear zone between quartzites and talcose chlorite schists known as the antimony line. Antimony minerals are sparsely distributed along this shear zone, but local concentrations occur in fissures associated with vertical drag folds. Massive and sheared dolomites or carbonate-chlorite schists associated with the folds have been favorable host rocks for replacement by stibnite, but the deposits are fissure fillings as well as replacements. Berthierite, $FeSb_2S_4$, is present with the stibnite in parts of the deposits.

Mineralization is thought to have been associated with the emplacement of granite of Precambrian age and to have proceeded in three stages: (1) formation of pyrite-arsenopyrite-gold-quartz reefs; (2) formation, at lower temperature, of stibnite-gold-quartz reefs; (3) formation of cinnabar-quartz reefs.

Bolivia

Bolivia is and has been a very important producer of antimony. Although it ranked well below China and the Republic of South Africa in 1962, the average production for the 15-year period 1945 to 1959 was almost equal to that of either China or the Republic of South Africa (Fig. 12-1). During the period 1945 to 1949 Bolivia furnished slightly more than 25 percent of the antimony of the world—nearly twice as much as China and South Africa combined.

The antimony deposits of Bolivia (1; 6, p. 87; 12, pp. 37–41) are widely scattered throughout the same general zone as the tin-silver deposits (Fig. 9-3) but are distinct from them, some occurring from 5 to as much as 15 miles away. In general they occur in rather narrow fissures in black Paleozoic shales in the eastern Andes Mountains and consist chiefly of stibnite in a quartz gangue. Some veins contain also galena, others contain gold. The most productive mines are in the Challapata district, department of Oruro, and the Tupiza district, department of Potosí. The major production comes from mines in the area around Tupiza near the south end of the antimony zone; second most important production comes from mines near Challapata. Perhaps as many as 150 mines exist within the antimony zone, but many of them are small; the number of major mines seems to be about a dozen.

The country rock is generally very fine-grained black shale or slate of Ordovician age. The deposits are associated with and localized in shear zones near the apices of anticlines; some shear zones are as much as a kilometer long. Individual veins generally are neither long nor wide, although some attain a width of 3 ft or more, and some mineralized zones are several hundred feet wide. Stibnite, the primary antimony mineral, may be accompanied by pyrite, ferberite, gold, quartz, and dickite, although in some places ferberite occurs in separate veins.

The antimony deposits were considered by Turneaure and Welker (51) to have been formed during part of the main period of mineralization of the area, perhaps belonging to the low-temperature end of the hydrothermal sequence, during which the tin-silver deposits were emplaced, although they stated that some doubt existed regarding the genetic relationship of the antimony deposits to the granodiorite and whether the deposits actually are a part of the metallogenetic province. They favored (p. 622) a leptothermal classification for the antimony deposits. Ahlfeld (1) stated that the antimony deposits show no evident connection with igneous rocks but that a connection with deep-lying, extensive granitic or granodioritic plutons was assumed. He classed the deposits as hypabyssal-epithermal and considered them to have been formed during the last phase of the activity that formed the deposits of the Bolivian tin province.

Mexico

Mexico furnished close to 20 percent of the antimony of the world in the five-year period 1945 to 1949, but only about 12 percent for 1950 to 1954, 8 percent for 1955 to 1959, and 8.7 percent for 1962. The actual tonnage supplied in 1962 was somewhat more than the average of that furnished in 1955 to 1959, but the total world production increased considerably in 1962 (Fig. 12-1).

Nearly all of the production of Mexico comes from a zone about 1,400 miles long and 10 to 100 miles wide that extends more or less through the central part of the country (10, 12). The concentration of deposits is especially great in the southern half of this zone, and the outstanding producing areas are, from north to south, the Antimonia district of the state of Sonora, the San José mines of the state of San Luis Potosí, the Pacheco district of the state of Zacatecas, the Soyatal district of the state of Querétaro, and the Tejocotes region of the state of Oaxaca. Among these the San José group of mines, San Luis Potosí, has been in operation since 1898 and is rated as the most productive in North America and perhaps in the Western Hemisphere. This group generally furnishes about one-third of the antimony of Mexico. The Tejocotes district, Oaxaca, however, has been the largest producer in Mexico since 1938. The Soyatel district, Querétaro, generally ranks third in production.

More than 70 percent of the production in 1942 (3) was from three states: San Luis Potosí, 31.3 percent; Oaxaca, 20.4 percent; and Querétaro, 19.6 percent. No other state furnished as much as 4 percent, but Zacatecas supplied 3.7 percent and Durango 3.2 percent. The reserves of Mexico are large but chiefly of low grade, and much of this material cannot be recovered economically at present (1963).

Most of the productive antimony deposits of Mexico are in pre-Tertiary rocks along the northeastern border and the ends of a large area of Tertiary lavas and are probably related in origin to Tertiary volcanic activity. They are generally associated with faults, fractures, and anticlinal axes, which were the chief structural controls of deposits. Generally the deposits occur in limestone and are most important adjacent to feeding channels where the limestone is overlain by shale. The original mineral of most deposits was stibnite, much of which has been oxidized to antimony oxides. Some mercury is present in deposits in limestone. In the San José group of mines, 7 miles east of Wadley, San Luis Potosí (14), the deposits are in four limestone beds overlain by shale and occur as disseminations, veinlets, and irregular masses. In the Tejocotes district (15), western Oaxaca, 13 miles west of Tlaxiaco, the deposits form high-grade masses in limestone near the axes of anticlines where the limestone is overlain by shale. In the Soyatal district (11), about 120 miles north of Mexico City in the north-central part of Querétaro, near Toliman, there is extensive mineralization in the upper part of a limestone overlain by limy shale. The major deposits are concentrated in the limestone adjacent to faults and fractures and also near the central parts of anticlines. In the Antimonia district (16) in Sonora, about 100 miles west of Santa Ana, most of the ore is in quartz and chalcedony veins that occur chiefly in siltstone but to a small extent in sandstone and Tertiary igneous rocks. The veins contain also a little silver and gold. The deposits of the Pacheco district, Zacatecas (12, pp. 32–33), are rather widespread but small. Ore bodies occur in limestone and are related to faults, fractures, and anticlinal axes.

United States

The United States never has been an outstanding producer of antimony. Although the average production in the period 1945 to 1949 was nearly 9 percent of the world total, it decreased in 1950 to 1954 to slightly less than 4 percent, dropped again in 1955 to 1959 to about 1.2 percent, and in 1962 to just a trifle more than 1 percent. Most of the antimony deposits (13) are in the western United States in a zone between 111 and 122°W longitude, which includes parts of Arizona, California, Nevada, Utah, Oregon, and Washington. Numerous deposits occur within this zone, but only a few of them are classed as of first importance (production plus resources greater than 10,000 short tons of antimony), and only about 10 as of second importance (between 1,000 and 10,000 short tons of antimony). Among those of first importance are the Stayton district, California, the Park City district, Utah, the Yellow Pine district, Idaho, and the Coeur d'Alene district, Idaho. The most important deposits of the United States are summarized in Table 12-1.

TABLE 12-1. Most Important Sources of Antimony in the United States*

State	Deposit First importance†	Deposit Second importance†	Summary of chief features
California	Stayton district, Quien Sabe (French mine), San Benito County		Chalcedony veins, pockets of stibnite in silicified breccia zones in Tertiary volcanic rocks; some mercury
		San Emigdio (Antimony Peak), Kern County	Antimony oxides and stibnite in quartz in shear zone in granodiorite and granitized metamorphic rock
		Wildrose Canyon, Inyo County	Antimony oxides and stibnite in quartz veins and lenses in schist
Idaho	Coeur d'Alene district, Shoshone County, Sunshine area		Ores contain tetrahedrite, galena, sphalerite, quartz, and carbonate in siliceous argillite of Precambrian Belt series. A Cu-Pb-Zn-Ag-As-Sb deposit
		Bunker Hill, Shoshone County	Galena-sphalerite-boulangerite-carbonate veins in Precambrian Belt series. A Pb-Zn-Ag-Sb deposit
	Yellow Pine, Valley County		Stibnite-scheelite-arsenopyrite-pyrite veinlets and disseminations in shear zone in granodiorite. A W-Au-Ag-As-Sb deposit
		Meadow Creek, Valley County	Same as above
Nevada		Bloody Canyon, Pershing County	Stibnite in quartz veins in rhyolite flow
		Southerland, Pershing County	Stibnite in quartz veins in calcareous shale near volcanic plug
Utah	Park City (Uintah), Summit County		Tetrahedrite, jamesonite, and bournonite are minor to abundant in veins and replacements in limestone, quartzite, and diorite. Recoverable metals are Pb, Zn, Cu, Ag, Au, As, Sb
		Bingham, Salt Lake County	Tetrahedrite is a minor mineral in copper and lead-silver ores. A Cu-Pb-Zn-Ag-Au deposit
		Tintic, Juab County	Complex sulfide ores, chiefly in Paleozoic carbonate rocks associated with Tertiary volcanic rocks. A Pb-Cu-Ag-Au-Zn-As-Sb deposit
		East Tintic, Juab County	Same as above
		Antimony Canyon (Coyote Creek), Garfield County	Pods and disseminations of stibnite and oxides in Upper Cretaceous sandstone overlying conglomerate; some fault control

* Compiled from *U.S. Geol. Survey Mineral Inv. Resource Map* MR-20, 1962.
† Deposits of the first importance are those in which the production plus resources is greater than 10,000 short tons of antimony; deposits of the second importance, between 1,000 and 10,000 short tons.

The deposits of Park City, Tintic, East Tintic, and Bingham, Utah, and of the Coeur d'Alene area, Idaho, have been described in connection with other metals.

The Yellow Pine district, Idaho, (2, 4, 9, 13) has furnished about 80 percent of the antimony produced in the United States since 1932. The district contains the outstanding reserves of the United States and is also one of the few areas that offers attractive possibilities for additional discoveries. Production of antimony, however, depends on a relatively high price for that metal, and if the price declines too much the major production is gold. The Yellow Pine mine, the most important mine of the district, is located on a branch fault that splits off from a major fault of the area. The deposits occur as veinlets and disseminations in a shear zone in granodiroite of the Idaho batholith. Stibnite and gold are closely associated, but gold occurs also with pyrite and arsenopyrite. On the west side of an open pit the chief ore mineral is stibnite, whereas on the east side of the same pit the main value is in gold, although some stibnite also occurs on that side.

Other antimony in the United States comes chiefly as a by-product from ores of silver, copper, lead, and zinc. In such deposits it generally is feasible to recover antimony even though the price of that metal is not especially high. Many of the deposits that must be worked chiefly or entirely for antimony cannot be mined profitably except when the price of antimony is high.

Other Countries

Although deposits of antimony are present in a number of other countries, perhaps the most important ones are those of Czechoslovakia, Yugoslavia, and the Soviet Union, although reliable information about production in those countries is lacking. Czechoslovakia furnished nearly 5 percent of the antimony of the world in the 15-year period 1945 to 1959 and 3 percent in 1962 (Fig. 12-1). The major part of the production comes from mines near Medzibrod, Poproc, and Spisska Bana, in Sovokia. Also, some recovery is a by-product of lead-silver ore from Pribram. Yugoslavia furnished somewhat less antimony than Czechoslovakia in 1945 to 1959 but somewhat more in 1962. Stibnite occurs in irregular pipes a few feet in diameter at the Stolica mines, Brzitsa, department of Drinska, and these mines furnished about three-fourths of the total production of Yugoslavia before 1940. Stibnite and antimony oxides occur in veins cutting limestone at the Lisanski mine near Drinska. Stibnite occurs in quartz veins at the Zajaca mines, about 9 miles northwest of Krupanj, and the output there in 1949 was stated to be one of the largest in Europe. Little information is available regarding antimony deposits of the Soviet Union. A summary of some material published by Shimkin (8) appears in Table 12-2.

Mercury

Italy and Spain have been the foremost producers of mercury in the world for many years. Together they furnished slightly more than 55 percent of the world total in the 15-year period 1945 to 1959 and about 43 percent

TABLE 12-2.   Principal Antimony Deposits of the Soviet Union*

Deposits	Location, latitude and longitude	Notes about deposit
Kadamzhai	39°58′N, 71°48′E	Largest in Soviet Union. Average content 2.1–2.8 percent finely disseminated stibnite. Ore occurs in siliceous breccia as fillings and some occasional more highly mineralized lenses. Ore body forms contact zone 5–7 m thick in Paleozoic limestone underlying shale.
Khaidarkan	39°58′N, 71°20′E	Mercury-antimony-fluorite deposit associated with breccia in the contact zone between Paleozoic limestones and shales, as at nearby Kadamzhai. Complex mineral composition. Average content: 19 percent CaF_2, 3 percent Sb_2S_3, 0.24 percent Hg. Stibnite and cinnabar increase with depth but fluorite decreases.
Razdol'noye	53°10′N, 94°10′E	Deposit consists of a quartz-antimony vein in highly metamorphosed Precambrian shales. Apparently contains high concentrations of stibnite; deposit may be very rich and fairly large.
Turgai	49°38′N, 63°30′E	May be second most important antimony deposit in Soviet Union. Other information not available.

* Reprinted by permission of the publishers from Demitri Shimkin, "Minerals—a Key to Soviet Power." Cambridge, Mass.: Harvard University Press, Copyright, 1953, by the President and Fellows of Harvard College.

in 1962. Other important producers have been the United States, probably the Soviet Union, Mexico, Yugoslavia, and China (Fig. 12-2). The greatest reserves are stated to be in Italy, Spain, Yugoslavia, the Soviet Union, and China (20).

The mercury deposits of the world are, in general, confined to a broad zone of late Tertiary orogeny and volcanism (21, p. 1). Although associated with volcanism, in many cases no close relationship between a volcanic source and a specific deposit can be demonstrated. Further, although occurring in the same general region as many antimony deposits, the mercury deposits not uncommonly are somewhat separated from them, apparently forming a lower-temperature marginal phase in the mineralized zone (13, p. 1; 33, p. 449).

Italy

The mercury deposits of Italy (20, pp. 22–23; 26; 41, pp. 467–477) at one time included the two important districts of Monte Amiata (Firenze or Florence) and Idria. The Idria mine near Trieste has been a possession of Yugoslavia since about 1947 and will be discussed with deposits of that country.

The Monte Amiata district, about 75 miles north of Rome, is in Toscana (Tuscany) and extends along the east slope of Monte Amiata and southward. The major deposits occur within a zone about 5 miles wide and 25 miles long, but other mines and prospects are known southwest of this zone so that the area

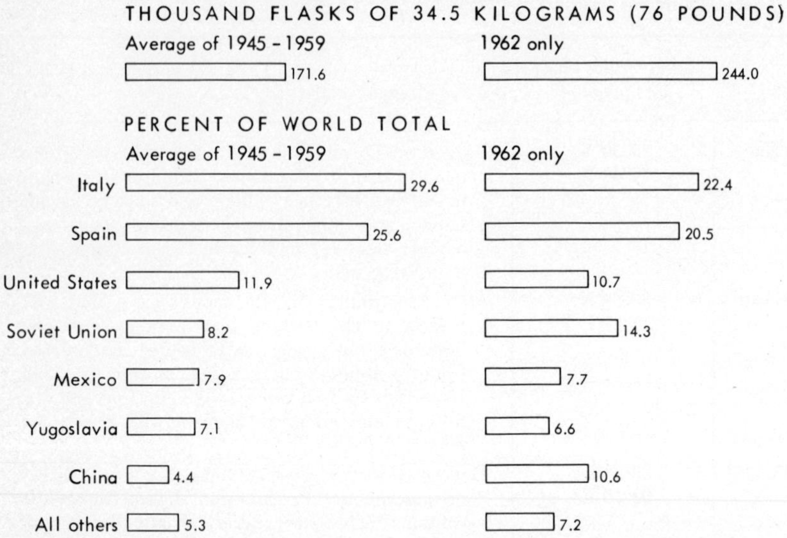

THOUSAND FLASKS OF 34.5 KILOGRAMS (76 POUNDS)

Average of 1945 – 1959   1962 only

[bar] 171.6   [bar] 244.0

PERCENT OF WORLD TOTAL

Average of 1945 – 1959   1962 only

Italy [bar] 29.6   [bar] 22.4

Spain [bar] 25.6   [bar] 20.5

United States [bar] 11.9   [bar] 10.7

Soviet Union [bar] 8.2   [bar] 14.3

Mexico [bar] 7.9   [bar] 7.7

Yugoslavia [bar] 7.1   [bar] 6.6

China [bar] 4.4   [bar] 10.6

All others [bar] 5.3   [bar] 7.2

Fig. 12-2 World mercury production. Compiled from *U.S. Bur. Mines Minerals Yearbook,* various years.

of the entire district is more than 500 sq miles. The ore contains from 0.6 to 3.0 percent mercury and the reserves are large. Some surface outcrops in the district were worked in the early centuries B.C., but modern history of mining dates from 1868. Production of late years has come from two major mines and several smaller ones, although Eckel (26, p. 287) in 1948 listed 42 mines and prospects, some of which were inactive, and discussed four principal active mines.

Most of the district is underlain by shale, sandstone, marl, and limestone that range in age from Triassic to Quaternary. The greater part of Monte Amiata itself is composed of trachyte, the only igneous rock of much extent in the district, apparently of late Tertiary age or younger. The structure of the district is incompletely known, but the rocks seem to have been gently folded and cut by two sets of major faults, one of which, at least 25 miles long, apparently afforded the principal control for ore deposition. Hot springs and gaseous emanations occur along this fault zone, some of them near the mines, but have not been regarded by mine operators as favorable prospecting areas.

Cinnabar is the principal ore mineral of all deposits of the Monte Amiata district. It is accompanied by pyrite and marcasite and by argillic alteration that seems to have resulted from hydrothermal activity. Stibnite is fairly common in the southern part of the mineralized zone. Most of·the large ore bodies occur in rocks of Eocene age. They occur in part in clay that fills spaces around fragments of unaltered limestone, shale, and sandstone, and in part as veinlets of cinnabar and as partial replacements of calcareous sandstone. Some of the richest ore is in clay around the borders of limestone blocks and in replacement bodies in calcareous sandstone. In at least one major mine the upward extent of the ore body is limited by overlying trachyte.

Spain

One of the oldest mercury producers of the world is Almadén, province of Ciudad Real, south-central Spain, about 140 miles southwest of Madrid. Perhaps the mines were exploited by the Celts and then by the Phoenicians, but it appears to be definite that excavations were undertaken during or at the end of the Punic Wars, about 150 B.C., and cinnabar was transported to Rome, reduced to vermilion, and used to decorate Roman villas and to make rouges and other beauty products.[1] Later the deposits were worked at different times by the Arabs (Almadén is Arabic for *the mine*) and by various other groups. Trustworthy data are available from 1499, and from then until 1949 production was about 7 million flasks of mercury. The reserves are not published, but apparently they are concentrated within a mass of rock which, at the surface, lies within a circle with a radius of 25 km with its center at the principal shaft. This is the extent of the deposits reserved by the Spanish government, by which they are controlled and exploited. One estimate is that the reserves are sufficient to support continuous production for 100 years at a rate of 82,000 flasks annually.

Accounts vary regarding the grade of ore that has been mined or that is now being mined. One estimate of recovery was about 6 percent mercury, others were about 8 percent. During the 16 years from 1901 through 1916 the recovery was about 8.6 percent, and losses in the furnaces indicate that the grade of ore was about 9.3 percent. Some estimates indicate that the grade of ore mined during the earlier years of work at Almadén was as high as 15 to 25 percent.

Deposits at Almadén (20, p. 23; 22; 25; 35; 36; 39; 41, pp. 477–482) apparently are associated with a major fault line that is known to be mineralized over a distance of 15 miles. A fracture zone coincides with the valley of the Rio Valdeazogue,[2] and along this zone there are numerous occurrences of cinnabar, especially at places 5, 8, and 18 km east of Almadén. The chief mine is located in nearly vertical interbedded quartzite and black slate of Silurian age, but the deposits are chiefly in the quartzite. The black slate, which is bituminous in part, contains disseminated pyrite, and both the slate and the quartzite are less resistant than surrounding ridges of unmineralized quartzite, so that the deposits lie in a valley. The deposits make three principal lodes or veins, which consist of much fractured and brecciated quartzite into which cinnabar was introduced through innumerable small fissures. The cinnabar then extensively replaced quartz and some associated sericite. The lodes, which trend nearly east-west and are separated by slate, are, from north to south, (1) the San Nicolas, average width about 10 ft, (2) the San Francisco, average width about 8 ft, and (3) the San Diego (eastern half) and San Pedro (western half), 17 to 36 ft wide with ore making about 60 percent of this width. Pos-

---

[1] See Bennett, (22) for a brief summary, in English, of the interesting early history of Almaden, and Sampelayo (39, pp. 69–79) for a more detailed account in Spanish.
[2] Rio Valdeazogue means essentially river of the valley of quicksilver.

sibly some of these lodes may unite in depth, at some place below the twelfth level, but this has not been determined.

United States

Mercury production in the United States to 1960 was a little more than 3,190,000 flasks, nearly one-sixth of the world's supply (21, p. 1). Most of this came from California, as shown by Table 12-3.

TABLE 12-3. Distribution of Mercury Production in the United States through 1960*

State	Percent of total production of the United States
California	86.3
Texas	4.6
Nevada	4.0
Oregon	3.2
Idaho	1.0
Arkansas	0.4
Arizona	0.2
Washington	0.2
Utah	0.1
Total	100.0

* From *U.S. Geol. Survey Mineral Inv. Resource Map* MR-30, 1962.
Alaska is not included, but production there has been chiefly since about 1954.

The mercury deposits of the United States generally occur 1,000 ft or less from the surface, but exceptionally, as at the New Almaden mine in California, they extend to a depth of 2,500 ft. They are present in a zone of late Tertiary volcanism and orogeny in the western part of the country, but occur in rocks of various ages. The principal ore mineral is cinnabar and the chief gangue is silica or carbonate minerals, although a considerable amount of diversity exists. Stibnite is present in some deposits. Pyrite and, more rarely, marcasite are present in some deposits but rare or absent in others.

California. The principal mercury deposits of California (21, pp. 2–3) extend through the Coast Ranges from Santa Barbara northward to Clear Lake, a distance of about 350 miles. In this zone there are 18 districts spaced at intervals of about 25 miles. Throughout the Coast Ranges highly deformed but un-metamorphosed graywacke, siltstone, and greenstone, along with minor limestone and chert of the Franciscan formation of Jurassic and Cretaceous age, have been intruded by masses of serpentine. These rocks are overlain by thick sequences of sedimentary rocks that range in age from Cretaceous to Pliocene, but in part of the area there are extensive flows of younger lavas.

Many of the ore bodies are in the margins of the intrusive serpentine where it was hydrothermally altered to a rock composed of silica minerals

TABLE 12-4. Some Features of the Major Mercury Districts of California*

District (and major references)	Location	Production through 1960	Notes about deposits
New Almaden (19; 41, pp. 411–417)	About 50 miles southeast of San Francisco; lat. 37°14′N, long. 121°51′W	More than 1,100,000 flasks, about 40 percent of total of United States	Discovered in 1845; ore bodies in silica-carbonate rocks, which cinnabar replaces
New Idria (28; 41, pp. 417–420)	About 140 miles southeast of San Francisco; lat. 36°21′N, long. 120°38′W	About 500,000 flasks from New Idria mine	Most ore occurs beneath a thrust fault near the margin of a plug of serpentine that intrudes shales
Mayacmas (41, pp. 420–423; 48)	About 60 miles north of San Francisco; lat. 38°45′N, long. 122°42′W	Oat Hill mine, 162,000 flasks; Great Western mine, 105,000 flasks; Aetna mine, 66,000 flasks; Mirabel mine, 42,000 flasks	Mineralized area about 25 miles long, 7 miles wide; most ore bodies in silica-carbonate rock, but at Oat Hill mine along faults in graywacke
Knoxville (18; 41, pp. 423–425)	About 70 miles north of San Francisco; lat. 38°51′N, long. 122°22′W	Knoxville mine, 121,000 flasks; Reed mine, 26,000 flasks; several less productive mines	All large ore bodies in silica-carbonate rock of sheared serpentine adjacent to a fault
Clear Lake (29; 41, pp. 425–426; 45; 46)	About 75 miles north of San Francisco; lat. 39°00′N, long. 122°41′W	Sulphur Bank mine, about 130,000 flasks; a few other mines with very small production	Ore deposits are Recent; thermal waters currently depositing some mercury and antimony
San Luis Obispo (27; 41, pp. 430–432)	About 175 miles southeast of San Francisco; lat. 35°41′N, long. 121°03′W	Oceanic mine, 41,000 flasks; Klau mine, 26,000 flasks; also a number of relatively small mines	Much of area underlain by Franciscan sedimentary rocks intruded by serpentine
Wilbur Springs	About 85 miles north of San Francisco; lat. 39°04′N, long. 122°26′W	Abbot mine, 43,000 flasks; other smaller mines, not producing in 1960	Small but some rich ore bodies in opaline silica-carbonate rock; cinnabar replaces silica-carbonate rock along fractures and fills cracks
Altoona (48)	About 230 miles north of San Francisco; lat. 41°08′N, long. 122°33′W	Altoona mine, 35,000 flasks; a few other mines with very small production	Mine workings are in an area of porphyritic diorite and minor serpentine; ore occurs in and near fault gouge

* Compiled from *U.S. Geol. Survey Mineral Inv. Resource Map MR-30*, 1962.

and magnesian carbonates, known as silica-carbonate rock. About 60 percent of the mercury produced in California has come from ore bodies in such silica-carbonate rock. Among the 18 districts in the mercury zone, 8 have been especially productive. Some features of those districts are shown in Table 12-4. Among these only the three that have been most productive will be described briefly—New Almaden, New Idria, Mayacmas—along with the one that shows clear relationships with hot-spring activity—Sulphur Bank. An interesting account of mercury in California before 1860, and also about mercury in general from perhaps as early as 300 B.C., is given by Egenhoff (49).

The New Almaden district (19; 41, pp. 411–417) has the distinction not only of having furnished the greatest amount of mercury of any district in the United States, but also of being the first place where mercury was discovered in North America. Legends indicate that Indians early used a cave in the area as a source of material for red paint for their bodies and early explorers reported that Indians from other areas obtained vermilion by trading with Indians of the New Almaden region. Recorded and well-authenticated discovery of the deposit was in 1824 by a Mexican, Luis Chaboya, who thought he had discovered a rich deposit of silver and abandoned it when he could obtain no silver from the ore. A second similar attempt was made by another in 1835. In 1845 a Mexican army officer, Don Andres Castillero, claimed that the deposit contained silver and a little gold but somewhat later that year demonstrated that the ore contained mercury by sprinkling some of it on hot coals and condensing some of the metal in an inverted tumbler. On December 30, 1845, he was awarded possession of the mine, which he had named Santa Clara, then formed a mining company and began production. By the use of crude methods he produced some 2,000 lbs of mercury, then obtained financial assistance for expansion from a group who optimistically renamed the mine New Almaden without seeing the property. No production is recorded until 1850, when it was 7,723 flasks, and for 1851 it was 27,779 flasks. This early production was stimulated by the discoveries of gold in California and the need for mercury in the amalgamation of gold.[1]

The most important ore of New Almaden was concentrated in silicate-carbonate rock formed by hydrothermal alteration of serpentine that intruded rocks of the Franciscan group (Jurassic?) and was localized especially near the contact of silica-carbonate rock and the Franciscan rocks, the richest bodies being within a few feet of the contacts. Cinnabar was introduced along fractures and replaced the silica-carbonate rock. Replacement generally extended only a few inches out from the fractures but was so complete that in many places more than half of the replaced rock was cinnabar. In many of the large ore bodies many steep fractures occurred very closely spaced and much of the intervening rock was rich ore. The largest ore body recorded was 200 by 15 ft and extended down dip for 1,500 ft. Ore supplied during the first 15 years of recorded production contained more than 20 percent mercury after cobbing and hand sorting. The average grade of all ore sent to the furnaces during the 100-year history of production was slightly less than 4 percent, although during the latter part of this time the grade of the ore

[1] For subsequent very interesting early history, see E. H. Bailey (19, pp. 264–268).

gradually declined to less than 1 percent, and by 1949 production had almost ceased.

The deposits of the New Idria district (28; 41, pp. 416–420), discovered in 1853, have been in continuous production since then except in the period 1921 to 1922. A large oval body of strongly sheared serpentine, as exposed at the surface, is rimmed by Franciscan (Jurassic?) and Upper Cretaceous rocks. The serpentine is more or less encircled by faults, and the cinnabar deposits are chiefly veins and stockworks that occupy fractures in altered rocks. Both rock alteration and ore deposition were localized chiefly at abrupt changes in strike of the New Idria thrust fault, beneath which the deposits of greatest commercial importance occur.

The Mayacamas district (41, pp. 420–423; 48) furnished 455,000 flasks of mercury from 1864 to 1948, about 92 percent of which came from the eastern half of the district. The period of greatest production was 1893 to 1895. The deposits occur along the flanks of an anticline, chiefly along or near contacts between intrusive serpentine and rocks of the Franciscan group. The deposits are irregular and veinlike masses that contain disseminated cinnabar, stockworks of cinnabar veinlets, and cinnabar-bearing veins. More than half of the deposits occur in silica-carbonate rocks, and the others occur in sandstone or chert of the Franciscan group or within and along the contacts of basalt dikes. All the ore bodies are associated with minor faults, and ore shoots were controlled by gouge zones, changes in dip and strike of faults and contacts, intersection of shears, and premineral brecciation.

Sulphur Bank (29; 41, pp. 425–426; 45; 46) has been stated to be the most productive mineral deposit in the world that is clearly related in origin to hot springs. Franciscan (Jurassic?) rocks overlain unconformably by Recent sedimentary rocks and andesite are cut by two major sets of faults. Along fault zones and at their intersections the rocks have been hydrothermally altered and there is a widespread distribution of small fumaroles and hot springs. The ore deposits were structurally controlled by fault zones and their intersections. In altered andesite they are high-grade stringers of cinnabar along jointing and sheeting planes. In Recent sediments fine-grained cinnabar occurs as fillings of open spaces in breccia and conglomerate, and the deposits commonly become wider and flatter beneath contacts with andesite. In the Franciscan rocks cinnabar coats blocks of sandstone in fault breccia and is finely disseminated in fault gouge. The deposits, discovered in 1856, were first worked in 1865 for sulfur which occurred nearly pure near the surface but were closed in 1868, partly because of the unfavorable selling price of sulfur, partly because the sulfur decreased in amount and became less pure somewhat beneath the surface as the mercury deposits were approached. The deposits were again opened in 1873 for mercury and have been worked at various intervals since then.

Nevada, Oregon, and Idaho.   Most of the mercury of Nevada (21, pp. 4–5; 41, pp. 438–443) has come from a zone in the central third of the state from small deposits of very rich ore and also from large bodies of low-grade ore. About half of the total production of about 120,000 flasks to the end of 1959 came from the Cordero mine. The Cordero mine (30, 31), discovered in 1929, is 11 miles by road southwest of McDermitt, Nevada, near the Nevada-Oregon boundary and is an extension of the Opalite district of southeastern Oregon. The ore bodies occur in rhyolite along two roughly parallel faults. The rhyolite has been brecciated and in part replaced or recemented by quartz and chalcedony near the surface but argillized beneath a shallow zone. Locally the silicified rock is designated *opalite*. Cinnabar and marcasite are the chief metallic minerals, the marcasite apparently earlier than the cinnabar. The ore body is a pipe about 50 ft wide and 200 ft long that extends downward to the deepest level worked in 1959, No. 8.

The chief sources of mercury in Oregon (21, pp. 3–4; 41, pp. 443–447) have been the Bonanza mine, 35,000 flasks, Black Butte mine, 17,000 flasks, Horse Heaven mine, 17,000 flasks, Opalite mine, 15,000 flasks, and the Bretz mine, 10,000 flasks. Some production comes from a number of small mines, also. The Bonanza and the Black Butte mines are in southwestern Oregon in Douglas and Lane Counties; the Horse Heaven mine is in central Oregon in Jefferson County; and the Opalite and Bretz mines are in southeastern Oregon in Malheur County near the Nevada border, and an extension of this district includes the Cordero mine of Nevada. The Bonanza (24) and Black Butte (44) mines are in a mineralized zone that extends from Medford to Cottage Grove, in which most of the deposits are in sedimentary rocks and lavas of Eocene age and occur along faults. At the Bonanza mine, fractures in argillized tuffaceous sandstone near its contact with overlying shale are mineralized with disseminated cinnabar along with quartz, chalcedony, various carbonates, and minor realgar and orpiment. At the Black Butte mine, andesitic lavas, tuffs, and breccias have been faulted and hydrothermally altered. Within and below the fault, cinnabar is disseminated through much of the rock and occurs in veinlets in the richer parts of the ore shoots. Associated minerals are calcite, opal, chlorite, sericite, pyrite, and marcasite.

The mercury of Idaho (21, p. 4) comes chiefly from the Cinnabar (Hermes) mine near Yellow Pine and the Idaho-Almaden mine near Weiser. The ore bodies of the Cinnabar mine (40) are in limestone and shale of Paleozoic (?) age that form part of a roof pendant in the granite of the Idaho batholith. These rocks were argillized, sericitized, and silicified along a broad fault zone and subsequently mineralized by cinnabar, which occurs as fracture fillings and disseminations, chiefly in limestone, accompanied by pyrite, stibnite, realgar, and orpiment. At the Idaho-Almaden mine (37, 38) fractures in an anticline where it is crossed by a pronounced sag have been mineralized by cinnabar, opal, and chalcedony. Cinnabar is also disseminated in the surrounding rock, which is silicified material of the opalite type and occurs as a blanket above feldspathic sandstone of the Payette formation of Miocene or Pliocene age.

Texas.   The Terlingua district of Texas (21; 41, pp. 432–437; 47) furnished more than 150,000 flasks of mercury between 1895 and 1946, when production ceased. The greatest period of activity was 1915 to 1919, with a supply of more than 4,000 flasks each year, a maximum of 10,791 flasks in 1917, and a total production of nearly 35,000 flasks during this period. Although there were about 20 mines and many prospects in the district, approximately 90 percent of the total production came from the Chisos-Rainbow, Mariposa, and Study Butte mines, which furnished, respectively, 100,000, 20,000, and 10,000 flasks.

The Terlingua district is a narrow zone about 20 miles long in the southern part of the Big Bend region, southwestern Texas, chiefly in Brewster County. The district is underlain by limestone and shale of Cretaceous age, above which are volcanic rocks of early Tertiary age, all of which are cut by basaltic to rhyolitic dikes, sills, and laccoliths. The rocks have been gently folded and faulted, and solution gave rise to the formation of collapse breccias. Mineralization was along faults and in breccias. At the Chisos-Rainbow mine cinnabar occurs in calcite veins that filled fissures along faults, in flat troughlike zones in limestone underlying clay where they were localized by faults, and in fault breccia zones in limestone and in a pipelike body of breccia. The pipe is a vertical cylinder containing blocks and fragments of mineralized limestone in a clay matrix. The breccia deposits were the richest and most productive of the mine. At the Mariposa mine mineralization was along faults where limestone just beneath clay was dissolved by hydrothermal solutions, clay collapsed into the limestone, and cinnabar mineralization produced deposits known as "cave fill zones." The deposits of the Study Butte mine were thin seams of cinnabar and pyrite along fractures in quartz syenite that intruded Cretaceous calcareous shales, and they were also impregnations and veinlets of pyrite, cinnabar, and calcite in the shale.

Alaska. Although Alaska (20, p. 12) has produced some mercury since 1926, the amount has been rather small until 1957. In that year Alaska accounted for almost 16 percent of the production of the United States, and in 1961 it furnished 13 percent of the total and ranked third—California about 60 percent and Nevada about 24 percent.

Cinnabar has been reported from various places in Alaska, but profitable mining has been essentially confined to the Kuskokwim region in the southeastern part of the state. The chief production has come from the Red Devil mine (34, 50), although some production has come also from the Alice and Bessie mine (or Parks property) about 3 miles to the northwest and from the Decoursey Mountain mines about 40 miles west-northwest of the Red Devil mine. Within the area of the Red Devil mine, interbedded graywacke and shale of Cretaceous age are cut by dikes and sills of basalt. The ore occurs in irregular veins and veinlets in fault zones and at the contact between silicified basalt and graywacke. Cinnabar is accompanied by much stibnite. The chief gangue mineral is quartz, but calcite and minor amounts of orpiment and realgar are present also.

Soviet Union

The principal mercury deposits of the Soviet Union (20, p. 24; 42, pp. 164–165) appear to be concentrated in three deposits, as shown by Table 12-5.

Much of the production has come from mines of the Nikitovka area, where ore occurs in sedimentary rocks of Carboniferous age. Cinnabar is present as disseminations in sandstones and conglomerates, especially beneath impervious rocks at the tops of domes, and in a breccia zone, in cracks, and as impregnations near the crest of an anticline. Some of the ore is in coal. Associated metallic minerals are stibnite, pyrite, and galena; the gangue is chiefly quartz. The ore bodies of the Khaidarkan area are localized in a breccia along the contact between highly folded and faulted Paleozoic limestones and shales.

TABLE 12-5. Principal Mercury Deposits of the Soviet Union*

Deposit	Location	Notes about deposit
Nikitovka	About 35 miles north of Stalino in the Donets basin area; 48°24′N, 38°03′E	Mined since 1885. Ore content: 0.37–0.40 percent Hg
Khaidarkan	About 50 miles southwest of Fergana; 39°58′N, 71°48′E	Mercury-antimony-fluorite deposit; large but low grade. Requires selective mining in richer blocks. Average content: 0.24 percent Hg, 3 percent Sb_2S_3, 19 percent CaF_2
Chagan-Uzon	Southeastern Altai Mountains; 50°07′N, 88°25′E	Mercury content of ore stated to be equal to that of Nikitovka (see above)

* Reprinted by permission of the publishers from Demitri Shimkin, "Minerals—a Key to Soviet Power." Cambridge, Mass.: Harvard University Press, Copyright, 1953, by the President and Fellows of Harvard College.

The mineralization brought in cinnabar, stibnite, fluorite, and quartz. The deposit is important for mercury, antimony, and fluorite. At Chagan-Uzan, stated to be about equally as important for mercury as Khaidarkan, ore is associated with a mineralized fault that cuts sedimentary rocks ranging in age from Cambrian to Tertiary. Cinnabar, stibnite, and pyrite occur in breccia veins cemented by carbonate. Silurian sandstones in contact with impervious Cambrian formations apparently were the sites of the richest ore bodies.

Mexico

The largest mercury deposits of Mexico (20, pp. 18–20; 32; 41, pp. 449–460) are in the states of Zacatecas, Guerrero, and Durango. Among these, in 1942 (32, p. 198), Zacatecas furnished 35.5 percent of the Mexican production, Guerrero supplied 11.3 percent, and Durango yielded 7.6 percent, altogether more than half of the total. Other significant but smaller producing states were Chihuahua with 4.3 percent, Guanajuato with 4.2 percent, San Luis Potosí with 2.9 percent, and Aguascalientes with 2.2 percent, bringing the total to about 68 percent.

In the Nuevo Mercurio district of northern Zacatecas there are many prospects, small mines, and 18 major mines. The deposits are in folded and faulted limestones, marls, and sandstones of Cretaceous age. Cinnabar, along with calcite and a little marcasite, occurs in the broken axial parts of anticlines and in zones of coarse, vuggy fault breccia. The deposits have been mined to a depth of about 500 ft but the grade of ore declined appreciably with depth. In the Sain Alto district, central Zacatecas, cinnabar forms impregnations in sandstone and cinnabar-calcite veinlets in joints in sandstone. Deposits are generally localized beneath shale or fault gouge. In the Canoas district, southern Zacatecas, mercury deposits occur in a dome of latite that was fractured and extensively altered. Very fine-grained cinnabar accompanied by abundant opal and chalcedony is present along fractures where the rock was extensively altered to montmorillonite. Most of the production has come from stockwork deposits within 100 ft of the surface.

The most important deposits of Guerrero are in the Huitzuco district and the Huahuaxtla district, northern part of the state. The Huitzuco district, which supplies both mercury and antimony, was discovered about 1870 and exploited soon afterward. The deposits differ somewhat from other types. Folded and faulted Cretaceous limestone was intruded by small bodies of granite porphyry, hydrothermally altered to dolomite and then to anhydrite and, near the surface, changed by weathering to gypsum. Cavities in the gypsum, made by surface waters, were partly filled by gravel and surface rubble. Subsequently hydrothermal solutions deposited livingstonite ($HgS \cdot 2Sb_2S_3$) and stibnite in some of the dolomite breccia and the rubble. Although enriched ore that is erratically distributed furnished some production, most production has come from primary ore, which has been followed to a depth of 820 ft in La Cruz mine, and the grade of which is about 1 percent Hg on the lower levels. At Huahuaxtla the ore bodies occur along a fault in Upper Cretaceous limestone and shale. The ore consists of cinnabar accompanied by some marcasite, pyrite, gypsum, and calcite, and much of it is in fault gouge about 30 ft thick.

In the Cuarenta district of northern Durango, granite has been intruded by diabase dikes, overlain by limestone conglomerate and rhyolite flows and tuffs, and all rocks cut by minor faults. The ore bodies are generally at the unconformity between the granite and the conglomerate and are localized especially under diabase dikes. The conglomerate contains some disseminated ore.

Yugoslavia

The Idria mine of Yugoslavia (20, pp. 24–25; 23; 41, pp. 463–467), second only to Almadén, Spain, in total production of mercury, has been worked since early in the fifteenth century. It is 25 miles north-northeast of Trieste, Italy, in the extreme northwestern corner of Yugoslavia. The geology of the area is complex and of the Alpine type. The ore occurs in sandstone, dolomite, and shale of late Paleozoic and younger age, along a highly broken fault zone for a length of 4,600 ft and a depth of at least 1,300 ft. The chief ore mineral, cinnabar, forms rich ore where it impregnates sandstone, good ore where it occurs in brecciated dolomite, and lower-grade ore in carbonaceous shale. The good ore bodies generally occur beneath a capping of dense shale, and the lower-grade ore bodies in the shale at the apices of tight folds. Sparse gangue minerals are pyrite, calcite, dolomite, chalcedony, quartz, gypsum, and locally fluorite. Replacement has been important.

China

The mercury deposits of China (20, p. 25; 33, pp. 447–450; 41, pp. 483–485) are closely associated with the antimony deposits, and production, just as in the case of antimony, has increased sharply since 1945. Thus the production of mercury in China was less than 0.6 percent of the world total for 1945 to 1949, rose to slightly more than 3 percent for 1950 to 1954, to almost 8 percent for 1955 to 1959, and was about 10 percent in 1962. Although deposits are widely scattered throughout the broad zone in which they occur, the most important ones are on the Kweichow-Hunan border (Fig. 9-5), among which are at least six productive deposits, two of which seem to be of especial importance—Wan-Shan-Ch'ang (Wanshanchang) and Ta-Tung-La (Tatungla). The ore bodies, in which cinnabar is the chief mineral, occur as irregular veins in brecciated layers of Paleozoic limestone underlying shale, especially in anticlines. Some stibnite, pyrite, calcite, quartz, and hydrocarbon may accompany the cinnabar. Although some rich ore shoots have yielded somewhat more than 4 percent mercury, the average yield is about 1 percent, and the ores are extensively sorted and upgraded prior to treatment.

SELECTED REFERENCES
CHIEFLY ANTIMONY

1. F. Ahlfeld, 1952, Die südbolivianische Antimonprovinz, *Neues Jahrb. Miner.*, Abh. vol. 83, no. 2-3, pp. 313–346.
2. J. R. Cooper, 1951, Geology of the tungsten, antimony, and gold deposits near Stibnite, Idaho, *U.S. Geol. Survey Bull.* 969-F, pp. 151–197.
3. J. González Reyna, 1956, Los yacimientos de antimonio de México, in "Riqueza Minera y Yacimientos Minerales de México," 3d ed., pp. 203–213, Internat. Geol. Cong., 20th, Mexico 1956, 497 pp.
4. J. B. Huttl, 1952, Yellow Pine is expanding its output of strategic minerals, *Eng. Mining Jour.*, vol. 153, no. 5, pp. 72–77.
5. V. C. Juan, 1946, Mineral resources of China, *Econ. Geology*, vol. 41, no. 4, pt. 2, supplement. Antimony, pp. 444–447.
6. B. L. Miller and J. T. Singewald, 1919, "The Mineral Deposits of South America," McGraw-Hill Book Company, New York, 598 pp.
7. E. W. Sahli, 1961, Antimony in the Murchison Range of the north-eastern Transvaal, Commonwealth Min. Metall. Cong., 7th, South Africa, *Trans.*, vol. 1, pp. 181–199.
8. D. B. Shimkin, 1953, "Minerals—a Key to Soviet Power," Harvard University Press, Cambridge, Mass., 452 pp. Antimony, pp. 153–154.
9. D. E. White, 1940, Antimony deposits of a part of the Yellow Pine District, Valley County, Idaho, a preliminary report, *U.S. Geol. Survey Bull.* 922-I, pp. 247–279.
10. D. E. White, 1945, The antimony deposits of Mexico [abs.], *Econ. Geology*, vol. 40, p. 96.
11. D. E. White, 1948, Antimony deposits of Soyatal district, state of Querétaro, Mexico, *U.S. Geol. Survey Bull.* 960-B, pp. 35–88.
12. D. E. White, 1951, Resources, chap. 3, 57 pp., in "Materials Survey, Antimony," U.S. Bureau of Mines.

13. D. E. White, 1962, Antimony in the United States, *U.S. Geol. Survey Mineral Inv. Resource Map* MR-20. Map and pamphlet of 6 pp.
14. D. E. White and J. González Reyna, 1946, San José antimony mines near Wadley, state of San Luis Potosí, Mexico, *U.S. Geol. Survey Bull.* 946-E, pp. 131–153.
15. D. E. White and R. Guiza, Jr., 1947, Antimony deposits of the Tejocotes region, state of Oaxaca, Mexico, *U.S. Geol. Survey Bull.* 953-A, pp. 1–26.
16. D. E. White and R. Guiza, Jr., 1949, Antimony deposits of El Antimonio district, Sonora, Mexico, *U.S. Geol. Survey Bull.* 962-B, pp. 81–119.
17. Anon., 1953, World's richest antimony mine, *Mining World,* vol. 15, no. 6, pp. 47–52.

CHIEFLY MERCURY

18. P. Averitt, 1945, Quicksilver deposits of the Knoxville district, Napa, Yolo, and Lake Counties, California, *California Jour. Mines and Geology,* vol. 41, no. 2, pp. 65–89.
19. E. H. Bailey, 1951, The New Almaden quicksilver mines, *California Div. Mines Bull.* 154, pp. 263– 270.
20. E. H. Bailey, 1959, Resources, chap. 3, 92 pp., "Mercury, a Materials Survey," *U.S. Bur. Mines Inf. Circ.* 7941.
21. E. H. Bailey, 1962, Mercury in the United States, *U.S. Geol. Survey Mineral Inv. Resource Map* MR-30. Map and pamphlet of 8 pp.
22. E. Bennett, 1948, Almaden, world's greatest mercury mine, *Mining and Metallurgy,* vol. 29, no. 493, pp. 6–9.
23. B. Berce, 1960, Methods and results of geochemical investigations of mercury, Internat. Geol. Cong., 21st, Copenhagen 1960, *Report,* pt. 2, pp. 65–74.
24. R. E. Brown and A. C. Waters, 1951, Quicksilver deposits of the Bonanza-Nonpareil district, Douglas County, Oregon, *U.S. Geol. Survey Bull.* 955-F, pp. 225–251.
25. C. DeKalb, 1921, The Almaden quicksilver mine, *Econ. Geology,* vol. 16, pp. 301–312.
26. E. B. Eckel, 1948, Mercury industry in Italy, *Am. Inst. Mining Engineers Tech. Pub.* 2292, 21 pp.; discussion, *Am. Inst. Mining Engineers Tech. Pub.* 2474, pp. 3–5.
27. E. B. Eckel, R. G. Yates, and A. E. Granger, 1941, Quicksilver deposits in San Luis Obispo County and southwestern Monterey County, California, *U.S. Geol. Survey Bull.* 922-R, pp. 515–580.
28. E. B. Eckel and W. B. Myers, 1946, Quicksilver deposits of the New Idria district, San Benito and Fresno Counties, California, *California Jour. Mines and Geology,* vol. 42, no. 2, pp. 81–124.
29. D. L. Everhart, 1946, Quicksilver deposits at the Sulphur Bank mine, Lake County, California, *California Jour. Mines and Geology,* vol. 42, no. 2, pp. 125–153.
30. E. L. Fisk, 1961, Cinnabar at Cordero, *Mining Eng.,* vol. 13, pp. 1228–1230.
31. J. E. Gilbert and V. P. Haas, 1959, Cordero—Nevada's largest Hg mine, *Eng. Mining Jour.,* vol. 160, no. 3, pp. 88–90.
32. J. González Reyna, 1956, Los yacimientos de mercurio de México, in "Ri-

queza Minera y Yacimientos Minerales de México," 3d ed., pp. 185–201, Internat. Geol. Cong., 20th, Mexico 1956, 497 pp.

33. V. C. Juan, 1946, Mineral resources of China, *Econ. Geology*, vol. 41, no. 4, pt. 2, supplement. Mercury, pp. 447–450.

34. E. M. Mac Kevett, Jr., and H. C. Berg, 1963, Geology of the Red Devil quicksilver mine, Alaska, *U.S. Geol. Survey Bull.* 1142-G, 16 pp.

35. L. Menendez y Puget, 1949, The riches of Almaden, *Mining World*, vol. 11, no. 6, pp. 34–36; no. 7, pp. 38–41; no. 8, pp. 35–37.

36. F. L. Ransome, 1921, The ore of the Almaden mine, *Econ. Geology*, vol. 16, pp. 313–321.

37. J. R. Reynolds, 1956, Idaho-Almaden mercury mine, mining and geology, *Mining Eng.*, vol. 8, no. 11, pp. 1096–1098.

38. C. P. Ross, 1956, Quicksilver deposits near Weiser, Washington County, Idaho, *U.S. Geol. Survey Bull.* 1042-D, pp. 79–104.

39. P. H. Sampelayo and others, 1926, Minas de Almaden, Internat. Geol. Cong., 14th, Madrid 1926, *Exc.* B-1, pp. 1–102.

40. F. C. Schrader and C. P. Ross, 1926, Antimony and quicksilver deposits in the Yellow Pine district, Idaho, *U.S. Geol. Survey Bull.* 780-D, pp. 137–167.

41. C. N. Schutte, 1931, Occurrence of quicksilver orebodies, *Am. Inst. Mining Metall. Engineers Trans.* (general volume), pp. 403–486; discussion, pp. 486–488.

42. D. B. Shimkin, 1953, "Minerals—a Key to Soviet Power," Harvard University Press, Cambridge, Mass., 452 pp. Mercury, pp. 164–165.

43. C. M. Swinney, 1950, The Altoona quicksilver mine, Trinity County, California, *California Jour. Mines and Geology*, vol. 46, no. 3, pp. 395–404.

44. F. G. Wells and A. C. Waters, 1934, Quicksilver deposits of southwestern Oregon, *U.S. Geol. Survey Bull.* 850, 58 pp.

45. D. E. White, 1955, Sulphur Bank, California, *Econ Geology*, Fiftieth Ann. Vol., pt. 1, pp. 117–120.

46. D. E. White and C. E. Roberson, 1962, Sulphur Bank, California, a major hot-spring quicksilver deposit, *Geol. Soc. America* (Buddington volume), pp. 397–428.

47. R. G. Yates and G. A. Thompson, 1959, Geology and quicksilver deposits of the Terlingua district, Texas, *U.S. Geol. Survey Prof. Paper* 312, 114 pp.

48. R. G. Yates and L. S. Hilpert, 1948, Quicksilver deposits of eastern Mayacmas district, Lake and Napa Counties, California, *California Jour. Mines and Geology*, vol. 42, no. 3, pp. 231–286.

MISCELLANEOUS REFERENCES CITED

49. E. L. Egenhoff, 1953, De argento vivo: Historic documents on quicksilver and its recovery in California prior to 1860, *California Jour. Mines and Geology*, vol. 49, no. 4, supplement, 144 pp.

50. R. F. Lyman, 1961, The Red Devil mercury mine, *Mining Eng.*, vol. 13, pp. 1337–1339.

51. F. S. Turneaure and K. K. Welker, 1947, The ore deposits of the eastern Andes of Bolivia, The Cordillera Real., *Econ. Geology*, vol. 42, pp. 619–620.

chapter 13

Uranium, Vanadium, Thorium, Rare Earths, Titanium, and Zirconium

Uranium, vanadium, thorium, rare earths, titanium, and zirconium occur in a number of minerals that are closely associated in some deposits, and commercial amounts of two or perhaps more of these elements may be obtained from a single deposit, yet there are some marked differences in occurrence of commercial deposits. Some of the similarities and differences among deposits are summarized in Table 13-1.

The mineral composition of some deposits, especially those containing uranium, is complex. Heinrich (1) lists about a hundred radioactive minerals that he designates main species and discusses many varieties of these as well as other species. A comparatively few minerals, however, along with some that are not radioactive, are important in commercial production. These are given in Table 13-2 along with some notes about them. This table shows, also, that some of the minerals contain several of the chemical elements with which we are presently concerned.

Uranium

Production and Reserves

Information regarding the world-wide production and reserves of uranium is generally unknown for countries of the Soviet bloc but fairly well known for other countries. Earlier, practically all information was kept secret, but as attention became focused more and more on the peaceful uses of atomic energy, and especially with the convening of the International Conference on the Peaceful Uses of Atomic Energy at Geneva in 1955, much more information was released. Consequently it is now possible to give a reasonably accurate picture of the production of the Free World (Table 13-3).

The average annual production of the Free World for the period 1958 to 1962 was about 38,000 short tons of uranium oxide (U_3O_8), and estimates of annual

TABLE 13-1. Some Relations among Deposits of Uranium, Vanadium, Thorium, Rare Earths, Titanium, and Zirconium

Chief kinds of deposits and their importance

Element or elements sought	Pegmatites	Veins	Sandstone and conglomerate	Placers	Supergene, oxidized	Carbonatites
Uranium	Chiefly by-product of other materials	Highly important; some of richest deposits	Highly important; furnish major production	Unimportant*	Highly important in some sandstone-conglomerate	
Uranium and vanadium	Vanadium generally not present	Vanadium generally not present with uranium	Highly important	Unimportant	May be important	
Vanadium without uranium	Unimportant	Unimportant	Usually unimportant	Unimportant	Important in some deposits	
Thorium, rare earths	Generally little production	Some are important	Unimportant	Highly important	Unimportant	Some are important
Titanium (ilmenite and rutile), zirconium	Generally not productive	Some are important	Unimportant	Highly important	Unimportant	

* If, that is, the Witwatersrand and Blind River deposits are not considered to be placer types. Uranium occurs in conglomerates, and the origin is a matter of debate.

TABLE 13-2. Some Important Minerals Present in Deposits of Uranium, Vanadium, Thorium, Rare Earths, Titanium, and Zirconium

Mineral	Approximate composition	Chief occurrences
Uraninite*	Uranium oxides	Pegmatites, especially; primary deposits, associated with pitchblende
Pitchblende*	Uranium oxides	Veins; sandstone and conglomerate; nearly all high-grade uranium ores
Coffinite	Uranium silicate	Some sandstone and conglomerate; some veins
Davidite	Rare earth iron-titanium oxide	Some veins, usually intergrown with ilmenite
Brannerite	Complex oxide of Ti, U, Ca, Fe, Y, Th	Some sandstone and conglomerate deposits
Uranophane	Calcium-uranium silicate	Common in oxidized deposits
Autunite and meta-autunite	Calcium-uranium phosphate	Some oxidized deposits if PO_4 is available
Torbernite and metatorbernite	Copper-uranium phosphate	Some oxidized deposits if both PO_4 and Cu are available
Schroeckingerite	Carbonate and sulfate of Na, Ca, U, plus F	Oxidized deposits; least important of uranium ore minerals
Carnotite	Potassium-uranium vanadate	Sandstone and conglomerate; oxidized parts of many deposits
Tyuyamunite	Copper-uranium vanadate	About same as carnotite
Vanadinite	Lead chloro-vanadate	Oxidized parts of some deposits
Descloizite	Lead-zinc vanadate	Oxidized parts of some deposits
Roscoelite	Vanadium-bearing mica	Some sandstone deposits; some gold veins
Thorianite-uranothorianite	Thorium oxide to thorium-uranium oxide	Pegmitites, granites, gneisses; placers
Thorite-uranothorite	Thorium silicate to thorium-uranium silicate	Pegmatites, granites, gneisses; veins; some placers; carbonatites
Monazite	Phosphate of Th, Ce, La, Y	Pegmatites, granites, gneisses; placers; some veins
Euxenite-polycrase series	Complex oxide of Nb, Ta, Th, rare earths, uranium	Pegmatites; commercial occurrence chiefly in placers
Fergusonite-formanite series	Complex oxide of Nb, Ta, Th, rare earths, uranium	Pegmatites; commercial ouccrrence chiefly in placers
Samarskite	Complex oxide of Nb, Ta, Th, rare earths, uranium	Pegmatites; commercal occurrence chiefly in placers
Gadolinite	Silicate of Be, Fe, Y	Pegmatites
Allanite	Silicate of Ca, Al, Fe, Mg, Ce, Th	Pegmatites
Pyrochlore-microlite series	Complex oxide of Na, Ca, Nb, Ta, plus F, Ce, Th, U	Carbonatites; some placers
Bastnaesite	$CeFCO_3$	Carbonatites
Rutile	TiO_2	Chiefly placers; some nelsonite masses
Ilmenite	$FeTiO_3$	Chiefly placers; some nelsonite masses; some titaniferous iron deposits
Zircon	$ZrSiO_4$	Pegmatites, granites, gneisses; placers

* Usage of the terms *uraninite* and *pitchblende* is somewhat variable. A consistent chemical distinction between the two does not seem to exist. Pitchblende is a variety of uraninite and is generally botryoidal, colloform, or massive.

production of Soviet-bloc countries range from 10,000 to 20,000 short tons. Production is thought to come from the Soviet Union, Czechoslovakia, East Germany, Hungary, and perhaps some others.

Estimates of reserves change as mining progresses. Recently (1963 and 1964) there has been a tendency toward downward revision of estimates made earlier.

United States

The uranium produced in the United States comes chiefly from deposits in sandstone, conglomerate, and limestone of the Colorado Plateau region, but some is supplied from adjacent areas, especially Wyoming (Table 13-4). Further, some of the production of states in the Colorado Plateau comes from veins.

The production for 1962, although it shows the present and most probable future sources of uranium in the United States for some years to come, does not show the interesting changes that have taken place between about 1950 and 1962 as a result of new discoveries and development of new ore bodies. This is

**TABLE 13-3.   Free-world Production of
Uranium Oxide (U₃O₈), 1958–1962 inclusive***

Country	Percent of free-world total
United States	43
Canada	32
Republic of South Africa	16
Republic of the Congo	3
France	3
Australia	3
Total†	100

* Compiled from *U.S. Bur. Mines Minerals Yearbook*, 1962, vol. 1, p. 1276.
† Estimated production of all other countries of the free world was but 0.7 percent of the total. Above figures rounded off to include this.

**TABLE 13-4.   Mine Production of Uranium
in the United States in 1962***

State	Percent of total production of the United States	Grade of ore, percent U_3O_3
New Mexico	46.2	0.23
Wyoming	18.7	0.25
Utah	16.1	0.35
Colorado	13.8	0.21
Arizona	2.1	0.26
South Dakota	0.3	0.18
All others	2.8	0.26
Total and grade	100.0	0.24

* Compiled from *U.S. Bur. Mines Minerals Yearbook*, 1962, vol. 1, p. 1267.

shown, in part, by Table 13-5, which summarizes information made available since 1957 after declassification of statistics regarding uranium. The large increase in the production from New Mexico reflects the development of the highly important deposits of the Ambrosia Lake area, and the increase in the production from Wyoming reflects the development of important deposits discovered in the Powder River and Wind River Basins.

The important reserves of uranium are contained in the states that are the major producers and are shown by Table 13-6.

TABLE 13-5.   Shifts in Production of Uranium in the United States, 1957–1962*

State	Approximate percent of total production in the United States					
	1957	*1958*	*1959*	*1960*	*1961*	*1962*
Utah	38	32	25	17	17	16
New Mexico	26	29	39	42	42	46
Colorado	19	19	16	15	15	14
Arizona	8	6	4	4	3	2
Wyoming	6	11	12	18	19	19
Total of above	97	97	96	96	96	97

* Compiled from *U.S. Bur. Mines Minerals Yearbook*, 1957–1962, vol. 1.

TABLE 13-6.   Estimate of Reserves of Uranium Ore in the United States, December 31, 1959*

State	Million tons of ore	Percent U_3O_8	Percent of total ore of the United States
New Mexico	55.7	0.26	64.7
Wyoming	15.8	0.34	18.3
Utah	5.3	0.33	6.2
Colorado	4.5	0.30	5.2
Arizona	1.2	0.35	1.4
Washington, Oregon, and Nevada	1.9	0.21	2.2
North Dakota and South Dakota	0.6	0.27	0.7
All others	1.1	0.24	1.3
Total and grade	86.1	0.28	100.0

* Compiled from *U.S. Bur. Mines Minerals Yearbook*, 1959, vol. 1, p. 1154.

General characteristics of many deposits.   The uranium deposits of the Colorado Plateau, of Wyoming, of South Dakota, as well as somewhat similar deposits in other parts of the world, show some features that are essentially common to the entire group. They have been designated by various names (17, p. 944), among which are *sandstone-type uranium deposits, sandstone-conglomerate uranium deposits, Plateau-type uranium deposits,* and a number of others. Finch (17) summarized their general features and proposed the name *peneconcordant uranium deposit* because one of their outstanding characteristics is that they almost conform to or are parallel to bedding or other sedimentary structures of sedimentary rocks. Finch stated:

> *Peneconcordant uranium deposits are tabular, lenticular, or irregularly-shaped masses of widely differing size that are, in general, concordant to the gross sedimentary structures of the enclosing rock but that in detail cut across sedimentary structures. This local discordance is a diagnostic characteristic indicating that the deposits are epigenetic rather than syngenetic. The long dimension of peneconcordant uranium deposits is commonly oriented parallel or subparallel to sedimentary features, such as bedding, lenses, channels, and fossil logs. The long dimension is rarely, and then probably coincidentally, parallel or subparallel to tectonic structures, such as joints, faults, or shear zones.*
>
> *The host rocks for peneconcordant uranium deposits are mainly arenaceous and argillaceous, and subordinately highly carbonaceous and carbonate rocks. The arenaceous and argillaceous rocks range from coasely clastic—conglomerate, sandstone, and arkose—to finely clastic—siltstone, mudstone, and tuff. Small amounts of carbonaceous materials are common in the arenaceous and argillaceous rocks. Carbonate host rocks are limestones that may be partly or completely clastic. . . . The environment of deposition of these host rocks was mainly fluvial and subordinately lacustrine, paludal, and marginal marine.*
>
> *The uranium and associated minerals principally fill the original sedimentary and diagenetic pore spaces in medium- to coarse-grained host rocks. These minerals, also, partially or wholly replace interstitial cement and detrital rock constituents, including plant remains and quartz grains.*[1]

The Colorado Plateau deposits.   The Colorado Plateau deposits (1, pp. 358–419; 2, pp. 208–221; 3, pp. 60–66; 4, pp. 274–311; 5, 338–345, 350–367, 525–529, 605–614; 8, pp. 143–168, 187–193, 243–280) are in adjoining parts of Colorado, Utah, Arizona, and New Mexico (Fig. 13-1). They are uranium-vanadium deposits that occur in most of the sedimentary rocks exposed in the Colorado Plateau region but are concentrated especially in the fluvial Shinarump conglomerate and the Chinle formation of Triassic age and the eolian Entrada sandstone, lacustrine Todilto limestone, and fluvial Morrison formation of Jurassic age. Among these the Morrison formation had been far more productive to 1955 than the others, followed by the Chinle and then the Shinarump (70). Dates given in 1963 (30) indicate that primary uranium mineralization in certain deposits of Utah and Arizona occurred about 210 million years ago and in certain deposits of Colorado about 110 million years ago.

Within the sedimentary rocks of the Colorado Plateau there are about a

[1] From W. I. Finch, *Econ. Geology,* vol. 54, pp. 945–946. Copyright 1959. Economic Geology Publishing Company. Used by permission.

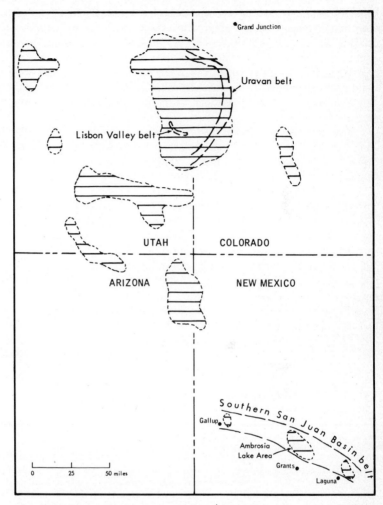

Fig. 13-1 Map showing generalized distribution of major uranium-vanadium areas of the Colorado Plateau, and the two major uranium belts. Simplified after R. P. Fischer, *U.S. Geol. Survey Prof. Paper* 300.

dozen deposits that contain more than 1 million tons each of material (production plus reserves) containing at least 0.1 percent U_3O_8, more than 50 that contain between 1,000 and 1 million tons each of the same grade, and many smaller ones (13).[1] The major deposits show a strong tendency to be grouped in elongate zones or belts (Fig. 13-1) (18; 34; 46) in which the deposits are more closely spaced, larger, and of higher grade than they are in adjoining areas. Two major

[1] An interesting popular, nontechnical, and illustrated pamphlet about the deposits and their development, titled *Mesa Miracle,* was published in 1952 by United States Vanadium Company, Division of Union Carbide and Carbon Corporation, Grand Junction, Colorado.

belts and a minor one have been recognized. The Uravan belt, about 70 miles long and 4 miles wide (average), is chiefly in western Colorado, and the ore deposits within it commonly occur in clulsters, athough some are isolated. The Southern San Juan Basin mineral belt, in northwestern New Mexico, is about 85 miles long and 20 miles wide. The small but sharply defined Lisbon Valley belt is in southeastern Utah, is about 20 miles long and less than a mile wide, and, although small, contains deposits of higher grade than the usual deposit in sandstone.

The deposits have been classified by Fischer (5, pp. 605–614; 8, pp. 143–154) on the basis of lithologic character into four types (Table 13-7). The deposits have also been classified according to the relative amounts of uranium, vanadium, and copper contained in them (Table 13-8).

The mineral composition of the deposits is complex (7, pp. 65–79) and varies with the amounts of uranium, vanadium, and copper present and with the degree of oxidation of the deposit. In vanadiferous ores that are essentially unoxidized the vanadium minerals are chiefly roscoelite, $K(Al,V)_2(Al,Si)_3O_{10}(OH,F)_2$, montroseite, $VO(OH)$, and paramontroseite, VO_2, along with vanadium-bearing chlorite and vanadium-bearing hydrous mica. Uranium minerals are uraninite or pitchblende and brannerite. The most common sulfides present are pyrite and marcasite, but the copper-bearing sulfides chalcopyrite, bornite, chalcocite, and covellite may be present also. Oxidation of such ores yields, from the vanadium minerals in the earlier stages, corvusite, $V_2O_4 \cdot 6V_2O_5 \cdot nH_2O$, and rauvite, $CaO \cdot 2UO_26V_2O_5 \cdot 20H_2O(?)$, and from combined vanadium and uranium minerals in the late stage, carnotite, $K_2(UO_2)_2V_2O_8 \cdot 3H_2O$, or tyuyamunite, $Ca(UO_2)_2V_2O_8 \cdot 5-8 \cdot 5H_2O$,

TABLE 13-7. Types of Uranium Deposits of the Colorado Plateau on the Basis of Lithologic Character*

Type of deposit	*General characteristics*	*Chief occurrence*
Sandstone-mudstone	Deposits in sandstone interbedded with mudstone and that contains mudstone lenses and pebbles; not enough finely divided argillaceous material to occupy all pore spaces between sand grains; plant fossils common	All deposits in the Morrison formation, and most deposits in the Shinarump and Chinle formations
Sandstone	Deposits in "clean" sandstone	In Entrada sandstone only
Limestone	Deposits in limestone	All deposits in the Todilto limestone
Mudstone, siltstone, and argillaceous sandstone	Deposits are in rocks that are more argillaceous than those of the sandstone-mudstone type; probably all or nearly all pore space between sand grains is occupied with argillaceous matrix; plant fossils are common	Some deposits in the Chinle formation

* Compiled from *U.S. Geol. Survey Prof. Paper* 300, pp. 605–614, 1956.

TABLE 13-8.   Types of Uranium Deposits of the Colorado Plateau on the Basis of the Relative Amounts of Contained Uranium, Vanadium, and Copper*

Type of deposit	General characteristics	Chief occurrence
Vanadium-uranium	Contain 1–20 parts V_2O_5 to 1 part U_3O_8	Central part of the Plateau region; most deposits in the Morrison formation and about one-third of those in the Shinarump and Chinle formations; includes nearly all sandstone-mudstone deposits
Vanadium (uranium)	Contain 20–40 parts V_2O_5 to 1 part U_3O_8	Eastern margin of the Plateau; all are in Entrada sandstone and are of the sandstone type
Copper-uranium	Copper content generally a few times greater than uranium-tent	Restricted to White Canyon area of central Utah; all are in the Shinarump formation and classed as sandstone-mudstone deposits
Uranium (vanadium-copper)	U_3O_8 is more abundant than V_2O_5 and Cu	Almost all along southern and western margins of the Plateau

* Compiled from *U.S. Geol. Survey Prof. Paper* 300, pp. 605–614, 1956.

which are generally very stable. Secondary copper minerals may be malachite, azurite, chrysocolla, and a number of others. In nonvanadiferous ores that are unoxidized the uranium appears to be chiefly in uraninite, and coffinite is scarce or lacking. Oxidation of such ores produces a great variety of secondary uranium minerals, chiefly sulfates or carbonates depending on the presence of sulfide or carbonate in the unoxidized ore.

The ore bodies (3, pp. 60–66; 5, pp. 41–49, 605–614, 338–345; 7, pp. 3–11; 8, pp. 143–154) are most common in the lenticular sandstone and mudstone facies of fluviatile sedimentary rocks, but some productive ore bodies occur in limestone, chiefly in the Grants and Laguna districts of New Mexico.

The ore bodies of the sandstone type are tabular, elongate, or podlike, range in thickness from a few inches to about 20 ft and, exceptionally, 50 ft, and range in length from a few feet to thousands of feet. Generally they are parallel to the bedding but in detail they are undulant, cut across the bedding gently or, locally, cut across the bedding in sharply curving forms that are designated *rolls*. The distribution of ore within a lens is somewhat variable. It may follow the bottom of a channel or a sandstone lens; it may follow the top of a lens or some sedimentary structure; or it may be midway in a permeable layer. The localization of primary deposits was controlled chiefly by sedimentary features, but some secondary minerals occur along joints and fractures. A very important factor in localization of ore bodies was the distribution of carbonaceous material (7, pp. 139–149). Deposits are associated with coalified wood, crude oil, carbonaceous shale, and carbonaceous material of unknown origin, but coalified plant debris is the most common type of carbonaceous material occurring on the

Colorado Plateau and the type associated with some high-grade ore bodies. Some logs 75 ft or more long were richly mineralized and in places yielded pockets of ore containing 50 percent U_3O_8. The ore minerals, which are typically fine grained, may fill pore spaces in sandstone, impregnate clay pellets and fossil wood, replace the cell walls of the wood, and replace interstitial clay. At the Happy Jack mine, San Juan County, Utah (7, pp. 185–195), about 75 miles west of Blanding, which is one of the most productive mines in the White Canyon district, ore occurs as replacements of woody "trash pockets,"[1] as replacements of larger wood fragments, and as bedded ores.

The characteristics of ore bodies of the sandstone type and also those of the limestone type are illustrated in the San Juan Basin area of New Mexico (5, pp. 338–345; 8, pp. 387–404; 20; 22). In that area the Ambrosia Lake district and the Laguna district (Fig. 13-1) together contain more than 50 percent of the known uranium reserves of the United States. "The total value of the uranium ore from these districts is expected to exceed $1 billion, placing them in combined strategic and economic value among the largest active districts in the United States."[2] More than 90 percent of the reserves of the area are in the Morrison formation although deposits in the Todilto limestone furnished a large part of the total production of the area from the time of their discovery in 1950 until 1956. The sandstone-type deposits will be considered briefly, after which attention will be turned to the limestone deposits.

In general the sandstone-type deposits are similar to those of other parts of the Colorado Plateau. The ore minerals in the Morrison formation are intimately mixed with carbonaceous material in ore bodies that are nearly parallel to the bedding but that occur at varying depths from the surface outcrops down to more than 2,000 ft, since the rocks dip under younger strata. The outstanding deposit of the area, in the Laguna district, is the Jackpile, stated probably to be the largest single uranium deposit in the United States and one of the great producers of the world. It consists of two ore bodies, the South, which is the smaller, and the North, both of which are mined by open pit. The North ore body is about 1,300 ft wide, several thousand feet long, and consists of at least two tabular layers. The thickness is as much as 50 ft in places but probably averages about 20 ft. The South ore body is oval in plan, about 350 ft wide and 750 ft long, consists of two tabular layers, and is as much as 50 ft thick.

The Todilto limestone (5, pp. 338–345; 8, pp. 387–407; 22), where it is ore-bearing in the central and eastern parts of the southern San Juan Basin, generally varies in thickness from about 20 to 30 ft. Commonly it contains as much as 1 percent organic carbon, has a fetid odor when freshly broken, and consists, from the base upward, of (1) platy almost lithographic limestone, (2) finely wrinkled very thin-bedded limestone, and (3) massive recrystallized limestone. In many places folds and faults are conspicuous and are confined almost entirely to the limestone. Uranium deposits occur in all three zones, but probably most deposits are in the two lower ones and are localized where the limestone has been deformed.

[1] Trash pockets are mixed accumulations of mudstone pebbles and carbonaceous material such as might form in stream channels at places of local turbulence.

[2] From H. C. Granger, E. S. Santos, B. G. Dean, and F. B. Moore, *Econ. Geology*, vol. 56, p. 1180. Copyright 1961. Economic Geology Publishing Company. Used by permission.

The major production has come from the Grants area, and minor production from the Laguna area. Most of the ore bodies are elongated, 20 to 30 ft wide and 100 to a few thousand feet long.

Wyoming deposits. Commercial deposits of uranium in Wyoming (1, pp. 420–426; 3, pp. 66–67; 5, pp. 392–402; 8, pp. 361–370; 31; 40; 45) were not discovered until 1951, although the occurrence of uranium in the state was known much earlier. Production was still relatively small in 1957 (Table 13-5) but increased rapidly thereafter. Uranium occurrences are known in 20 of the 23 counties of Wyoming, but only 10 or 12 of them yield production, and only 5 have been of much importance. Actually, Fremont County has consistently furnished most of the production. The production from Campbell County was particularly important between 1955 and 1957 but has been relatively low since then. The production from Carbon County became of much importance in 1961 and 1962 and was about a fifth of the total of the state. The most important area has always been the Gas Hills field of Fremont County, but the Crooks Gap field of the same county has been a good contributor. The early production of Campbell County was from the Pumpkin Buttes field, but commercial operations there have since declined very much. Production from Carbon County has been from the Boggs-Poison Basin and the Shirley Basin fields.

Wyoming is characterized by a number of mountains and intermountain basins. Sedimentary rocks that flank the mountains and fill the basins range in age from Cambrian to Pliocene, but the known productive uranium deposits are chiefly in fluviatile rocks of Tertiary age, especially in the Wind River and Wasatch formations, both of Eocene age, with some production from the Browns Park formation of Miocene age.

The Gas Hills area, chiefly in Fremont County but in part in Natrona County, in the Wind River Basin, discovered in September 1953, is the most important producer of the state. The deposits are chiefly in the upper part of the Wind River formation, which consists of coarse-grained cross-bedded feldspathic sandstone along with some mudstone, carbonaceous shale, and siltstone. Some carbonaceous fossil plant fragments are disseminated throughout the sandstone. Ore bodies are irregular in shape, typically follow the bedding and sedimentary structures but locally depart from such features, and mineralization may follow fractures. Ore bodies occur at the surface but also at several hundred feet below the surface. Lenses of primary minerals, chiefly pitchblende and coffinite, along with pyrite, are 1 to 15 ft wide, as much as several feet thick, and generally pinch and swell. Oxidation has formed a considerable number of secondary minerals, some of which surround the primary ores or are disseminated throughout the oxidized sandstone, where they occur as interstitial fillings in irregular blankets, and in pockets. Apparently structural control was of direct or indirect importance in bringing about localization of the deposits. Wilson (45, pp. 16–18) summarized previous work, which indicated that groundwater solutions may have been impeded by faults, and that, in general, faults may have been instrumental in bringing about secondary enrichment.

The production at the Crooks Gap area of Fremont County, about 65 miles north of Rawlins, comes chiefly from the Wasatch formation. The ore bodies are similar to those of the Gas Hills area, but carbonaceous fossil plant material is more abundant and is associated with the best ore.

The Pumpkin Buttes area, chiefly in Campbell County in the Powder River Basin, was discovered in October, 1951, and the first ore was shipped from the district in 1953. All ore bodies are in the Wasatch formation, especially in a red sandstone zone near the center of the formation. The deposits are associated with and occur in lenses of red sandstone dispersed throughout siltstone, claystone, and carbonaceous shale. The major part of the mining has been from small high-grade concretionary deposits that contain 1 percent and up to 5 percent U_3O_8. These concretionary deposits consist of pitchblende associated with oxides of iron and manganese and are surrounded by uranophane and in places some carnotite. Other deposits, much larger but of lower grade, consist of irregular zones and impregnations in which the chief ore minerals are carnotite and tyuyamunite.

Production in Carbon County has been from the Boggs–Poison Basin area, where the deposits occur in the Brown Park formation of Miocene age, and from the Shirley Basin, where the deposits are in the Wind River formation. The Brown Park formation consists of conglomerate, cross-bedded sandstone, tuffaceous sandstone, and quartzite. The unoxidized ores contain pitchblende, coffinite, and pyrite; the oxidized ores contain uranophane and some other secondary minerals. In the Shirley Basin area pitchblende coats, cements, and replaces detrital grains in lenticular feldspathic sandstones of the lower part of the Wind River formation. The average thickness of the deposits is 4 to 5 ft, and the reserves are stated to be several million tons of ore containing 0.5 percent U_3O_8. In 1959 it was reported (76) that new ore reserves, stated to average 1 percent or more U_3O_8, had been developed in the Wind River formation of the Shirley Basin.

Some production in Wyoming has come from older rocks, such as the Madison limestone (Mississippian) in Big Horn County, and from Lower Cretaceous rocks of Crook County (Black Hills area).

Origin of peneconcordant deposits.   The origin of the sandstone-type or peneconcordant deposits has been a matter of debate for many years. The problem involves the source of the uranium, which is not obvious in many deposits, the manner of its transportation, and the cause of its deposition. In general two major concepts have been proposed regarding these deposits, (1) escape of uranium from granitic or other magmas as late residual fluids and subsequent concentration of the uranium in the sedimentary rocks and (2) leaching of uranium from detrital or igneous rocks by groundwater and subsequent deposition in sedimentary rocks. McKelvey, Everhart, and Garrels (29) reviewed discussions that had been presented to 1955 and suggested, regarding the deposits of the Colorado Plateau and the Witwatersrand that

> *they were formed from deepseated solutions, probably originating from igneous rocks that, for the most part, did not reach the surface. These solutions rose along fractures to permeable beds which they then followed. . . . We believe that this interpretation fits the facts of the occurrence of similar deposits in other important areas, such as the Black Hills area of South*

Dakota and Wyoming, and the newly discovered deposits in the Blind River area of southwestern Ontario.[1]

Since 1955 the published results of experimental work (21) on the solution of uranium showed that much uranium and vanadium could be taken into solution by groundwater. Other experimental work on sulfur isotopes (23) indicated that hydrogen sulfide of organic origin may have been an important precipitant of uranium in deposits of the Colorado Plateau. Suflur isotopes were investigated by numerous people, and this work has been summarized briefly by Miller and Kulp (30, p. 611).

The concept of origin by groundwater was discussed by Nininger and others in 1960 (33). They suggested (p. 48) that strong evidence for deposition by groundwater is presented by the basin or fluviatile environment, the fossil plant material, and the sulfur-isotopes ratios which indicate bacterial origin of the sulfides, and that these things are opposed to direct deposition from hydrothermal solutions of magmatic origin. Further, they stated that study of fracture systems on the Colorado Plateau and elsewhere have not proved that fractures served as passageways for uranium-bearing magmatic solutions.

In the same year (1960), Page (35) discussed the source of uranium in ore deposits, called attention to the presence of mafic dikes closely related in age to the uranium deposits in most uranium districts of the world, and stated that such dikes could indicate the presence of subjacent igneous bodies from which uranium-bearing solutions could have come. He also stated that spatial and geochemical correlations suggest that emanations from subjacent igneous masses may have been incorporated into groundwater and may have migrated to suitable sites of deposition. Further, because of the probable composition of waters at depth compared with those near the surface, he suggested that leaching by groundwater would be restricted to a near-surface environment. Surface waters, therefore, would function only as a means of leaching and redistributing hypogene uranium and associated metals in deposits of the Colorado Plateau type.

Opinion still remains divided. Perhaps the one idea that is generally accepted for deposits of the Colorado Plateau is that the presence of organic debris was an important factor in bringing about precipitation of uranium. Adler, in 1963 (9), discussing the oxidized sandstone-type deposits, stated: "The ores are in virtually all cases associated with organic debris that has been assumed to play a dominant role in the precipitation of the uranium as well as accompanying sulfide minerals."[2] Strong support for an origin by circulating groundwaters was presented in 1963 by Miller and Kulp (30). On the basis of isotopic evidence from samples of most of the typical Colorado Plateau ores they concluded, in part (p. 627), that the sulfur isotope data indicate that deposition took place at less than 90°C and that in general the deposits may be accounted for by deposition from extremely dilute solutions in the presence of H_2S at surface temperature. Two extracts from their paper follow:

[1] From V. E. McKelvey, D. L. Everhart, and R. M. Garrels, *Econ. Geology, Fiftieth Ann. Vol.*, pt. 1, p. 509. Copyright 1955. Economic Geology Publishing Company. Used by permission.
[2] From H. H. Adler, *Econ. Geology*, vol. 58, p. 840. Copyright 1963. Economic Geology Publishing Company. Used by permission.

The theory of origin of these deposits that is most probable and also consistent with the isotopic data involves deposition of uranium by ground water in H_2S-rich locations. This generally occurs shortly after sedimentation but may also occur later, if circulatory conditions are favorable. . . .

The initial source of uranium could have been hydrothermal solution from the basement, surface rock and soil, or the permeable strata of the present ore deposits. The low concentration of uranium needed for the formation of the deposits, however, does not require any more complex assumption than derivation from the permeable sedimentary rocks through which the water passed.[1]

Vein deposits. Uranium-bearing vein deposits in the United States (5, pp. 275–278, 283–287; 8, pp. 97–103; 13), although apparently relatively common, have not generally furnished much production, but a few of them have been consistent producers, among which are the veins of the Marysvale district, Utah, the most productive vein deposits in the United States.

Several types of vein deposits of uranium have been noted in various parts of the world. Although these differ in detail, Everhart and Wright (16) have attempted to compare and classify them on the basis of mineral composition and other features and arrived at the conclusion that they could be placed in two broad groups, as follows: (1) the uranium-bearing nickel-cobalt-native silver type, which contains a wide variety of metals, in which the gangue minerals are mostly carbonates, and in which the host rocks are typically meta-sediments; (2) the siliceous-pyrite-galena type, in which uranium is the chief metal present, in which the gangue minerals are fluorite and various forms of silica, and in which the host rocks are felsic intrusives. The second type is the one that has furnished most of the uranium from vein deposits in the United States.

Uranium-bearing veins of the United States are in the western part of the country. In a general way the greatest concentration of veins is around the margins of the Colorado Plateau and the Wyoming basins, although they extend outward from these areas for a considerable distance. Practically all of these were formed during Tertiary time and probably were closely related in time to Tertiary igneous activity. Characteristically the veins contain primary pitchblende, some iron and copper sulfides, galena, and commonly sphalerite, and some may contain molybdenum minerals. Fluorite and quartz are generally abundant, and hematite staining is characteristic. Oxidized zones may be conspicuous and contain secondary "sooty pitchblende," uranophane, autunite, torbernite, and various other secondary minerals.

The Marysvale district (1, pp. 350–351; 3, pp. 53–54; 5, 283–287; 8, pp. 123–129; 28) is the most productive area of uranium vein deposits in the United States. The principal deposits are about 4 miles northeast of Marysvale, Piute County, Utah, in and near the western edge of the High Plateaus division of the Colorado Plateaus, and are in a zone about a mile long and half a mile wide. The rocks of the immediate area are volcanics of early Tertiary age, intrusive monzonite, and rhyolite of late Tertiary age. The rocks were subjected to two stages of hydrothermal alteration, an alunitic stage prior to uranium mineralization and an argillic stage associated with uranium mineralization. The dominant host rock is the monzonite, in which faults, fractures, and breccia zones were mineralized by filling; however, some mineralization took place in fractures of early Tertiary volcanics and the late Tertiary rhyolite. Some veins are known to extend throughout a vertical distance of 1,500 ft. Veins pinch and swell and are a few inches to several feet thick. The primary uranium mineral is pitchblende, which is associated chiefly with pyrite, dark fluorite, quartz, and adularia. Umohoite, a newly discovered hydrous uranyl molybdate, formed later than the pitchblende, occurs in crosscutting veinlets in the main vein of one of the mines (Freedom No. 2) and is also regarded as of hypogene origin. Near-surface oxidized parts of veins may contain ore-grade concentrations of secondary uranium minerals— autunite, torbernite, uranophane, and others—as well as other secondary minerals.

Other vein deposits worthy of note, but which can be mentioned only briefly, are those of the Front Range, Colorado (8, pp. 97–103; 16, p. 88), the Coeur d'Alene district, Idaho (27), and the Thomas Range, Utah (5, pp. 275–278; 8, pp. 131–137). The pitchblende of the deposits of the Front Range commonly is a local constituent of veins that contain galena, sphalerite, copper minerals, gold, and silver. Uranium mineralization in the Coeur d'Alene district occurs as veinlets of uraninite associated with pyrite and jasper in the deeper levels of the Sunshine mine. It differs in age from the usual deposits of the western United States, being Precambrian. The uranium of the Thomas Range, Utah, is associated with the fluorspar deposits of that area, which can best be stated as uraniferous fluorspar from which uranium has not been recovered.

Canada

Production of uranium in Canada (6, p. ·70) has come from Port Radium and Raycock in the Northwest Territories, Beaverlodge in extreme northwestern Saskatchewan, and Blind River and Bancroft in Ontario. Distribution of uranium claims and mining properties of all types throughout Canada covers a much wider area, but about 80 percent of all properties are in the three provinces stated. Among them in 1961 and 1962 about 77 percent of the total production in Canada was from Ontario, about 20 percent was from Saskatchewan, and about 3 percent from the Northwest Territories.

The deposits of Canada have been grouped into three main types, conglomeratic, pegmatitic, and veins and related deposits, and reserves of each type estimated to the end of 1957 (6, p. 97). The reserves are shown by Table 13-9, along with estimates of the U_3O_8 content of the ore. The productive conglomeratic ore is entirely in the Blind River area, Ontario; the pegmatitic ore is chiefly in Saskatchewan, Quebec, and Ontario; and the vein and related deposits are chiefly in Saskatchewan, the Northwest Territories, and Ontario. Deposits similar to those of the Blind River area are known at intervals throughout the district of Sudbury, Ontario. They are of the same type and occur at the same stratigraphic horizon as those of the Blind River area but were of insufficient size and of too low grade to be minable under the marketing conditions of 1958 and 1959 (74).

TABLE 13-9.   Estimate of Reserves of Uranium Ore of Canada at the End of 1957*

	Short tons of ore				
Type of ore	Measured†	Indicated	Inferred	Total	Approximate percent U_3O_8
Conglomeratic	31,300,000	279,200,000	45,200,000	355,700,000	0.1
Pegmatitic	300,000	9,600,000	980,000	10,880,000	0.1
Veins and related deposits	180,000	10,000,000	28,000	10,308,000	0.2
Total and grade	31,780,000	298,900,000	46,208,000	376,888,000	0.103

* From A. H. Lang, J. W. Griffith, and H. R. Steacy, *Geological Survey of Canada, Econ. Geol. Series No.* 16, 1962, p. 97–98. Used by permission.

† The terms *measured, indicated,* and *inferred* are roughly equivalent to *proven, probable,* and *possible.*

The Blind River area, Ontario.   The Blind River area, Ontario (6, pp. 127–144; 25; 26; 39; 41), which is also referred to as the Algoma area or the Elliot Lake area, extends over about 200 sq miles immediately north of Lake Huron (Fig. 13-2). The area is one that has been developed very recently, since diamond drilling was not done until 1953 and production started in 1955 (73, 75). Development of new mines and a rapid rise in production followed until a peak was reached in 1959 (77). Since then production has decreased because of a large surplus of uranium in the western world. A shift to the greater use of atomic energy for peaceful purposes might be expected to again stimulate production some years in the future.

The area of the Blind River deposits is underlain by rocks of Precambrian age, which consist of an older, pre-Huronian complex of granite and granite-gneiss, along with some metavolcanic and metasedimentary rocks, and a younger, Huronian succession of sedimentary rocks that are only slightly metamorphosed in the area of the deposits. The Huronian rocks consist of a lower series, the Bruce, and an upper series, the Cobalt. The Huronian and older rocks are intruded by dikes, sills, and small bodies that vary in composition from silicic to mafic but that are chiefly the latter (quartz diabase and olivine diabase) and, in the southeastern part of the Blind River area, by a large body of granite. The Huronian rocks of the Blind River area have been folded into a broad northern syncline, commonly designated the Quirke Lake basin or trough (or the Algoma basin), and a southern broad anticline that terminates to the south against a regional thrust fault. Erosin has stripped off the core of the anticline, so that the contact between the Huronian and the pre-Huronian rocks forms a large Z in outline.

Most of the producing deposits known to date are in the Bruce series of the Quirke Lake basin. The Bruce series consists chiefly of quartzite, graywacke, argillite, and conglomerate, along with some limestone. Formerly it was subdivided into two groups, each of which was again subdivided into two formations. The commercial uranium deposits all occur in the lower division of the Bruce

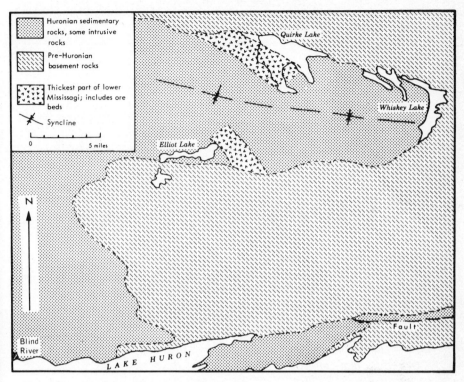

Fig. 13-2 Map showing generalized geology and location of uranium deposits, Blind River area, Ontario, Canada. Modified after the Geological Survey of Canada, from *Economic Geology Series No.* 16, 2d ed., 1962, p. 134. Used by permission.

series; the names used are shown in Table 13-10. Although the most recent work suggests the division of the Mississagi into two groups and four formations, many descriptions of the deposits refer only to the Mississagi formation, which is used with the older meaning. This older terminology will be used here and attention called to what would be the equivalent in the newer terminology.

Commercial uranium deposits in the Blind River area are known to occur only in the lower part of the Mississagi formation (Matinenda formation), within about 100 ft of the underlying basement rocks. This part of the Mississagi (Matinenda)

> *is composed of coarse-grained, poorly bedded clastic rocks, including quartz grit, feldspathic quartzite, arkose, and quartz pebble conglomerate; locally, pebbles and cobbles of greenstone and, more rarely, of granite are present in quartz pebble conglomerate at or near the base of the formation, constituting a basal conglomerate. The formation shows pronounced local variations in thickness as well as a general regional thickening from north to south (0 to 700 feet). . . . The formation is evidently of fluvial origin, deposited by streams which flowed in a southeasterly direction.*

TABLE 13-10. Nomenclature of the Lower Division of the Bruce Series, Blind River Area, Ontario, Canada*

Older terminology		Newer suggested terminology		Common terminology
Bruce Series	Mississagi formation	Mississagi group	Ten-mile formation	Upper Mississagi
			Whiskey formation	Middle Mississagi
		Elliot group	Nordic formation	Lower Mississagi
			Matinenda formation	

* From D. S. Robertson and N. C. Steenland, *Econ. Geology*, vol. 55, p. 664. Copyright 1960. Economic Geology Publishing Company. Used by permission.

The Matinenda formation is distinctly radioactive. . . . Closely packed quartz pebble conglomerate is particularly pyritic and radioactive with the radioactivity commonly due to disseminated high grade uranium minerals— "brannerite", uraninite, and "thucolite"—as well as to thorium in the more ubiquitous monazite and zircon. The thickest, coarsest-grained, most closely packed, and most uraniferous conglomerate beds are found in relatively thick parts of the formation, that is, within the overlying pre-Huronian "valleys." Two such "valley" structures contain most of the uranium ore deposits discovered to date in the area.[1]

The two "valley" structures to which Roscoe alludes are the two shown in Fig. 13-2 as the thickest part of the Lower Mississagi.

Typical ore bodies are composed chiefly of subrounded pebbles of vein quartz from about 0.25 to 2 in. in diameter, are 7 to 35 ft thick but commonly 10 ft, and vary considerably in length, although some have been traced for hundreds of feet and, by intermittent drill holes, apparently thousands of feet. Commercial ore throughout the area as a whole contains an average of about 0.12 percent U_3O_8. The chief uranium minerals present are brannerite and uraninite. Numerous other minerals are present, including uranothorite, coffinite, monazite, and zircon, but the most abundant mineral except the ordinary constituents of the conglomerate is pyrite, which generally constitutes less than 6 percent of the rock by weight but locally forms as much as 20 percent.

The Bancroft area, Ontario. The Bancroft area of Ontario (6, pp. 175–186), about 100 miles northeast of Toronto, is the only area in Canada where uranium is produced from pegmatitic deposits. Although the presence of uraninite in this area was known from at least 1922, production did not start until 1956. Within the area that contains producing mines there are three somewhat circular areas

[1] From "The Proterozoic in Canada," pp. 55–56, Stratigraphy Quirke Lake–Elliot Lake Sector, Blind River Area, Ontario, by S. M. Roscoe. Copyright 1957. University of Toronto Press. Used by permission.

of granitic complexes of Precambrian age, which have been designated the *Cheddar granite,* the *Cardiff plutonic complex,* and the *Faraday granite;* each of which is surrounded by metasedimentary and meta-igneous Grenville rocks. Radioactive deposits occur in dikes and lenses of pegmatitic granite and syenite, in zoned granite pegmatites, in various metasomatic bodies in the Grenville rocks, and in calcite-fluorite-apatite veins. Production has come only from pegmatitic granite and syenite around the margins of the three granitic masses at the Dyno mine (Cheddar granite), the Bicroft mine (Cardiff plutonic complex), and the Faraday and Greyhawk mines (Faraday granite). Commonly the producing bodies are complex, cut the wallrocks at some places and replace them at others, and contain irregularly distributed ore shoots, which are usually associated with concentrations of mafic minerals and magnetite. The ore minerals are uraninite and uranothorite and are generally accompanied locally by a variety of other minerals.

The Beaverlodge area, Saskatchewan. The Beaverlodge area, Saskatchewan (1, pp. 312–315; 3, pp. 48–50; 6, pp. 145–175; 12; 24; 44), was the leading producer of Canada before the exploitation of the Blind River deposits. The area is midway along the north shore of Lake Athabasca, about 450 miles by air north of Edmonton, Alberta. Gold was known in the area since about 1910, but mining did not start until 1934. Pitchblende was discovered in 1935. Extensive prospecting and drilling followed, and production began about 1953. Throughout the area there are roughly 3,000 vein-type occurrences of uranium, in many of which pitchblende has been identified.

The rocks of the area consist of the older Tazin group of metasedimentary and metavolcanic rocks, generally considered to be of Archean age, granitic and related rocks intrusive into or derived from those rocks, and the younger Athabasca group of relatively unmetamorphosed sedimentary and volcanic rocks, generally thought to be of Proterozoic age although originally classed as Cambrian. All of these rocks are cut by dikes and sills of gabbro and basalt. Faults are common throughout the area. Pitchblende deposits occur in the Tazin, the granitic, and the Athabasca rocks, but the major deposits have been found in the Tazin rocks, and two deposits of the area have been the most productive. One, the Gunnar, is a large disseminated deposit, and the other, the Ace-Verna, consists of relatively closely spaced stringers, lenses, small veins, and disseminations. Most productive deposits of the area are related spatially to prominent faults but generally occur in secondary structures close to them.

The ore body of the Gunnar mine is pipelike, plunges about 45°S, and extends at least 1,400 ft from the surface along the plunge. The ore minerals, pitchblende and uranophane, are disseminated throughout a mass of rock that approaches syenite in composition but contains much carbonate and apparently is a much-crushed, porous, alteration product of the granitic gneiss in which it is located. Other metallic minerals include hematite and traces of pyrite, chalcopyrite, and galena. Ore reserves at the end of 1957 were reported as 1,375,810 tons of ore averaging 3.4 lb of U_3O_8 a ton (0.17 percent U_3O_8) that could be mined from open pit and 1,800,000 tons of ore averaging 4 lb a ton (0.20 percent) that could be recovered by underground mining to the seventh level.

The Ace-Verna property consists of two deposits, the Ace and the Verna, which are 2 miles apart and connected by underground workings. The workable material at the Ace deposits consists of tabular and irregular masses that have been classed as breccia ore bodies and as vein ore. The breccia bodies are composed of breccia cemented by pitchblende, quartz, calcite, and chlorite, along with pitchblende-bearing stringers in fractures. One of the larger breccia bodies is 650 ft long, 5 to 30 ft wide, and has been partly explored to a depth of 1,800 ft. The vein ore bodies consist of networks and stringers of small pitchblende-bearing veins. The workable material at the Verna deposit consists of tabular, branching bodies that contain pitchblende in fractures from microscopic size to 4 in. wide.

The Northwest Territories.   The uranium deposits of the Northwest Territories (1, pp. 275–279; 3, pp. 46–48; 6, pp. 186–224; 14) are in the Great Bear Lake area, about 30 miles south of the Arctic Circle and 950 miles north of Edmonton. The Eldorado mine (Port Radium), discovered in 1930 and brought into production in 1933, was the oldest producer of Canada and for a number of years the world's second most important source of radium, the Shinkolobwe mine of Katanga, the Congo, being first. After a long period of production, however, the reserves became practically exhausted and the mine was closed in 1960. The Contact Lake mine, 9 miles southeast of the Eldorado mine, produced native silver and minor amounts of pitchblende between about 1936 and 1950. Production also has come from the Rayrock mine, 100 miles northwest of Yellowknife, from about 1957 to 1959, when the mine was closed.

The rocks of the Great Bear Lake area are of Precambrian age, and those exposed in and around the Eldorado mine are members of the Echo Bay group along with intrusive rocks. The Echo Bay group consists of tuffs, flow rocks, and sedimentary rocks, chiefly thinly banded cherty rocks along with bedded tuffs and coarser fragmental rocks. The rocks have been complexly deformed and are cut by numerous faults. The ore bodies occur almost entirely in the stratified rocks, especially the cherty ones, or where fracture zones follow the contacts of such rocks and early diabase. The ore bodies include narrow high-grade veins and lower-grade stockworks up to 40 ft wide. They occur irregularly within eight roughly parallel shear and fracture zones. Minable shoots have been 50 to 700 ft long.

These deposits are typical of the nickel–cobalt–native silver type (16) and contain, in addition to pitchblende, native silver and a large number of minerals, including various ones that contain nickel and cobalt.

Africa

The uranium of Africa (1, pp. 289–297, 319–324; 2, pp. 97–104, 132–138; 3, pp. 43–44, 56, 58, 83–84; 4, pp. 320–345, 350–368; 5, pp. 94–128; 10, pp. 299–300; 11; 15; 32; 38) occurs chiefly in the gold deposits of the Witwatersrand, Republic of South Africa, and the copper deposits of the Shinkolobwe mine, Katanga, Republic of the Congo. The gold and the copper deposits have been discussed previously. Here we will be concerned only with a few additional details regarding uranium.

All of the uranium produced in the Republic of South Africa is a by-product of gold mining except in a very few places where uranium in the conglomerate is a more valuable constituent than gold. The amount of uranium in the gold ores mined in 1959 (38) ranged from 0.31 to 1.51 lb of U_3O_8 per ton of ore, that is, about 0.015 to 0.075 percent U_3O_8. Known information (32) suggests that the reserve ore amounts to more than 1,100 million tons of indicated ore that should contain about 370,000 tons of U_3O_8. Recently published figures (early 1960s) of uranium-producing mines show 47,879,000 tons of ore reserves blocked out and available for mining, with an average value of 0.575 lb of U_3O_8 per ton (about 0.029 percent). In addition, there are 21,679,000 tons of ore in mine pillars, tailings dumps, and slimes deposited prior to the commencement of uranium production, which contain an average of 0.463 lb of U_8O_3 per ton (about 0.023 percent) and are being treated for recovery of uranium.

The Shinkolobwe mine is in the central part of the Katanga synclinorium, about 20 km (12.4 miles) west of Jadotville (Fig. 11-7). Although it is one of the Katanga copper deposits, it is the most westerly of the group and relatively unimportant as a copper producer. However, uranium was discovered there in 1915, and operations for the production of radium were begun in 1921. Open-pit mining to 1936, when operations were suspended, had yielded about 100,000 tons of radioactive minerals. Open-pit operations began again in 1944, underground operations in 1945, and production has been continuous since then.

The formation at the Shinkolobwe deposit, according to Derriks and Vaes (5, pp. 94–128), forms a spur of the Mine series (see Table 11-14), which is wedged into a fold-fault. Certain layers of siliceous dolomite and alternate dolomitic and graphitic schist apparently were favorable beds for mineralization. The structure is complex, since the rocks have been folded and faulted. The deposits are close to main faults but such faults are generally masked by a clay-talc breccia and are not mineralized. The uranium mineralization occurs in part as veins that follow the stratification. They have little continuity, however, and make a deposit resembling a stockwork. In addition to following bedding planes, mineralization extends into fissures, cavities, and along some secondary faults. Pitchblende marks the centers of the veins, and it is surrounded by gummite,[1] torbernite, and sklodowskite. Cobalt and nickel sulfides and their oxidation products are common associates, and a great many other minerals are present. The principal metals in the order of importance are uranium, cobalt, and nickel.

After very careful work, many of the details of which having been omitted from the foregoing description, Derriks and Vaes (5, p. 127) concluded, in part, that the deposit is of the vein type and is of pneumatolytic and hydrothermal origin; that the association of copper and uranium is not obvious, but that cobalt and nickel are intimately associated and nickel shows a special affinity for uranium. Deposition of uraninite was followed, in order, by (1) molybdenite, monazite, and some other minerals, (2) cobalt and nickel sulfides, then dolomite, and (3) copper.

Derriks and Oosterbosch (15) called attention to a small uraniferous deposit at Swambo, recently discovered west of Shinkolobwe. Mineralization, which is of

[1] Gummite is a fine-grained mixture of various secondary uranium minerals. Sklodowskite is a complex urano-silicate.

the vein type, is localized in the breccia and walls of a fault and is similar to that at Shinkolobwe, and it has been suggested that the Swambo deposit is perhaps an extension of the one at Shinkolobwe.

Uranium is known to occur at various mines in the Rhodesian Copperbelt (11), but prospecting to date has not been particularly successful.

Soviet Union

Several summaries of information available before 1950 about the uranium deposits of the Soviet Union were published between 1949 and 1958 (1, pp. 438–451; 2, pp. 187–199; 3, pp. 67, 80–81; 4, pp. 242–249; 10, pp. 310–313; 36; 42). The Fergana Valley area (roughly around 40°N, 72°E) and the area around Lake Baikal (roughly 52°N, 105°E) probably are the two most favoráble places for production of uranium ore. The most important present source of supply apparently is the general Fergana Valley area.

In the Fergana Valley area commercial production began at Tyuya Muyun (40°21′N, 72°00′E) in 1908, and between then and 1913 the amount of ore mined was 2,088,000 lb, reported to contain an average of 0.97 percent U_3O_8, 2.36 percent V, and 3.73 percent Cu. The deposit, as known in 1933, consisted of at least five ore-bearing veins of barite and over 30 barren veins of barite. The productive veins are near the center of the deposit and extend outward´ from the center as much as 1,500 m. The main vein was known to extend to a depth of 500 m. Ore bodies within the productive veins are a few centimeters to 1.5 m thick. The grade of run-of-the-mine ore is 1.5 percent U_3O_8 and ranges from 0.6 to 4.0 percent. The veins occur in metamorphosed Paleozoic limestone that was extensively modified by solution; some ore occurs in solution cavities. Some of the uranium-free barite veins contain radium and have the composition $(Ba,Ra)SO_4$. The ore minerals present are secondary ones of copper, vanadium, and uranium. The principal uranium mineral is tyuyamunite, which received its name from this deposit.

Uranium deposits are present in other parts of the Fergana Valley also. A deposit discovered in 1923 at Uigar-sai or Atbash (41°02′N, 71°12′E), on the northern side of the Fergana Valley, is stated to resemble closely the carnotite deposits of the Colorado Plateau. These deposits are reputed to be extensive. Other deposits in sandstone are known throughout the Fergana area.

Near Slyudyanka (51°40′N, 103°35′E), in the Lake Baikal area, the productive deposits contain niobium, tantalum, and uranium in the mineral betafite (locally designated *mendelyeevite* or *mendeleyevite*). This actually is a deposit of phlogopite mica, the betafite apparently occurring only in pegmatite veins. The area might be relatively unimportant for uranium, but niobium-tantalum-uranium ores associated with phlogopite deposits are known in Madagascar, and a number of other phlogopite deposits have been reported in the Soviet Union.

A number of other possible productive areas are known in northern and southern Siberia and central Kamchatka. The actual resources of the Soviet Union are not known by the Western world.

Other Countries

Australia and France each produce about 3 percent of the uranium of the world, so only brief mention of the deposits there will be made. Also, because of its historical significance, a brief discussion of the deposits at Joachimsthal, Czechoslovakia, will be presented. A recent report from Prague claims that a newly discovered deposit could make Czechoslovakia into the largest uranium producer of the world (79).

The known commercial deposits of Australia (1, pp. 253–254, 317–319; 2, pp. 107–112; 3, pp. 55–58; 4, pp. 380–394; 5, pp. 21–24, 91–92; 10, pp. 319–320; 19; 37) are in the Northern Territory and in South Australia. The Rum Jungle deposits of the Darwin-Katherine region, North Australia, about 60 miles south of Darwin, occur in folded and faulted carbonaceous shale, slate, and schist of Precambrian age. The deposits consist of replacements and quartz veinlets, in some of which uraninite is associated with chalcopyrite, pyrite, and a few other sulfides, and in others saleeite (a hydrous magnesium-uranium phosphate designated *autunite* in some reports) is the chief mineral. Secondary oxidized minerals are present in the upper parts of the deposits. At Radium Hill, South Australia, about 60 miles southwest of Broken Hill, several lodes containing davidite occur along fractures in gneiss and other rocks of Precambrian age. Associated minerals in the lodes are quartz, ilmenite, rutile, hematite, and biotite.

The uranium deposits of France (1, pp. 336–342; 2, pp. 158–168; 3, p. 54; 4, pp. 143–211; 5, pp. 152–161) occur chiefly in three areas in the Massif Central (south-central France), the Massif of Britanny (west-central France), and the Vosges (northeastern France). Among the producing deposits (4, p. 146), six are in the Massif Central and two are in the Massif of Brittany. In the Vendee district, Massif of Brittany, most of the deposits occur in the Mortagne granite close to its margins or along granite-schist contacts. In the Clisson area (4, pp. 192–196), about 22 km southeast of Nantes, at the l'Écarpière mine, the deposit consists of veins of fluorite and chalcedony that contain very small nodules and veinlets of pitchblende. Another group (4, pp. 196–198), the La Chapelle-Largeau, about 35 km southeast of the Clisson area and about 12 km east and slightly south of Mortagne, consists in part of pitchblende mineralization in siliceous veins and crushed zones. In the Limousin region of the western part of the Massif Central the most important deposits appear to be those of the LaCrouzille area about 18 km north of Limoges. At LaCrouzille (4, pp. 178–183), pitchblende is localized in fractures in lamprophyre (minette) dikes that cut granulitic granite. The pitchblende is in concretionary and spherulitic bodies, some of which are several cm across, and is accompanied by small amounts of various sulfides, especially pyrite and marcasite. In the Bois Noirs region (4, pp. 168–174) near St. Priest, about 33 km southeast of Vichy, in the Forez Mountains of the Massif Central, pitchblende occurs in silicified and pyritized breccia. About 10 km to the northwest of these deposits, in the Lachaux area (4, pp. 162–168), some of the deposits are chalcedonic pitchblende veins and others are the unusual parsonite[1] veins which, although interesting, appear to be of limited commercial importance.

[1] Parsonite is $Pb_2(UO_2)(PO_4)_2 \cdot 2H_2O \cdot UO_3$. See 1, pp. 536–538, for a description of these deposits, in English.

The Joachimsthal deposit, Czechoslovakia (1, pp. 284–286; 2, pp. 201–208; 3, pp. 44–46; 5, p. 31; 10, pp. 303–304), is of historic interest because the veins there furnished the material used by Pierre and Marie Curie when they discovered radium, and this deposit was an early source of radium, although the veins were mined chiefly for other metals. The deposit is of the nickel–cobalt–native silver type.

Vanadium

Most of the vanadium of the world is obtained as a by-product of the treatment of uranium ores, and the United States generally has supplied a little more than 70 percent of the total. Almost all of the rest has come from South-West Africa, Finland, and the Republic of South Africa (Fig. 13-3). Peru at one time was the chief producer, but rich reserves there were practically depleted in 1955. The vanadium resources of the world have been summarized very recently (47, 51).

United States

The vanadium deposits of the United States (47, 48) are chiefly in the Colorado Plateau, associated with the uranium deposits, but some production comes also from Wyoming and South Dakota from the uraniferous deposits there, and some comes from Idaho, where vanadium is obtained from ferrophosphorous that results from the electric production of elemental phosphorous. Production has come chiefly from southwestern Colorado, southeastern Utah, and northeastern Arizona. The known reserves of those areas are about 4,350,000 tons of ore that contains an average of 1.64 percent V_2O_5, or nearly 80 million lb of vanadium. Some deposits of Colorado that contain little uranium have been mined solely

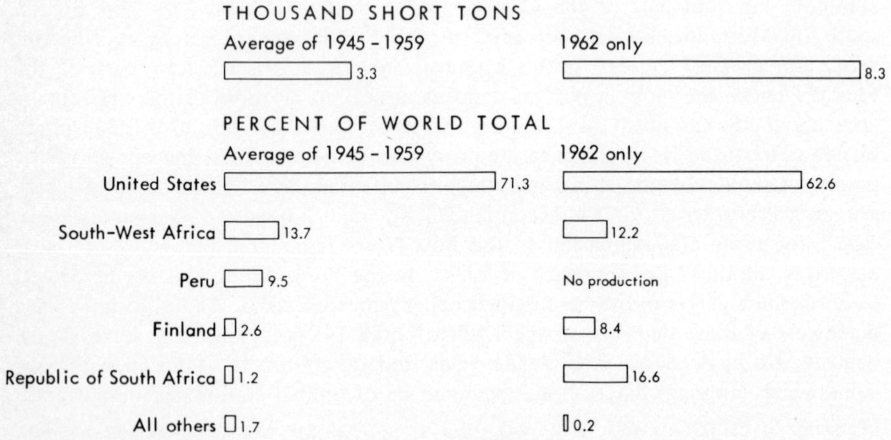

THOUSAND SHORT TONS

Average of 1945 - 1959	1962 only
3.3	8.3

PERCENT OF WORLD TOTAL

	Average of 1945 - 1959	1962 only
United States	71.3	62.6
South-West Africa	13.7	12.2
Peru	9.5	No production
Finland	2.6	8.4
Republic of South Africa	1.2	16.6
All others	1.7	0.2

Fig. 13-3 World vanadium production. Compiled from *U.S. Bur. Mines Minerals Yearbook*, various years.

for vanadium and those reserves, which average 1.5 to 2.0 percent V_2O_5, are stated to contain nearly 10 million lb of vanadium. Some of the other uranium ores, including the large reserves of New Mexico, average only about 0.13 percent V_2O_5. No effort is now made to recover vanadium from those ores, but the known deposits contain about 75 million lb of vanadium.

Other potential sources of vanadium in the United States are titaniferous magnetities, nontitaniferous magnetites, titanium-bearing veins, certain gold-quartz veins, certain base-metal deposits, vanadiferous phosphate and vanadiferous shale. Very little production has come from any of these sources.

Africa

The vanadium of Africa comes chiefly from South-West Africa, Zambia (Northern Rhodesia), and the Republic of South Africa.

The vanadium production of South-West Africa has come from mines near Otavi and Tsumeb to the west and northwest of Grootfontein, from the Abenab and Abenab-West mines to the north of Grootfontein (Fig. 16-1), and from the Berg-Aukus mines to the northeast of Grootfontein (47, 52). Recent production has been almost entirely from the Abenab-West and the Berg-Aukus mines. Production began in 1920 and has been continuous since then.

The deposits occur chiefly in dolomite of the Otavi system of Precambrian age, which rests on a basement complex composed of nearly vertical metasedimentary and metaigneous rocks that were extensively invaded by granite before the deposition of the rocks of the Otavi system. The rocks of the Otavi system in the southern part of the area have been intensely folded and faulted, those in the northern part only gently folded and not faulted. The deposits consist of vanadates of lead, zinc, and copper, intimately associated with oxidized sulfide ores of those metals. The chief vanadium minerals are vanadinite, descloizite, and mottramite. They occur in collapse breccias, tectonic breccias, solution cavities, and "sand-sacks." Many of the deposits, especially those in the sand-sacks, were shallow and were impoverished at slight depth, but some, as at the Abenab mine, extended to a depth of as much as 700 ft. The general nature of the deposits is exemplified by the occurrences at the Abenab and the Abenab-West mines, the Tsumeb and the Tsumeb-West mines, and the Berg-Aukus mine.

The ore body at the Abenab mine is a roughly cylindrical, steeply dipping mass of collapse breccia situated within platy limestone near its contact with massive dolomite. The interstices of the breccia are filled with the vanadates, calcite, and red clay. The ore body at Abenab-West is somewhat similar. The Tsumeb mine (72), active since 1900, is important chiefly for the production of lead, zinc, and copper, and has produced more than 2 million tons of those metals from hydrothermal deposits in the Precambrian dolomite. The deposit is a cylindrical pipe containing dolomite breccia and intrusive pseudoaplite and is the largest producer of lead in Africa. Oxidation of galena, sphalerite, tennantite, and several other hypogene minerals extended to about 1,200 ft, below which sulfides are dominant. In the upper part of the ore body there was very high-grade oxidized ore, especially rich in copper. The vanadium minerals—descloizite, mottramite, pyromorphite, mimetite, and vanadinite—occur in the upper part of the oxidized zone. By-product mottramite was a source of vanadium at the Tsumeb mine for many years. The ore bodies at Tsumeb-West are of the sand-sack type—highly irregular and oddly shaped sand-filled depressions in the surface of the dolomite. Vanadium ore occurs within the sand as concretionary nodules, either of pure, massive ore or as coatings on dolomite boulders contained in the sand. This type of deposit is the one most widely distributed throughout the area, but such deposits are shallow and ore is confined to the sand and walls of the depressions, although minor amounts have occurred in breccias or fractures in the limestone. The Berg-Aukas deposit apparently is a modification of the sand-sack type of ore body. An irregular pipe of smithsonite and calamine was wholly or partly surrounded by a calcareous clay that contained numerous boulders of dolomite, some of which were encrusted with vanadium ores. Locally vanadium minerals line the dolomite walls of the cavity. Away from that deposit there are a number of typical sand-sack deposits.

Various concepts of the origin of these vanadium deposits have been presented. The vanadium minerals are rather obviously of supergene origin, and the problem really concerns the source of the vanadium. Vanadium has not been found in the hypogene minerals at depth in the Tsumeb mine, nor were vanadium minerals found with the smithsonite and calamine in the central pipe at Berg-Aukas. Analyses of the Otavi sedimentary rocks, made on samples of fresh material collected some distance from known mineral deposits, show that the shales contain about 0.5 percent V_2O_5, and the dolomites about 0.2 percent. The total vanadium content of the Ovati sedimentary rocks probably is millions of tons. It has been suggested, therefore, that the formation of the vanadium deposits resulted from a combination of surface waters leaching vanadium from the Otavi sedimentary rocks and depositing it in suitable places around sulfide deposits of lead, zinc, and copper.

At Broken Hill, Zambia (Northern Rhodesia), vanadium is associated with lead and zinc in massive dolomite of the Roan series. Deposits are localized by intersecting fractures but are associated with collapse breccias and cavities in the dolomite. The vanadium minerals are chiefly vanadinite and descloizite.

Near Steelpoort in the Lyndenburg district, Republic of South Africa, vanadium is recovered from weathered material of the Bushveld complex (53). Stratiform sheets of vanadium-bearing titaniferous magnetite of the complex are stated to contain more than 25,000 short tons of recoverable vanadium. The weathered material that is mined contains about 1.6 percent V_2O_5. Production, which started in 1958, was slightly more than doubled in 1960 and much increased since then, so that in 1962 South Africa furnished 16.6 percent of the world production. An expansion program planned by the Transvaal Vanadium Co., Ltd., is aimed at making this the world's largest single producer of V_2O_5.

Finland

Vanadium ore is produced in Finland at Ontanmäki, a titaniferous iron mine, where titaniferous magnetite is reported to contain about 1 percent V_2O_5. The ore, a mixture of magnetite and ilmenite, is enclosed within an assemblage of amphibolites and gabbros. An expansion in production of about 40 percent is planned.

Soviet Union

Vanadium is contained in low-grade disseminated titaniferous magnetites of the Soviet Union, which occur in the central Ural Mountains. The average vanadium content of these ores is 0.1 to 0.15 percent V_2O_5. On Kachkanar Mountain, in the Urals, the largest open-pit mine in the Soviet Union was under development in 1962 and had begun operation, although a concentrating plant was far from completed. The ore consists of low-grade titaniferous magnetite that contains not more than 16 or 17 percent iron, 0.1 percent V_2O_5, and 1 percent TiO_2. Concentrates will contain 53 percent Fe, 0.4 percent V_2O_5, and about 4 percent TiO_2. This deposit contains about 80 percent of the entire iron ore reserves of the Urals and apparently is about the only commercial source of vanadium. The capacity of the plant, which should be reached in two or three years from 1962, is stated to be 6 million to 7 million metric tons of concentrate annually from 25 million metric tons of ore.

Peru

Vanadium-bearing asphaltite deposits are widespread throughout the Peruvian Andes (49, 50). These deposits were placed in three groups by Hernández Aquije (49), one on the east side of the Andean cordillera, one on the west side, and one that is intermediate between these two. Fissures, or in some cases openings between beds, chiefly Lower Cretaceous limestone, contain asphalt. The vanadium content of the asphalt ranges from about 0.3 to 1.0 percent, the highest values in general being in the group of deposits on the east side of the Andes. Within this eastern group was the world's richest vanadium deposit, Minasragra (Fig. 10-10), in the province of Pasco, which has furnished about a quarter of the total world production but was closed in 1955 because of the exhaustion of the ore body.

The Minasragra ore body consisted essentially of a mass resembling an inverted cone that occupied an area 850 by 180 ft at the surface and diminished at depth. The central part of the body was natural coke that enveloped shoots of patronite, a vanadium sulfide. The coke was surrounded by quisqueite, a hydrocarbon, and all of this material, in turn, by an envelope of shale impregnated with vanadium sulfide and some quisqueite, locally designated *bronce*. This shale, which is generally gray, grades outward into red shale of the surrounding area. Oxidation formed vanadates and oxides in the upper parts of the deposit, which was mined to a depth of 210 ft. It yielded many thousand tons of material that contained more than 6 percent vanadium.

Thorium, rare earths, titanium, and zirconium

General Relations

Most of the thorium, rare earths, titanium, and zirconium of the world are obtained from placer deposits, especially beach sands (57). The principal minerals from which these elements are obtained are monazite (thorium and rare earth constituents of the cerium group), ilmenite and rutile (titanium), and zircon (zirconium). These four minerals may all be present in a single deposit, but usually one or two of them are much more abundant than the others. Ilmenite is perhaps the most abundant one in most deposits. Most of the monazite produced up to 1903 came from beach sands of Brazil, and since then it has come from those sands and beach sands of India. Good concentrates of Brazilian monazite contained 5 to 6 percent ThO_2 whereas those of India contained 9 to 10 percent. Ilmenite is recovered from both of those areas—indeed, ilmenite is the main product, monazite a by-product. Ilmenite sands of Ceylon contain large amounts of ilmenite, lesser amounts of rutile and zircon, and some monazite. More than half of the total world output of rutile and zircon comes from beach and dune sands of Australia. Beach sands of the United States, in Florida, are mined for ilmenite but yield some by-product monazite, rutile, and zircon. Potential by-product sources of some of the minerals being considered are alluvial tin deposits of Nigeria, Malaya, and Indonesia and placer gold deposits at various places. Important future production of ilmenite, along with lesser amounts of rutile and zircon, should come from Africa.

During recent years the Malagasy Republic (Madagascar) has become an important producer of thorium from the mineral uranothorianite. Deposits were discovered there in 1953 and 1954 (4, pp. 358–367) and are known throughout a zone about 150 km long and 30 to 40 km wide in the southern part of the country, extending northward from Fort Dauphin and occupying, in part, the basin of the Mandrare River. The country rock of the area consists, in general, of Precambrian schists containing, along with some other rocks, lenses and masses of pyroxenite and "werneritite"—rocks composed to a large extent of wernerite. The uranothorianite deposits are concentrated almost wholly in the werneritites and pyroxenites, especially the latter, which appear to favor the axes of anticlines. The lenses of pyroxenite are generally a few meters thick, exceptionally 80 to 100 m, and from 10 m to several hundred meters long. Uranothorianite crystals, generally a few millimeters in size, rarely a centimeter, are disseminated in and make lenses in the pyroxenite and are associated with phlogopite, which is the chief mineral produced in the area. Weathering has released the uranothorianite in parts of the area, and both eluvial and alluvial deposits have been formed.

Vein or lode deposits, and carbonatites, supply important amounts of thorium and rare earths at some places, notably in Cape Province, South Africa (monazite), and at Mountain Pass, California (bastnaesite).

Two potentially important sources of thorium, if proper metallurgical processes of recovery can be devised, are (1) the Blind River, Ontario, uranium deposits, which possibly contain about one-third as much thorium as uranium in the minerals uranothorite and brannerite and (2) pyrochlore-bearing carbonatite complexes, as the pyrochlore may contain interesting amounts of thorium.

Brazil

Beach sands that contain more than usual amounts of heavy minerals occur along more than 1,000 miles in the states of Rio de Janiero, Espirito Santo, Bahia, Parabyba, and Rio Grande do Norte in Brazil (1, pp. 506–509; 3, pp.

92–94; 5, pp. 134–139; 57). The material comes from the erosion of lower Tertiary rocks, which in turn received it from Archean rocks. The most extensive reserves are in elevated or "fossil" bars up to 100 m wide. About 20 workable deposits are known, the largest of which are at Comaxatiba and Guaratiba in Bahia, at Guarapary in Espirito Santo, and at Barra do Itapopoana in Rio de Janiero. The heavy mineral content of the sands is roughly between 20 and 40 percent, of which 2 to 5 percent (exceptionally 20 percent) is monazite. Zircon is abundant but rutile is generally not present in recoverable amounts. The economic value of the sands depends on monazite and ilmenite (78), the ilmenite being about ten to fifteen times more abundant than the monazite.

Recently a thorium–rare earth deposit was discovered at Morro do Ferro in the Pocos de Caldas zirconium district of Brazil (68), about 230 miles west and somewhat north of Rio de Janeiro. Magnetite veins of a stockwork were enriched along their borders by thorium-bearing solutions that also deposited material in the highly fractured adjacent rocks. Weathering produced clayey material to a depth of as much as 100 m. The central part of the deposit is the richest and contains an average of more than 1 percent ThO_2 and more than 5 percent total rare-earth oxides. Much of the thorium apparently is present as thorogummite. Rare-earth elements occur chiefly in allanite.

India and Ceylon

Monazite-bearing sands occur along many parts of the coast of India (1, pp. 511–513; 3, pp. 94–95; 5, pp. 163–166; 57), but the only commercial production has come from the beaches of Travancore (80) where both high-grade beach sands and lower-grade dune deposits have been exploited. The two principal deposits are at Manavalakurichi, near Kovilam, and at Chavara, north of Quilon. At Manavalakurichi, Archean rocks weathered to laterites have supplied heavy minerals throughout a beach about a mile long. Ilmenite containing 55 percent TiO_2 and monazite containing 9.5 percent ThO_2 are the chief products. At Chavara, material derived from Tertiary sediments is ilmenite that contains 60 percent TiO_2 and monazite that contains 8.5 percent ThO_2. The dominant mineral is ilmenite, but zircon, rutile, and monazite are all recovered and marketed.

Along the west coast of Ceylon there are a number of small, rich beaches, especially at Kaikawela and Beruwela. In the northeast part of Ceylon, at Pulmoddai, there is a large deposit. The commercial heavy minerals are about 72 percent ilmenite (54.75 percent TiO_2), 18 percent zircon and rutile combined, and 0.3 to 0.4 percent monazite (8.9 to 9.2 percent ThO_2).

Australia

Very large deposits of heavy-mineral sands occur on beaches and dunes between Southport and Yamba, New South Wales, Australia (1, pp. 515–517; 57; 69). These sands are especially rich in zircon and rutile, and in many deposits each of these is as abundant as ilmenite. Monazite is rather rare and averages not more than 0.3 percent of the commercial heavy minerals. It contains about 6.6 percent ThO_2. The commercial products are chiefly rutile and zircon, with some monazite as a by-product. A recent report (81) states that about 90 percent of the world supply of rutile comes from Australia. The ilmenite generally has not been marketed because it contains about 1 percent Cr_2O_3 and is not acceptable to manufacturers of titanium pigments.

Africa

Along the Natal coast of Africa is a heavy mineral deposit known as Ilmenite Hill, which is a large dune (58) about 1.25 miles long and 0.25 mile wide. The top layer of the dune, 5 to 20 ft thick, consists of loamy soil, beneath which are yellow and white sands containing lenses of hard red clay. These sands extend downward for 100 ft or more. The heavy minerals are concentrated especially in the surface capping, but lesser concentrations do occur in lower horizons. A plant has been constructed 25 miles south of Durban to recover materials from this deposit. The annual production is expected to be 100,000 long tons of ilmenite, 6,500 long tons of rutile, and 9,000 long tons of zircon.

An important monazite vein is present at Steenkampskrall, northwestern Cape Province, about 50 miles north of Van Rhynsdorp and 200 miles north of Cape Town (1, pp. 325–326; 57). The deposit occurs along a mineralized shear zone in granite of Precambrian age. The vein pinches and swells, changing thickness from an inch or two to as much as 6 ft. The highest-grade material contains 60 to 75 percent monazite, which assays about 4 to 5 percent ThO_2.

United States

Thorium and rare earths in the United States (3, pp. 97–100; 5, pp. 562–567; 62) come chiefly from lode deposits of the Mountain Pass district, California (1, pp. 241–243; 61), first in 1962 production; from beach sands of Florida, second in 1962 production, and North and South Carolina (1, pp. 497–505; 59; 63); and from placers in Idaho (55, 56, 67). Some of these areas, notably Florida, also furnish ilmenite, rutile, and zircon. Much ilmenite comes from titaniferous magnetites of New York, first in 1962 production, and both ilmenite and rutile come from Virginia (64, 65, 66).

Bastnaesite, $(Ce,La)FCO_3$, that generally contains thorium, is present in a large mass of carbonate in the Mountain Pass district, San Bernardino County, California, very near the Nevada boundary. A complex of metamorphic and igneous rocks of Precambrian age was intruded by potash-rich igneous rocks, also probably of Precambrian age. Among those igneous rocks is a composite shonkinite-syenite body 6,300 ft long, in and near the southwest side of which are abundant carbonate veins, most of which are less than 6 ft thick. One huge mass of carbonate rock, however, near the old Sulphide Queen gold mine and mill, has a maximum width of 700 ft and a length of 2,400 ft. This has been designated the *Sulphide Queen carbonate body*, not because of its sulfide content but because of its proximity to the old Sulphide Queen mine. Both this body and the veins, about 200 of which have been mapped, consist of about 60 percent carbonates, chiefly calcite, dolomite, ankerite, and siderite. The other constituents are barite, bastnaesite, parisite, $(Ce,La)_2Ca(CO_3)_3F_2$, quartz, and variable amounts of many other minerals. The rare-earth mineral content of the carbonate generally is from 5 to 15 percent, but locally it is as much as 40 percent. This Sulphide Queen carbonate body, discovered in April, 1949, is stated to be the greatest concentration of rare-earth minerals known in the world. It is a carbonatite, apparently formed as a carbonate-rich end product of the differentiation of alkaline magma and intruded into the surrounding rocks.

Mining of beach sand at Jacksonville, Florida (59), and of raised barrier beaches at Trail, Florida, yields ilmenite, rutile, monazite, and zircon. Ilmenite is the most abundant constituent recovered at both places, but considerable amounts of rutile and zircon are present, along with some monazite. Mining of these deposits brought Florida into first place in 1962 for the production of rutile and zircon and into second place for ilmenite and rare-earth minerals.

Monazite mined from placers in the Carolina Piedmont (8, pp. 497–601; 63) caused that area to be a major world source of monazite from about 1886 to 1895, before competition from Brazilian monazite began. Only a very small amount of mining was done in the region from 1911 until 1954, when one dredge began operation. The deposits in general were not minable under the prices that prevailed in 1958, but they are an important potential source of monazite and zircon. The deposits occur (63, pp. 709, 712) in alluvial sediments on the floors of shallow, narrow valleys. Recent work has indicated that 84 floodplains of streams between the Savannah River, South Carolina, and the Catawba River, North Carolina, can be classed as placers in which the average tenor is 1.3 lb of monazite and 0.6 lb of zircon per cu yd. The monazite may contain an average of 5.67 percent ThO_2 and 0.38 percent U_3O_8. The zircon may contain about 0.01 percent ThO_2 and 0.04 percent U_3O_8.

Virginia was second in production of rutile in the United States in 1962 and third in the production of ilmenite. The deposits have been an important source of rutile for many years and have also furnished a considerable amount of ilmenite. These deposits (64, 66), which are present in eight counties of Virginia but are most important in Amherst and Nelson Counties, occur in nelsonite dikes associated with anorthosite. These dikes are composed chiefly of ilmenite and apatite, with rutile the third most abundant consistuent. Other minerals present in varying amounts are amphibole, pyroxene, biotite, and, uncommonly, magnetite. Ilmenite, and especially rutile, also occur disseminated throughout rock that has been designated anorthosite and several other types. The rutile, present in concentrations up to 10 percent but more commonly 4 or 5 percent, is generally associated with apatite in the more feldspathic parts of the rock.

Thorium-rich veins are present at various places in Colorado, Idaho, and Montana, but one of the most important areas is around the Idaho batholith, where both veins and placers occur. In that area a number of placer deposits are present but have been worked most extensively at Bear Valley and Cascade, in Valley County. In general the Idaho placers are monazite-bearing but some, especially at Bear Valley, contain euxenite and columbite also (1, pp. 501–504; 67). Vein and lode deposits occur especially in Lemhi County in the Lemhi Pass district, which extends into adjoining Beaverhead County, Montana, but they also occur in other parts of Lemhi County (54, 55, 56, 60). The deposits in the Lemhi Pass area occur in veins and lodes that traverse rocks of the Precambrian Belt series and are localized along simple to complex shear and fracture zones. The veins of the area, all of which contain quartz, have been classed as those that contain (1) copper-bearing sulfides, (2) hematite, (3) barite, hematite, and thorite, and (4) copper-bearing sulfides and thorite. Recent discoveries of greatest promise are not associated with the copper-bearing veins but occur along independent zones that are essentially mineralized shear and fracture zones along which both filling and replacement occurred. Thorite is the mineral of economic importance and the one that is most widespread, but some monazite and allanite are present also. The thorite apparently contains a considerable amount of rare-earth elements and a small amount of uranium. Production has started from some of the deposits, and the area apparently contains a major concentration of thorium.

SELECTED REFERENCES
MAJOR PUBLICATIONS OF WIDE SCOPE

1. E. W. Heinrich, 1958, "Mineralogy and Geology of Radioactive Raw Materials," McGraw-Hill Book Company, New York, 654 pp.
2. E. Kohl, 1954, Uran, "Die Metallischen Rohstoffe," no. 10, Ferdinand Enke, Stuttgart, 234 pp.
3. R. D. Nininger, 1956, "Minerals for Atomic Energy; a Guide to Exploration for Uranium, Thorium, and Beryllium," 2d ed., D. Van Nostrand Company, Inc., Princeton, N.J., 399 pp.
4. M. Roubault, 1958, "Géologie de l'uranium," Masson et Cie, Paris, 462 pp.
5. Various authors, 1955, Geology of uranium and thorium, Proc. Internat. Conf. on the peaceful uses of atomic energy, vol. 6, Geneva, Switzerland, 1955, 825 pp. United Nations, New York, 1956.
6. A. H. Lang, J. W. Griffith, and H. R. Steacy, 1962, Canadian deposits of uranium and thorium, *Canada Geol. Survey Econ. Geology Ser. No. 16*, 2d ed., 324 pp. Contains summary of production and reserves of the world as well as discussion of Canadian deposits.
7. R. M. Garrels and E. S. Larsen, 3d, 1959, Geochemistry and mineralogy of the Colorado Plateau uranium ores, *U.S. Geol. Survey Prof. Paper* 320, 236 pp.
8. L. R. Page, H. E. Stocking, and H. B. Smith, 1956, Contributions to the geology of uranium and thorium by the United States Geological Survey and Atomic Energy Commission for the United Nations international conference on peaceful uses of atomic energy, Geneva, Switzerland, 1955, *U.S. Geol. Survey Prof. Paper* 300, 739 pp.

CHIEFLY URANIUM

9. H. H. Adler, 1963, Concepts of genesis of sandstone type uranium ore deposits, *Econ. Geology*, vol. 58, pp. 839–852.
10. G. W. Bain, 1950, Geology of the fissionable materials, *Econ. Geology*, vol. 45, pp. 273–323.
11. S. H. U. Bowie, 1959, Note on uranium and thorium occurrences in the Federation of Rhodesia and Nyasaland (Notes sur les gites d'uranium et de thorium dans la Federation des Rhodesie et du Nyasaland), *Chronique Mines Outre-Mer*, vol. 27, no. 279, pp. 304–308, (English and French).

12. B. S. W. Buffam, D. D. Campbell, and E. E. N. Smith, 1957, Beaverlodge mines of Eldorado Mining and Refining, Ltd., in "Structural Geology of Canadian Ore Deposits, Congress Volume," pp. 220–240, Min. Metall. Cong., 6th, Canada 1957, Canadian Institute of Mining and Metallurgy, Montreal.

13. A. P. Butler, Jr., W. I. Finch, and W. S. Twenhofel, 1962, Epigenetic uranium deposits in the United States, *U.S. Geol. Survey Mineral Inv. Resource Map* MR-21. Map and pamphlet of 42 pp.

14. D. D. Campbell, 1957, Port Radium mine, in "Structural Geology of Canadian Ore Deposits, Congress Volume," pp. 177–189, Min. Metall. Cong., 6th, Canada 1957, Canadian Institute of Mining and Metallurgy, Montreal.

15. J. J. Derriks and R. Oosterbosch, 1956, Les gites de Swambo et de Kalongwe comparés à Shinkolobwe (The Swambo and Kalongwe deposits compared with that of Shinkolobwe), *Chronique Mines Outre-Mer*, vol. 27, no. 279, pp. 300–303 (French and English).

16. D. L. Everhart and R. J. Wright, 1953, The geologic character of typical pitchblende veins, *Econ. Geology*, vol. 48, pp. 77–96.

17. W. I. Finch, 1959, Peneconcordant uranium deposit—a proposed term, *Econ. Geology*, vol. 54, pp. 944–946.

18. R. P. Fischer and L. S. Hilpert, 1952, Geology of the Uravan mineral belt, *U.S. Geol. Survey Bull.* 988-A, pp. 1–13.

19. N. H. Fisher and C. J. Sullivan, 1954, Uranium exploration by the Bureau of Mineral Resources, Geology and Geophysics in the Rum Jungle Province, Northern Territory, Australia, *Econ. Geology*, vol. 49, pp. 826–836.

20. H. C. Granger, E. S. Santos, and others, 1961, Sandstone-type uranium deposits at Ambrosia Lake, New Mexico—an interim report, *Econ. Geology*, vol. 56, pp. 1179–1210.

21. J. W. Gruner, 1956, Concentration of uranium in sediments by multiple migration-accretion, *Econ. Geology*, vol. 51, pp. 495–520.

22. L. S. Hilpert and R. H. Moench, 1960, Uranium deposits of the southern part of the San Juan Basin, New Mexico, *Econ. Geology*, vol. 55, pp. 429–464.

23. M. L. Jensen, 1958, Sulfur isotopes and the origin of sandstone-type uranium deposits, *Econ. Geology*, vol. 53, pp. 598–616.

24. A. W. Jolliffe and E. P. Evoy, 1957, Gunnar mine, in "Structural Geology of Canadian Ore Deposits, Congress Volume," pp. 240–246, Min. Metall. Cong., 6th, Canada 1957, Canadian Institute of Mining and Metallurgy, Montreal.

25. F. R. Joubin and D. H. James, 1956, Uranium deposits of the Blind River district, Ontario, *Mining Eng.*, vol. 8, pp. 611–613.

26. F. R. Joubin and D. H. James, 1957, Algoma uranium district, in "Structural Geology of Canadian Ore Deposits, Congress Volume," pp. 305–316, Min. Metall. Cong., 6th, Canada 1957, Canadian Institute of Mining and Metallurgy, Montreal.

27. P. F. Kerr and R. F. Robinson, 1953, Uranium mineralization in the Sunshine mine, Idaho, *Mining Eng.*, vol. 5, pp. 495–511.

28. P. F. Kerr, G. P. Brophy, and others, 1957, Marysvale, Utah, uranium area, *Geol. Soc. America Spec. Paper* 64, 212 pp.

29. V. E. McKelvey, D. L. Everhart, and R. M. Garrels, 1955, Origin of uranium deposits, *Econ. Geology*, 50th Ann. Vol., pt. 1, pp. 464–533.

30. D. D. Miller and J. L. Kulp, 1963, Isotopic evidence on the origin of the Colorado Plateau uranium ores, *Geol. Soc. America Bull.*, vol. 74, pp. 609–630.

31. V. A. Mrak, 1958, Uranium deposits in the Tertiary sediments of the Powder River Basin, Wyoming, 13th Ann. Field Conf., Powder River Basin 1958, "Wyoming Geological Association Guidebook," pp. 233–240.

32. L. T. Nel, 1959, Uranium and thorium in the Union of South Africa (Uranium et thorium en Union Sud Africaine), *Chronique Mines Outre-Mer*, vol. 27, no. 279, pp. 325–331, (French and English).

33. R. D. Nininger, D. L. Everhart, and others, 1960, The genesis of uranium deposits, Internat. Geol. Cong., 21st, Copenhagen 1960, pt. 15, Genetic problems of uranium and thorium deposits, pp. 40–50.

34. E. A. Noble, 1960, Genesis of uranium belts of the Colorado Plateau, Internat. Geol. Cong., 21st, Copenhagen 1960, pt. 15, Genetic problems of uranium and thorium deposits, pp. 26–39.

35. L. R. Page, 1960, The source of uranium in ore deposits, Internat. Geol. Cong., 21st, Copenhagen 1960, pt. 15, Genetic problems of uranium and thorium deposits, pp. 149–164.

36. J. Paone, 1957, Uranium, *U.S. Bur. Mines Minerals Yearbook*, vol. 1, pp. 1234–1251.

37. L. W. Parkin and K. R. Glasson, 1954, Geology of the Radium Hill uranium mine, South Australia, *Econ. Geology*, vol. 49, pp. 815–825.

38. E. T. Pinkney and R. J. Westwood, 1961, Uranium in South Africa, Commonwealth Min. Metall. Cong., 7th, South Africa 1961, *Trans.*, vol. 3, pp. 1023–1038.

39. D. S. Robertson and N. C. Steenland, 1960, On the Blind River uranium ores and their origin, *Econ. Geology*, vol. 55, pp. 659–694.

40. C. S. Robinson and G. B. Gott, 1958, Uranium deposits of the Black Hills, South Dakota and Wyoming, 13th Ann. Field Conf., Powder River Basin, "Wyoming Geological Association Guidebook," pp. 241–244.

41. S. M. Roscoe, 1957, Geology and uranium deposits, Quirke Lake-Elliot Lake, Blind River area, Ontario, *Canada, Geol. Survey Paper* 56-7, 21 pp.

42. D. B. Shimkin, 1949, Uranium deposits in the U.S.S.R., *Science*, vol. 109, January 21, pp. 58–60.

43. South Africa Department of Mines, and Geological Survey, 1959, Uraninite in the Witwatersand and associated systems, pp. 321–329, in "The Mineral Resources of the Union of South Africa," 4th ed., Pretoria, South Africa.

44. L. P. Tremblay, 1957, Ore deposits around Uranium City, in "Structural Geology of Canadian Ore Deposits, Congress Volume," pp. 211–220, Min. Metall Cong., 6th, Canada 1957, Canadian Institute of Mining and Metallurgy, Montreal.

45. W. H. Wilson, 1960, Radioactive mineral deposits of Wyoming, *Wyoming Geol. Survey Rept. Inv. No. 7*, 41 pp.

46. H. B. Wood and M. A. Lekas, 1958, Uranium deposits of the Uravan mineral belt, "Guidebook to the geology of the Paradox Basin," pp. 208–215, Intermountain Association of Petroleum Geologists.
47. R. P. Fischer, 1963, chapters on Geochemistry and Geology of Vanadium, and Resources, in Vanadium, a materials survey, *U.S. Bur. Mines Inf. Circular* 8060, 95 pp.
48. R. P. Fischer, 1962, Vanadium in the United States, *U.S. Geol. Survey Mineral Inv. Resource Map* MR-16. Map and pamphlet of 8 pp.
49. S. Hernández Aquije, 1954, 1955, El vanadio en el Perú, *Mineria*, Lima, vol. 2, no. 5, pp. 19–30; no. 6, pp. 31–49; vol. 4, no. 8, pp. 21–22; no. 9, pp. 23–29.
50. H. E. McKinstry, 1957, Review of El Vanadio en el Perú, *Econ. Geology*, vol. 52, pp. 324–325.
51. P. Nicolini, 1960, Recherches bibliographiques sur la géologie du vanadium, *Chronique Mines Outre-Mer*, vol. 28, no. 291, pp. 1–36.
52. C. M. Schwellnus, 1946, Vanadium deposits in the Otavi Mountains, South West Africa, *Geol. Soc. South Africa Trans.*, vol. 48, pp. 49–73.
53. C. M. Schwellnus and J. Willemse, 1944, Titanium and vanadium in the magnetite iron ores of the Bushveld Complex, *Geol. Soc. South Africa Trans.*, vol. 46, pp. 23–38.

**CHIEFLY THORIUM, RARE EARTHS,
TITANIUM, ZIRCONIUM**

54. A. L. Anderson, 1958, Uranium, thorium, columbium, and rare earth deposits in the Salmon region, Lemhi County, Idaho, *Idaho Bur. Mines and Geology, Pamphlet No.* 115, 81 pp.
55. A. L. Anderson, 1961, Thorium mineralization in the Lemhi Pass area, Lemhi County, Idaho, *Econ. Geology*, vol. 56, pp. 177–197.
56. A. L. Anderson, 1961, Geology and mineral resources of the Lemhi Quadrangle, Lemhi County, Idaho, *Idaho Bur. Mines and Geology, Pamphlet No.* 124, 111 pp.
57. C. F. Davidson, 1956, The economic geology of thorium, *Mining Mag. (London)*, vol. 94, no. 4, pp. 197–208.
58. G. Langton and E. J. Jackson, 1961, Recovery of ilmenite, rutile and zircon at Umgaba, Commonwealth Min. Metall. Cong. 7th, South Africa 1961, *Trans.*, vol. 3, pp. 1073–1091.
59. J. B. Mertie, Jr., 1958, Zirconium and hafnium in the southeastern Atlantic states, *U.S. Geol. Survey Bull.* 1082-A, pp. 1–28.
60. J. Newton, D. LeMoine, and others, 1960, Study of two Idaho thorite deposits, *Idaho Bur. Mines and Geology, Pamphlet No.* 122, 53 pp.
61. J. C. Olson and others, 1954, Rare-earth mineral deposits of the Mountain Pass district, San Bernardino County, California, *U.S. Geol. Survey Prof. Paper* 261, 75 pp.
62. J. C. Olson and J. W. Adams, 1962, Thorium and rare earths in the United States, exclusive of Alaska and Hawaii, *U.S. Geol. Survey Mineral Inv. Resource Map* MR-28. Map and pamphlet of 16 pp.
63. W. C. Overstreet, P. K. Theobald, Jr., and J. W. Whitlow, 1959, Thorium and uranium resources in monazite placers of the western Piedmont, North and South Carolina, *Mining Eng.*, vol. 11, pp. 709–714.

64. A. A. Pegau, 1950, Geology of the titanium-bearing deposits in Virginia, in "Symposium on Mineral Resources of the Southeastern United States," pp. 49–55, University of Tennessee Press, Knoxville, Tenn.
65. C. L. Rogers and M. C. Jaster, 1962, Titanium in the United States, exclusive of Alaska and Hawaii, *U.S. Geol. Survey Mineral Inv. Resource Map* MR-29. Map and pamphlet of 18 pp.
66. C. S. Ross, 1941, Occurrence and origin of the titanium deposits of Nelson and Amherst counties, Va., *U.S. Geol. Survey Prof. Paper* 198, 59 pp.
67. C. N. Savage, 1960, Nature and origin of central Idaho blacksands, *Econ. Geology,* vol. 55, pp. 789–796.
68. H. Wedow, Jr., 1961, Thorium and rare earths in the Pocos de Caldas zirconium district, Brazil, *U.S. Geol. Survey Prof. Paper* 424-D, pp. 214–216.
69. Anon., 1962, The beach sand deposits of Queensland, *Mining Jour. (London),* vol. 259, no. 6640, pp. 488–489, 491.

MISCELLANEOUS REFERENCES CITED
70. Y. W. Isachsen, T. W. Mitcham, and H. W. Wood, 1955, Age and sedimentary environments of uranium host rocks, Colorado Plateau, *Econ. Geology,* vol. 50, pp. 127–134.
71. S. M. Roscoe, 1957, Stratigraphy, Quirke Lake–Elliot Lake sector, Blind River area, Ontario, in "The Proterozoic in Canada," pp. 55–56, *Royal Soc. Canada Spec. Pub. No.* 2, University of Toronto Press.
72. Staff of the Tsumeb Corporation, 1961, Geology, mining methods and metallurgical practice at Tsumeb, Commonweath Min. Metall. Cong., 7th, South Africa 1961, *Trans.,* vol. 1, pp. 159–179.
73. H. S. Strouth, 1955, Canada's new uranium camp at Blind River, *Mining Eng.,* vol. 7, pp. 462–463.
74. J. E. Thomson, 1960, Uranium and thorium deposits at the base of the Huronian system in the district of Sudbury, *Ontario Dept. Mines, Geol. Rept. No.* 1, 40 pp.
75. Anon., 1955, Canada taps Blind River's uranium, *Eng Mining Jour.,* vol. 156, no. 6, pp. 95–97.
76. Anon., 1959, Huge uranium ore find is reported, *Eng. Mining Jour.,* vol. 160, no. 3, p. 114.
77. Anon., 1959, At Elliot Lake in Canada, *Mining World,* vol. 21, no. 7, pp. 39–40.
78. Anon., 1962, Ilmenite and monazite in Brazil, *Mining Jour. (London),* vol. 258, no. 6597, p. 91.
79. Anon., 1963, Czechs to become biggest uranium producers?, *Mining Jour. (London),* vol. 261, no. 6679, p. 173.
80. Anon., 1963, Minerals from the beach sands of Travancore, *Mining Jour. (London),* vol. 260, no. 6647, p. 29.
81. Anon., 1963, Markets for Australian beach sand producers, *Mining Jour., (London),* vol. 261, no. 6688, p. 397.

chapter **14**

Iron, Magnanese, and Aluminum

Deposits of iron, manganese, and aluminum show many features that are similar. The major commercial deposits of each metal are of sedimentary origin, but the major deposits of aluminum were formed as residual accumulations whereas those of iron and manganese were formed not only as residual accumulations but also as chemically deposited sedimentary beds, some of which were later modified by weathering and the circulation of groundwater. The major similarities and differences among the deposits of these three metals are shown by Table 14-1.

Iron

Some Features of Iron and Steel Production

Iron is the most important metal in modern economic development, even though we are living in the Atomic Age. Iron is the foundation on which the steel industry has been built and without which it would collapse. "The importance of iron ore is emphasized by the fact that approximately 100 million long tons of steel is consumed annually in the United States, as compared with a total consumption of about 5 million long tons of copper, lead, zinc, and aluminum." (14, p. 62). Further, steel manufacturing not only uses iron ore but also various metals in making alloys and large quantities of limestone and fluorite in processing operations. Among metals other than iron used in making steel of various types, manganese is of first importance, followed by silicon, chromium, and nickel. Many other chemical elements are employed, designated *secondary alloying elements*. These, in the general order of tonnage used, are aluminum, molybdenum, copper, phosphorous, tungsten, titanium, vanadium, cobalt, zirconium, niobium, tantalum, selenium, boron, calcium, and cerium.

The production of steel generally parallels the production of iron ore throughout the world, as indicated by Figs. 14-1 and 14-2, which show that for the 15-year period 1945 to 1959 the United States and the Soviet Union led the world in the production of iron ore and also in the production of steel. Further, France the United Kingdom, and West Germany all were important producers of iron ore

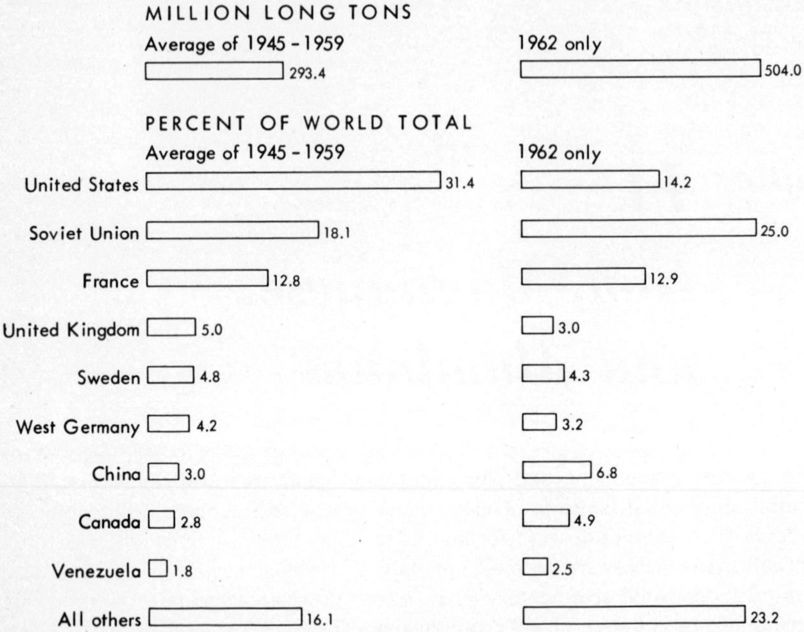

MILLION LONG TONS

Average of 1945 – 1959	1962 only
293.4	504.0

PERCENT OF WORLD TOTAL

	Average of 1945 – 1959	1962 only
United States	31.4	14.2
Soviet Union	18.1	25.0
France	12.8	12.9
United Kingdom	5.0	3.0
Sweden	4.8	4.3
West Germany	4.2	3.2
China	3.0	6.8
Canada	2.8	4.9
Venezuela	1.8	2.5
All others	16.1	23.2

Fig. 14-1 World iron ore production. Compiled from *U.S. Bur. Mines Minerals Yearbook,* various years.

MILLION SHORT TONS

Average of 1945 – 1959	1962 only
232.9	397.3

PERCENT OF WORLD TOTAL

	Average of 1945 – 1959	1962 only
United States	40.5	24.7
Soviet Union	16.3	21.1
United Kingdom	8.1	5.7
West Germany	6.7	9.0
France	4.7	4.8
Japan	3.3	7.6
China	1.3	3.3
All others	19.1	23.8

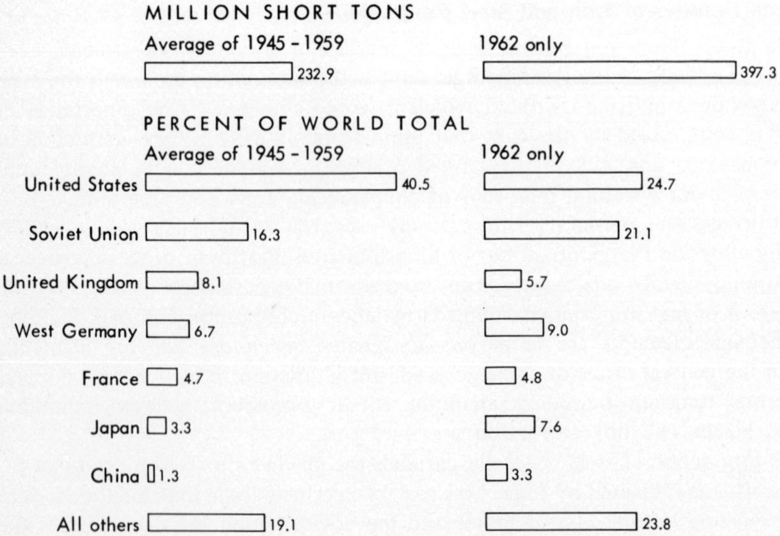

Fig. 14-2 World steel production. Compiled from *U.S. Bur. Mines Minerals Yearbook,* various years.

TABLE 14-1. Major Similarities and Differences among Deposits of Iron, Manganese, and Aluminum

Deposit	Major minerals	Origin and importance				
		Sedimentary beds, little modified	Sedimentary beds modified by weathering, groundwater	Residual	Magmatic	Contact-metasomatic
Iron	One or more: hematite, limonite,* siderite, chamosite, other silicates	Important as concentrating ores	Highly important source of rich ore; carbonates, silicates changed to oxides	Important; carbonates, silicates changed to oxides	Perhaps some minor deposits of hematite	Minor deposits of hematite
	Magnetite	Uncertain; probably minor		Perhaps some deposits	Some important deposits	Some important deposits
Manganese	Rhodochrosite	Generally too low grade	Important source of enriched ore	Highly important		
	Cryptomelane, psilomelane, pyrolusite, wad,*manganite	Highly important	Highly important			
Aluminum	Bauxite*			Highly important		

* Limonite, wad, and bauxite are not true mineral species but mixtures.

355

and also of steel, although the rank in production differed. Important steel industries do not always develop near the source of iron ore, however, because other materials, such as coal and limestone, which are bulky and expensive to transport, may not be available.

Certain noteworthy changes have been taking place during late years, in the production of both iron ore and steel, as indicated by the production in 1962. This has resulted in part from the effort of certain countries, such as the Soviet Union, China, and Japan, to bring about a marked increase in the production of steel. Steel-producing capacity has increased to such an extent that the American Iron and Steel Institute concluded that there would be a substantial world-wide surplus of steel-making capacity for some time to come, which would result in a seriously unbalanced supply-demand position. A long-range projection by the United Nations Steel Committee had indicated a substantial increase in steel consumption by 1972 to 1975, but this forcast had not proved to be realistic in 1963. World production of iron ore, however, has continued to rise and reached 504,012,000 gross tons in 1962 compared with 497,610,000 tons in 1961. Much of the 1962 increase came from the Soviet Union, the 1962 production being about 25 percent of the world total (208).

Interesting changes in the production of iron ore have taken place in the 50 years from 1910 to 1960 and are summarized in Table 14-2. The increase in Canadian production since 1950 reflects the development of the large Labrador deposits, and the increases in production of Brazil, Peru, and Venezuela reflect the development of some of the rich South American deposits. Other conspicuous increases in production were made by China and the Soviet Union.

Another important development in the iron and steel industry during comparatively recent years has been increasing beneficiation of ores at the sites of mining, so as to ship a more concentrated product, and the ever-increasing use of agglomerated material, especially pelletized ore.[1] In the United States, 12.3 percent of the iron ore shipped was beneficiated in 1920, whereas 55.5 percent was so treated in 1960 and 67.1 percent in 1962.

Pelletizing consists essentially of agglomerating fine material (173, 195). The process makes use of various ores that formerly were not marketable. Pelletizing yields a product that is mechanically strong and can be shipped without serious disintegration. Moreover, the iron content of the pelletized material is considerably higher than that of ores previously utilized, and research is consistently improving the grade. The use of pellets has brought about an iron ore revolution, and the pellet is becoming the preferred material for the blast furnace charge (196).

---

[1] Iron-bearing material may be classed as (1) direct-shipping ore, which can be sent directly to the furnaces; (2) intermediate or concentrating ore, which is improved by rather simple processes before shipping; and (3) low-grade material, including unenriched iron formation, which must undergo extensive beneficiation before shipping, usually fine grinding and agglomeration. Direct-shipping ore is difficult to define, as the type of material that is acceptable will depend on the practice currently in use at the blast furnaces of a particular area. Fine materials must undergo agglomeration so that they will not be blown out of the blast furnace. Agglomeration may be accomplished by sintering, nodulizing, or pelletizing. Sintering and nodulizing are generally done in plants at blast furnaces, from finer ores received, whereas pelletizing is done in plants at the mines and the pellets are then shipped to the blast furnaces. See 90, Chap 3, pp. 7–16, for a discussion of beneficiation and pp. 16–27 for a discussion of smelting practices.

**TABLE 14-2.   Production of Iron Ore at 10-year Intervals,
1910–1960, Selected Countries, and World Total***

Country	*Percent of world production*					
	1910	*1920*	*1930*	*1940*	*1950*	*1960*
Australia	0.1	0.5	0.6	1.2	0.9	0.8
Brazil	—	—	—	0.1	0.8	1.0
Canada	0.2	0.1	—	0.2	1.3	3.8
Chile	—	—	0.9	0.9	1.2	0.9
China	0.1	1.2	1.0	?	0.8	10.7
France	10.3	11.4	27.4	6.8	12.0	13.0
Germany	20.3†	5.2	3.2	8.6	—	—
Germany, West	—	—	—	—	4.4	3.7
India	—	0.5	1.1	1.7	1.2	2.1
India, Portuguese	—	—	—	—	0.1	0.9
Luxembourg	—‡	3.0	3.7	?	1.5	1.4
Malaya	—	—	0.4	1.0	0.2	1.1
Peru	—	—	—	—	—	1.4
Spain	6.1	3.9	3.1	1.4	0.8	1.1
Sweden	3.9	3.7	6.3	6.0	5.4	4.1
United Kingdom	10.9	10.5	6.6	9.6	5.3	3.4
United States	40.9	56.0	33.4	39.8	39.8	17.5
Soviet Union (Russia)	—	0.1	5.7	14.7	17.6	20.8
Venezuela	—	—	—	—	0.1	3.8
Total of above	92.8	96.1	93.4	92.0	93.4	91.5
	Million long tons					
World	139.5	122.6	178.0	188.0	250.0	507.0

* Compiled from "Mineral Resources of the United States," U.S. Geological Survey through 1923, U.S. Bureau Mines 1924 through 1931, and *Minerals Yearbook*, U.S. Bur. Mines, thereafter.
† Includes Luxembourg.
‡ Included with Germany.

Pelletizing has increased considerably the grade of material fed into the furnaces. Before pellets were used in quantity, early in the 1950s, the average for all ores from North American mines was a ratio of iron to silica of 4.9:1, and the natural iron content of the ores averaged about 50 percent. Pellet shipments in the early 1950s had a ratio of iron to silica of about 7:1 and an iron content around 61 percent, but by 1961 the ratio had changed to 8:1 and the iron content had reached 62 percent. The forecast for 1965 is pellets with a ratio of iron to silica of 11:1 and an average iron content of 63.5 percent. The tonnage of pellets then produced is expected to be more than 42 million tons which, with a content of 63.5 percent iron, would be equivalent to 54 million tons of 50 percent iron ore.

    World reserves of iron ore are difficult to estimate, partly because pelletizing has changed the picture regarding material that might be classed as ore, partly because insufficient work has been done on known important deposits that have been discovered or have come into use lately. A detailed tabulation of reserves (48) published in 1962 indicates that nearly 60 percent of the total reserves of iron contained in ore of all types is in four countries—the Soviet Union, Brazil, India, and the United States; that slightly less than 19 percent is in three other countries—China, Canada, France; and that nearly 10 percent is in five other countries—Sweden, West Germany, Guinea, Australia, and the United Kingdom. About 40 percent of the total ore in these countries is explored ore, about 60 percent is potential ore (Table 14-3).

This table of reserves may not give a complete picture of the probable amount of iron that may be available within the next 10 to 15 years. Some of the deposits are remotely situated (Soviet Union) or in an awkward geographic location (Bolivia). Others, as those of Canada and of South America generally, could well be much larger than the amounts stated. The Canadian reserves might be as much as 30,000 million tons instead of the 4,566 million tons indicated by the table. The reserves of South America may well be larger than the amount indicated. It has been stated (3, p. 32) that the potential medium-grade ore there will not be used soon because of the very large proven reserves. Exploration and development of deposits in Africa (158, 161, 209) may well increase estimates of reserves there. Further, new discoveries are made from time to time, as the recently found deposits of Malaya (197, 198) and the recently announced deposits on the Ivory Coast of Africa and on Baffin Island in the Arctic.

An interesting example of the manner in which material that formerly could

TABLE 14-3. Some of the Most Important Iron Ore Reserves of the World*

| | Iron in iron ore | | | | | |
| | Millions of metric tons | | | Approximate percent of total iron of world reserves | | |
Country	Explored ore	Potential ore	Total ore	Explored ore	Potential ore	Total ore
Soviet Union	9,246.0	5,571.2	14,817.2	31.9	13.2	20.8
Brazil	1,797.0	8,487.5	10,284.5	6.2	20.1	14.4
India	3,393.6	6,555.0	9,948.1	11.7	15.5	12.5
United States	2,554.2	6,042.2	8,596.4	8.8	14.3	12.0
China	1,375.9	3,303.3	4,679.2	4.7	7.8	6.5
Canada	1,753.6	2,812.8	4,566.4	6.0	6.6	6.4
France	2,181.8	1,880.0	4,061.8	7.5	4.4	5.7
Sweden	1,393.3	507.9	1,901.2	4.7	1.2	2.6
West Germany	445.7	1,021.6	1,467.3	1.5	2.4	2.0
Guinea	260.0	1,170.0	1,430.0	0.9	2.5	2.0
Australia	300.6	907.1	1,207.7	1.0	2.1	1.7
United Kingdom	675.8	325.8	1,001.1	2.3	0.8	1.4
Total of above	25,377.5	38,584.4	63,961.4	87.2	90.9	88.0
World total	29,041.1	42,128.9	71,170.0			

* Compiled from R. W. Hyde, B. M. Lane, and W. W. Glaser, *Engineering and Mining Journal,* vol. 163, no. 12, p. 84–88. Copyright 1962. McGraw-Hill Publishing Company. Used by permission.

not be regarded as iron ore has become available and is being used is the iron sands that lie under Ariake Bay in Jyushu, Japan (191, 194). These iron sands, which are high in TiO_2, were needed to provide more domestic raw material for Japan's growing steel industry, because about 77 percent of her iron ore was being imported. A new method of treatment and the development of undersea mining have made possible the economic development of pig iron entirely from these iron sands, which are divided into two groups, a rich deposit containing 27.5 million metric tons of refinable material and a medium-grade deposit containing 8.9 million metric tons.

United States

The production of iron ore in the United States comes from four general regions: Lake Superior, Southeastern states, Northeastern states, and Western states. Some scattered production comes also from states not included in these four regions (Table 14-4).

The Lake Superior region not only was the most important producing area of the United States in 1962 but also has had that distinction almost from the beginning of production there. Further, in 1955 the reserves there constituted 73.4 percent of the total reserves of the United States, and the potential reserves 77.3 percent. The second most important region is the Southeastern, with 11.2 percent of the reserves and 17.3 percent of the potential reserves.[1] In the Lake Superior region the major production has come from the Mesabi district, Minnesota; in the Southeastern region, from the Birmingham district, Alabama; in the

TABLE 14-4. Production of Iron Ore in the United States, 1962*

Region and states	Thousand long tons†	Approximate percent of U.S. total
Lake Superior (Minnesota, Michigan, Wisconsin)	118,829	78.0
Northeastern States (New York, New Jersey, Pennsylvania)	11,232	7.9
Southeastern States (Alabama, Georgia)	7,357	5.1
Western States (Arkansas, Colorado, Idaho, Missouri, Montana, Nevada, New Mexico, South Dakota, Utah, Wyoming)	6,350	4.4
Undistributed	6,535	4.6
Total	143,303	100.0

* *U.S. Bur. Mines Minerals Yearbook,* 1962, vol. 1, p. 660.
† Does not include ore that contained more than 5 percent manganese.

[1] Reserves include "direct-shipping ore and concentrates usable under present technologic and economic conditions" (1959); potential reserves, those "probably usable, partly as direct-shipping ore but mostly after beneficiation, under more favorable technologic and economic conditions." (14, p. 65)

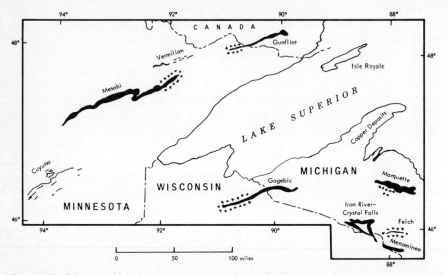

Fig. 14-3   Map showing location of the major Lake Superior iron ranges. Areas of moderate-grade to high-grade metamorphism shown by crosses. Shapes of iron ranges generalized and positions approximate. After H. L. James, *Econ. Geology,* vol. 49, p. 238. Copyright 1954. Economic Geology Publishing Company. Used by permission.

Northeastern region, from the Adirondack district, New York; in the Western region, from Missouri, Utah, and Wyoming.

Mining of iron ore in the United States started as early as 1682 in New Jersey, 1692 in Pennsylvania, 1785 in New York, and probably about 1818 in Alabama. Iron ore was discovered in the Lake Superior region in 1844 and the first shipment made from there in 1850 (22, p. 402).[1]

The Lake Superior region.   The Lake Superior region (14, 21, 22, 30, 33, 49, 50, 51, 70, 76, 83, 84, 90, 91, 95) is generally divided into six iron-bearing areas known as ranges (Fig. 14-3), some of which are further subdivided into districts. The six ranges and the percent of the total iron ore production of the Lake Superior region, 1854 to 1962, are shown by Table 14-5. More than two-thirds of the total production has come from the Mesabi range, even though active mining there started 38 years later than it did on the Marquette range.

The iron ranges are so designated in part because, to a considerable extent, the iron formation makes prominent ridges, although this is not always the case and the ridges may be made chiefly from flanking quartzites or other resistant rocks and the iron formation may occupy lowlands. In general the iron ranges are long and relatively narrow and mark the sites of troughs that contain sedimentary and associated volcanic rocks, which generally were complexly folded and faulted and subsequently eroded. The structure varies from range to range, and from place to

[1] An interesting summary of the early history of the iron and steel industry in the United States is given in reference 22, pp. 399–402.

TABLE 14-5. Iron Ore Produced in the Lake Superior Region, by Ranges, 1854–1962*

Range	Date production is first shown in tables	Production, million long tons	Percent of total Lake Superior production
Mesabi, Minn.	1892	2,375	68.9
Marquette, Mich.	1854	317	9.2
Gogebic, Mich.-Wis.	1884	316	9.2
Menominee, Mich.-Wis.	1877	275	8.0
Vermilion, Minn.	1884	98	2.8
Cuyuna, Minn.	1911	66	1.9
Total		3,447	100.0

* Compiled from *U.S. Bur. Mines Minerals Yearbook* and *U.S. Geol. Survey Mineral Resources of the United States,* various years.

place within a range, and for the most part it is complex, but some idea of the general conditions may be obtained from the examples shown in Figs. 14-4, 2-12, and 2-13. The rocks of most of the ranges are steeply dipping, but those of the Mesabi range dip more gently than those of the others, and in places they approach horizontality.

The rocks that contain the iron formations and the iron ores are of Precambrian age. In general throughout the Lake Superior region the rocks may be subdivided into the Lower Precambrian basement, composed chiefly of gneisses, schists, and amphibolites but including also sedimentary rocks and some iron formation; the Middle Precambrian, consisting of quartzites, dolomites, the major iron formations, graywacke, slate, and basic volcanic rocks; and the Upper Precambrian, consisting of sandstones, conglomerates, lava flows, and intrusive rocks. The earlier subdivision of the Middle Precambrian rocks of the major iron ranges (91, pp. 118, 159, 212, 225, 251–252, 291, 330) classed them as Lower Huronian, Middle Huronian, and Upper Huronian, and further classed the Upper Huronian as the Animikie group. Later usage has been somewhat different. Roughly, usage of the United States Geological Survey has been as follows:

	Earlier usage	*Later usage in Michigan**	
Middle Precambrian	Upper Huronian (Animikie group)	Paint River group	Animikie series
		Baraga group	
	Middle Huronian	Menominee group	
	Lower Huronian	Chocolay group	

* H. L. James, 1958, Stratigraphy of Pre-Keweenawan Rocks in Parts of Northern Michigan: *U.S. Geol. Survey Prof. Paper* 314-C, p. 30.

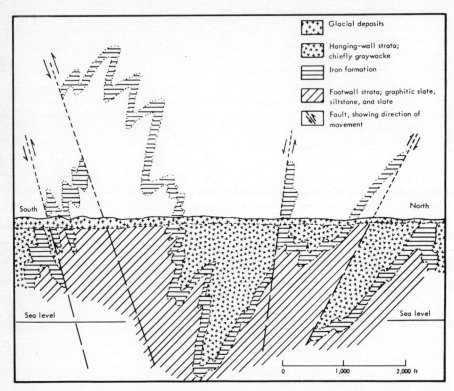

Glacial deposits

Hanging-wall strata; chiefly graywacke

Iron formation

Footwall strata; graphitic slate, siltstone, and slate

Fault, showing direction of movement

South

North

Sea level   Sea level

0   1,000   2,000 ft

Fig. 14-4 Generalized section showing structure of part of the Iron River district, Iron County, Michigan. After H. L. James and C. E. Dutton, *U.S. Geol. Survey Circ.* 120.

The usage of the Minnesota Geological Survey differs somewhat from either of the designations shown here. Major iron formations and ores are present in the Middle and Upper Huronian rocks of the older usage and in the Menominee and Paint River groups of the later usage.

Stratigraphic terminology throughout the area tends to be confusing to one not familiar with both early and recent geological work that has been done there. No attempt is made here to present the details but merely to state that throughout the region as a whole the major iron formations and iron ores are in rocks of Middle Precambrian age[1] regardless of the position within those rocks. Some of the features of the six ranges are shown in Table 14-6.

The iron formations of the region consist of one or more iron minerals, either alone in layers and lenses or associated with chert, jasper, or slate. The most usual iron minerals in the areas where the grade of metamorphism is low are hematite (soft, hard, or specular), magnetite, limonite, siderite, and various silicates such as greenalite and minnesotaite. Where the grade of metamorphism is moderate to high (Fig. 14-3) the more usual iron minerals are specular hematite, magnetite, grune-

[1] Terminology of the United States Geological Survey

rite, and fayalite. Much of the iron formation is composed of alternating layers of chert or jasper and iron minerals, some of which are granular or oolitic. Much of the iron formation causes magnetic anomalies, which vary from weak to intense and may be mapped by use of a magnetometer where not too strong and by use of a dip needle where very strong. They have proved invaluable as a means of tracing the iron formation under cover and making structural interpretations.

Iron ore throughout the region has, in general, accumulated from the iron formations as a result of weathering and the circulation of groundwater.[1] Siderite and iron silicates were attacked, and in places residual concentrations of iron oxides were formed. The leaching of silica, however, along with the oxidation of the iron minerals not already in the oxide state, seems to have been the key to much of the concentration of the ore. In part, however, iron also must have migrated downward in solution and have been later precipitated in favorable places against impermeable barriers, aiding in the formation of ore bodies along faults (Fig. 2-12), in the troughs of synclines (Fig. 2-13), at the intersections of dikes and iron formations, and above sills. The source of the iron contained in the iron formations, as distinct from the iron ores, is not as evident as the source of the iron in the ores. This was discussed briefly in connection with the origin of sedimentary iron deposits.

Perhaps the most exciting recent development in the history of the Lake Superior region is the shift from the use of high-grade direct-shipping ores containing 50 to 60 percent or more iron, which were rapidly becoming depleted, to low-grade ores containing less than 30 percent iron, which need concentrating and pelletizing. In 1953 the statement was made: "A revolution is underway in the Lake Superior district. A new group of active, technically-trained young men is moving in. These men do not believe that the Iron Country is 'running out of ore.' They believe that is simply 'running out' of the type of ore that has been mined in such constantly increasing quantities over the past 50 years. It is now time to learn how to use a new type."[2]

The new type of material to which the foregoing statement alludes is generally designated, in mining parlance, *taconite* in Minnesota and *jasper* in Michigan. *Taconite*[3] means essentially unenriched oxide iron formation in Minnesota; *jasper* or *jaspilite* has the same meaning in Michigan. The taconites and the jaspers may be further divided into the magnetic and the nonmagnetic ones, which require different processes to recover the iron. Generally the taconites and jaspers contain less than 30 percent iron.

The pioneer attempt to use taconite was at Babbit, Minnesota, at the eastern end of the Mesabi range, where the taconite contains much magnetite as a result of metamorphism caused by intrusion of the Duluth gabbro. Between 1915 and 1924 experimental work was done, a plant was constructed, some ore mined, con-

---

[1] Some people have ascribed part of the ore accumulation to ascending hydrothermal solutions, but, in general, opinion and field relations strongly favor the effects of weathering and downward circulating groundwater.

[2] From E. W. Davis, *Engineering and Mining Journal*, vol. 154, no. 7, p. 103. Copyright 1953. McGraw-Hill Publishing Company. Used by permission.

[3] The original term, which was *taconyte,* was proposed by H. V. Winchell "for the cherty or jaspery, but at times calcareous or more or less quartzitic rock, that incloses the soft hematites of the Mesabi Range, Minn." [(5, p. 670) of Chap. 2] The type locality was the eastern end of the Mesabi range.

TABLE 14-6. Some Features of the Iron Ranges of the Lake Superior District of the United States*

Range	Extent of range	Thickness of iron formation	General type of ore	Some characteristics of ore bodies
Cuyuna, Minnesota	Two parallel parts, North range (main deposits) and South range; about 65 miles long, 1–2 miles wide	50–400 ft	Limonite and hematite, generally manganiferous on North range	Maximum size of ore bodies: 2 miles long, $\frac{1}{4}$ mile wide
Mesabi, Minnesota	About 110 miles long, about 1.5 miles wide	400–750 ft	Limonite and hematite about equal amounts on main range; magnetite associated with metamorphic iron silicates, eastern part	Ore bodies up to $3\frac{1}{4}$ miles long, $\frac{1}{2}$–1 mile wide, 500 ft thick
Vermilion, Minnesota	Two areas of iron ore about 25 miles apart, approximately 5 miles long and 0.5 mile wide	Possibly several hundred feet	Hard, dense, high-grade hematite	Lenticular to tabular bodies; some are 1,500 ft long, 100 ft thick, extend downward 2,500 ft vertically
Gogebic, Michigan-Wisconsin	About 70 miles long, 800–1,000 ft wide	400–1,000 ft	Hematite	Steeply inclined tabular bodies and plunging masses along intersection

District	Dimensions	Iron formation thickness	Ore minerals	Remarks
Menominee, Michigan-Wisconsin	Disconnected segments over length of 50 miles; in places 15 miles wide	650 ft, without ore in middle part, in eastern area; 200–600 ft in western area	Specular hematite and magnetite in eastern part; nonspecular hematite and limonite in western part	of dikes and iron formation. Maximum size of ore bodies: 3 miles in pitch length, 400 ft thick, and 1,500 ft parallel to dip of inclined strata. Eastern part: maximum dimensions before any mining, but not in a single ore body, 150 by 2,500 ft horizontally, 2,000 ft vertically. Western part: maximum dimensions before mining, about 1½ miles long, 300 ft wide, 1,000–2,000 ft deep
Marquette, Michigan	About 30 miles long, 1–6 miles wide	Main iron formation, 2,000 ft maximum; minor iron formation, 200 ft maximum	Chiefly hematite	Main ore formation: tabular masses parallel to stratification of enclosing rock; cylindrical masses along intersections of dikes; irregular masses above sills; ore bodies as much as 250 ft thick, mined to vertical depths of about 3,000 ft. Minor iron formation: small, shallow, and irregular ore bodies

* Compiled from *U.S. Geol. Survey Bull.* 1082-C, pp. 75–78, 1959.

centrated, and shipped, but the venture had to be abandoned because material could not be produced at a cost low enough to compete with direct-shipping ores. Between 1929 and 1947 other investigations were conducted, climaxed by a decision by the Reserve Mining Company, in 1947, to erect a large concentrating plant, which resulted finally in the construction of such a plant at Silver Bay and a large mine and crushing plant at Babbit (168, 169, 172, 178, 179, 186). The deposit controlled by the Reserve Mining Company is a wedge-shaped mass about 9 miles long, 2,800 ft wide, and 175 ft deep at its thickest part. It was estimated in 1956 to contain about 1.5 billion tons of magnetic taconite containing 25 to 30 percent iron (168).

The processes from mining to finished product involve a great many interesting details which cannot be given here. The operations start with the mining of extremely hard material—the only thing tougher is said to be an old-time hard-rock miner—for which jet piercing[1] is used, and follow with much crushing, screening, magnetic concentrating, and making into pellets.

After the pioneer work at Babbitt, other plants came into existence: in Minnesota successful operations began on lands west of those controlled by Reserve Mining (175); in Michigan (160, 173, 180) they were undertaken successfully at the Groveland mine (Menominee range) and at the Republic mine and the Humboldt mine (Marquette range). Much of the Michigan ore is not magnetic and is concentrated by flotation. All these plants have undergone expansion. In 1963 the annual capacity of six pelletizing plants in Michigan was about 5,500,000 long tons of pellets, and the annual capacity of three pelletizing plants in Minnesota was about 16,500,000 long tons of pellets. The content of iron in the pellets is much higher than it is in the direct-shipping ores or in the concentrates, as indicated by Table 14-7.

TABLE 14-7.   Iron Content of Ore Shipped from the Lake Superior Region, 1962*

	Thousand long tons		
	Direct-shipping ore	*Concentrates*	*Agglomerates*
Ore	18,068	20,387	16,307
Iron in ore	9,360	11,162	10,003
Approximate percent iron	51.8	54.7	61.3

* From *U.S. Bur. Mines Mineral Yearbook,* 1962, vol. 1, p. 666.

The Northeastern region.   The iron deposits of the Northeastern region (14; 90, chap. 4, pp. 21–24, chap. 5, pp. 38–48) are localized in three principal areas: (1) the Adirondack district in northern New York; (2) the Dover district in the highlands of northern New Jersey; and (3) the Cornwall deposit near Lebanon,

[1] Jet piercing makes use of a blowpipe in which kerosene and oxygen are burned. The blowpipe rotates against the rock face, and the flame is followed by a jet of water. Rock fragments spall off and are blown out by steam that is formed.

Pennsylvania. These are among the oldest producing deposits of the United States and have furnished iron ore since the early part of the eighteenth century.

The deposits of New Jersey and New York are similar (2; 14, pp. 72–73; 44; 45; 71; 81; 82; 90, chap. 4, p. 21, chap. 5, pp. 42–45). In general the major deposits consist of lath-shaped, pod-shaped, veinlike masses of magnetite that are steeply inclined and conform essentially to the gneissic structure of the enclosing rocks, which are a variety of gneisses, skarns, and igneous rocks of Precambrian age. In the best deposits the magnetite is massive, especially in the central parts of the ore bodies, and may become disseminated toward the outer parts. Some ore bodies occur in granite, in which they generally are of the disseminated type. In general throughout both states the more massive magnetite makes some lump ore, which is cobbed after coarse crushing and used as direct-shipping ore. Much material is further crushed and concentrated by magnetic separation. Martite (hematite after magnetite), however, which is present in varying amounts, remains in the tailings from the magnetic separation and must be further concentrated by other processes. The best ore bodies contain 35 percent to more than 60 percent iron, but some of the mined bodies of disseminated magnetite contain 25 to 30 percent iron, whereas others contain still smaller amounts and can be regarded only as possible sources of iron ore.

In the Dover district, New Jerseey (81, 82), which has furnished about 70 percent of the total production of that state, the deposits are concentrated in two belts, generally on the limbs of the major folds. Some of the lath-shaped ore bodies, which are the most important type, are as much as 2 miles long at the surface, average 10 to 20 ft thick, are somewhat less than 100 to as much as 2,400 ft broad, and extend indefinitely along the pitch length—some have been mined for more than a mile along the pitch length and more than 2,000 ft vertically.

In the Lyon Mountain area, New York (2, 71), the magnetite ore bodies occur on the limbs and in the keels of synclines in complexly folded rocks of Precambrian age. The deposits at Mineville, New York, are somewhat similar. At the Benson mine, New York, the deposit consists of magnetite disseminated throughout granitic rock and, although the ore is of relatively low grade, it can be recovered by low-cost open-pit mining.

Various ideas have been expressed regarding the origin of the iron deposits of New York and New Jersey. The more recent work shows that the magnetite clearly replaced other minerals. In the Dover area, New Jersey, the concept expressed by Sims (82) is that the magnetite deposits are high-temperature replacements formed by magmatic emanations, possibly late in Precambrian time after the orogeny that deformed the enclosing rocks. In the Lyon Mountain area, New York, Postel (71) expressed the concept that the bulk of the ore was introduced by pneumatolytic emanations but that some magnetite was introduced by hydrothermal solutions during the waning stage of mineralization.

The chief deposits of Pennsylvania are those at Cornwall (14, p. 73; 27; 90, chap. 5, pp. 46–47), which have furnished ore continuously since 1742. They are contact-metasomatic ore bodies in Cambrian limestone above its contact with a Triassic diabase sheet, which is partly concordant and partly cross-cutting. Two major ore bodies and a minor one consist chiefly of magnetite and iron-rich amphibole, although chalcopyrite, pyrite, diopside, and various other gangue minerals are present. Magnetite replaced not only limestone but also silicates and quartz. The ore averages about 40 to 42 percent iron, but magnetic concentrates contain about 62 percent. By-products recovered are copper, gold and silver (contained in chalcopyrite), cobalt (contained in pyrite), and sulfuric acid from the roasting of pyrite. Another deposit in Pennsylvania, recently discovered by airborne magnetometer surveys, is in the vicinity of Morgantown, Berks County. It consists of a large, kidney-shaped mass 1,500 to 3,000 ft below the surface and furnished ore from the Grace mine.

Southeastern region. The Birmingham district, Alabama, is the most important ore-producing area in the Southeastern region, although many small deposits that have been mined are present in other places. The Birmingham district, about 75 miles long and 40 miles wide, ranks next to the Mesabi range, Minnesota, in ore reserves (14, p. 73). The deposits (9) are beds of the Red Mountain formation[1] of Silurian age. This formation consists in general of shale and sandstone but contains 60 to 100 ft of iron-bearing strata throughout the active mining area in the vicinity of Birmingham. The uppermost of these beds are ferruginous sandstone and shale that average 20 to 30 percent iron but are very high in silica. Beneath this are beds designated the *Hickory Nut Seam,* the *Ida Seam,* the *Big Seam* (the chief producer), and the *Irondale Seam.* A generalization of the major features of these seams is shown in Table 14-8.

TABLE 14-8. Generalized Features of the Iron-ore Seams of the Red Mountain Formation, Birmingham Area, Alabama*

Material	Approximate thickness	General character
Hickory Nut Seam	3–5	Sandy ore characterized by great abundance of *Pentamerus oblongus,* named from resemblance of fossils to hickory nuts incased in partly open outer shucks
Less ferruginous rock	12–20	
Ida Seam	2–6	Siliceous ore. Generally only soft ore in surface exposures is mined, where it contains 35–44 percent iron, 45–52 percent silica
Less ferruginous rock	35–50	
Big Seam	16–40; good ore generally 10–12	Good hematite; has produced most of the ore of the district; contains about 35–39 percent iron, 17–29 percent silica, 18–23 percent calcium carbonate; generally in two benches, the upper one usually most important
Less ferruginous rock	2–4	
Irondale Seam	4–6	Similar to lower bench of Big Seam

* Compiled from *U.S. Geol. Survey Bull.* 315-D, 1906.

[1] Formerly designated the *Clinton formation* and the *Rockwood formation.*

The ore has been designated *hard, semihard,* and *soft.* The hard ore is the unaltered material; the semihard ore is the partly altered material; and the soft ore is the altered material that is present at the outcrops and some distance down the dip, which is 15 to 45°. The hard ore contains less iron than the soft ore, but it also contains less silica and more lime. Obviously there is gradation from hard to soft types, depending on the amount of leaching by groundwater, which removes calcium carbonate. This gradation is illustrated by the following analyses (9, p. 135):

	Hard ore	*Intermediate ore*		*Soft ore*
Fe	37.00	45.70	50.44	54.70
SiO₂	7.14	12.76	12.10	13.70
CaO	19.20	8.70	4.65	0.50

The lower content of iron in the hard ore is offset by the high lime content, which is present as calcium carbonate and makes a considerable part of the ore self-fluxing.[1]

Less important iron deposits in the Southeastern region are the brown ores (limonite) which are rather widespread throughout the area but which are generally mined as iron ore only in Alabama, Georgia, and Tennessee. They are also used as an important source of mineral-earth pigments. Generally the deposits are small residual accumulations and the ores need some type of concentration before shipping. The two most important areas are the Russelville district, Alabama (9), and northwestern Georgia (42).

Western region. Among the states of the Western region[2] Utah is the outstanding producer, and California, Texas, and Wyoming are important producers.

The most important deposits of Utah are in the Iron Springs district near Cedar City (6; 14, p. 79; 57; 62; 90, Chap. 5, pp. 60–61), which covers an area of about 60 sq miles. Quartz-monzonite of early Tertiary age, intrusive into limestone of Jurassic age, occurs in three masses about which are ore bodies of mixed hematite and magnetite, some of which replace the limestone and some of which fill joints and other fractures. Ore bodies are from a few hundred to more than a thousand feet long, as much as 250 ft thick, and generally contain more than 50 percent iron.

Numerous small contact-metasomatic iron deposits are present in California, but the most productive are those of the Eagle Mountains (14, p. 79; 35; 90, Chap. 5, pp. 51–52), northern Riverside County, about 10 miles north of Desert Center and a few miles south of the southern boundary of Joshua Tree National Monument. Contact-metasomatic replacements of limestone or dolomite occur throughout an area 6 miles long and 1½ to 2 miles wide near intrusive quartz-monzonite. The major ore bodies, 600 to 1,500 ft long, 70 to 300 ft thick, occur

[1] Calcium carbonate is needed to flux the silica present in the ores, which then goes into the slag. If enough calcium carbonate is present the ores are self-fluxing and are equivalent, from an economic viewpoint, to much richer ores to which limestone must be added as a flux.
[2] The Western region as used here includes all states west of the Mississippi River except Minnesota.

in two beds that are separated by quartizite and have been designated the *North ore body* and the *South ore body*. Magnetite and hematite are associated with metamorphic silicates. The best grade of ore contains more than 50 percent iron, poorer grades as low as 30 percent.

Iron ore is present in several places in Wyoming (4; 5; 14, pp. 79–80; 72; 90, Chap. 5, p. 62). Production has taken place in the Hartville district, about 90 miles north of Cheyenne, since 1898, where irregular lenses of high-grade hematite occur in schist of Precambrian age. The lenses are a few feet to a few hundred feet wide, and some are more than a thousand feet long. The most recent operations have been in deposits of the southern part of the Wind River Range, Fremont County, near Atlantic City. The deposits consist of sedimentary quartz-magnetite taconite about 150 ft thick in a sequence of metamorphosed sedimentary and volcanic rocks of Precambrian age. The iron formation occurs in the limbs of steeply pitching synclines. The magnetite content is 40 to 50 percent and, although the size of the grains is small, the material is suitable for magnetic concentration after fine grinding. Ore is mined by open pit, crushed, concentrated, and agglomerated. Reserves are estimated at more than 73 million tons of ore (199).

The iron deposits of Texas (14, p. 78; 69; 90, Chap. 5, pp. 59–60) are chiefly in the eastern and northeastern part of the state and lie in a shallow structural trough in two areas designated the *North basin* and the *South basin*. The deposits of the North basin are chiefly in Cass and Morris Counties, which are adjoining and are north of the town of Marshall; those of the South basin are in several counties east of Marshall. Greensand of early Tertiary age about 25 ft thick, composed of a mixture of glauconitic sand, quartz sand, and clay, contains almost all the ore, which consists of limonite and siderite. The deposits are close to the surface, and in general a surficial weathered zone containing limonite gives way to siderite below. Iron silicates of the greensand are thought to have been altered to siderite and, in turn, to have been converted into limonite by oxidation and hydration.

A very recent mining operation (early 1960s) is the Meramec Mining Company project at Pea Ridge, Missouri. Drilling to a body that produced a strong magnetic anomaly disclosed a tabular mass of magnetite, standing in a nearly vertical position, up to 400 ft wide, half a mile long, and of unknown depth. Drilling indicated the possibility of 100 million tons of ore, and shaft sinking began in 1957. The ore body lies 1,400 ft below the surface and was being developed on four levels in 1963, the lowest one 2,275 ft below the surface. The ore is concentrated and pelletized (192, 210).

Canada

Canada produced only insignificant quantities of iron ore before 1939, but in 1962 it furnished almost 5 percent of the total production of the world. The major increase in production began about 1954 when, on July 31, the first commercial shipment of iron ore was made by the Iron Ore Company of Canada from the Labrador-Quebec area (181, 182, 183). Since then production has increased more or less steadily, with some declines because of world-market conditions, and future production should increase very much because, in addition to direct-shipping and concentrating ore, great advances have been made in the establishment of pelletizing plants. Production was taking place at 13 properties in Canada at the end of 1961, and by-product iron was being obtained by three companies. Some of the major features of production as they were in 1963, when significant advances had been made, are shown in Tables 14-9 and 14-10. Summaries of some developments appear in various mining publications (163, 184, 185, 187, 188, 189, 190, 200, 201). The major production and developments are in the Quebec-Labrador region and in Newfoundland, with much activity in Ontario and a considerable amount of activity in British Columbia.

TABLE 14-9.   Some Major Features of Production, Iron Deposits of Canada, 1963*

Company and location of property	Product mined and average natural grade	Product shipped and average natural grade	Shipments, million long tons
Iron Ore Company of Canada: (a) Schefferville, Quebec	Hematite-goethite from open-pit mines, 54.89 percent Fe	Direct-shipping ore, 54.89 percent Fe	6.753
(b) Labrador City, Newfoundland	Specular hematite from open-pit mine, 36.1 percent Fe	Specular hematite concentrates, 64 percent Fe	2.217
Quebec Cartier Mining Co.: Gagnon, Quebec	Specular hematite from open-pit mine, 31.3 percent Fe	Specular hematite concentrate, 64.4 percent Fe	6.353
Caland Ore Co., Ltd.: East arm of Steep Rock Lake, north of Atikokan, Ontario	Hematite-goethite from open-pit mine, 53.69 percent Fe	Direct-shipping ore, 53.69 percent Fe	2.003
Carol Pellet Co.: Labrador City, Newfoundland	Processes concentrate of I.O.C.C.	Pellets, 65.0 percent Fe	1.835
The Algoma Steel Corp., Ltd.: mines and sinter plant near Wawa, Ontario	Siderite from open-pit and underground mines, 33.32 percent Fe	Ore beneficiated by sink-float and sintered, 50.86 percent Fe, 2.84 percent Mn	1.618
Dominion Steel and Coal Corp., Ltd., Wabana Mines Div.: Bell Island, Newfoundland	Hematite-chamosite from underground and open-pit mines, 48.39 percent Fe	Heavy-media concentrate, 50.54 percent Fe	1.168
Steep Rock Iron Mines, Ltd.: Steep Rock Lake, north of Atikokan, Ontario	Hematite-goethite from open-pit and underground mines, 50.93 percent Fe	Direct-shipping ores and gravity concentrate, 54.0 percent Fe	0.963

* From Mineral Resources Division, Department of Mines and Technical Surveys, 1963 *Canadian Minerals Yearbook.* Used by permission.

TABLE 14-10. Chief Projects under Development and Announced Plans for Production, Iron Deposits of Canada, 1963 Data*

Company, location, expected date of production	Product to be mined	Product to be shipped	Expected annual production, long tons
Wabush Mines, Pickands Mather & Co., managing agent, Wabush Lake, near Labrador City, Lab., 190 miles north of Sept-Iles (1965)	Specular hematite iron formation from open-pit mine, 37 percent Fe	Concentrate and pellets, 64–65 percent Fe	4,900,000 pellets 400,000 concentrates
Caland Ore Co., Ltd.,† East arm of Steep Rock Lake, north of Atikokan, Ontario (1965)	Hematite and goethite from open-pit mines, 53.76 percent Fe	Pellets, plus 60 percent Fe Lump ore, 54 percent Fe	1,000,000 1,500,000
Jones & Laughlin Steel Corp., near Kirkland Lake, Ontario (1964)	Magnetite iron formation from open-pit mine, 25 percent Fe	Pellets, 65–66 percent Fe	1,000,000

* From Mineral Resources Division, Department of Mines and Technical Surveys, 1963 *Canadian Minerals Yearbook.* Used by permission.
† Company presently produces 2 million tons of natural ore (54 percent Fe) a year.

The Quebec-Labrador-Newfoundland region.

The Labrador trough region. The Quebec-Labrador iron region (25, 28, 29, 34, 38, 39) extends southward from the west side of Ungava Bay to Mt. Reed and slightly beyond (Fig. 14-5), a distance of about 600 miles from north to south but considerably more than that throughout the extent of the iron-bearing rocks. The iron formations, which are of the Lake Superior type, are in the western part of a broad zone of folded Precambrian (Proterozoic) rocks that comprise the Labrador trough, which is 60 miles wide in the central part but less than that both to the north and to the south. The central part of the trough or geosyncline is best known. It contains Late Precambrian (Proterozoic) rocks that unconformably overlie Early Precambian (Archean) ones. The Early Precambrian rocks consist chiefly of the Ashuanipi complex of granitic and other gneisses, whereas the Late Precambrian rocks consist of about a dozen formations or groups composed chiefly of clastic sedimentary and, especially toward the top, basic volcanic rocks, except for the iron formations. The principal iron-bearing rocks occur near the central part of this succession, and the major formation, the Sokoman, consists of various types of iron formation—silicate, jasper, cherty siderite.[1]

Structures throughout the region are complex (28, p. 3). Doubly plunging isoclinal folds overturned to the southwest are characteristic, and these have been traversed by numerous thrust faults. Southwest of Wabush Lake very complex local structures have resulted from folding of at least two ages and several sets of faults.

[1] Details of this succession throughout the Labrador trough, which are still incompletely known and about which there is some disagreement, can be found in references 25, 28, 34, 38.

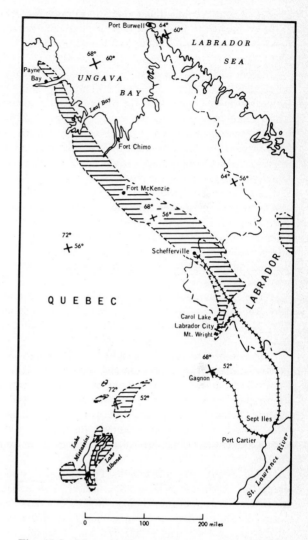

Fig. 14-5 Map showing extent of sedimentary and volcanic rocks (horizontal ruling) of the Quebec-Labrador area, Canada. Modified after the Geological Survey of Canada from Map 900 A, 13th ed., 1963. Used by permission.

Metamorphism is variable throughout the region, but in general along the western side of the trough it is of low grade in the central part and of higher grade both to the north and to the south. Generally, also, metamorphism is of high grade to the east throughout the entire trough.

The major iron formation varies in thickness from less than 100 ft to more than 500 ft. In most places a ferruginous, slaty argillite underlies the iron formation and is transitional upward through interbedded chert and argillite to a cherty magnetite-silicate-carbonate facies of iron formation, above which is a thin-banded, red, jasper-magnetite-hematite facies or cherty metallic iron formation (28, p. 2). Other types are present locally.

Gross (28, p. 3) divided the Labrador geosyncline into three divisions based on economic considerations. In the central division, extending from Sawbill Lake in the south to Finger Lake in the north, and including the main Schefferville–Knob Lake mining area, the rocks are not highly metamorphosed. There the iron formation has been leached and high-grade goethite-hematite ores have been formed. In this area some 45 ore deposits are known (174). In the southern division, metamorphism has produced coarse-grained iron formations that contain specular hematite, magnetite, and quartz, which can be easily concentrated and are the basis of the new developments in the Wabush Lake–Mount Wright and Jeannine Lake area. In the northern division the rocks are considerably metamorphosed and the recrystallized iron formation provides potential deposits suitable for beneficiation. In the area south of Payne Bay (29) a total of more than 1,000 million tons of crude ore than contains 24.5 to 35.5 percent iron has been proven, another 1,000 million tons is indicated, and additional large tonnages may be inferred.

The southern section of the trough has been undergoing active development. (See Tables 14-9, 14-10.) The Carol Lake project and the Wabush project (43, 171, 211, 212) both utilize iron formation and concentrate it to 64 to 65 percent iron.

In addition to the iron deposits of the Labrador trough, substantial reserves of fine-grained iron formation are present in the Lake Albanel region about 200 miles southwest of Mt. Reed in Mistassini Territory, Quebec (68, 73, 94). There the Temiscamie iron formation, about 700 ft thick, consists of cherty siderite, hematite, magnetite, and iron silicates, and contains about 29 percent iron. This iron formation is known throughout a distance of about 100 miles and is thought to be the equivalent of some of the iron formation of the Labrador trough.

Bell Island, Newfoundland. The Wabana[1] iron-ore deposits (40, 41, 60) along the northwest coast of Bell Island, Conception Bay, Newfoundland, near St. John's, have been mined since 1895. For many years after 1910 they supplied between half a million and two million tons of ore annually. The total production to the end of 1955 was 58,588,516 long tons of ore that contained 51.5 percent iron (dry analysis), 11.8 percent silica, 0.9 percent phosphorous, and 1.5 percent moisture (60, p. 506). In 1949 Newfoundland became a part of Canada and production thereafter was included in the total for Canada. The Wabana deposits are rated among the largest of the world (90, Chap. 5, p. 74). Minable reserves have been estimated at 3,500 million tons of ore (90, Chap. 4, p. 30). The ores cannot be used in many furnaces of the United States, however, because of the high phosphorous content. Present American smelting practice requires ores of relatively low phosphorous content, with ores containing more than 0.18 percent phosphorous being classed as high-phosphorous ores.

[1] *Wabana* is an Indian name that means "the place where daylight first appears," and was given by Mr. Thomas Cantley as appropriate for this far eastern part of North America (40, p. 5).

The deposits, of Lower Ordovician age, are beds of hematite, chamosite, and siderite that dip 8 to 10° to the northwest and pass beneath Conception Bay. Surface and onshore mines have been almost worked out and mining is now confined to submarine operations that extend as much as 2 miles from shore, the deepest working being 1,850 ft below sea level (60, p. 506). Safety in mining operations makes mandatory the leaving of at least 200 ft of rock cover, and this has made a considerable amount of good ore unminable in part of the area.

The deposits consist of three major zones of minable ore within sedimentary rocks composed of shale and sandstone in which cross-bedding and ripple marks are common. These zones are designated the *Lower Bed,* the *Middle Bed,* and the *Upper Bed.* The Lower Bed, which is the major source of iron ore, is a zone 15 to 40 ft thick that contains several large and thick lenses of hematite. Beneath this zone for about 70 ft there are numerous beds of hematite, rarely more than 2 ft thick, interlayered with thinly bedded shales and sandstones. The clean ore of the Lower Bed contains 45.00 to 57.60 percent iron, 20.00 to 7.50 percent silica. The Middle Bed, which lies about 215 ft above the Lower Bed, is rather sharply defined, contains 5 to 15 ft of minable ore, and is characterized by a persistent bed of hematite 4 to 6 ft thick at the top. The clean ore contains 51.50 to 59.60 percent iron, 12.00 to 6.40 percent silica, and is the richest ore in the area. The Upper Bed, which starts about 30 ft above the top of the Middle Bed, is marked by weak and erratic deposits of hematite-chamosite-siderite, many of the layers of which are so interbedded with shale that the grade is too low for mining. A zone about 20 ft above the base contains minable ore 7 to 12 ft thick in limited areas, but little mining of it has been done.

Selected areas in Ontario.   Some of the major producing mines and projects of Ontario are shown by Tables 14-9 and 14-10. A number of smaller operations are present, including by-product recovery from pyrrhotite flotation concentrates in the Sudbury area, the shipping product being iron oxide containing about 68 percent iron.

The Steep Rock Lake deposits (24; 52; 74; 75; 90, Chap. 5, pp. 67–68; 101), near Atikokan, are probably the most important ones in Ontario. Several large but local masses of hematite are known. The reserves of two of these are estimated at 21 million tons of measured ore and an additional 50 million tons of indicated ore, all expected to average about 54 percent iron, 4.62 percent silica, 0.024 percent phosphorous, and 0.03 percent sulfur. The potential reserves may amount to hundreds of millions of tons (52, p. 374). Ore has been mined by open-pit and underground methods.

The story of exploring the Steep Rock Lake deposits and bringing them into production is a fascinating one. Early in 1800 prospectors noted iron-ore fragments along the south side of Steep Rock Lake, and a map published in 1897 indicated that an iron-bearing horizon appeared to lie beneath the lake. Early exploratory diamond drilling from the shore of the lake revealed little of importance, but in 1937 there was inaugurated a winter exploration program of diamond drilling through the ice of the lake. During the winter of 1937 and 1938 7 of 12 holes disclosed good ore, and drilling during the next few years outlined enough ore to warrant production. Before production could be undertaken, however, the eastern part of Steep Rock Lake had to be drained, and this necessitated, first, the diversion of the Seine River, which flowed into the lake. Diversion of the river was begun in 1943, followed by dewatering of the lake, with the result that open-pit mining began at the Errington pit in October 1944. Explorations have indicated that ore extends to a depth of at least 2,000 ft.

The origin of the Steep Rock Lake ore bodies is an unsettled problem. An early concept (75), based on the results of drilling before production began, attributed the deposits to hydrothermal replacement, chiefly along an unconformity between underlying limestone and overlying volcanic rock. Information obtained from subsequent drilling, from exposures in two open pits, and from underground mining led Jolliffe (52) to suggest, in 1955, that the ore bodies consist of brecciated sedimentary limonite of Steeprock age and that local migration of iron might have resulted from recrystallization or hydrothermal modification of the finer matrix of this brecciated material. This suggestion was based in part on the apparent restriction of ore bodies to a stratigraphic member of the Steep Rock group and the almost-certain sedimentary origin of the nonore parts of the ore zone. Evidence indicates, however, that hydrothermal solutions did move through the rocks, and it is these solutions that may have caused later modification. These are but two of a number of suggestions that have been made.

Two other producing areas in Ontario deserve brief discussion (90, Chap. 4, p. 31, Chap. 5, pp. 68–69). The Michipicoten district, near Michipicoten Harbor at the northeast end of Lake Superior, furnishes concentrating material from steeply inclined tabular masses, chiefly of siderite. A more recently developed area is at Marmora (184, 185), about 120 miles east of Toronto, near the shore of Lake Ontario. There a body of magnetite, discovered by aeromagnetic surveys, furnishes material that is concentrated and pelletized.

The Northwestern region.   British Columbia has been furnishing slightly more than a millon long tons of ore a year (1960 and 1961), roughly 6 percent of the total Canadian production. Numerous deposits are known, but production in 1961 was from only three of them, two of which are on Vancouver Island and one on Texada Island. All three produced magnetite from open-pit mines (roughly 41 to 48 percent iron) and shipped magnetite concentrate (roughly 58 to 61 percent iron). Three projects were under development and expected to start production in 1962, two of which are on Vancouver Island, one on Moresby Island. These three are expected to produce 1.6 million long tons of ore annually, all from open-pit mines in magnetite ore (48 to 52 percent iron), and to ship magnetite concentrate (more than 60 percent iron). Much of the material from British Columbia is shipped to Japan.

The iron ore of British Columbia comes from contact-metasomatic deposits within sedimentary and volcanic rocks of Mesozoic age. The ore bodies, which are commonly in limestone, occur near or at the contact of the host rock and granitic rocks of the Coast Range batholith (165).

In Alberta, low-grade iron deposits have been investigated in the Crowsnest Pass region (170) and the Clear Hills–Swift Creek area (213), southwestern part and northwestern part of the province, respectively. The deposits of the Crowsnest Pass region are low-grade sedimentary titaniferous magnetite in beds up to 3 ft thick, with reserves apparently less than 8 million tons of material containing 25 to 30 percent iron. The deposits of the Clear Hills–Swift Creek area (63) are oolitic goethite, siderite, or iron silicate ores in Late Cretaceous sandstones and shales. Exploration indicates about 200 million tons of material containing 34 percent iron in the Swift Creek area and about 34 million tons containing 33 percent iron in the Clear Hills area immediately to the south.

Near the Yukon–Northwest Territories border a recent discovery (202) indicates the presence of hematite-jasper deposits containing 50 percent iron. The thickness of the deposits, which are in beds that dip about 10° and crop out at the surface, is 150 to 200 ft. They are known over an area of at least 146 sq miles.

South America

Production of iron ore in South America in 1962 is shown by Table 14-11. The largest reserves of iron ore appear to be in Brazil, the next largest in Venezuela. Although no single country of South America is among the world leaders in production of iron ore, the reserves and potential future production are so great that somewhat detailed description of the deposits seems desirable.

TABLE 14-11. Production of Iron Ore in South America in 1962*

Country	Long tons of iron ore	Approximate percent of world total
Venezuela	13,057,000	2.5
Brazil	9,842,000	1.9
Chile	7,874,000	1.5
Peru	6,569,000	1.3
Total	37,342,000	7.2

* From *U.S. Bur. Mines Minerals Yearbook*, 1962, vol. 1, p. 674.

Venezuela. The iron deposits of Venezuela (56; 58; 65, pp. 534–538; 77; 78; 86; 90, Chap. 4, pp. 31–32, Chap. 5, pp. 86–88; 96) are south of the Orinoco River along the northern border of the Guiana Highlands in the state of Bolivar. The region is underlain by rocks of Precambrian age, which constitute the northernmost part of the Guiana Shield. Both Lower Precambrian and Upper Precambrian (?) rocks are present, the Lower Precambrian consisting of a complex of gneisses, schists, granites and other intrusive rocks, the Upper Precambrian (?) consisting of two series, the Imataca and the Pastora.[1] The Imataca series, the one that contains the iron deposits, consists of ferruginous quartzites (iron formations) and gneisses that are strongly folded, with dips nearly vertical. Generally the trend of the rocks is east-northeast and the resistant iron formations form distinct ridges that rise above lateritic clay formed by weathering of the gneisses. The

[1] The age relations of these rocks are not clearly established. Some workers have considered the Pastora series to be much younger than the Imataca series.

iron formations or ferruginous quartzites, which possibly are recrystallized ferruginous cherts, are fine grained and generally thinly laminated and composed of iron-rich layers that alternate with iron-poor ones. The unweathered iron formation is composed largely of hematite, magnetite, and quartz. Weathering and leaching by groundwater have produced the iron ores.

Iron ores, which were used locally as early as 1750, occur at a number of places throughout the extent of the iron formation (167, 176), but two of the deposits are outstanding, El Pao and Cerro Bolivar. El Pao, discovered in 1926, finally was brought into production in 1950 after many difficulties and interruptions (177). The main ore body being mined is about 2,600 ft long and 1,700 ft wide, and a few feet to about 400 ft thick. It is somewhat bowl-shaped and the central part is filled with overburden, which is as much as 425 ft thick in places and must be removed before mining. The ore is of high-grade, direct-shipping quality, containing 63 to 66 percent iron.

Cerro Bolivar, which may contain seven or eight times as much iron ore as El Pao, is the largest-known iron deposit in the area and is about the size of the largest of the ore bodies of the Mesabi range, Minnesota. Reserves here have been estimated at 500 million tons of ore, the average grade of which may be 63.5 percent iron, dry analysis, or about 58 percent natural. Ruckmick (78) stated that the ores of Cerro Bolivar may be classed as hard, "crustal" ores and soft, friable ores. The crustal ores are residual accumulations that mantle the ore bodies to an average depth of about 50 ft. The friable ores, which are the ones of chief importance, occur to depths as much as 800 ft below the top of the hill before mining began. They were formed by leaching of silica by meteoric water, although some migration of iron occurred also.

Brazil. Iron deposits are known in Brazil in the states of Minas Gerais, Mato Grosso, Bahia, Coias, and the territory of Amapá (Fig. 14-6), but the most important ones are those of Minas Gerais, Mato Grosso, and Amapá. Among these, only the ones of central Minas Gerais seem likely to be suitable for the establishment of a large steel industry in Brazil. They contain an enormous tonnage of ore that is of high grade and exceptional purity, and they are close to major industrial centers (16). The deposits of Minas Gerais are suitable for exportation, also, and those of Mato Grosso and Amapá may prove to be. Most of the production so far has come from Minas Gerais.

The deposits of Minas Gerais (3; 18; 19; 20; 31; 32; 36; 37, pp. 358–362, 385, 404; 65, pp. 166–176; 89; 90, Chap. 5, pp. 76–78, Chap. 4, pp. 33–35) are in the south-central part of the state, where iron formation crops out in long ridges within an area of about 7,000 sq km in the Central Highlands. The deposits occur in severely folded and locally faulted rocks of Precambrian age known as the Minas series, which Guild (32) has divided into three groups. The middle group consists of chemical precipitates—iron formation and dolomite—whereas the lower and upper groups are chiefly clastic rocks. Slight metamorphism has changed the material of the middle group into quartz (recrystallized chert), specular hematite and some magnetite, and crystalline dolomite which, locally, contains some tremolite and other silicates. The iron-bearing material has been referred to as the Itabira iron-formation (37, p. 358), so named from Itabira Peak, a striking

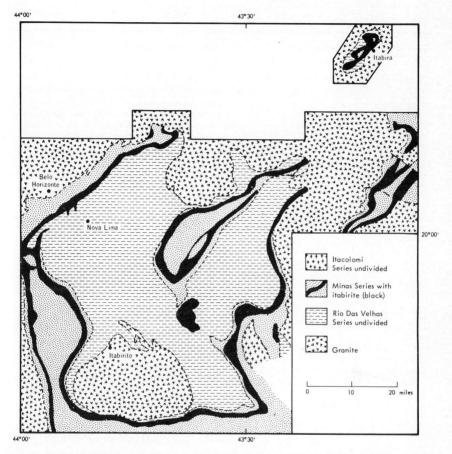

Fig. 14-6 Map showing distribution of the Itabira group in part of Minas Gerais, Brazil. Generalized from J. Van N. Dorr, 2d., and A. L. de M. Barbosa, *U.S. Geol. Survey Prof. Paper* 341-C, p. 7.

mountain of specular hematite near the town of Itabira do Campa. The thickness of the iron formation varies from 5 to 1,200 m.

Itabirite is the name generally used for the iron formation, but this term has been employed with different meanings: high-grade hematite ore; siliceous iron formation containing less than 50 percent iron; hard ore and iron formation, as opposed to soft varieties of both ore and iron formation. The original term (32, p. 14) seems to have been adapted from an aboriginal word meaning whetstone. The general characteristics of itabirite, as defined by Guild (32, p. 14), are (1) a simple mineral composition, chiefly granular quartz and flakes of specular hematite; (2) lamination, with hematite and quartz forming alternating dark and light layers and giving the rock a conspicuously banded appearance. Thus it is very similar to the taconite and jaspilite in Minnesota and Michigan of the Lake Superior region, United States.

TABLE 14-12.   Types of Iron Ore, Minas Gerais, Brazil

Type of ore	Variety of ore	General characteristics
Hematite ore	Hard massive ore	Iron content 67–69 percent, silica usually less than 1 percent. Best ore and type most widely used. Fine-grained specular hematite, some magnetite. Some masses up to 150 m thick, more than 1,000 m long. Masses may be tabular and parallel to bedding or may cut across bedding.
	Soft and intermediate ore	Iron content 50–69 percent. Leached material near surface may be friable mass of specular hematite (powdery ore). Some material may break into plates parallel to lamination (laminated or thin-bedded ore), and grade into itabirite.
Itabirite ore		Iron content generally at least 35 percent, average may be 40–45 percent. Original iron formation similar to taconite or jaspilite, needs concentration.
Surficial ore	Canga	Iron content 55–60 percent for much material. Fragments cemented by limonite. Best type, "canga rica," composed almost entirely of blocks of hard hematite with minimum limonite cement; other types contain varying amounts of hematite, itabirite, and other fragments. Canga forms capping several meters thick over most leached itabirite. Contains variable amount of aluminum hydroxide and may grade into laterite.
	Chapinha	Residue of thin iron-rich fragments mixed with soil. (*Chapinha* is *little plate* in Portuguese.) Easily screened material may average as much as 64 percent iron. Desired for local use.
	Rubble	Cobbles and blocks of hard hematite on flanks of itabirite ridges.

Several types of iron ore are recognized throughout the area, the name used depending in part on physical characteristics and in part on position with respect to the surface of the ground. These are summarized in Table 14-12.

General agreement seems to exist that the original iron formation was deposited as a chemical sediment and that the surficial types of ore and intermediate to soft ores were formed by circulation of surface water and groundwater. Some have considered the hard hematite to have been precipitated as unusually pure lenses of iron oxide in the original sediments; others have thought that the hard hematite resulted from replacement after the original sediments had been deposited.

The extent of the reserves throughout the entire area of the iron formation in Minas Gerais is unknown. The reserves within the Congonhas district, a relatively small part of the area, are shown by Table 14-13, and those within the Itabira district by Table 14-14.

TABLE 14-13.   Reserves of Iron Ore within the Congonhas District, Minas Gerais, Brazil*

Type of ore	Metric tons above depth of 1,000 m	Iron content, percent
Shipping ore		
Hematite	153,900,000	68
Surficial ore	103,950,000	50–60 plus
Itabirite	39,045,000,000	40 (?)

* From *U.S. Geol. Survey Prof. Paper* 290, 1957.

TABLE 14-14.   Reserves of Iron Ore within the Itabira District, Minas Gerais, Brazil*

Type of reserve	Metric tons	Average iron content, percent	General characteristics
Reasonably assured	374 million	More than 67	About one-third lump ore
Inferred	638 million	More than 67	Probably better physical characteristics than assured ore
Possible	2.5 billion	—	Probably better physical characteristics than surface ore, but much ore is too deep for economical mining

* From *U.S. Geol. Survey Prof. Paper* 341-C, 1963.

The deposits of Mato Grosso (19; 90, Chap. 5, p. 36, Chap. 6, p. 77) occur in the vicinity of Corumbá and extend into Bolivia. The iron formation there is typical banded jasper-hematite rock that may average 50 to 55 percent iron but that contains as much as 20 percent silica, making it unattractive except as a concentrating ore.

Bodies of magnetic iron ore are present at various places in Bahia, Ceará, Minas Gerais, São Paulo, Santa Catarina, and Paraná. Many of these appear to be highly metamorphosed inliers of material similar to the iron formation of the Minas series, surrounded by older gneisses. Some, however, such as those in São Paulo, may be magmatic deposits.

Chile.   Numerous iron deposits are present in Chile (3; 7; 65, pp. 263–272; 67; 87), but the workable ones appear to be concentrated chiefly in the provinces of Atacama and Coquimbo, although large low-grade deposits are known much farther south in the provinces of Arauco and Malleco. Production has come chiefly from El Tofo and nearby El Romeral in the province of Coquimbo, a short distance north of La Serna, but important production started in 1962 at

El Algarrobo in the province of Atacama, southeast of the port of Huasco (1, 104).

The deposits throughout the provinces of Coquimbo and Atacama occur near the contact of Mesozoic (?) granodiorite that intrudes older rocks, chiefly igneous but including some sedimentary ones. The major ore bodies, which consist chiefly of magnetite and hematite, occur as lenses that may be as much as 500 m wide and several km long. In many places smaller, veinlike masses of magnetite and hematite occur. The larger bodies appear to be of magmatic origin, the veins to be of hydrothermal origin.

The deposits at El Tofo and El Romeral (7; 65, pp. 267–270; 90, Chap. 5, pp. 77–78), department of La Serena, province of Coquimbo, furnished most of the production before 1962. The ore bodies, composed of hematite and some magnetite, are associated with igneous rocks and appear to be of magmatic origin although this is not clear because of some late alteration. El Tofo contained about 45 million tons of high-grade ore and was worked by Bethlehem Chile Iron Mines from 1913 to 1956, when the ore body was exhausted and activity was shifted to the nearby similar but smaller El Romeral deposit. A plant for the treatment of fines that had accumulated from El Tofo and the upgrading of ore from El Romeral was under construction in 1962.

The deposits at El Algarrobo (1; 7; 65, p. 267; 104), province of Atacama, southeast of the port of Huasco, although known since 1907 and explored to some extent between then and 1946, were not intensively explored until 1959. Exploration has now indicated that El Algarrobo contains the largest drilled reserve of iron ore in Chile, and large-scale open-pit production began in 1962 by Compania de Acero del Pacifico, South America, the sole producer of steel in Chile. The deposits lie within two hills, Algarrobo and Penoso, and consist of more than 25 apparently separate bodies of high-grade magnetite and specular hematite that extend throughout a north-south distance of over 2.5 miles in a zone about 600 ft wide. The deposits occur in porphyritic igneous rocks, apparently of Lower Cretaceous age, that seem to be a roof pendant in an intrusion of granodiorite. The surface is mantled with a residue of ore fragments (small pebbles to boulders) mixed with waste rock, enclosed within a claylike matrix that appears to be a result of weathering of andesitic country rock. This detrital material is up to 80 ft thick. Reserves, which include the detrital covering, are estimated to be 70 million metric tons of ore containing 55 to 68 percent iron and 2 to 5 percent silica.

The material in the provinces of Arauco and Malleco (7, 67), known as the Relún deposits, discovered about 1942, consists of fine granular magnetite and quartz, iron-rich zones alternating with those much higher in silica. The deposits occur in a zone of micaceous slates that generally dip 10 to 15° but in places are highly folded. The iron-rich material seems to be an iron formation similar to taconite and contains 35 to 40 percent iron. The thickness of the iron formation is about 60 to 70 ft. The deposits lie about 25 miles inland from the port of Quidico and extend in a northerly direction from the Relún River for many miles into the province of Malleco, but the most important zone is about 10 miles long and 5 miles wide. They constitute an important reserve of low-grade, concentrating ore.

Peru.   Although iron deposits are known in various places in Peru (3; 17; 90, Chap. 5, p. 85), production has been almost entirely from the Marcona deposits, 15°10′S, 75°10′W, province of Nazca, about 17 miles from the port of San Juan. The deposits, thought to be of contact-metasomatic origin, consists of numerous ore bodies in metamorphosed sedimentary rocks of Paleozoic age that were intruded by granodiorite and granite and were strongly metamorphosed. The upper parts of the mineralized bodies, from the surface to 25 to 45 m below, are hematite with some limonite, but at greater depth the hematite gradually gives place to magnetite, apparently the original mineral of the deposits. The iron content appears to be about 55 to 60 percent. A concentrating plant at San Nicolas Bay was completed late in 1961 to furnish high-grade concentrates for shipping, and a pelletizing plant was scheduled for completion in 1963 to furnish pellets for shipment.

Soviet Union

Iron ore production in the Soviet Union has made great strides in the period since 1945 and was 25 percent of the world total in 1962. Further, the reserves may be larger than those of any country in the world, though of lower grade (54, 97). Extensive beneficiating and sintering plants have been established, and these convert much low-grade ore into material that is satisfactory for use in the Russian blast furnaces. Larger facilities are planned for the future.

Iron ore occurs in various places throughout the Soviet Union, but the deposits may be placed in a few regional groups, as shown in Table 14-15. Many other deposits are known in addition to those cited in this table. These include Precambrian magnetite deposits of Yeno (67°40′N, 31°10′E), northwestern Russia, 23 to 48 percent iron; the Khopher deposits (50°40′N, 41°50′E), of Cretaceous age, Volga region, 35 to 44 percent iron; a number of contact-metasomatic and other magnetite deposits in the Ural Mountains; and various deposits throughout Siberia, many of which are reported to be large.

The Krivoi Rog deposits extend throughout a zone 130 miles long and 30 miles wide from 20 miles southeast of Krivoi Rog (about 48°N, 33°30′E) slightly west of north to 25 miles northwest of Kremenchug (about 49°N, 33°30′E). These deposits are similar to those of the Lake Superior region of the United States. Intensive work is being done in this area, both in mining and concentrating the ores, and it is expected that the content of iron from underground mines will be raised from 60 to 64 or 65 percent by concentrating and that from open-pit mines will be raised from 55 to 63 percent. These deposits were expected to furnish more than half of the iron ore of the Soviet Union in 1965.

The Kursk magnetic anomaly covers one of the very large iron deposits of Russia. According to Tomkeiff (88), the Soviet Minister of Geology stated in 1958 that the iron deposits of this anomaly are unrivaled in the world. Magnetic anomalies in this region were known as early as 1783, but detailed surveys between 1919 and 1926 were rather disappointing. Much more recent investigations, however, disclosed rich ore bodies. Beneath a cover of 40 to 500 m or more of Paleozoic, Mesozoic, and Tertiary sedimentary rocks is a mass of schists, ferruginous quartzites, and gneisses of Precambrian age. These rocks occupy a series of folds along a northwest-southeast belt that extends over an area of about 150 km by 250 km. The rich deposits (88) are secondary ore immediately above the Proterozoic floor

TABLE 14-15.   Some of the Largest Iron Deposits of the Soviet Union*

General area	Deposit	Location, latitude and longitude	Notes about deposit
Vicinity of the Black Sea	Kerch	45°05′–45°27′N, 36°05′–36°30′E	Iron phosphorites of Pliocene age; 33–42 percent iron; open-pit mining; phosphorous fertilizer to be produced from slag.
	Krivoi Rog	47°40′–48°25′N, 33°25′E	Principal iron producer; jaspilite type iron formation of Precambrian age; 55–63 percent iron.
	Kursk Belgorod	Near 52°N, 36°E Near 50°40′N, 36°20′E	Part of Kursk magnetic anomaly; reserves apparently greatest in Soviet Union; jaspilite type of iron formation of Precambrian age.
Ural Mountains	Magnitogorsk (Mount Magnitnaya)	53°24′N, 59°05′E	Contact-metasomatic deposit in Carboniferous limestone; 42–62 percent iron.
Kazakhstan	Kustanay	53°40′N, 61°30′E	Magnetite. Reserves may be larger than those of Kursk magnetic anomaly.

* Reprinted by permission of the publishers from Demitri Shimkin, "Minerals—a Key to Soviet Power." Cambridge, Mass.: Harvard University Press, Copyright, 1953, by the President and Fellows of Harvard College.

and resulted from weathering and leaching of silica from ferruginous quartzites. The rich ores generally contain more than 53 percent iron and average 61.4 percent iron, 5 percent silica, 0.1 percent sulfur, and 0.02 percent phosphorous. The total extraction of rich iron ore in this region may reach 70 to 80 million tons a year. In addition to the rich iron ore there is a vast amount of taconite-type material that contains 25 to 35 percent iron but may furnish concentrates that contain 60 percent iron. The total reserves of the Kursk magnetic anomaly have been estimated at 1,113 million metric tons containing 57 percent iron (54, p. 281).

France, Luxembourg, Belgium

The major iron deposits of France (15), which extend somewhat into Luxembourg (59) and less so into Belgium, are those of the Lorraine region. Less important deposits are present in the Armorican area of western France (Brittany and Normandy), and minor deposits in the south—the Canigou district, Pyrennes. More than 90 percent of the production has come from the Lorraine region, generally less than 1 percent from the Pyrennes region (15). Similarly, the major reserves are in the Lorraine region. Mikami (64) gave the reserves in 1944 as: (1)

Lorraine, 4,000 million tons of actual ore containing 30 to 40 percent iron, potential reserve about the same, with the extension into Luxembourg adding about 150 million tons of actual ore; (2) Brittany and Anjou, 150 million tons, 48 to 52 percent iron, Normandy, well over 300 million tons, 45 percent iron, with good potential reserves in each case; (3) Canigou district, Pyrennes, about 50 million tons containing 50 to 55 percent iron. More detailed estimates were given in 1952 (15, pp. 181–189). The deposits of the Pyrennes and the Armorican regions are in rocks of Paleozoic age, those of the Lorraine region are in rocks of Jurassic age. Only the latter deposits will be discussed.

The Lorraine region (15, 89) has been the most important producer of iron ore in Europe. Production began as early as 1833 in the Moselle area and expanded rapidly after the discovery, about 1882, of an important extension of the deposits. The principal area of deposits is a composite basin extending slightly northwest from Metz; the smaller area is a basin to the south, around Nancy, the two being separated by a barren anticlinal area (the Pont-à-Moussan anticline). The northern basin actually consists of a series of gentle folds that trend northeast, and throughout the rocks are cut by a number of normal faults that have about the same trend as the folds. Dips generally are but 2 to 3° to the southwest but carry the ore beds, which crop out in the east, to depths of as much as 3,000 ft in the southwest.

The deposits, generally designated the *minette* ores, are ferruginous oolitic beds of Jurassic age (Upper Lias and Dogger), associated with limestones. The beds in the upper part of the ore-bearing formation are more calcareous and less siliceous than those in the lower part, and thus two general types of ore are recognized, calcareous and siliceous, the calcareous one being the more important commercially as it is self-fluxing. The iron minerals also vary from the upper to the lower parts of the beds. In the higher beds the iron mineral generally is goethite, which is associated with calcite. In the beds toward the middle, several iron minerals may be present—goethite, hematite, magnetite, siderite, chlorite—associated with calcite. In the lower beds the iron minerals generally are siderite and chlorite (11, 12). A number of beds are workable, and generally they are designated by color—green, black, brown, gray, yellow, red, from the bottom upward. The gray, yellow, and red calcareous beds generally have been the most productive. Thickness of workable beds varies but generally must be 3 to 6 m with a content of 31 to 36 percent iron. Analyses made in 1948 and 1950 show the following: Fe, 31.4 to 37.1 percent; CaO, 5.2 to 19.6 percent; SiO_2, 6.37 to 24.50 percent; P, 0.61 to 0.74 percent; S, 0.08 to 0.23 percent.

The origin of the deposits seems most likely to be deposition as sedimentary beds, with differences in mineralogy reflecting differences in environment of deposition (11, 12).

Sweden

Iron deposits occur in two general areas of Sweden: (1) the northern area, north of the Arctic Circle and mostly in Lapland, and (2) the central area, between 59 to 61°N latitude and really well south of the geographic center of Sweden. The northern area is by far the more important, as it contains the very large deposits of Kiirunavaara and Gällivare (26). The central area contains a great number of deposits but most of them are of only moderate size. The deposits of both areas are chiefly magnetite bodies of several types. The date at which the magnetites were first mined is unknown, but it may have been long before 1354, and original crude furnaces were beginning to be replaced by blast furnaces during the fifteenth century; in 1740 Sweden produced not less than 40 percent of the pig iron of the world. Reserves (64) of the northern area are estimated at 1,100 million tons of actual ore containing 62 percent iron, with the same amount of potential ore; those of the central area, 150 million tons of actual ore containing 55 to 60 percent iron, and a similar amount of potential ore.

The great Kiirunavaara magnetite ore body at Kiruna is 4,400 m long, about 90 m wide horizontally, originally stood 240 m above the level of a neighboring lake, and dips about 55°E as a tabular body between syenite porphyry and quartz-bearing porphyry of Precambrian age. Dikelike bodies of magnetite intrude the rocks of both the hanging wall and the footwall on a small scale, making "ore breccia." Apatite occurs with the magnetite in varying amounts. The smaller Gällivare ore body is somewhat similar, but the enclosing rocks have been somewhat strongly metamorphosed and are leptites[1] and gneisses. The ore body shows evidence of deformation and the apatite, about 4 percent, is present in bands. Other smaller, similar ore bodies are present in this northern area, along with some ore bodies of other types. The Kiirunavaara and Gällivare deposits are generally considered to be magmatic in origin and probably injections.

The ore deposits of the central area were worked much earlier than those of the northern area and are the ones on which the iron industry of Sweden was established. All the deposits occur in the Precambrian (Archean) leptite series, which consists of volcanic and sedimentary rocks intruded by granites of three different age groups, all Archean. Three principal types of deposits are present: (1) quartz-banded ores, (2) skarn and limestone ores, and (3) apatitic ores, the first two being very low in phosphorous. The quartz-banded ores represent Precambrian iron formations. They contain hematite, magnetite, or both, the richer, shipping ores containing 47 to 52 percent iron, the leaner, concentrating ores down to 34 percent iron. Some of them contain, in addition to quartz, some metamorphic silicates. The skarn and limestone ores, which grade into each other, are magnetite bodies associated with varying amounts of metamorphic silicates. Some show the characteristics of contact-metasomatic deposits formed in limestone; others show somewhat different features. Geijer and Magnusson (26, p. 481) divided the skarns into two types, reaction skarns and primary skarns. The first type they regarded as formed through reactions within a deposit, the second type through addition of iron, magnesia, and silica during replacement at high temperature. Not all deposits can be placed definitely in one type or the other. The apatitic ores are represented by only a few deposits, the largest of which is Grängesberg. They are lenticular bodies of magnetite, some hematite, and apatite, show intrusive relations to the surrounding leptites, and apparently had the same origin as the bodies in the Kiruna area.

[1] *Leptite* as used in Sweden refers to recrystallized acid volcanics in which the lower limit of the groundmass grain size is between 0.03 and 0.05 mm and the upper limit is 0.5 mm. (26, p. 478).

China

China produced 6.8 percent of the iron ore of the world in 1962 and contains relatively large reserves, especially of low-grade material. The reserves were given by Mikami (64, pp. 14, 22) in 1944 as 500 million tons of actual ore containing 40 percent iron and 700 million tons of potential ore containing 35 percent iron; but Juan (53, p. 427) gave the total reserves as between 1,400 and 1,800 million metric tons of various grades (26 to 65 percent iron). Possibly 60 percent of the total reserves are in Liaoning Province of Manchuria (Fig. 9-5), much of which is low-grade ore, and 13 percent in Chahar (eastern Inner Mongolia) and Shansi Provinces of North China, with smaller reserves in the Yangtze Valley and other places. The types of ore have been classed as (1) metamorphosed deposits, (2) sedimentary deposits, (3) magmatic deposits (actually partly contact-metasomatic, partly hydrothermal), and (4) detrital and residual deposits. The metamorphosed deposits, of Precambrian (Proterozoic) age, constitute 65 percent of the total reserves and consist of banded lean ore (less than 30 percent iron, more than 50 percent silica) with which is intercalated rich ore (60 to 70 percent iron) composed chiefly of magnetite. The material apparently is an original sedimentary iron formation that has been enriched in places as a result of leaching by ascending hydrothermal solutions. The deposits are especially widespread in Liaoning Province. The sedimentary deposits, which constitute about 18 percent of the reserves, include Precambrian (Sinian) bedded hematites (48 to 60 percent iron), best known in Chahar Province; Upper Devonian oolitic hematites (50 to 63 percent iron), present in Hunan, Kiangsi, and eastern Hupeh Provinces of the Yangtze Valley; Lower Carboniferous hematite concretions; and Jurassic siderite nodules. Some contact-metasomatic deposits of hematite and magnetite occur in Hupeh and Anhwei Provinces of the Yangtze Valley. The hydrothermal deposits are tabular bodies or veins composed chiefly of hematite. The detrital and residual deposits are of little importance.

Other Countries

Slightly more than 76 percent of the iron ore produced in the world during 1962 came from the countries that have been discussed; nearly 9 percent came from three other countries, West Germany (3.2 percent), the United Kingdom (3.0 percent), and India (2.5 percent); nearly 5 percent from five other countries, Malaya (1.3 percent), Portuguese India (1.0 percent), Australia (0.9 percent), the Republic of South Africa (0.8 percent), and Liberia (0.7 percent). Only brief comments will be made regarding the deposits of selected countries.

The reserves of iron ore in Germany (47; 64, pp. 8–9; 92, p. 2) are chiefly in (1) the Ilsede-Peine-Salzgitter area of Niedersachsen, southeast of Hanover and (2) in the Württemberg-Baden area of southern Germany in the Rhine Valley along the west side of the Black Forest and the upper waters of the Donau (Danube) River. The deposits of the Ilsede-Peine-Salzgitter area are in rocks of Cretaceous age and contain but 23 to 30 percent iron. The deposits of the Württemberg-Baden area in the Rhine Valley, near Freiburg, contain about 20 percent iron but more lime than silica, whereas those of the Donau (Danube) River area contain about 22 percent iron and more silica than lime.

The iron deposits of the United Kingdom (64, p. 9; 93, 98) are of two main types, hematite ores and bedded ores. The hematite ores, mainly in Carboniferous limestone, contain 56 to 60 percent iron, but these have been mostly worked out and reserves are extremely limited. The bedded ores are of Ordovician, Carboniferous, Jurassic, and Cretaceous age, but the principal ones are of Jurassic age and occur in eastern and central England. The ore minerals are siderite, chamosite, and limonite, and the iron content generally is about 30 percent. The reserves of the Jurassic ores were estimated at 2,000 million tons of actual ore in 1944, and possibly an equal amount of potential ore.

Large reserves of ore containing 60 percent iron are reported to be present in India (55; 64, pp. 14–15; 99). Iron-ore mining and smelting have been carried on for perhaps 3,000 years, and Indian steel furnished the famous Damascus sword blades. Iron ores are widely distributed and are of four general types; hematite ores, quartz-magnetite rock, sideritic and limonitic ores, and lateritic ore. The hematite ores, which are by far the most important, have been derived from sedimentary banded iron formations of Precambrian age. Large deposits of such ores are present in Bihar, Orissa, and Central Provinces; lesser deposits in Bombay and Mysore. The quartz-magnetite rock, apparently metamorphosed hematite ore, occurs especially in Salem and Trichinopoly districts of Madras and in parts of Mysore. The sideritic and limonitic ores occur in the Iron Shale of the Gondwana system, especially in the Raniganj coal field of Bengal. The lateritic ores are rather widely distributed. The major reserves consist of hematite ores.

Australia has furnished only a small amount of the iron ore of the world, but recent work there indicates that newly developed deposits may be very large (46, 100, 102, 103, 105). The deposits worked in 1963 were those of the Middleback Ranges, South Australia, and the islands of Cockatoo and Koolan, Yampi Sound, Western Australia. Other important areas are the recently discovered deposits in the Pilbara region, Western Australia, stated to contain 3 billion tons of ore, and several other deposits farther south in Western Australia. Moreover, magnetometer surveys in northwestern Tasmania have shown the presence of two large magnetic zones, Savage River and Long Plains. The reserves there appear to be enough for the establishment of an iron and steel industry in Tasmania.

The deposits of the Middleback Ranges of South Australia (23, 79) are in the vicinity of Whyalla on Spencer Gulf and consist of a rather typical iron formation of Precambrian age, the Middleback banded quartzites composed of alternating layers of iron oxides and silica. The deposits generally are in the eroded keels of pitching synclines and consist of high-grade hematite ores, 65 percent iron, and rather large reserves of low-grade hematite-quartzite type, possibly 30 percent iron. The high-grade ores apparently were formed by leaching silica from the hematite-quartzite type.

The Cockatoo and Koolan deposits of Western Australia (13), in Yampi Sound, are about 160 miles northeast of Broome. The ore occurs in hematite-quartzite believed to be part of the Mosquito Creek series of Precambrian age. The hematite-quartzite is the dominant rock of the two islands and varies from a weakly ferruginous rock to one almost wholly composed of hematite where the ore bodies occur. The reserves of high-grade ore are high: 74.2 million tons of ore containing 64 to 69 percent iron and 10.5 million tons containing 54 percent iron. The reserves here and elsewhere in Western Australia indicate that Australia now is one of the four major sources of high-grade iron ore in the world.

Manganese

More than 40 percent of the manganese of the world generally is produced by the Soviet Union and about 40 percent by six other countries—India, Ghana, the Republic of South Africa, Brazil, Morocco (Southern Zone), and China (Fig. 14-7). Less important production in 1962 came from many countries: the Congo, Japan, and British Guiana, each about 2 percent of the world total; Gabon, Roumania, and Mexico, each slightly more than 1 percent; and 31 other countries furnished the rest. The major ore reserves of high and intermediate grade are shown by Table 14-16. In addition to these reserves, it has been estimated that there are, in the world, something more than 1 million thousand tons of low-grade manganese materials, from which the amount of manganese that could be recovered economically is unknown. Exclusive rights to a process to obtain high-purity manganese dioxide from low-grade ores was obtained by Sterling Oil of Oklahoma, Inc. (193), and a five-year program planned. It was stated that by this process materials containing as little as 15 percent manganese by dry weight could be economically treated. If such treatment should become feasible on a large scale, much of the low-grade material of the world might ·become usable.

Most of the manganese ore of the world is used in making steel; other uses include the making of dry-cell batteries (battery ore), manufacture of chemicals, and some others. Manganiferous material is classified by the United States Bureau

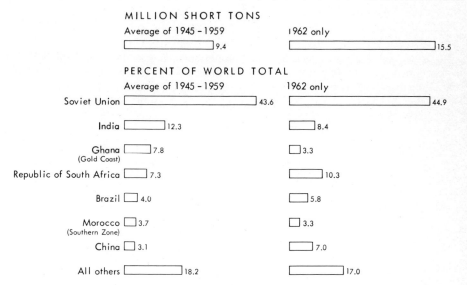

Fig. 14-7 World manganese ore production. Compiled from *U.S. Bur. Mines Minerals Yearbook,* various years.

TABLE 14-16.   Approximate Major World Reserves of Manganese Ore*

Country	Thousands of tons	Mn content, percent
Soviet union	625,000	20–50
India	92,000	47–52
Brazil	60,000	38–50
Republic of South Africa	60,000	30–54
Morocco (Southern Zone)	47,400	40–50
China	29,000	20–50
Ghana (Gold Coast)	12,000	50 plus
Total of above	925,400	
World total	980,250	High and intermediate grade

* Compiled from U.S. Bur. Mines, Materials Survey, Manganese, Chap. 2, pp. 7–11, 1952.

of Mines as manganese ore (35 percent or more manganese), ferruginous manganese ore (10 to 35 percent manganese), and manganiferous iron ore (5 to 10 percent manganese). In general, in the United States material used as metallurgical, battery, and chemical ore contains 35 percent or more manganese. Nearly 75 percent of the ore used in producing manganiferous pig iron in 1962 contained 5 to 10 percent manganese, nearly all of the rest contained 10 to 35 percent manganese, and a negligible amount contained more than 35 percent manganese.

Soviet Union

The manganese deposits of the Soviet Union (80, pp. 65–68; 107, pp. 185–203; 123, pp. 29–30; 136, Chap. 10, pp. 24–37) occur chiefly in two areas, Nikopol of the southern Ukraine and Chiaturi (Chiatura, Tschiaturi, Chiatury) in the Transcaucasus, both near the Black Sea. Less important deposits occur at Laba and at Bely Kluch (about 190 miles north and 70 miles south, respectively, of Chiaturi); at Polunochnoye (60°50′N, 60°25′E) and Marsyaty (Marsiata) (60°04′N, 60°28′E) in the northern Ural Mountains; in the Abzelilovo (Beloretsk) group (53°50′N, 58°50′E) and the Baimak group (52°35′N, 58°20′E) of the southern Ural Mountains; at Mangyshalk (43 to 44°33′N, 50 to 53°E) in Kazakhstan, near the northeast coast of the Caspian Sea; and at Mazul'skoye (56°15′N, 90°35′E) in western Siberia. Still less important deposits occur at a number of places, especially in the Ural Mountains. A new manganese deposit in the Ural Mountains, stated to contain up to 55 percent manganese and to be suitable for open-pit mining, was reported recently.

The Chiaturi deposit (80, pp. 66–67; 89, p. 70; 136, Chap. 10, pp. 24–25), on the south slope of the Caucasus Mountains (42°17′N, 43°18′E), is reputed to be the largest manganese deposit in the world and also the largest single producer of high-grade ore—42 to 55 percent manganese with 0 to 1.8 percent iron. It consists of a nearly horizontal bed of marly sand of Oligocene age, 6.5 to 10 ft thick, that contains oolites of pyrolusite and psilomelane, along with some braunite and wad. This bed, which is known for a length of 19 miles in a strip 5 to 6 miles wide, is deeply buried, but erosion along the Kvirila River has exposed the ore on the slopes of deep valleys and permitted access by adits (tunnels), some of which are beneath 300 ft of overlying rock.

The Nikopol district consists of two ore fields about 10 to 15 miles apart (47°32'N, 34°40'E and 47°32' to 47°40'N, 34°05' to 34°30'E). The ore bed is sandy clay of Oligocene age, 3 to 12 ft thick, that contains oolitic pyrolusite along with some psilomelane, wad, and manganite. It is covered by 50 to 200 ft of overburden and is reached by shafts. The ore is of lower grade than that of Chiaturi and averages about 33 percent manganese and possibly 1 to 3 percent iron, but it is concentrated into three grades, manganese content about 50 percent, 42 percent, and 36 percent respectively.

According to Maksimov (125) hundreds of manganese deposits are present in central Kazkhstan, most of which have received very little study. One of the better-known deposits, at Dzhedzy (about 49°N, 67°E), consists of blanketlike ore bodies along the bedding of Middle Devonian sedimentary rocks and, near faults, veinlike or lens-shaped bodies that wedge out above and below and consist chiefly of braunite, apparently formed by metamorphism of original psilomelane.

India

Manganese deposits are rather widely distributed throughout India (107, pp. 66–81; 111, pp. 204–221; 123, pp. 27–29; 132; 136, Chap. 10, pp. 3–14; 137), but the principal deposits are in the states of Madhya Pradesh (Central Provinces), Bombay, Bihar and Orissa, and Andhra. Estimates in 1956 (132) indicated that by far the largest-known reserves of India are in Madhya Pradesh—20 million tons of measured and indicated ore, possibly 60 million tons of inferred ore. Other reserves of considerable importance are in the states of Bihar and Orissa, Bombay, Andhra, Mysore, and Goa (Portuguese).

Recorded production of manganese in India began in 1891 in the state of Orissa, although the first recorded occurrence of manganese was in 1829 in the Nagpur district of the state of Madhya Pradesh (135, p. 67). Total production of India up to 1955 was 38,890,000 tons of ore, of which about 66 percent came from Madhya Pradesh, about 12 percent from Madras and Andhra, and about 11 percent from Orissa and Bihar. During the 55-year period 1900 to 1954, Madhya Pradesh furnished nearly 25 million long tons of ore that averaged about 48 percent manganese.

The manganese deposits of India, according to Sondhi (132, p. 9), may be divided broadly into two general types, conformable tabular deposits and lateritic or superficial deposits. The first type was probably deposited as sediments in Precambrian rocks, the second type was formed as a result of weathering of manganiferous rocks. The first type is by far the more important one.

The manganese deposits of Madhya Pradesh (130, 131, 135) extend throughout a zone about 130 miles long and up to 16 miles wide, which is situated in central India about half way between Bombay and Calcutta. Although about 200 individual deposits within this zone have yielded some manganese ore, about 20 of them have furnished over 90 percent of the total production. Among these 20 deposits the length ranges from a few hundred feet to nearly 4 miles; seven of them show a continuous length of more than 1 mile each. The largest one is reported to be the Balaghat deposit (18°30′N, 76°00′E) of the Bharweli mine (137), with a continuous length of 8,800 ft. Generally the thickness of deposits is from 3.5 to 50 ft. Deposits have been mined as much as 600 ft down dip from the outcrops and have been proved to extend as much as 1,000 ft down dip.

The deposits of Madhya Pradesh occur in intensely deformed and metamorphosed sedimentary rocks of the Precambrian Sausar series, which are overlain by the Cretaceous Lameta formation (sandstones and limestone) and the overlying Deccan traps (plateau basalts) of Late Cretaceous–Eocene age. The lodes are closely associated with one formation, the Mansar, but occur in three zones. The manganese-bearing rock of the Sausar series has generally been referred to as "gondite."[1] As originally used, gondite referred to metamorphosed manganiferous rocks composed of spessartite and quartz as essential minerals, but containing also other manganese silicates. As later applied it apparently was restricted to a quartz-spessartite rock, generally a very minor type. Straczek an others (135, p. 67) use gondite in the broader sense for a manganiferous rock that is not ore but contains quartz and variable amounts of manganese silicates in addition to spessartite and commonly some braunite. They thus distinguish between ore deposits and gondite deposits, either one of which may be oxidized and enriched.

The deposits of the region have been classified (135, p. 72) as (1) lode deposits, subdivided into (*a*) hypogene or primary deposits and (*b*) supergene deposits; (2) detrital or "boulder ore" deposits. The hypogene deposits are subdivided into (1) hypogene oxide (braunite) deposits with or without subordinate manganese silicates and (2) gondite deposits, which contain subordinate braunite. The supergene deposits may be either oxidized braunite deposits or oxidized gondite deposits. Intense oxidation generally extends only to a few feet or a few tens of feet below the surface; exceptionally to as much as 600 ft. In the early days of mining probably the detrital deposits were important, but now the lode deposits furnish more than 90 percent of the ore. In general the lodes are tabular, conformable lenses.

Origin of the deposits as original sediments that were subsequently metamorphosed has been supported by a recent comprehensive field and laboratory study in the Bharweli mine area, Balaghat district (131).

The principal deposits of the state of Bombay (130) are in the Panch Mahals and Baroda districts. They occur as tabular to irregular lenses which generally are parallel to the foliation and apparently the bedding of highly folded phyllites, quartzites, and cherts of Precambrian age. Individual deposits attain a thickness of 50 ft and a length of several thousand feet. The ore minerals, which are chiefly pyrolusite and psilomelane-type oxides, are known to extend to a depth of 500 ft below the surface. Apparently they replace the rocks in which they occur, and ore bodies grade into phyllites and quartzites. Roy (130, p. 57) is inclined to attribute the concentration of the ore bodies to the circulation of meteoric waters.

The deposits of the states of Orissa and Bihar (129) are chiefly in Orissa. They occur in four principal areas: (1) Singhbhum-Keonjhar-Bonai, about 200 miles west and slightly south of Calcutta; (2) Gangpur, about 50 miles farther west, inactive since 1948; (3) Patna, west-central part of Orissa, about 120 miles

---

[1] Named after the Gonds of the Central Provinces (Madhya Pradesh State) of India [Chap. 2 (6, p. 311)].

southwest of the Gangpur area; and (4) Koraput-Kalahandi, about 100 miles south of the Patna area. The major deposits of the Singhbhum-Keonjhar-Bonai area are tabular and lenticular bodies up to 1,000 ft long and 20 ft thick in highly folded rocks of the Precambrian Iron Ore series. Some of them have been mined to as much as 400 ft down the dip. Most of the ores contain less than 40 percent manganese and many contain much iron. They apparently were formed by oxidation and enrichment of some primary material that has not yet been reached by mining. The Gangpur deposits were oxidized and enriched zones of gondite. The deposits of the Patna and the Koraput-Kalahandi areas (134) are similar and occur in deeply weathered, isoclinally folded Precambrian garneti-ferous quartzites, schists, and minor intercalated rocks of other types. The deposits of the Patna area occur in a zone about 15 miles long and 4 to 6 miles wide; those of the Koraput-Kalahandi area in a zone about 20 miles long and 1 to 2 miles wide. The most important ore bodies are tabular, conformable lenses that contain pyrolusite, psilomelane, and some other manganese oxides in clay and weathered schist. They are as much as 2,500 ft long and 35 ft wide. The ores contain 25 to 54 percent manganese, but only about a quarter of them contain more than 40 percent manganese. The richer, central parts of ore bodies character-istically grade outward into ferruginous manganese ore or manganiferous iron ore.

Africa

Manganese deposits occur at many places in Africa, and that continent furnished 22 percent of the manganese produced in the world in 1962. More than nine-tenths of this production came from five countries: South Africa, 47 percent; Morocco, 15 percent; Ghana, 15 percent; the Congo, 10 percent; Gabon, slightly less than 7 percent. The major evaluated reserves are in South Africa and Morocco, with considerably less in Ghana (Table 14-16). Only the deposits of those three countries will be discussed.

Republic of South Africa. The manganese deposits of South Africa (107, pp. 109–116; 108; 114; 115; 119; 120, pp. 260–274; 123, pp. 30–31; 124; 136, Chap. 12, pp. 16–18) that have furnished most of the production are in the Postmasburg field, somewhat more than 100 miles northwesterly from Kimberly. Another field, the Kalahari or Kuruman, which is chiefly sand-covered, is marked by a large outcrop (Black Rock) about 90 miles north of Postmasburg and approximately 50 miles northwest of the village of Kuruman (Fig. 16-1). Recent investigations indicate that the reserves of the Kalahari field may be far greater than those of the Postmasburg field. Manganese ores, which are associated with iron ores, occur throughout this region from about 60 miles south of Postmasburg to several miles north of Black Rock, a total distance of about 180 miles; substantial barren gaps are present, however.

Throughout the Postmasburg and Kalahari (Kuruman) fields various manga-nese deposits are present and are closely associated with a zone of thrust faulting that caused much fracturing of the rocks. The deposits occur in several formations of Precambrian age, as well as in cataclastic rocks that came into existence during an orogeny that followed the deposition of the Precambrian formations. The stratigraphic occurrences and some features of the ore bodies are shown by Table 14-17.

The Postmasburg field consists of an eastern and a western zone, which merge both to the north and to the south. The ore bodies of the eastern zone are highly irregular; those remaining are of small size, since the large bodies have been worked out, and are widely scattered throughout extensive masses of siliceous breccias, fault gouge, and residual and slump material from the dolomite of the Dolomite series. Although the ore contains but a small amount of iron and is of a very desirable type, the scattered occurrence and small size of the deposits make mining conditions generally unprofitable. The ore bodies of the western zone occur along a distance of 25 miles throughout which the basal Gamagara beds (Table 14-17) were thrust over the beds of the Dolomite series. Generally the base of the ore bodies lies on or closely adjacent to dolomite, which forms a highly irregular floor. Locally mining has shown that pockets of ore have subsided into irregular solution channels to depths of 30 ft or more, whereas nearby pillars of dolomite extend upward nearly to the surface (115, p. 47).

In the Kalahari (Kuruman) field, manganese ore is present in two and possibly three persistent layers in the Upper Griquatown banded ironstones (Table 14-17). The two well-known manganese-bearing layers are separated by 70 to 100 ft of banded ironstones. The upper ore body, about 20 feet thick so far as it has been exposed, contains 45 to 55 percent Mn and 5 to 15 percent Fe. The lower ore body, about 50 ft thick, contains 30 to 38 percent Mn, 3 to 5 percent Fe, 5 to 8 percent SiO_2, the rest being chiefly $CaCO_3$. This layer is persistent under cover, but as it goes deeper beneath the surface down the dip of the beds the manganese content decreases to about 25 percent and the calcium carbonate content increases to as much as 30 percent. Apparently this is about the character of the original unweathered material, which seems to have been a well-laminated manganiferous limestone. The upper parts of such beds, near the surface, have been leached of some calcium carbonate and enriched in manganese as well as in iron. The manganese-bearing layers of this field are known intermittently for 35 miles along the strike and, in places, for 1,000 ft down the dip.

Various estimates have been made of the reserves of the Postmasburg and the Kalahari (Kuruman) fields. One estimate (120, p. 266) for the Postmasburg field was 20 million tons of ore, distributed as follows: 8 million tons, 25 to 34 percent Mn; 7 million tons, 35 to 42 percent Mn; 4 million tons, 43 to 47 percent Mn; 1 million tons, more than 48 percent Mn. Another estimate (115, p. 101) was 50 million tons of ore close to the surface. Estimates for the Kalahari (Kuruman) field are more difficult to make, because of limited exploration, but there is general agreement that the reserves are large. At Black Rock alone the reserves have been estimated to be at least 8 million tons of ore that probably contains 45 to 47 percent Mn, 6 to 10 percent Fe, and 4 to 8 percent SiO_2. An estimate of the probable reserves over the entire field was many hundred million tons, although locally the ore is covered by thick overburden (115, p. 3).

Various suggestions have been made about the origin of the Postmasburg and the Kalahari deposits, and these have been reviewed by de Villiers (114, pp. 84–90, 155–157). The concept that seems to be accepted by the largest number of investigators is that the ores resulted from infiltartion by manganese-rich

TABLE 14-17. Occurrences of Manganese Ore, Postmasburg and Kalahari (Kuruman) Fields, South Africa*

Age	Materials	Geographic and stratigraphic occurrence of manganese ore	Types of ore bodies and ores
Tertiary-Recent and Perno-Carboniferous	Formations overlying manganese deposits		
UNCONFORMITY			
Pre-Perno-Carboniferous and post-Matsap-Gamagara	Cataclastic rocks, fault gouges, silicified breccias	Postmasburg, eastern zone; below and in cataclastic rocks, fault gouges, breccias, where they lie on dolomite of Dolomite series	Siliceous type; largest and richest ore bodies occur on or near contact of dolomite and siliceous breccias. Ore minerals chiefly braunite and some low-iron psilomelane and pyrolusite. Irregular and widely scattered bodies of small size.
OROGENIC PERIOD			
Precambrian Loskop system	Matsap and Gamagara beds (quartzites, shales, conglomerates)	Postmasburg, western zone; along base of Gamagara shales where they lie on dolomite of the Dolomite series	Ferruginous type; iron content upward from 10 percent to that of manganiferous iron ore. Ore minerals are braunite, bixbyite, ferruginous psilomelane, jacobsite, some hausmannite. Ore bodies furnish major production of Postmasburg field.
UNCONFORMITY			
Transvaal system	Upper Griquatown banded ironstones	Kalahari (Kuruman) field; near base of the ironstones	Ore bodies apparently conformably intercalated in the banded ironstones; persistent long distances down dip. Deposits some distance below surface contain 25 percent Mn and 30 percent $CaCO_3$, but near surface are enriched to 40–55 percent Mn with 5–15 percent Fe.
	Middle Griquatown lavas	About 50 miles south of Postmasburg; in shaly facies near top of ironstones	Somewhat ferruginous ore.
	Lower Griquatown banded ironstones	Various places in Postmasburg field; may occur in deep solution cavities	Unweathered dolomite may contain as much as 3 percent MnO_2.
	Dolomite series		

* Compiled from Commonwealth Mining and Metallurgical Congress, 7th, South Africa, vol. 1, pp. 201–215, Manganese in the Union of South Africa, by L. G. Boardman; Copyright 1961. Used by permission.

meteoric waters. The source of the manganese, at least in the Postmasburg field, is thought to have been the dolomite of the Dolomite series, which is known to contain as much as 3 prcent MnO_2, which crops out over large areas, and which was much fractured by faulting in many places. This explanation seems to apply well to the Postmasburg field but offers more difficulty in the Kalahari field (114, pp. 62–63; 119, pp. 593–595; 124, pp. 85–86). The character of material in the Kalahari field suggests that some original layers in the ironstones were essentially manganiferous limestones and that these were enriched by weathering and leaching near the surface. These beds are stratigraphically about 5,000 ft above the dolomite of the Dolomite series.

Morocco.   The chief manganese deposits of Morocco (107, pp 154–159; 109; 110; 123, pp. 24–26; 136, Chap. 12, pp. 9–13) are those of Imini, Bou Arfa, and Tioune and vicinity. The deposits of Imini are by far the most important and have supplied most of the production. Both those deposits and the ones at Tioune and around Ouarzazate are in the same general region as the cobalt deposits of Morocco (Fig. 11-8) and are unfavorably situated regarding transportation, which must be by truck to Marrakech, the nearest railway station.

The Imini deposits are known throughout a distance of about 25 km and occur in dolomites of Middle Cretaceous age. Three manganiferous beds are present, but only the lower two have been mined. The thickness of the beds is 1 to 2.5 m, and in some places they join. A red clay associated with the ore contains pyrolusite and appears to be related to solution of the dolomite and partial replacement by manganese oxide.

The Bou Arfa deposits of eastern Morocco are somewhat similar to those of Imini, as they occur in limestone and dolomite, but they are of Middle Liassic age and have been much disturbed by folding and faulting. Two mineralized beds are present, the lower one mined at Hamaraouet, the upper one at Ain Beida. In general, solution has formed cavities and canal-like passages in the dolomite, which in places contains thin layers of manganese oxide. Solution has been especially notable along fractures. The deposits are pocketlike and lenticular masses composed chiefly of pyrolusite and psilomelane, although hausmannite is abundant in some places. Goethite is also abundant. The dolomite walls of the deposits, as well as blocks of dolomite in the ore, are covered with red clay, apparently connected with solution of the dolomite and its partial replacement.

The deposits at Tioune and around Ouarzazate are of Precambrian age and of two types. Those at Tioune are lenticular beds in continental formations associated with volcanic rocks. The ore minerals are chiefly braunite and psilomelane and show evidence of at least some replacement of the beds in which they occur. The deposits around Ouarzazate are chiefly in volcanic rocks. The ore minerals are mainly braunite, hollandite, and hausmannite; the gangue of some is dolomite, of others it is barite and quartz.

Ghana (Gold Coast).   The manganese ore of Ghana (123, pp. 26–27; 133; 136, Chap. 12, pp. 13–14) comes chiefly from the Nsuta area near the town of Nsuta. The deposits occur for 2.5 miles along a ridge composed of highly folded phyllites, apparently of Precambrian age, which contain beds of quartz-spessartite rock (gondite). Within these rocks are lenticular ore bodies that may be original beds that have been enriched. Much detrital ore is present also, down to depths of as much as 40 ft. The detrital material is washed to remove clay and quartz. The average content of manganese is 48 percent; of iron, 4 to 5 percent; of silica, 4 to 5 percent. Proved reserves are estimated to be something more than 10 million tons of ore.

A recent study by Sorem and Cameron (133) indicates that three genetic types of ore are present: replacement, cavity-filling, and detrital or residual. The chief minerals are various manganese oxides, with which goethite is closely associated in some ore. Some disagreement exists regarding the origin of the deposits, but it seems clear that weathering and circulation of meteoric water played an important part in bringing about deposition of some of the material.

Brazil

The major manganese production of Brazil has come from the state of Minas Gerais. Important reserves in addition to those in Minas Gerais are in Mato Grosso and the Federal Territory of Amapá (106; 107, pp. 66–81; 116; 117; 118; 123, p. 14; 126; 127; 128; 136, Chap. 9, pp. 4–5).

Minas Gerais. Manganese ore was first mined in Minas Gerais in 1894, although the presence of ore was known as early as 1871. During the earlier history of manganese mining, production came from numerous small and medium-sized deposits, many of which were completely exhausted. Production from the Lafaiete (Quéluz) district has been greater than that of any other district and probably exceeds that of all of the other deposits of Brazil combined, but the reserves of the Amapá deposits, as well as those of the Urucum deposit in Mato Grosso, apparently are greater than those of the Lafaiete district (118, p. 326). Early producing mines of the district are inactive, and production now is from the Merid (Morro da Mina) mine, which began operation in 1902 and has furnished more than half of the manganese ore of Brazil. Between 1902 and 1956 the total production from that mine was 6,074,238 metric tons of ore, the greatest production of any manganese mine in the Western Hemisphere.

Unfossiliferous metamorphosed rocks that have been classed as Precambrian have been the protores for most of the high-grade manganese deposits of Minas Gerais. Rocks of especial importance are metasedimentary ones and possibly some pyroclastic and other metaigneous rock (118, p. 265). Original protores of three types have been recognized (118, p. 289): (1) manganese silicate-carbonate rock; (2) the chemical sediments that now form itabirite and marble (marble-itabirite protore); and (3) the clastic sediments that now form phyllite, quartzite, and metatuffs (?) (clastic protores). These protores were weathered and were leached by circulating groundwater, and commercial-grade ore was formed from the protores by supergene enrichment. The manganese silicate-carbonate rock has been by far the most important protore and probably has been the source of three-fourths of the total production. Millions of tons of ore that was formed by enrichment of the various protores averaged about 48 percent manganese, but most of this rich ore has been mined, so that in 1954 the average grade of the ore exported was about 44 percent manganese, and the grade of that used for domestic supplies was probably not much greater than 40 percent manganese.

Amapá. The manganese deposits of the territory of Amapá (117, 126, 127, 128; 136, Chap. 9, pp. 7–10), discovered in 1946, lie within an area about 10 km long and probably 200 m wide about in the center of Amapá, known as the Serra do Navio district. Production began in 1957 at the rate of more than 600,000 metric tons of ore a year, and in 1961 it reached 750,000 tons. The shipping-grade ore averages more than 48 percent manganese. The rocks in which the deposits occur are part of the crystalline complex that forms the Guianan Highlands; in the Serro de Navio district the rocks apparently consist of a series of metamorphosed sedimentary and volcanic rocks, within which are small nodules to large lenses of manganese oxide. Outcrops of ore, which generally are elongate, characteristically are composed of hard, massive manganese oxides that show no trace of bedding. Typically the outcrops are vuggy and cavernous, and many display solution chimneys in the ore. These chimneys are nearly circular, up to 1 m across and 2 m deep. They apparently result from the accumulation and decomposition of vegetation in low spots, with the formation of organic acids that dissolve small amounts of mineral and are flushed out during heavy rains. Remnants of altered garnetiferous rock occur in the outcrops of ore, and these may form small patches or masses several meters in diameter. Boulders of ore, and areas of such boulders, occur near outcrops, and small concretionary nodules of earthy manganese oxide, locally designated *granzon*, are widely distributed in the soils above and around the outcrops. The ore has been divided into four types (126): (1) hard ore, which is of relatively high grade and composed of botryoidal and colloform cryptomelane mixed with pyrolusite; (2) schistose ore, of widely varying grade, composed of a mixture of oxide ore and clay; (3) gondite ore, 15 to 45 percent manganese, which contains a high concentration of garnet pits, preserved garnet crystals, or both; (4) lateritic ore, which consists of boulders of float ore and pellets of granzon in red clay.

Proved reserves are greater than 14 million metric tons of ore containing 46 percent manganese. Probably they are much larger. The largest known ore body contains more than 3 million tons of ore, and five others contain more than 1 million tons each.

Insufficient detailed work has been done to give a clear picture of the origin of the ore. Some features indicate that weathering and leaching by meteoric waters had some part in the formation of the ore, but doubt has existed regarding the nature of the original material (127, pp. 370–374). Nagel (126, pp. 496–497) regarded the ores as residual products that formed by weathering and enrichment of the protores, which he described as both manganese carbonate and gondite and believed to be sediments of Precambrian age derived from an unknown source. Subsequent deformation and regional metamorphism were followed by erosion, renewed uplift, and deep weathering. High-grade massive manganese deposits were formed from carbonate protores, whereas somewhat lower-grade deposits were formed from silicate protores.

Mato Grosso. The manganese deposits of Mato Grosso (106; 107, pp. 76–77; 128) are 22 to 30 km south of Corumbá, which is on the bank of the Paraguay River close to the Brazil-Bolivia border. These deposits, as well as iron deposits, crop out in the Urucum Mountains, which rise from the plain around Corumbá and thus are generally designated the Urucum deposits. The iron deposits were explored in 1876, the manganese deposits in 1894. Mining was sporadic until about 1955 but increased thereafter. The reserves are very large; one estimate published in 1956 (106, p. 271) was 50 million metric tons of indicated ore and 20 million metric tons of inferred ore; another one in 1945 (116, pp. 37–38) was 33,670,000 metric tons of measured, indicated, and inferred ore containing 45.6 percent manganese and 11.1 percent iron, coupled with a statement that the geologically possible ore might be 53,200,000 metric tons containing about 45.6 percent manganese and 11.0 percent iron. In addition to the manganese ore there is present an estimated 1,310,000 metric tons of banded hematitic iron formation containing 55 to 56 percent iron and 18 to 20 percent silica (116, p. 43).

The manganese and iron deposits occur as beds in the Banda Alta formation, thought to be of early Paleozoic age. The formation consists chiefly of alternating layers of hematite and chert—a typical iron formation similar to those of the Lake Superior region—but it contains two beds of manganese oxides that are of commercial importance. The lower bed, designated Bed No. 1, occurs at the base of the Banda Alta formation, and the upper bed, No. 2, occurs 30 to 45 m above it. A third bed of no economic importance is present also. Bed No. 1, which is the principal one, generally varies in thickness from 2 to 6 m but averages about 3 m. It is known throughout a distance of 7.5 km around the cliffs of Morro do Urucum, about 500 to 600 m above the surrounding plain. Bed No. 2 varies in thickness from 1 to 2.2 m and averages about 1.2 m. The manganese-bearing beds consist of an intimate mixture of very fine-grained hematite and cryptomelane. Generally the ore is dense and hard, but a minor amount of it is vuggy and some is nodular.

The manganese and iron deposists of Urucum have been classed as original sedimentary beds that have changed but little since deposition (128, pp. 4–6). The presence of clastic beds associated with the manganese beds indicates an abrupt but temporary change in the environment of deposition, possibly with an attendant change in pH which favored rapid deposition of manganese oxide. Also, it has been suggested (106, pp. 268–271) that certain microorganisms may have been a contributing factor in the deposition of the manganese.

China

China produced about 7 percent of the manganese of the world in 1962. Reserves may be about 29 million tons of ore containing 20 to 50 percent MnO_2 (136, Chap. 10, p. 3).

The most important manganese deposits (53, pp. 433–436; 107, pp. 122–125; 112; 136, Chap. 10, p. 3) occur chiefly in South China, south of the Yangtze River, in Kiangsi, Hunan, Kwangtung, and Kwangsi Provinces (Fig. 9–5). Other deposits are present in Shantung, Hopeh, and Liaoning Provinces. Most of the ores are sedimentary beds of several ages: Precambrian (Sinian), Devonian, Carboniferous, and Permian. Many of the original beds consisted of manganese-bearing carbonate that contained 10 to 25 percent manganese; others apparently contained primary manganese oxides and were of higher grade. Oxidation and enrichment of some of the lower-grade ores brought the manganese content up to 40 to 50 percent. Surficial red clays in Kwangsi and Kwangtung Provinces contain residual boulders and masses of psilomelane and concretions of pyrolusite and psilomelane, which can be sorted to make good ore containing more than 40 percent manganese.

United States

Production of manganese ore in the United States was only 2 percent of the world total during the 15-year period 1945 to 1959, and since then it has been less than that. Production through 1959 was stimulated by government contracts, which were terminated at the end of 1959, with the result that in 1960 the production was less than 40 percent of that supplied in 1959. Montana furnished practically all of the manganese ore shipped in 1962 and generally has been the leading state since 1916. Other production during relatively recent years has come chiefly from Nevada, Arizona, and New Mexico. Managiferous iron ore (5 to 10 percent manganese) is produced almost entirely in the Cayuna iron district of Minnesota. Manganiferous residuum is recovered from the treatment of manganiferous zinc ores of New Jersey. Early production of manganese ore came chiefly from Arkansas, Virginia, Georgia, and California. Total production of these states and of Montana through 1925 follows, expressed as the percentage of the total production of the United States:

Montana	42.2
Virginia	26.4
Georgia	7.9
Arkansas	6.6
California	5.4
Total	88.5

Production began in 1850 in Arkansas; in 1859 in Virginia; in 1866 in Georgia; in 1867 in California; and in 1916 in Montana (122, pp. 209–212). The late date of the start of production in Montana, coupled with a production of more than 42 percent of the total of the United States, emphasizes the great importance of Montana as a source of manganese.

Various estimates of reserves in the United States show that, although a considerable amount of manganiferous material is available, the percent of manganese contained in various deposits is generally low. The major reserves, with some comments about the deposits, are shown in Table 14-18.

The deposit at Butte, Montana, is the most important commercial source of manganese in the United States. Rhodochrosite occurs in the peripheral zone of the copper deposits, previously discussed. Base and precious metals are separated from the rhodochrosite by flotation, after which the rhodochrosite concentrate is roasted and made into nodules that contain about 58 percent manganese.

Among the other deposits, those of Aroostook County, Maine, of Chamberlain, South Dakota, and of Artillery Peak, Arizona, although extensive, have not

TABLE 14-18. Reserves of Manganese in the United States in 1955*

District	Raw material, long tons	Average percent manganese	Contained manganese, long tons	Type of material
Cayuna Range, Minnesota	450,000,000	5.0	22,500,000	Precambrian manganiferous iron ore, plus large low-grade unoxidized and oxidized iron formation
Aroostook County, Maine	280,000,000	9.0	25,200,000	Bedded Paleozoic ferruginous and manganiferous marine sediments
Chamberlain, South Dakota	69,000,000	15.5	10,700,000	Manganese carbonate concretions in Cretaceous Pierre shale
Artillery Peak, Arizona	156,000,000	4.0	6,240,000	Bedded oxides in Pliocene continental deposits
Butte, Montana	4,460,000	14.0	624,000	Rhodochrosite from peripheral zone of hydrothermal copper deposits
Three Kids, Nevada	4,460,000	10.0	446,000	Wad and manganese oxides in Pliocene (?) Muddy Creek formation; derived from hot-spring sources (?)
Pioche, Nevada	3,570,000	10.0	357,000	Manganoan siderite replacement deposits in Cambrian limestone; peripheral to precious and base-metal hydrothermal deposits
Philipsburg, Montana	710,000	22.5	160,000	Rhodochrosite replacements in Lower Paleozoic carbonate rocks, associated with hydrothermal vein deposits (Ag-Pb-Zn); some supergene enrichment
Leadville, Colorado	3,570,000	15.0	536,000	Oxides from manganosiderite replacements of Paleozoic limestone associated with lead-zinc-silver deposits; hydrothermal origin
Other small deposits	—	—	3,000,000	

* Compiled from *U.S. Geol. Survey Mineral Resource Inv. Resource Map* MR-23, 1962.

furnished any production because at present the manganese cannot be recovered economically.

The Three Kids district, Nevada, about 15 miles southeast of Las Vegas and near Lake Mead, furnished some 15,000 to 20,000 tons of manganese ore to the end of World War I (136, Chap. 7, p. 26) and then little until 1952 when a concentration plant was completed whereby the relatively low-grade ore of the deposit was converted into nodules containing 45 percent manganese. This, together with government contracts, stimulated production so that in 1954 the production from Nevada exceeded that of Montana, and Nevada remained in first place in all but one year through 1960, but after the termination of a government contract production stopped and the concentration plant was sold at auction in 1961. Although the manganiferous beds, 10 to 75 ft thick, seem to have an extent of about 6 miles, they contain numerous partings of sand and silt; individual beds of manganese oxide generally cannot be traced more than a few feet, and the rocks have been cut into a series of small fault blocks.

A manganiferous zone extends along the Appalachian Mountains from Pennsylvania to Alabama (122, pp. 182–184), in which Paleozoic rocks have undergone prolonged weathering and erosion. As a result, small amounts of manganese have been concentrated by supergene action, and manganese oxide nodules, along with some veins and disseminated oxides, have formed in clay that accumulated from the decomposition of carbonate rocks. Most of the residual clay masses lie on terraces on the slopes of the higher ridges, within which the ore bodies make lenticular masses a few feet to 20 ft thick. Within this zone are two deposits of especial note because they supplied much of the manganese of the United States before 1916: Crimora, Virginia, and Cartersville, Georgia.

Manganese deposits are present at several places in Arkansas and Oklahoma, the most notable of which are in the Batesville district, Arkansas (122, pp. 184–186). The history of the deposits, which are associated with the Ordovician Fernvale limestone and overlying Cason shale, is somewhat complex. It involves weathering and erosion with the accumulation of nodules of manganese oxide in residual clay, the manganese probably derived from the Fernvale limestone, and mild hydrothermal action that dissolved manganese and redeposited it. The deposits in residual clay have furnished most of the production.

The manganese ore of California (122, pp. 190–191) comes from a number of small deposits that extend for some miles throughout the Coast Ranges. These deposits occur in the Franciscan formation (Jurassic and Cretaceous age) within lenticular masses of radiolarian chert. Most of the ore bodies are lenses 1 or 2 ft up to 35 ft thick and a few feet to several hundred feet long. Among these the Ladd mine of San Joaquin County is the most important. There, rhodochrosite and bementite deposits in the chert are overlain, near the surface, by high-grade supergene oxides.

Aluminum (bauxite)

Bauxite is the principal ore of aluminum, and from it most of the aluminum of the world is recovered. Bauxite, however, also is an important industrial material, as it is used in the manufacture of artificial abrasives, a number of

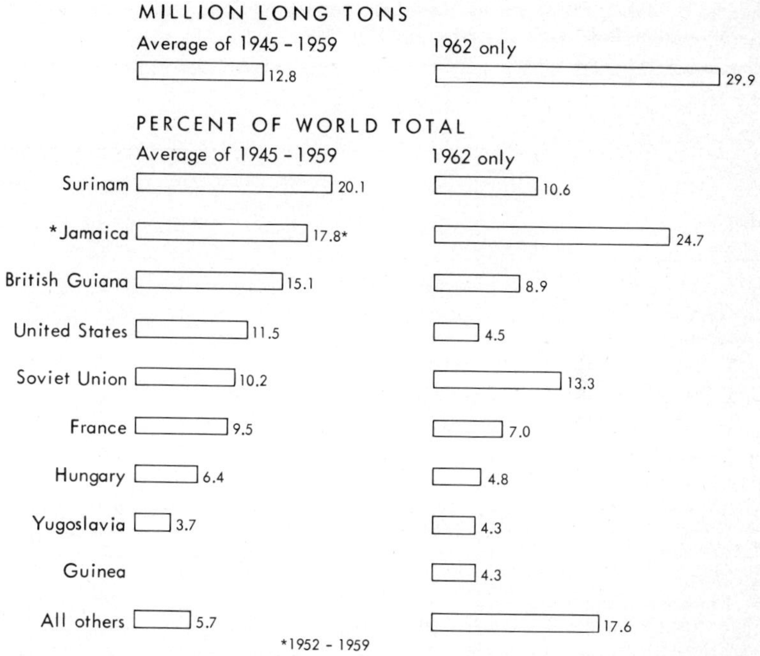

MILLION LONG TONS

Average of 1945 - 1959	1962 only
▭ 12.8	▭ 29.9

PERCENT OF WORLD TOTAL

	Average of 1945 - 1959	1962 only
Surinam	▭ 20.1	▭ 10.6
*Jamaica	▭ 17.8*	▭ 24.7
British Guiana	▭ 15.1	▭ 8.9
United States	▭ 11.5	▭ 4.5
Soviet Union	▭ 10.2	▭ 13.3
France	▭ 9.5	▭ 7.0
Hungary	▭ 6.4	▭ 4.8
Yugoslavia	▭ 3.7	▭ 4.3
Guinea		▭ 4.3
All others	▭ 5.7	▭ 17.6

*1952 - 1959

Fig. 14-8 World bauxite production. Compiled from *U.S. Bur. Mines Minerals Yearbook,* various years.

chemicals and refractories, and for other purposes. Further, bauxite is a potential source of by-product titanium and thorium (157, 166).

Slightly more than 84 percent of the world production of bauxite during the 15-year period 1945 to 1959 came from six countries—Surinam, Jamaica, British Guiana, the United States, the Soviet Union, and France (Fig. 14-8). Those same countries produced 69 percent of the world total in 1962, but the relative amounts furnished were considerably changed, Jamaica supplying almost 25 percent of the world total. Other important producing countries were Hungary, Yugoslavia, and Guinea, each furnishing something more than 4 percent of the world total. Estimates of the reserves of bauxite are presented in Table 14-19. From this table it will be noted that the two countries containing the greatest reserves of the world, Australia and Guinea, are not included among the major producers for the 15-year period 1945 to 1959, and only one of them, Guinea, is included among the major producers for 1962. Some production from Guinea has been recorded for many years, but development of important deposits has been in progress only recently. Some difficulties were encountered by the major operating company after Guinea became a republic, and work on an important project that was scheduled for completion in 1964 was suspended in 1961 because of inability to reach a satisfactory financial agreement with the government. In 1963 the government signed a new agreement with another company to undertake development of the major deposit.

Although bauxite has been the chief material from which aluminum has been

TABLE 14-19.   Estimate of the Most Important Bauxite Reserves and Marginal and Submarginal Bauxite Resources of the World*

| Country | Millions of long tons† | | Approximate percent of world total | |
	Reserves	Marginal and submarginal resources	Reserves	Marginal and submarginal resources
Australia	2,060	1,190	35.7	13.6
Guinea	1,100	2,400	20.0	27.4
Jamaica	600	400	10.4	4.5
Hungary	300	—	5.2	—
Yugoslavia	290	—	5.2	—
Surinam	250	150	4.3	1.7
Ghana	254	—	4.3	—
British Guiana	150	1,000	2.6	11.4
China (Communist)	150	1,000	2.6	11.4
Soviet Union	100‡	—	1.7	—
Greece	84	100	1.4	1.1
France	70	190	1.2	2.2
India	58	200	1.0	2.3
Dominican Republic	60	40	1.0	0.4
United States	50	300	0.8	3.4
Brazil	40	200	0.7	2.3
Cameroon	—	985	—	11.3
Total of above	5,616	8,155	98.1	93.0
World total	5,760	8,740	100.0	100.0

* Compiled from *U.S. Geol. Survey Prof. Paper* 475-B, 1963.
† Some estimates did not specify the kind of ton used.
‡ Includes a considerable amount of material that may be marginal or submarginal.

obtained, shale has been considered as a possible source for some years. In 1956 the Anaconda Company reported that, after nearly two years of experimental work, commercial alumina could be produced successfully from clay. Experimental work has been in progress also in Canada and elsewhere (203). In 1962 a further announcement by Anaconda stated that material was being processed successfully at the Anaconda reduction plant, Columbia Falls, Montana, and that the process appeared to be successful and could well compete with alumina produced from imported bauxite. If this proves to be an economical source of supply and alumina can be produced successfully from high-alumina clays, it could be a highly important development for the United States. In 1963 (152) it was estimated that, at the 1961 rate of consumption in the United States, the domestic reserves of bauxite would be exhausted in less than six years if foreign sources of supply were cut off and domestic deposits became the only material available.

United States

Bauxite occurs in two principal areas of the United States: (1) Arkansas and (2) Georgia, Alabama, and Tennessee (144, pp. 891–892; 146; 154, pp. III,

4–10). Some deposits occur also in Mississippi, Virginia, Missouri, and Pennsylvania. The deposits are gibbsite except those of Missouri and Pennsylvania, which are diaspore. About 90 percent of the production of bauxite in the United States has come from Arkansas, and most of the rest has come from Georgia and Alabama. The diaspore deposits have been used in the manufacture of aluminous refractories.

In addition to the bauxite deposits, large deposits of high-alumina clay (20 to 25 percent Al_2O_3) are present in Idaho, Washington, and Oregon (153). These could be a source of material for production of alumina from clays.

Arkansas.   The bauxite deposits of Arkansas (142; 143; 154, pp. III, 4–7) are near the center of the state in Pulaski and Saline Counties. The deposits are in two districts, a more important one in Saline County about 25 miles southwest of Little Rock, a less important one in Pulaski County about five miles south of Little Rock. The reserves are chiefly in Saline County and in 1950 were estimated at 70.7 million long tons that averaged 50 percent Al_2O_3 and 9 percent SiO_2. Bauxite was identified in Arkansas in 1887, but production did not begin until 1898 because bauxite mining was well advanced in Georgia and Alabama. By 1903, however, Arkansas was the leading producer in the United States.

Within the bauxite area, Paleozoic sedimentary rocks were intruded by nepheline syenite and related rocks that form bosses of a large batholith. After a period of prolonged weathering and erosion, much of the area was buried beneath Tertiary sediments of the Coastal Plain. Bauxite occurs as discontinuous blanketlike deposits over the weathered surface of the nepheline syenite and as lenticular deposits that lie unconformably on the Tertiary Wills Point formation. The bauxite formed almost entirely from the nepheline syenite in place or from detritus derived from it. Four types of deposits are recognized (Fig. 14-9): (1) residual deposits on the upper slopes of partly buried hills of nepheline syenite; (2) colluvial deposits

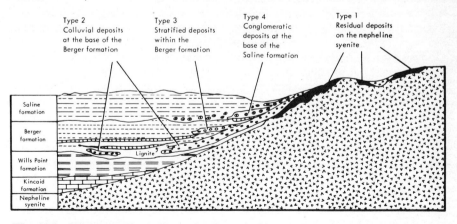

Fig. 14-9 Diagrammatic section showing principal types of bauxite deposits of Arkansas. Generalized after M. Gordon, Jr., J. I. Tracey, Jr., and M. W. Ellis, *U.S. Geol. Survey Prof. Paper* 299, p. 102.

within the Berger formation, which grade laterally and vertically into kaolin and consist of material washed downslope from exposed nepheline syenite and weathered in place; (3) bedded deposits in the Berger formation, which characteristically truncate, overlie, and fill channels in the colluival (type 2) deposits and which consist of grains, pisolites, and pebbles of bauxite that collected as alluvial material; (4) conglomeratic deposits at or near the base of the Saline formation, which consist of rubble derived chiefly from the residual (type 1) deposits. The first two types are the most extensive and the most important commercially.

Southeastern area. The principal deposits of the southeastern United States (139; 145, pp. 47–51; 154, pp. III, 7–9) are in Georgia and Alabama, but smaller deposits occur in Tennessee, Virginia, and Mississippi. Mining of bauxite began in 1888, and from 1889 to 1898 the entire production of the United States came from the Hermitage and Bobo districts of northwestern Georgia and the Rock Run district of northeastern Alabama. Now production is chiefly from the Andersonville district, south-central Georgia, and the Eufaula district, southeastern Alabama.

The deposits occur in (1) the Ridge and Valley province and (2) the eastern Gulf Coastal Plain. The deposits of the Ridge and Valley province are the more northerly ones and include the early productive Hermitage and Bobo districts, Georgia, and the Rock Run district, Alabama. Such deposits occur in limestone and dolomite, commonly of the Knox group (Cambro-Ordovician), and generally are circular or elliptical in plan outline with a maximum horizontal diameter usually not greater than 100 to 200 ft and a thickness that may, in some cases, equal the diameter. These deposits apparently accumulated in sinkholes. The deposits of the Coastal Plain are the more southerly ones and include those of the Andersonville district, Georgia, and the Eufaula district, Alabama. The deposits in Georgia are a few hundred to more than a thousand feet in diameter, generally not more than 10 ft thick; those of Alabama usually are not more than 200 ft in diameter and locally are as much as 50 ft thick. These Coastal Plain deposits are associated with and commonly enclosed in kaolinitic clay, apparently of fluviatile or lacustrine origin, and generally are overlain by nonmarine strata of Tertiary age.

According to Bridge (139) the deposits of the Ridge and Valley province and also those of the Coastal Plain were formed during early Tertiary time as a result of deposition of finely divided claylike material carried by streams from the crystalline rocks of the Blue Ridge and Piedmont areas and deposited in sinkholes and in shallow depressions. Although complete agreement does not exist, it seems likely that the bauxite was then formed by prolonged weathering of these clays.

Surinam and British Guiana

Bauxite deposits of similar type occur throughout Surinam and British Guiana within a zone about 30 miles wide that lies about 15 to 50 miles inland from the coast, and they may be present nearer the coast under heavy overburden (144, pp. 893–894; 146; 151; 154, pp. III, 14–16). The important commercial deposits of Surinam are those of (1) the Surinam River area and (2) the Cottica River area; the deposits of British Guiana are (1) along the Essequibo River, (2) the Demerara River, and (3) the Berbice River. Production began in Surinam in 1922 and in British Guiana about five years earlier. During the 30-year period 1921 to

1950 these two countries furnished 31 percent of the bauxite produced in the world, in about equal amounts.

In both countries the deposits are in a hilly area on the tops and along the flanks of the hills. Ore bodies in Surinam vary from very small ones to those several miles long, a thousand feet wide, and 30 ft thick. The ore bodies in British Guiana may be a few miles long and 15 to 35 ft thick. The overburden in Surinam varies from almost nothing to 60 ft, that in British Guiana from a few feet to 100 ft. The bauxite is the gibbsite type and is generally of high grade. The ores shipped from Surinam contain 54 to 57 percent Al_2O_3, 8 to 12 percent Fe_2O_3, 2 to 3 percent SiO_2; those shipped from British Guiana contain 57 to 61 percent Al_2O_3, 1 to 3 percent Fe_2O_3, and 4 to 6 percent SiO_2.

The Mediterranean-Adriatic region: France, Hungary, Yugoslavia

A large region of bauxite deposits extends from Spain through the Pyrenees, along the north shore of the Mediterranean Sea, across into Hungary, along the western shore of the Adriatic Sea in Yugoslavia, and southwesterly into Greece. Within this region are the important deposits of France, Hungary, and Yugoslavia, and the less important deposits of Italy and Greece. Throughout this entire region the deposits are similar and generally occur in solution cavities in limestones or dolomites of Triassic to Eocene age (138, pp. 231–235; 144, pp. 889–891; 145, pp. 41–44; 148, pp. 244–247; 150, p. 93; 154, pp. III, 24–31). In general they are thought to have developed by prolonged weathering of the clayey residue derived from the decomposition of the limestones and dolomites. Three countries of this region, France, Hungary, and Yugoslavia, produced nearly 20 percent of of the bauxite of the world in the period 1945 to 1959 (Fig. 14-8).

France. Bauxite received its name from Les Baux, Bouches du Rhône, France, where it was discovered in 1821. The most important deposits of France are those of Var and Herault; less important deposits are in Ariege and Bouches du Rhône (the original bauxite area). The largest individual deposits as well as the largest reserves are in Var, in the vicinity of Brignoles. Most of the French bauxites contain boehmite as the chief aluminous mineral and average 55 to 58 percent Al_2O_3, 20 to 25 percent Fe_2O_3, 3 to 6 percent SiO_2, and 2.5 to 3.5 percent TiO_2. Several varieties are present, however, some of which contain 60 to 70 percent Al_2O_3, 4 to 12 percent Fe_2O_3, 8 to 16 percent SiO_2, and 2 to 3 percent TiO_2. (146, p. 71).

Hungary. The major bauxite deposits of Hungary (146, pp. 71–72; 154, pp. III, 30–31) occur in an area extending from north of Lake Alaton to about 40 miles southwest of Budapest. Other deposits occur about 35 miles north of Budapest and about 100 miles south of Budapest near the border of Yugoslavia. Much of the production has been from the northeastern part of the major area—the Gant deposits in the Vertes Mountains. The deposits, of the boehmite type, averaged, according to information available before World War II, 57 to 60 percent Al_2O_3, 14 to 20 percent Fe_2O_3, 3 to 7 percent SiO_2, 2.5 to 3.5 percent TiO_2. Much low-grade material must be mined along with the commercial ore. Production began in 1925.

Yugoslavia. The bauxite deposits of Yugoslavia (141; 146, pp. 72–73; 154, pp. III, 29–30) occur along the coast of the Adriatic Sea and extend inland as far as 90 miles in some places. Some of the deposits have been mined since 1915, the early work being especially in Istria, south and southeast of Trieste, which was formerly a part of Italy; production in that area before World War II is credited to Italy. These Istrian deposits are high grade and some of them are similar to the French deposits. A red variety of bauxite contains 55 to 58 percent Al_2O_3, 22 to 25 percent Fe_2O_3, 2 to 5 percent SiO_2, 2.5 to 3.5 percent TiO_2; a yellow variety contains 57 to 65 percent Al_2O_3, 12 to 20 percent Fe_2O_3, 5 to 9 percent SiO_2, 3 to 4 percent TiO_2. Deposits, generally of lower grade, also are mined in Herzegovina, Montenegro, and Dalmatia. The Dalmatian deposits were stated to be the most important and productive ones in 1955 (141), although they contain an average of only 51 to 53 percent Al_2O_3. Production from deposits around Drnis began about 1915 and has been increasing since World War II. The ore bodies are associated with steeply dipping beds and consist of scattered pockets and pipes of irregular form and varying thickness—1 to 25 m or more. The steep dip makes underground mining necessary. In the Maslenica area of northern Dalmatia there are many large outcrops of low-grade bauxite—47 to 48 percent Al_2O_3. Generally the bauxites of Yugoslavia contain both boehmite and gibbsite and are high in iron—21 to 26 percent Fe_2O_3.

Jamaica

One of the most outstanding cases of rapid development of bauxite deposits is that of Jamaica. Atlhough near-surface residual material overlying limestone was known to have a high content of alumina as early as 1869 (149, p. 63), attention was not focused on the importance of this information until analysis of a soil sample collected in 1938 and subsequent work regarding lack of the fertility of the soil as pasture land (147). A very small amount of bauxite for large-scale testing was mined in 1943. This material was low grade and refractory and development of the deposits was postponed (159). Important production began in 1952. Since then it has increased rapidly until, in 1962, Jamaica furnished about 25 percent of the bauxite mined in the world. Production from a new property began in 1963, and another project is expected to be in operation before 1969. This project, planned by Kaiser Bauxite Company, is expected to cost about 30 million dollars, and will involve dredging a channel through a coral reef, development of an ocean shipping port, railroads, and other facilities (164).

Bauxite occurrences are known throughout much of the island of Jamaica (144; 146, pp. 69–70; 147; 148; 154, pp. III, 10–13) but the chief ore bodies are in the north-central part (St. Ann and Trelawny parishes) and the southwestern part (Manchester and St. Elizabeth parishes). The bauxite lies beneath a cover of but 1 or 2 ft of soil, consists chiefly of gibbsite and hematite, and occupies solution depressions and troughs in limestone of Oligocene and Miocene age on which it rests in very sharp contact. Individual ore bodies are pipelike masses that may pass into long tabular bodies or blankets. Thickness is from a few inches to more than 100 ft, with the average mined thickness being 25 ft and bodies too thin for economical mining being less than 5 ft. Some blanketlike bodies are 10 to 20 miles long. The average commercial ore body generally contains about 500,000 long tons of bauxite. The ore generally contains 46 to 50 percent Al_2O_3, 16 to 20 percent Fe_2O_3, 1 to 2 percent SiO_2. The unusually low silica content is one of its striking characteristics.

At least three major concepts have been advanced regarding the origin of the Jamaican deposits: (1) residual accumulation of insoluble constituents in the limestone that underlies the deposits; (2) alteration of argillaceous material derived from lateritic weathering of Upper Cretaceous or in part early Eocene andesitic rocks that were trapped in solution pockets and depressions in the limestone surface, and (3) alteration of volcanic ash deposited on the limestone. Although opinion is divided, a combination of residual accumulation from insoluble constituents in the limestone, which contains but 0.036 percent Al_2O_3, plus material washed into the depressions may be the most reasonable explanation.[1]

Australia

Nearly half of the total reserves of bauxite, including marginal or submarginal resources, are in Australia, and 35 percent of this material is good bauxite. The major reserves (140) are in the Weipa deposits of Cape York Peninsula, Queensland, the deposits of the Gove Peninsula, Northern Territory, and deposits in the Darling Range, Western Australia (204, 205, 206, 207). Analyses of ores from these three areas average about as follows:

	Al_2O_3	SiO_2	Fe_2O_3
Weipa, Cape York Peninsula	55.0	6.2	7.0
Gove Peninsula	49.0	4.0	18.0
Darling Range	43.3–51.0	6.0–17.2	5.8–28.5

Most of the deposits are gibbsitic, but those of Weipa are a mixture of gibbsite and boehmite.

In addition to the deposits of these three main areas, deposits are present in New South Wales, Victoria, and Tasmania. Further, discovery of very rich deposits in northwestern Arnhem Land was reported recently.

The Weipa bauxite deposit, Queensland, is stated to be one of the largest single deposits in the world. It is known over an area of about 200 sq miles, and extensive exploration indicates a possible reserve of 3 billion tons. The deposit forms a capping of flat-lying Tertiary sedimentary rocks. It consists of an upper layer some 10 to 15 ft thick composed of pisolites 10 to 200 mm across, chiefly uncemented, that contain 55 to 57 percent Al_2O_3 and 3 to 4 percent SiO_2, and a lower layer composed of pisolites and nodules that are cemented and in which the content of Al_2O_3 decreases sharply and the content of SiO_2 increases sharply in

---

[1] Recent discussions of the origin of these deposits are contained in various issues of *Econ. Geology*, 1955 through 1963, in references 147 and 149, and in the *American Jour. Sci.*, 1961, pp. 288–294.

its upper part but, through a thickness of 10 to 15 ft, passes gradually into un-
laterized sedimentary rocks below. Overburden is but 10 to 15 ft.

The Gove deposits, Northern Territory, are similar to those at Weipa, although not as large. The upper layer of pisolites attains a thickness of 20 ft in
some places; the lower nodular layer is about as thick as the upper layer. Overburden is but 2 to 3 ft.

Deposits on the Wessel Islands are similar to the Weipa and the Gove deposits.

The deposits of the Darling Ranges, Western Australia, a short distance east
of Perth, extend throughout a distance of about 120 miles. The deposits make
cappings on the rocks of the Darling Ranges and, although they contain 10
percent or more silica, they are favorable for the production of alumina because
the silica is present chiefly as quartz, not as a silicate, and the ore can be
treated without difficulty since the quartz does not react with the soda used in
the process.

Other Countries

Among other countries that were important producers of bauxite in 1962
are the Soviet Union and Guinea; potentially important deposits exist in Communist China, in India, and in Brazil. Although the production from Guinea in
1962 was but 4.3 percent of the world total, contrasted with an estimated production of 13.3 percent by the Soviet Union, the reserves of Guinea are among the
highest in the world. Further, a considerable amount of bauxite is present in
other countries of West Africa, especially Cameroon and Ghana, so that West
Africa could assume considerable importance in the future.

Soviet Union.   Deposits of bauxitic material in the Soviet Union, a considerable
amount of which is of low grade, occur in northwest Russia, in the Ural Mountains, in Siberia, and in Kazakhustan (66, pp. 15–16, 36, 57, 78; 80, pp. 98–109;
146, p. 72; 154, pp. III, 33–34). Most of the commercial deposits are of Paleozoic
age and contain diapore (144, p. 888). Deposits that are generally noncommercial
at present are of Mesozoic age, contain gibbsite, and are characterized by a high
content of iron oxide or silica.

Early mining of bauxite was in the Tikhvin district of northwest Russia
(59°23′–59°32′N, 33°36′E), near Leningrad. These deposits rest on Upper Devonian sand and clay and are overlain by Lower Carboniferous strata (144, p.
888). Generally the grade of the ore is low and variable—35 to 55 percent
Al_2O_3, 15 to 25 percent Fe_2O_3, 10 to 25 percent SiO_2.

Perhaps three-fourths of the bauxite reserves of the Soviet Union are in
the Ural Mountains. The major deposits are in the northern Urals and extend
from Vyiya (58°35′N, 59°59′E) to Talitsa (61°06N′, 60°30′E). Other deposits
occur in the central and southern Urals. The deposits of the northern Urals generally consist of bauxite, probably of Devonian age, associated with clays and
limestones. They are reported to contain 49 to 59 percent Al_2O_3 and 1.5 to 6 percent SiO_2. Some of the deposits of the southern Urals, of Paleozoic age, are small
but contain 52 to 78 percent Al_2O_3. Moderately large deposits of the central
Urals, of Mesozoic age, are of low grade, 23 to 37 percent Al_2O_3, and contain
much iron, 31 to 53 percent Fe_2O_3.

West Africa.   The deposits of the Republic of Guinea (144, p. 896; 146, p. 73; 154, pp. III, 23–24) are very extensive on the mainland in the areas around Boke, Fria, and Tougue. Production began, however, on the Los Islands near Conakry in 1952. Bauxite on the islands rests on nepheline syenite, is 15 to 25 ft thick, and is generally covered by several feet of lateritic material. It is gibbsitic and contains 48 to 54 percent Al_2O_3, 12 to 14 percent Fe_2O_3, and 4 to 7 percent SiO_2. The deposits on the mainland, which is to a large extent covered by lateritic material, form local concentrations in ferruginous laterite.

In Cameroon discovery of a large bauxite deposit in the Minim-Martap area was confirmed in 1959. The deposit is on the Adamaoua Plateau, about 80 km southwest of Ngaoundere. It is estimated to contain at least 500 million tons of ore and possibly as much as 2 billion tons. The average content of Al_2O_3, however, is but 40 percent. Other deposits at Fongo-Tongo are reported to contain 45 million tons of material averaging 45 percent Al_2O_3.

Bauxite has been mined in Ghana since 1941 (144, p. 896; 146, p. 74; 154, pp. III, 20–23). Deposits of medium to low grade are known in an east-west zone 90 to 130 miles inland from the coast, the most extensive ones being north of Yenahin, 35 to 50 miles west of Kumasi. Production has come only from Mount Ejuanema, north of the seaport of Accra. Development of the bauxite deposits of Ghana to any great extent is unlikely before completion of the Volta River power project, as a result of which an eventual large annual production of alumina is expected. Considerable progress has been made, and power on a limited basis might become available about 1966.

China.   The bauxite deposits of China (53, pp. 462–463; 144, p. 888; 154, pp. III, 38–41; 155) occur in the provinces of Yunnan, Kweichow, Szechwan, of southwestern China, Fukien of southern China, and Shantung and Liaoning of eastern China (Fig. 9-5). The deposits near Nanting (36°44′N, 118°01′E), Shantung Province, have been the most productive; those at Penhsihu (41°19′N, 123°54′E), Liaoning Province, seem to rank second at present; and perhaps the Kweiyand deposits (26°35′N, 106°43′E) in Kweichow Province rank third. Extensive reserves seem to have been discovered recently near Sian in Shensi Province (36°16′N, 108°54′E).

The deposits are of Paleozoic age and are associated with sedimentary rocks, except those of Fukien Province, which are of Quaternary age and overlie basalts from which they were derived. In Yunnan and Kweichow Provinces deposits 10 to 15 m thick contain 2 to 3 m of good ore—57 to 77 percent Al_2O_3, up to 11 percent SiO_2, and from very little to as much as 16 percent Fe_2O_3, These deposits apparently contain boehmite and diaspore. The deposits of Shantung and Liaoning Provinces are similar. The better grade material, which may constitute about 40 percent of the deposits, contains about 50 to 60 percent Al_2O_3, 13 to 18 percent SiO_2, and 8 to 18 percent Fe_2O_3. The deposits of Fukien Province are small and consist chiefly of gibbsite masses and nodules in red clay.

India.   The bauxite deposits of India (111, pp. 257–264; 144, pp. 895–896; 146, p. 74; 154, pp. III, 35–38) of the greatest potential importance are in the states of Bihar and Madhya Pradesh, which possibly contain 75 percent of the reserves. Other deposits of some importance occur in Bombay, Kashmir, and Madras. Deposits consist of pockets and lenses in limestone and clay, or cappings of high, flat-topped plateaus, many of which are related to the Deccan basaltic flows. Most of the deposits are gibbsitic, but some contain diaspore also. Average composition is 54 to 58 percent Al_2O_3, 1 to 3 percent SiO_2, 3 to 8 percent Fe_2O_3, and unusually high TiO_2, which is 6 to 11 percent, probably because of derivation from basalts.

Brazil.   The best known deposits of Brazil (144, pp. 896–897; 146, p. 74; 154, p. III, 17; 156) are in the Poços de Caldas district of the western part of the state of Minas Gerais and near Ouro Preto in the central part of the same state. In the Poços de Caldas area, near the boundary with the state of São Paulo, bauxite as much as 25 ft thick rests on phonolite and related rocks. In the Ouro Preto area, low-grade ferruginous bauxitic laterite rests on residual clays in an area underlain by Precambrian schists.

SELECTED REFERENCES
IRON

1. G. E. Aiken and R. C. Temps, 1962, El Algarrobo: New open pit iron mine in the Atacama desert, *Mining Eng.*, vol. 14, no. 3, pp. 52–57.

2. H. L. Alling, 1942, The Adirondack magnetite deposits, in "Ore Deposits As Related to Structural Features," pp. 143–146, Princeton University Press, Princeton, N.J., 280 pp.

3. S. Anderson and H. T. Reno, 1960, Iron ore industry is young, vigorous and growing fast, *Mining World,* vol. 22, no. 5, pp. 32–35.

4. S. H. Ball, 1906, The Hartville iron ore range, Wyoming, *U.S. Geol. Survey Bull.* 315-D, pp .190–205.

5. R. W. Bayley, 1963, A preliminary report on the Precambrian iron deposits near Atlantic City, Wyoming, *U.S. Geol. Survey Bull.* 1142-C, 23 pp.

6. D. P. Bellum and L. Nugent, 1963, Iron mining, healthy Utah industry, *Mining Eng.,* vol. 15, no. 11, pp. 34–38.

7. F. Benitez, 1957, The iron mines of Chile, *Eng. Mining Jour.*, vol. 158, no. 7, pp. 90–93.

8. F. Blondel and L. Marvier, editors, 1952, "Symposium sur les Gisements de Fur du Monde," Internat. Geol. Cong., 19th, Algiers 1952, vol. 1, 638 pp, vol. 2, 594 pp.

9. E. F. Burchard, 1906, The Clinton or red ores of the Birmingham district, Alabama, *U.S. Geol. Survey Bull.* 315-D, pp. 130–151.

10. E. F. Burchard, 1906, The brown ores of the Russelville district, Alabama, *U.S. Geol. Survey Bull.* 315-D, pp. 152–160.

11. S. Caillère and F. Kraut, 1953, Considérations sur la genèse des minerais de fer oolithiques Lorrains, Internat. Geol. Cong., 19th, Algiers 1952, comptes rendus, sec. 10, f. 10, pp. 101–117.

12. S. Caillère and F. Kraut, 1960, Les minerais de fer Siluriens de la Peninsule Armoricaine, Internat. Geol. Cong., 20th, Copenhagen 1960, pt. 16, sec. 16, Genetic problems of ores, pp. 171–180.

13. F. Canavan, 1953, The iron ore deposits of Yampi Sound, W.A., in *Fifth Empire Min. Metall. Cong.*, Australia and New Zealand, pp. 276–283, Australasian Institute of Mining and Metallurgy, Melbourne.

14. M. S. Carr and C. E. Dutton, 1959, Iron-ore resources of the United States, including Alaska and Puerto Rico, *U.S. Geol. Survey Bull.* 1082-C, pp. 61–134.

15. F. Cusset, F. de Torcy, and P. Maubeuge, 1952, Etudes sur les gisements de fer Français, Internat. Geol. Cong., 19th, Algiers 1952, "Symposium sur les Gisements de Fer du Monde," vol. 2, pp. 129–222.

16. L. J. DeMoraes, 1952, The iron ore deposits of Brazil, Internat. Geol. Cong., 19th, Algiers 1952, "Symposium sur les Gisements de Fer du Monde," vol. 1, p. 285.

17. Divisiones de geología y minas, Instituto nacional de investigación y fomento mineros (I.N.I.F.M.), 1952, El fierro en el Perú, Internat. Geol. Cong., 19th, Algiers 1952, "Symposium sur les Gisements de Fer du Monde," vol. 1, pp. 455–460.

18. J. Van N. Dorr, II, 1954, Comments on the iron deposits of the Congonhas district, Minas Gerais, Brazil, *Econ. Geology,* vol. 49, pp. 659–662.

19. J. Van N. Dorr, II, P. W. Guild, and A. L. de M. Barbosa, 1952, Origin of the Brazilian iron ores, Internat. Geol. Cong., 19th, Algiers 1952, "Symposium sur les Gisements de Fer du Monde," vol. 1, pp. 286–298.

20. J. Van N. Dorr, II, and A. L. de M. Barbosa, 1963, Geology and ore deposits of the Itabira district of Minas Gerais, Brazil, *U.S. Geol. Survey Prof. Paper* 341-C, 110 pp.

21. C. E. Dutton, 1949, Geology of the central part of the Iron River district, Iron County, Michigan, *U.S. Geol. Survey Circ.* 43, 9 pp., 6 map sheets.

22. C. E. Dutton, 1952, Memorandum on iron deposits in the United States of America, Internat. Geol. Cong., 19th, Algiers 1952, "Symposium sur les Gisements de Fer du Monde," vol. 1, pp. 371–411.

23. A. B. Edwards, 1953, Mineralogy of the Middleback iron ores, in *Fifth Empire Min. Metall. Cong.*, Australia and New Zealand, pp. 464–472, Australasian Institute of Mining and Metallurgy, Melbourne.

24. M. S. Fotheringham, 1952, Steep Rock's huge reserves, an ace for Canada's future, *Eng. Mining Jour.*, vol. 153, no. 4, pp. 82–85.

25. G. Gastil, R. Blais, D. M. Knowles, and R. Bergeron, 1960, The Labrador geosyncline, Internat. Geol. Cong., 21st, Copenhagen 1960, pt. 9, sec. 9, Precambrian stratigraphy and correlations, pp. 21–38.

26. P. Geijer and N. H. Magnusson, 1952, The iron ores of Sweden, Internat. Geol. Cong., 19th, Algiers 1952, "Symposium sur les Gisements de Fer du Monde," vol. 2, pp. 477–499.

27. C. Gray and D. M. Lapham, 1961, Guide to the geology of Cornwall, Pennsylvania, *Pennsylvania Geol. Survey,* B. G. 35, 18 pp.

28. G. A. Gross, 1961, Iron-formations and the Labrador geosyncline, *Geol. Survey Canada, Dept. Mines and Tech. Surveys, Paper* 60-30, 7 pp., geologic map.

29. G. A. Gross, 1962, Iron deposits near Ungava Bay, Quebec, *Geol. Survey Canada, Bull.* 82, 48 pp.

30. F. F. Grout and J. F. Wolff, Sr., 1955, The geology of the Cuyuna district, Minnesota, *Minnesota Geol. Survey Bull.* 36, 144 pp.

31. P. W. Guild, 1953, Iron deposits of the Congonhas district, Minas Gerais, Brazil, *Econ. Geology,* vol. 48, pp. 639–676.

32. P. W. Guild, 1957, Geology and mineral resources of the Congonhas district, Minas Gerais, Brazil, *U.S. Geol. Survey Prof. Paper* 290, 90 pp.

33. J. N. Gunderson and G. M. Schwartz, 1962, The geology of the metamorphosed Biwabik iron-formation, eastern Mesabi district, Minnesota, *Minnesota Geol. Survey Bull.* 43, 139 pp.

34. J. K. Gustafson and A. E. Moss, 1953, The role of geologists in the development of the Labrador-Quebec iron ore districts, *Mining Eng.,* vol. 5, pp. 593–602.

35. J. B. Hadley, 1948, Iron ore deposits in the eastern part of the Eagle Mountains, Riverside County, California, *California Dept. Nat. Resources, Div. Mines, Bull.* 129, pt. A, p. 3–24.

36. E. C. Harder, 1914, The "itabirite" iron ores of Brazil, *Econ. Geology,* vol. 9, pp. 101–111.

37. E. C. Harder and R. T. Chamberlin, 1915, The geology of central Minas Geraes, Brazil, *Jour. Geology,* vol. 23, pp. 341–378, 385–424.

38. J. M. Harrison, 1952, The Quebec-Labrador iron belt, Quebec and Newfoundland, *Geol. Survey Canada, Dept. Mines and Tech. Surveys, Paper* 52-20, 20 pp., map.

39. J. M. Harrison, 1953, Iron formations of Ungava Peninsula, Canada, Internat. Geol. Cong., 19th, Algiers 1952, comptes rendus, sec. 10, f. 10, pp. 19–33.

40. A. O. Hayes, 1924, Wabana iron ores of Newfoundland, *Geol. Survey Canada Mem.* 78, 155 pp.

41. A. O. Hayes, 1931, Structural geology of the Conception Bay region and the Wabana iron deposits of Newfoundland, *Econ. Geology,* vol. 26, pp. 44–64.

42. C. W. Hayes and E. C. Eckel, 1903, Iron ores of the Cartersville district, Georgia, *U.S. Geol. Survey Bull.* 213, pp. 233–242.

43. B. R. Henderson, 1962, Carol—IOCC's big new iron project, *Eng. Mining Jour.,* vol. 163, no. 11, pp. 74–80, 102.

44. P. E. Hotz, 1953, Magnetite deposits of the Sterling Lake, N.Y.–Ringwood, N.J., area, *U.S. Geol. Survey Bull.* 982-F, pp. 153–244.

45. P. E. Hotz, 1954, Some magnetite deposits in New Jersey, *U.S. Geol. Survey Bull.* 995-F, pp. 201–253.

46. T. A. G. Hungerford, 1963, B H P to double output of Australian 'iron' sands, *Eng. Mining Jour.,* vol. 164, no. 3, pp. 118, 127.

47. J. B. Huttl, 1959, Salzgitter brown iron ores basis for a second Ruhr, *Eng. Mining Jour.,* vol. 160, no. 11, pp. 82–86.

48. R. W. Hyde, B. M. Lane, and W. W. Glaser, 1962, Iron ore resources of the world, *Eng. Mining Jour.,* vol. 163, no. 12, pp. 84–88.

49. H. L. James and C. E. Dutton, 1951, Geology of the northern part of the Iron River district, Iron County, Michigan, *U.S. Geol. Survey Circ.* 120, 12 pp., 10 map sheets.

50. H. L. James, C. E. Dutton, F. J. Pettijohn, and K. L. Weir, 1959, Geologic map of the Iron River–Crystal Falls district, Iron Country, Michigan, *U.S. Geol. Survey Mineral Inv. Field Studies Map* MF 225.

51. H. L. James, L. D. Clark, C. A. Lamey, and F. J. Pettijohn, 1961, Geology of Central Dickinson County, Michigan, *U.S. Geol. Survey Prof. Paper* 310, 176 pp.

52. A. W. Jolliffe, 1955, Geology and iron ores of Steep Rock Lake, *Econ. Geology,* vol. 50, pp. 373–389.

53. V. C. Juan, 1945, Mineral resources of China, *Econ. Geology,* vol. 41, pt. 2, pp. 399–474.

54. J. Kowalewski, 1963, Russia's iron ore reserves, *Mining Jour. (London),* vol. 261, no. 6684, pp. 279–282.

55. M. S. Krishnan, 1952, The iron ores of India, Internat. Geol. Cong., 19th, Algiers 1952, "Symposium sur les Gisements de Fer du Monde," vol. 1, pp. 503–532.

56. M. O. Lake, 1950, Cerro Bolivar—U.S. Steel's new iron ore bonanza, *Eng. Mining Jour.,* vol. 151, no. 8, pp. 72–83.

57. C. K. Leith and E. C. Harder, 1908, The iron ores of the Iron Springs district, Southern Utah, *U.S. Geol. Survey Bull.* 338, 102 pp.

58. T. W. Lippert, 1950, Cerro Bolivar, saga of an iron ore crisis averted, *Mining Eng.,* vol. 2, pp. 178–192.

59. M. Lucius, 1952, Etude sur les gisements de fer du Grand-Duché de Luxembourg, Internat. Geol. Cong., 19th, Algiers 1952, "Symposium sur les Gisements de Fer du Monde," vol. 2, pp. 349–387.

60. J. C. Lyons, 1957, Wabana iron ore deposits, in "Structural Geology of Canadian Ore Deposits, Congress Volume," pp. 503–516, Commonwealth Min. Metall. Cong., 6th, Canada 1957, Canadian Institute of Mining and Metallurgy, Montreal, 524 pp.

61. R. D. Macdonald, 1960, Iron deposits of Wabush Lake, Labrador, *Mining Eng.,* vol. 12, pp. 1098–1102.

62. J. H. Mackin, 1947, Some structural features of the intrusions in the Iron Springs district, *Utah Geol. Soc. Guidebook to the geology of Utah, No.* 2, 62 pp.

63. G. B. Mellon, 1962, Petrology of Upper Cretaceous oolitic iron-rich rocks from northern Alberta, *Econ. Geology,* vol. 57, pp. 921–940.

64. H. Mikami, 1944, World iron-ore map, *Econ. Geology,* vol. 39, pp. 1–24.

65. B. L. Miller and J. T. Singewald, 1919, "The Mineral Deposits of South America," McGraw-Hill Book Company, New York, 598 pp.

66. D. V. Nalivkin, 1960, "The Geology of the U.S.S.R., a Short Outline," Pergamon Press, New York, 170 pp., geologic map of the U.S.S.R.

67. E. Nef, 1962, Los yacimientos de fierro en el Chile, Internat. Geol. Cong., 19th, Algiers 1952, "Symposium sur les Gisements de Fer du Monde," vol. 1, pp. 353–359.

68. J. M. Neilson, 1953, Albanel area, Mistassini Territory, Quebec, *Quebec. Dept. Mines, Geol. Surveys Branch, Geol. Rept.* 53, 35 pp., map.

69. J. M. Perkins and J. T. Lonsdale, 1955, Mineral resources of the Texas Coastal Plain (preliminary report), *Texas Univ. Bur. Econ. Geol. Rept. for Bur. of Reclamation,* 49 pp.

70. F. J. Pettijohn, 1952, Geology of the northern Crystal Falls area, Iron County, Michigan, *U.S. Geol. Survey Circ.* 153, 17 pp., 8 map sheets.

71. A. W. Postel, 1952, Geology of Clinton County magnetite district, New York, *U.S. Geol. Survey Prof. Paper* 237, 88 pp.

72. W. F. Pruden, 1961, Taconite—new iron ore for the west as U.S. Steel builds at Atlantic City, *Mining World*, vol. 23, no. 5, pp. 18–22.

73. T. T. Quirke, Jr., S. S. Goldich, and H. W. Kaueger, 1960, Composition and age of the Temiscamie iron-formation, Mistassini Territory, Quebec, Canada, *Econ. Geology*, vol. 55, pp. 311–326.

74. H. C. Rickaby, 1943, Steep Rock Lake, Canada's first big iron mine, *Mining and Metallurgy*, vol. 24, October, pp. 436–439.

75. H. M. Roberts and M. W. Bartley, 1943, Replacement hematite deposits, Steep Rock Lake, Ontario, *Am. Inst. Mining Metall. Engineers, Tech. Pub.* 1543, 26 pp.; discussion, various authors, *Tech. Pub.* 1632, 15 pp.

76. S. Royce, 1942, Iron ranges of the Lake Superior district, in "Ore Deposits as Related to Structural Features," pp. 54–63, Princeton University Press, Princeton, N.J., 280 pp.

77. E. Rubio, C. M. Bellizzia, A. Bellizzia, and R. Laforest, 1952, Geologia paragenesis y reservas de los yacimientos de hierro de Imataca en Venezuela, Internat. Geol. Cong., 19th, Algiers 1952, "Symposium sur les Gisements de Fer du Monde," vol. 1, pp. 477–498.

78. J. C. Ruckmick, 1963, The iron ores of Cerro Bolivar, Venezuela, *Econ. Geology*, vol. 58, pp. 218–236.

79. E. A. Rudd and K. R. Miles, Iron ores of the Middleback Ranges, South Australia, in *Fifth Empire Min. Metall. Cong.*, Australia and New Zealand, pp. 449–463, Australasian Institute of Mining and Metallurgy, Melbourne.

80. D. B. Shimkin, 1953, "Minerals—a Key to Soviet Power," Harvard University Press, Cambridge, Mass., 452 pp.

81. P. K. Sims, 1953, Geology of the Dover magnetite district, Morris County, New Jersey, *U.S. Geol. Survey Bull.* 982-G, pp. 245–305.

82. P. K. Sims, 1958, Geology and magnetite deposits of Dover district, Morris County, New Jersey, *U.S. Geol. Survey Prof. Paper* 287, 162 pp.

83. R. G. Schmidt, 1958, Bedrock geology of the southwestern part of the North Range, Cuyuna district, Minnesota, *U.S. Geol. Survey Mineral Inv. Field Studies Map* MF 181.

84. R. G. Schmidt, 1959, Bedrock geology of the northern and eastern parts of the North Range, Cuyuna district, Minnesota, *U.S. Geol. Survey Mineral Inv. Field Studies Map* MF 182.

85. R. G. Schmidt and C. E. Dutton, 1957, Bedrock geology of the south-central part of the North Range, Cuyuna district, Minnesota, *U.S. Geol. Survey Mineral Inv. Field Studies Map* MF 99.

86. J. C. Stam, 1963, Geology, petrology, and iron deposits of the Guiana Shield, Venezuela, *Econ. Geology*, vol. 58, pp. 70–83.

87. R. C. Temps, 1963, Iron mining in Chile: 1962, *Mining Eng.*, vol. 15, no. 1, pp. 38–39.

88. S. I. Tomkeieff, 1959, Iron ore in the Soviet Union, *Nature*, vol. 183, no. 4667, Apr. 11, p. 1028.

89. P. M. Tyler, 1952, Iron ore and Manganese, in W. Van Royan and O. Bowles, "The Mineral Resources of the World," vol. 2, pp. 56–71. Prentice-Hall, Inc., Englewood Cliffs, N.J., 181 pp.

90. U.S. Bureau of Mines, 1956 "Materials Survey, Iron Ore," 337 pp.

91. C. R. Van Hise and C. K. Leith, 1911, The geology of the Lake Superior region, *U.S. Geol. Survey Mon.* 52, 641 pp. Contains a condensed report of each range in the Lake Superior region at the time of publication.

92. Various authors, 1952, Die Eisenerz-Lagerstätten Deutschlands, Internat. Geol. Cong., 19th, Algiers 1952, "Symposium sur les Gisements de Fer du Monde," vol. 2, pp. 1–38.

93. Various authors, 1952, The iron ore deposits of Great Britain, Internat. Geol. Cong., 19th, Algiers 1952, "Symposium sur les Gisements de Fer du Monde," vol. 2, pp. 405–471.

94. W. G. Wahl, 1953, Temiscamie River area, Mistassini Territory, Quebec, *Quebec Dept. Mines, Geol. Surveys Branch, Geol. Rept.* 54, 32 pp., map.

95. D. A. White, 1954, The stratigraphy and structure of the Mesabi range, Minnesota, Minnesota *Geol. Survey Bull.* 38, 92 pp.

96. G. Zuloaga, 1935, The geology of the iron deposits of the Sierra de Imataca, Venezuela, *Am. Inst. Mining Engineers Trans.*, vol. 115, pp. 307–340; discussion, pp. 341–345.

97. Anon., 1958, Report on Soviet iron ore and steel, *Mining World*, vol. 20, no. 10, pp. 36–37.

98. Anon., 1961, U.K. ironstone workings and land restoration, *Mining Jour. (London)*, vol. 257, no. 6578, pp. 255, 257, 259.

99. Anon., 1961, India's iron ore deposits, *Mining Jour. (London)*, vol. 257, no. 6590, pp. 599, 601.

100. Anon., 1962, Enormous iron ore discoveries spur Western Australia boom, *Eng. Mining Jour.*, vol. 163, no. 12, pp. 80–83.

101. Anon., 1962, Caland open pits—Mink and Lime—new major source of Inland's iron ore, *Mining World*, vol. 24, no. 7, pp. 22–26.

102. Anon., 1963, Australia's Pilbara iron ore deposit, *Mining Jour. (London)*, vol. 260, no. 6646, p. 11.

103. Anon., 1963, Western Australia's iron islands, *Mining Jour. (London)*, vol. 261, no. 6675, p. 81.

104. Anon., 1963, Iron ore at El Algarrobo, *Mining Jour. (London)*, vol. 261, no. 6682, pp. 230–231.

105. Anon., 1963, Iron ore in Australia, *Mining Jour. (London)*, vol. 261, no. 6685, pp. 311–315.

MANGANESE

106. O. Barbosa, 1956, Manganese at Urucum, State of Mato Grosso, Brazil, Internat. Geol. Cong., 20th, Mexico 1956, "Symposium sobre Yacimientos de Manganeso," vol. 3, pp. 261–274.

107. G. Berg and F. Friedensburg, 1942, "Die Metallischen Rohstoffe," no. 5, "Mangan," Ferdinand Enke, Stuttgart, 235 pp.

108. L. G. Boardman, 1961, Manganese in the Union of South Africa, Commonwealth Min. Metall. Cong., 7th, South Africa, *Trans.*, vol. 1, pp. 201–215.

109. J. Bouladon and G. Jouravsky, 1952, Manganese, Internat. Geol. Cong., 19th, Algiers 1952, *Mon. Régionales,* ser. 3, Maroc, no. 1, pp. 45–80.

110. J. Bouladon and G. Jouravsky, 1956, Les gîtes de manganèse du Maroc (suivi d'une description des gisements du Précambrian III), Internat. Geol. Cong., 20th, Mexico 1956, "Symposium sobre yacimientos de manganeso," vol. 2, pp. 217–248.

111. J. C. Brown and A. K. Dey, 1955, "India's Mineral Wealth," 3d ed., Oxford University Press, London, 761 pp.

112. C. Chia-hsiang and L. You-hsin, 1960, Preliminary study of the manganese deposits of China, *Internal. Geol. Review,* vol. 2, no. 10, pp. 833–850. (Translated by E. C. T. Chao.)

113. M. S. Crittenden and L. Pavlides, 1962, Manganese in the United States, *U.S. Geol. Survey Mineral Resourrces Inv. Resource Map* MR-23. Map and pamphlet of 8 pp.

114. J. de Villiers, 1956, The manganese deposits of the Union of South Africa, Internat. Geol. Cong., 20th, Mexico 1956, "Symposium sobre Yacimientos de Manganeso," vol. 2, pp. 39–71.

115. J. de Villiers, 1960, The manganese deposits of the Union of South Africa, *South Africa Geol. Survey, Handbook* 2, 271 pp.

116. J. V. N. Dorr, II, 1945, Manganese and iron deposits of Morro do Urucum, Mato Grosso, Brazil, *U.S. Geol. Survey Bull.* 946-A, pp. 1–47.

117. J. V. N. Dorr, II, C. F. Park, Jr., and Glycon de Paiva, 1949, Manganese deposits of the Serra do Navio district, Territory of Amapá, Brazil, *U.S. Geol. Survey Bull.* 964-A, pp. 1–45.

118. J. V. N. Dorr, II, I. P. Coehlo, and A. Horen, 1956, The manganese deposits of Minas Gerais Brazil, Internat. Geol. Cong., 20th, Mexico 1956, "Symposium sobre Yacimientos de Manganeso," vol. 3, pp. 279–346.

119. J. J. Frankel, 1958, Manganese ores from the Kuruman district, Cape Province, South Africa, *Econ. Geology,* vol. 53, pp. 577–597.

120. Geological Survey Division and Department of Mines, Union of South Africa, 1959, "The Mineral Resources of South Africa," 4th ed., 622 pp.

121. J. González Reyna (ed.), 1956, "Symposium sobre Yacimientos de Manganeso," Internat. Geol. Cong., 20th, Mexico 1956; vol. I, "El manganeso en general"; vol. 2, "Africa"; vol. 3, "America"; vol. 4, "Asia y Oceania"; vol. V, "Europa."

122. D. F. Hewett, M. D. Crittenden, L. Pavlides, and G. L. DeHuff, Jr., 1956, Manganese deposits of the United States, Internat. Geol. Cong., 20th, Mexico 1956, "Symposium sobre Yacimientos de Manganeso," vol. 3, pp. 169–230.

123. J. N. Hoffman, 1958, Manganese, its minerals, deposits and uses, *Pennsylvania State Univ., Min. Ind. Exper. Sta., Circ. No.* 49 *(rev.),* 116 pp.

124. W. Kupferburger, L. G. Boardman, and P. R. Bosch, 1956, New considerations concerning the manganese ore deposits of the Postmasburg and Kuruman areas, South Africa, Internat. Geol. Cong., 20th, Mexico 1956, "Symposium sobre Yacimientos de Manganeso," vol. 2, pp. 73–87.

125. A. A. Maksimov, 1960, Types of manganese and iron-manganese deposits in central Kazakhstan, *Internat. Geol. Rev.,* vol. 2, no. 6, pp. 508–521.

126. R. H. Nagell, 1962, Geology of the Serra do Navio manganese district, Brazil, *Econ. Geology,* vol. 57, pp. 481–498.

127. C. G. Park, Jr., 1956, Manganese ore deposits of the Serra do Navio district, Federal Territory of Amapá, Brazil, Internat. Geol. Cong., 20th, Mexico 1956, "Symposium sobre Yacimientos de Manganeso," vol. 3, pp. 347–376.

128. C. G. Park, Jr., J. V. N. Dorr, II, P. W. Guild, and A. L. M. Barbosa, 1951, Notes on the manganese ores of Brazil, *Econ. Geology,* vol. 46, pp. 1–22.

129. G. H. S. V. Prasada Rao and Y. G. K. Murty, 1956, Manganese-ore deposits of Orissa and Bihar, India, Internat. Geol. Cong., 20th, Mexico 1956, "Symposium sobre Yacimientos de Manganeso," vol. 4, pp. 115–131.

130. B. C. Roy, 1956, Manganese-ore deposits of Bombay State, India, Internat. Geol. Cong., 20th, Mexico 1956, "Symposium sobre Yacimientos de Manganeso," vol. 4, pp. 41–61.

131. S. Roy, 1962, Study of the metamorphic manganese ores of Bharweli mine-area, Madhya Pradesh, India, and their genesis, *Econ. Geology,* vol. 57, pp. 195–208.

132. V. P. Sondhi, 1956, Manganese ores in India, Internat. Geol. Cong., 20th, Mexico 1956, "Symposium sobre Yacimientos de Manganeso," vol. 4, pp. 9–23.

133. R. K. Sorem and E. N. Cameron, 1960, Manganese oxides and associated minerals of the Nsuta manganese deposits, Ghana, West Africa, *Econ. Geology,* vol. 55, pp. 278–310.

134. A. Sriramadas, 1963, The manganese ore deposits of Kutingi mines, Koraput district, Orissa, India, *Econ. Geology,* vol. 58, pp. 569–578.

135. J. A. Straczek an dothers, 1956, Manganese ore deposits of Madhya Pradesh, India, Internat. Geol. Cong., 20th, Mexico 1956, "Symposium sobre Yacimientos de Manganeso," vol. 4, pp. 63–96.

136. U.S. Bureau of Mines, 1952, "Materials Survey, Manganese," 535 pp.

137. Anon., 1952, India ups manganese production to supply "free world" market, *Mining World,* vol. 14, no. 8, pp. 34–37.

ALUMINUM (BAUXITE)

138. V. T. Allen, 1960, Comparison of bauxite deposits of Europe with those in the U.S.A., Internat. Geol. Cong., 21st, Copenhagen 1960, pt. 16, sec. 16, Genetic problems of ores, pp. 230–236.

139. J. Bridge, 1950, Bauxite deposits of the southeastern United States, in "Symposium on Mineral Resources of the Southeastern United States," pp. 170–201, University of Tennessee Press, Knoxville, Tenn., 236 pp.

140. S. B. Dickinson, 1962, Aluminum in Australia. Pt. 1, Australia's bauxite resources, *Australian Chem. Proc.,* vol. 15, no. 10, pp. 8–15.

141. D. Franotovic, 1955, Dalmatia leads Yugoslavia's growing bauxite industry, *Eng. Mining Jour.,* vol. 156, no. 12, pp. 78–84.

142. M. Gordon, Jr., J. I. Tracey, Jr., and M. W. Ellis, 1958, Geology of the Arkansas bauxite region, *U.S. Geol. Survey Prof. Paper* 299, 268 pp.

143. M. Gordon, Jr., and J. I. Tracey, Jr., 1952, Origin of the Arkansas bauxite deposits, in "Problems of Clay and Laterite Genesis," pp. 12–34, American Institute of Mining and Metallurgical Engineers, New York, 244 pp.

144. E. C. Harder, 1949, Stratigraphy and origin of bauxite deposits, *Geol. Soc. America Bull.*, vol. 60, pp. 887–908.
145. E. C. Harder, 1952, Examples of bauxite deposits illustrating variations in origin, in "Problems of Clay and Laterite Genesis," pp. 35–64, American Institute of Mining and Metallurgical Engineers, New York, 244 pp.
146. E. C. Harder and E. W. Greig, 1960, Bauxite, in "Industrial Minerals and Rocks," 3d ed., pp. 65–85, American Institute of Mining, Metallurgical, and Petroleum Engineers, New York, 934 pp.
147. V. C. Hill, 1955, The mineralogy and genesis of the bauxite deposits of Jamaica, B.W.I., *Am. Mineralogist,* vol. 40, pp. 676–688.
148. H. R. Hose, 1960, The genesis of bauxite, the ores of aluminum, Internat. Geol. Cong., 21st, Copenhagen 1960, pt. 16, sec. 16, Genetic problems of ores, pp. 237–247.
149. H. R. Hose, 1963, Jamaica type bauxites developed on limestones, *Econ. Geology,* vol. 58, pp. 62–69.
150. H. F. Kurtz, 1952, Bauxite, in W. Van Royen and O. Bowles, "The Mineral Resources of the World," vol. 2, pp. 91–95, Prentice-Hall, Inc., Englewood Cliffs, N.J., 181 pp.
151. J. H. Moses and W. D. Mitchell, 1963, Bauxite deposits of British Guiana and Surinam in relation to underlying unconsolidated sediments suggesting two-step origin, *Econ. Geology,* vol. 58, pp. 250–262.
152. S. H. Patterson, 1963, Estimates of world bauxite reserves and potential resources, *U.S. Geol. Survey Prof. Paper* 475-B, pp. 158–159.
153. I. G. Sohn, 1952, Geologic environment of alumina resources of the Columbia River basin, *U.S. Geol. Survey Mineral Inv. Resource Appraisals Map* MR-1.
154. U.S. Bureau of Mines, 1953, "Materials Survey, Bauxite," 296 pp.
155. K. P. Wang, 1959, Vast expansion of aluminum-alumina is planned by Chinese Communists, *Eng. Mining Jour.*, vol. 160, no. 7, pp. 75–77, 123.
156. B. N. Webber, 1959, Bauxitization in the Pocos de Caldas district, Brazil, *Mining Eng.*, vol. 11, pp. 805–809.

MISCELLANEOUS REFERENCES CITED

157. J. A. S. Adams and K. A. Richardson, 1960, Thorium, uranium and zirconium concentrations in bauxite, *Econ. Geology,* vol. 55, pp. 1653–1675.
158. J. V. Beall, 1962, The little shift in the big picture, *Mining Eng.*, vol. 14, no. 12, pp. 35–39.
159. S. Bracewell, 1958, Jamaican bauxite in the West Indies economy, *Mining Eng.*, vol. 10, pp. 1079–1080.
160. R. Cochran, 1954, New plant successfully floats Michigan jasper, *Eng. Mining Jour.*, vol. 155, no. 8, pp. 100–104.
161. C. Danielsson and S. Iversson, 1963, Iron ore developments surge in West Africa, *Mining Eng.*, vol. 15, no. 3, pp. 32–38.
162. E. W. Davis, 1953, Iron ore for the future, *Eng. Mining Jour.*, vol. 154, no. 7, p. 103.
163. W. H. Durrell, 1954, Development of Quebec-Labrador iron ore deposits, transportation is major factor, *Mining Eng.*, vol. 6, pp. 387–390.

164. M. Eigo, 1964, Jamaica keeps growing—two new projects will boost bauxite shipments to the U.S., *Eng. Mining Jour.*, vol. 165, pp. 50–57.

165. Geology and economic minerals of Canada, 3d ed., 1947, pp. 294–295, *Canada Dept. Mines and Resources, Mines and Geol. Branch, Geol. Survey, Econ. Geol. Ser. No.* 1, 357 pp.

166. J. A. Hartman, 1959, The titanium mineralogy of certain bauxites and their parent materials, *Econ. Geology*, vol. 54, pp. 1380–1403.

167. J. E. Kelly, 1950, More iron ore in Venezuela, *Mining World*, vol. 12, no. 10, pp. 34–37.

168. A. W. Knoerr and G. P. Lutjen, 1956, Reserve's new taconite project, *Eng. Mining Jour.*, vol. 157, no. 12, pp. 75–102.

169. O. Lee, 1954, Taconite beneficiation comes of age at Reserve's Babbit plant, *Mining Eng.*, vol. 6, pp. 484–490.

170. G. B. Mellon, 1961, Sedimentary magnetite deposits of the Crowsnest Pass region, southwestern Alberta, *Research Council of Alberta, Bull.* 9, 98 pp.

171. K. E. Merklin and F. D. Devaney, 1961, The Wabush pilot plant: stepping stone to $200-million development, *Eng. Mining Jour.*, vol. 162, no. 4, pp. 84–90.

172. R. H. Ramsey, 1955, Teamwork on taconite, *Eng. Mining Jour.*, vol. 156, no. 3, pp. 71–93.

173. F. W. Starratt, 1956, Iron agglomerates from the Marquette Range, *Mining Eng.*, vol. 8, pp. 1100–1102.

174. R. J. Traill and others, 1962, Raw materials of Canada's mineral industry, *Geol. Survey Canada, Dept. Mines and Tech. Surveys, Paper* 62-2, p. 33.

175. Various authors, 1963, The story of Erie Mining Company, *Mining Eng.*, vol. 15, no. 5, pp. 39–54.

176. Anon., 1950, Venezuela—iron ore aplenty, *Mining World.*, vol. 12, no. 5, pp. 35–36.

177. Anon., 1951, After 15 years—El Pao iron, *Mining World*, vol. 13, no. 5, pp. 36–39.

178. Anon., 1952, Reserve Mining Company's taconite program underway, *Mining World*, vol. 14, no. 12, pp. 28–32.

179. Anon., 1952, Reserve Mining Co. starts taconite plant at Babbit, *Eng. Mining Jour.*, vol. 153, no. 11, pp. 72–79.

180. Anon., 1952, Specular hematite pilot plant, *Mining World*, vol. 14, no. 8, p. 51.

181. Anon., 1954, Historic first shipment of Labrador iron ore opens supply line to 500-million ton deposit, *Eng. Mining Jour.*, vol. 155, no. 8, pp. 124–125.

182. Anon., 1954, Ore in '54, Quebec-Labrador ships iron ore a year ahead of schedule, *Mining Eng.*, vol. 6, p. 884.

183. Anon., 1954, First Quebec-Labrador iron ore in '54 fitting climax to $250,000,000 project, *Mining World*, vol. 16, no. 9, pp. 66–67.

184. Anon., 1955, Marmora: Bethlehem beneficiates open pit ore, pelletizes concentrates, at new iron producer, *Mining Eng.*, vol. 7, pp. 634–637.

185. Anon., 1955, Marmora mine and plant ships pellets to feed Bethlehem's blast furnaces, *Eng. Mining Jour.*, vol. 156, no. 7, pp. 75–79.

186. Anon., 1956, Taconite milestone, the story of the Reserve project, *Mining World*, vol. 18, no. 10, pp. 52–59.

187. Anon., 1958, Mesabi to the north, *Mining Eng.*, vol. 10, pp. 678–679.

188. Anon., 1958, Canadian iron-ore—and where it will go in the next 25 years, *Mining Eng.*, vol. 13, pp. 680–682.

189. Anon., 1959, Canadian iron ore survey, *Mining Eng.*, vol. 11, pp. 202–203.

190. Anon., 1959, St. Lawrence seaway offers new avenue for output of Canadian and U.S. mines, *Eng. Mining Jour.*, vol. 160, no. 8, pp. 82–88.

191. Anon, 1961, Japan's big iron source—bottom of Tokyo Bay, *Mining Eng.*, vol. 13, p. 1303.

192. Anon., 1961, Meramec—new underground Missouri iron mine, *Mining World*, vol. 13, no. 5, pp. 30–31.

193. Anon., 1961, Low grade manganese to MnO_2; Sterling announces new process, *Eng. Mining Jour.*, vol. 162, no. 11, pp. 18–19.

194. Anon., 1962, Japanese steel from undersea sand iron, *Mining Jour. (London)*, vol. 258, no. 6603, p. 241.

195. Anon., 1962, Pellets for steelmaking, *Mining Jour. (London)*, vol. 259, no. 6633, pp. 314–315.

196. Anon., 1962, Iron ore pellets, *Mining Jour. (London)*, vol. 259, no. 6640, p. 484.

197. Anon., 1962, Malayan iron ore, *Mining Jour. (London)*, vol. 258, no. 6615, pp. 559, 561.

198. Anon., 1962, Major iron ore find, *Mining Jour. (London)*, vol. 259, no. 6620, p. 3.

199. Anon., 1962, Atlantic City: West's first taconite operation ships on schedule, *Mining World*, vol. 14, no. 10, pp. 16–18.

200. Anon., 1962, Iron ore in Canada, *Mining Jour. (London)*, vol. 258, no. 6604, pp. 261, 263.

201. Anon., 1962, Canada's iron ore industry, *Mining Jour. (London)*, vol. 259, no. 6637, pp. 403–405

202. Anon., 1962, Major iron discovery in Canadian north, *Mining World*, vol. 24, no. 9, p. 63.

203. Anon., 1962, Alumina from low-grade bauxites and clay, *Mining Jour. (London)*, vol. 259, no. 6632, pp. 280–283.

204. Anon., 1962, Report on Australia's integrated aluminum industry, *Mining Jour. (London)*, vol. 259, no. 6631, pp. 260–263.

205. Anon., 1963, Development of Australia's bauxite resources, *Mining Jour. (London)*, vol. 261, no. 6689, p. 415.

206. Anon., 1963, Australian aluminum, *Mining Jour. (London)*, vol. 261, no. 6693, pp. 515–516.

207. Anon., 1963, Here's Weipa, the world's largest known bauxite deposit, *Mining World*, vol. 25, no. 8, p. 41.

208. Anon., 1963, Iron and steel, *Mining Jour. (London)*, vol. 261, no. 6685, pp. 303, 305.

209. Anon., 1963, Iron ore from Liberia, *Mining Jour. (London)*, vol. 260, no. 6667, p. 540.

210. Anon., 1963, Meramec's Pea Ridge iron project nearing completion, *Mining World,* vol. 15, no. 5, p. 42.

211. Anon., 1963, New iron ore project in Canada, *Mining Jour. (London),* vol. 260, no. 6652, p. 153.

212. Anon., 1963, The Carol Lake project, *Mining Jour. (London),* vol. 260, no. 6657, pp. 270–271.

213. Anon., 1963, Winning low-grade iron ore, *Mining Jour. (London),* vol. 260, no. 6669, pp. 592–593.

part III

Some Important Industrial Deposits

chapter 15

Survey of
the Industrial
Mineral Field

The materials that form the industrial minerals and rocks make a diverse group
that includes not only materials used in various industrial operations but also a
number that are used chiefly as sources of metals and have already been discussed.
The industrial minerals and rocks have been discussed in considerable detail in
three good recent books, two published in 1960 (4, 45) and one in 1961 (18).
Consequently, the present book considers somewhat in detail the metallic deposits,
which have not been summarized for 20 years or more, and treats more briefly
the industrial minerals and rocks.

The chief industrial materials and some of their uses are listed in the following pages. This listing includes some minerals used chiefly as a source of metals
but also generally discussed with industrial materials. Other materials used as a
source of metals that might have been included because of their rather wide application industrially are gold, silver (in photographic emulsions), platinum, and
some others. These, along with many uses of industrial materials, have been omitted
so that the major industrial materials might be shown to better advantage.

The chief industrial materials

Andalusite, dumortierite, kyanite, sillimanite: Used in the manufacture of
refractories.
Asbestos: Used for making asbestos textile products, asbestos paper, sheets, and
similar material; asbestos shingles and asbestos-cement products; heat-insula-
tion cements.
Barite and witherite: Barite is used chiefly as oil-well drilling mud. Some other
uses of barium minerals are the making of various chemicals and lithopone pigment and the making of glass and rubber products.

Beryl and other beryllium minerals: Used as gem stones, especially beryl, and in the manufacture of some ceramic materials; source of beryllium.

Bauxite: Used in the manufacture of artificial abrasives; making of chemicals and refractories; source of aluminum.

Borax and borates: Used for making ceramic glazes, porcelain enamels, and various types of glass; making of soaps, detergents; manufacture of certain leather and textile products.

Chromite: Used in manufacture of chemicals and refractories; source of chromium.

Clays: Used in the manufacture of various ceramic products, brick, tile, etc., and refractories; as filler in manufacture of paper, rubber products, insecticides, and other products.

Diamond: Industrial diamonds are used extensively as abrasives; gem quality used as gems.

Diatomite: Used as a filtering material in purification of a variety of liquid products; as a filler in paints, papers, and other products; in making insulating materials; in making polishing materials.

Feldspar: Used in the manufacture of glass, pottery, enamel.

Fluorspar: Used in the manufacture of steel, hydrofluoric acid, and ceramic products.

Gem stones: Used as gems and semiprecious stones.

Graphite: Used in the manufacture of crucibles and other refractory products, especially in the brass, aluminum, and titanium industries; making lubricants; making brushes for electric motors; making foundry facings, pigments, batteries, pencils.

Gypsum: Used in the manufacture of portland cement, making plasters, lath, wallboard; as a soil conditioner.

Limestone: Used in the manufacture of lime, which is widely used in construction and in agriculture; manufacture of steel; making many chemical and industrial products.

Lithium minerals (spodumene, amblygonite, lepidolite, etc.): Used in the manufacture of lithium stearate greases; making ceramic and glass materials; in air conditioning; some varieties used as gem stones; source of lithium.

Magnesite, brucite, and other magnesium materials: Used in the manufacture of refractories, especially for use in the steel and copper industries; manufacture of certain cements used in flooring and wallboard; making paper pulp, rayon, fertilizers.

Manganese ore minerals: Used in the manufacture of dry cells and chemicals; source of manganese.

Mica: Best quality used largely by the electrical industry in making radios and other instruments; lower quality used for insulation in electrical appliances, insulating cloth, tape, etc.; ground mica used in making roofing, paints, and some other products.

Monazite and other rare-earth minerals: Used in making cores of carbons for motion picture projection, floodlights, searchlights; making polishing materials for lenses, prisms, mirrors; source of thorium, cerium, and others.

Nitrogen minerals: Used in the manufacture of fertilizers, chemicals, explosives.

Perlite: Used in the making of expanded perlite, which is used in the manufacture of lightweight aggregates for plaster and insulating.

Phosphate rock and apatite: Used in the manufacture of fertilizers and phosphoric acid.

Potash minerals: Used in the manufacture of fertilizers; making of soaps, detergents, some ceramic products, textiles, dyes, chemicals, drugs.

Pyrophyllite: Used in making wall tile, refractories, wallboard, paints, insecticides.

Quartz crystal: Used in making electronic and optical instruments.

Rutile and ilmenite: Used in the manufacture of white pigment; source of titanium.

Salt (halite): Used in the manufacture of chlorine; manufacture of soda ash, which is used in making glass, soap, paper, chemicals, paints, and various other products.

Sand and gravel: Used in building and paving industries; as railroad ballast; certain sands used in making glass and in foundry work.

Slate: Used for roofing, floor tiles, and numerous interior types of work.

Sodium carbonate minerals: Used in the production of soda ash.

Sodium sulfate minerals: Used in the production of kraft pulp, which is used in the manufacture of brown paper; also used in the glass and chemical industries.

Stone: Used as building stone, crushed and broken stone for concrete and road work; manufacture of cement.

Strontium minerals: Used for making signal flares; in manufacture of sugar from sugar beets; in the ceramic industry.

Sulfur and pyrites: Used in the manufacture of fertilizers; making paper pulp; in the textile and rubber industries; in the chemical industries.

Talc and soapstone: Used in the ceramic and paint industries; in the manufacture of insecticides, roofing, rubber, and other materials.

Vermiculite: Used in making expanded vermiculite, which is used in making insulation, plaster aggregate, and concrete aggregate.

Zircon: Used in the manufacture of refractories, foundry sand and facings, ceramics; some varieties used as gem stones; source of zirconium.

Production

World-wide production statistics are available for about half of the diverse materials listed in the foregoing summary. This information has been condensed, chiefly from the *United States Bureau of Mines Minerals Yearbook,* and assembled in Tables 15-1 through 15-4, which show the importance of various countries as producers of some industrial minerals and rocks. Generally if a country produced less than 3 percent of the world total for the period considered, which is the average of the 15 years 1945 to 1959 for most materials, it was omitted from the tables. Production from some Communist countries was not available. Deposits of the materials listed in these tables will be discussed briefly in subsequent chapters. Deposits of some others are summarized in this chapter.

TABLE 15.1.   Production of Certain Industrial Minerals and Rocks in Africa, Average of 1945–1959

Country	Material	World rank and percent of world production	
		Rank	*Percent*
Angola	Diamond (gem)*	5	10.0
Congo, Republic of the	Diamond (industrial)*	1	66.2
	Diamond (gem)*	6	8.5
Ghana	Diamond (industrial)*	2	9.6
	Diamond (gem)*	3	13.2
Malagasy Republic (Madagascar)	Graphite	4	5.7
Morocco	Phosphate rock	2	17.1
Sierra Leone	Diamond (industrial)*	4	4.5
	Diamond (gem)*	4	11.3
South Africa, Republic of	Asbestos	3	6.5
	Corundum	1	25.8
	Diamond (industrial)*	3	8.1
	Diamond (gem)*	1	20.7
Rhodesia (Southern Rhodesia)	Asbestos	4	5.7
	Corundum	2	14.4
South-West Africa	Diamond (gem)*	2	13.3
Tunisia	Phosphate rock	4	7.0

* Average of 1958–1962.

TABLE 15-2.   Production of Certain Industrial Minerals and Rocks in Asia and Australasia, Average of 1945–1959

Country	Material	World rank and percent of world production	
		Rank	*Percent*
Ceylon	Graphite	5	4.4
China	Salt	3	9.1
India	Corundum	3	4.8
	Mica	2	14.6
	Salt	6	5.0
Japan	Pyrite	1	15.2
	Pyrophyllite, talc, soapstone	2	20.2
Korea, Republic of	Graphite	1	22.0
Nauru Island	Phosphate rock	5	3.5

TABLE 15-3.   Production of Certain Industrial Minerals and Rocks in Europe, Average of 1945–1959

Country	Material	World rank and percent of world production	
		Rank	Percent
Austria	Graphite	3	8.0
	Magnesite	1	19.7
Cyprus	Pyrite	4	7.2
France	Feldspar	3	6.6
	Gypsum	3	10.5
	Potash	4	17.4
	Salt	5	5.1
	Talc, soapstone, pyrophyllite	3	8.3
Germany, East	Potash	3	20.5
Germany, West	Barite	2	14.5
	Feldspar	2	11.8
	Fluorspar	3	9.3
	Graphite	6	4.0
	Potash	2	22.3
Italy	Feldspar	6	3.4
	Pyrite	3	7.5
	Sulfur (native)	3	3.3
	Talc, soapstone, pyrophyllite	4	5.2
Norway	Feldspar	5	3.8
	Pyrite	5	5.2
	Talc, soapstone, pyrophyllite	5	4.5
Spain	Gypsum	6	5.6
	Pyrite	2	13.2
Sweden	Feldspar	4	4.6
United Kingdom	Barite and witherite	5	4.5
	Fluorspar	4	6.4
	Gypsum	4	10.5
	Salt	4	7.6
Soviet Union	Asbestos	2	20.9
	Barite	4	4.9
	Fluorspar	6	6.2
	Gypsum	5	7.2 (?)
	Phosphate rock, apatite	3	14.2
	Potash*	5	10.0
	Salt	2	10.2
Yugoslavia	Magnesite	3	3.1

* 1950–1959 world production.

**TABLE 15-4. Production of Certain Industrial Minerals
and Rocks in North America, Average of 1945–1959**

| Country | Material | World rank and percent of world production | |
		Rank	Percent
Canada	Asbestos	1	55.2
	Barite	3	7.9
	Fluorspar	5	6.2
	Gypsum	2	13.3
Mexico	Barite	6	4.1
	Fluorspar	2	13.4
	Graphite	2	12.4
	Sulfur (native)	2	5.4
United States	Barite	1	41.3
	Feldspar	1	55.9
	Fluorspar	1	23.6
	Gypsum	1	29.5
	Magnesite	2	11.9
	Mica	1	57.2
	Phosphate rock	1	45.6
	Potash	1	26.7
	Pyrite	5	7.1
	Salt	1	31.7
	Sulfur (native)	1	84.5
	Talc, soapstone, pyrophyllite	1	37.8

Summary of selected materials

Borates

Generally more than 95 percent of the world production of boron comes from the United States, but some comes also from Argentina, Chile, Peru, Bolivia, Turkey, the Soviet Union, and Italy (18, p. 90; 22, pp. 107–117; 41; 46).

The most important borate deposits of the United States are in California; deposits of minor importance are in Nevada and Oregon (1, 42). The two most productive deposits of California are at Kramer in Kern County and at Searles Lake in San Bernardino County. These two deposits furnish most of the borates now produced in the United States.

The deposit at Kramer, which is the largest and most productive one in the country, is interlayered with limy shales and siltstones of Tertiary age. Coarsely crystalline borax is the chief ore mineral, some of which has been replaced by kernite. The deposit at Searles Lake (22, pp. 111–112; 41; 42) consists of brines that are pumped from two very porous layers of saline minerals. Borax and other products are extracted from the brines. Some solid borax is present also, but as yet it has not been mined.

The deposits of other countries are somewhat varied. Most of those of South America are of the marsh type, in which thin crusts of borate minerals are mixed with other evaporites at or near the surface of desert basins. Generally the deposits are difficultly accessible, and economical production depends on location with respect to transportation. Production in Turkey has come chiefly from deposits in Balikesir and Bursa Provinces, but more recently discovered deposits in Kutahya Province are reported to be larger than those previously known. Some of the deposits are associated with gypsum, others are associated with marly rocks. The deposits of the Soviet Union, which are thought to be extensive, are about 100 miles north of the Caspian Sea near the Ural River. They occur along fracture zones in a salt dome. The boron of Italy is obtained as boric acid from volcanic emanations (soffioni) at steam vents.

Natural Nitrates

The nitrate deposits of Atacama and Taracapa, Chile (17; 18, pp. 429–432; 22, pp. 360–368; 27, pp. 283–302; 35; 36; 37; 38), are the only ones of commercial importance in the world, although deposits are known in Peru, Colombia, Argentina, Mexico, Egypt, South Africa, and the United States. The Chilean deposits occur as part of the cement of beds of gravel close to the surface of the desert. Several different layers occur locally, but three are rather general. The cemented material generally is immediately overlain by loose soil, locally designated *chuca,* below which occurs soil tightly cemented by mixed salts, chiefly chlorides and sulfates, the *costra,* which grades downward into the nitrate deposits, which are less consolidated material designated *caliche.* The caliche contains 5 to 30 percent sodium nitrate, along with 0.1 to 5 percent salts of potassium, calcium, and magnesium. Iodates are present also and iodine is recovered as a by-product.

Natural Sodium Carbonate

Deposits of natural sodium carbonate minerals—trona and natron—are relatively small, so most of the world's sodium carbonate is obtained from soda ash, which in turn is produced from salt. Two important producers of natural sodium carbonate are Kenya, Africa, and the United States. Deposits are known also in China, the Soviet Union, South Africa, and a number of other places (18, pp. 538–543).

The production of Kenya, second to that of the United States, comes from a large deposit of trona at Lake Magadi in the Rift Valley. The lake, about 18 miles long and 2 or 3 miles wide, is covered by a crust of trona as much as 6 ft thick in places, below which is saturated brine that contains 40 to 50 percent salts. Hot springs feed the lake and increase the supply of soda faster than it is removed by dredging.

Production of sodium carbonate in the United States (2; 4, pp. 387–390; 18, p. 540; 22, pp. 496–498; 25) was chiefly from Owens Lake and Searles Lake, California, until 1948; since then important production has come from Green River, Wyoming. Production at Owens Lake and at Searles Lake is from brines, whereas that at Green River is from bedded trona in the Green River formation of Eocene age at depths from about 650 ft to 1,500 ft. Production began in 1948 from a depth of about 1,500 ft (47). Subsequent drilling (5) disclosed an important deposit at Big Island, 20 miles north of Green River, at depths of 650 to 1,100 ft. Throughout the area there are seven beds of trona of commercial thickness and known minable reserves of more than 2 million tons of trona. The area is rated as the world's largest known source of trona. Production at the Big Island mine began in May, 1962. The beds represent deposits laid down in a former lake.

Natural Sodium Sulfate

Deposits of natural sodium sulfate, although widespread, generally have been little exploited because much of the world's sodium sulfate was obtained as a by-product from the manufacture of hydrochloric acid, sulfuric acid, and other products. Changes in technology, however, have caused the increasing use of natural sodium sulfate—mirabilite, thenardite, glauberite, blödite, and a few other minerals (2; 14; 18; 22, pp. 499–501; 25). Production of natural material comes chiefly from the United States and Canada, but production is known to occur also in Argentina, Chile, Egypt, Germany, the Soviet Union, and China.

The production of the United States comes from Searles Lake, California; from underground brines in Texas; and from solid deposits in Wyoming and California. Many deposits have been worked from time to time throughout a number of western states.

The deposits of Canada are chiefly in southern Saskatchewan. Most of the deposits consist of bedded sulfate, but some production comes from brine.

Strontium Minerals

Celestite furnishes most of the strontium used for industrial operations, but strontianite supplies a limited amount of it (3; 12; 18, pp. 547–552; 22, pp. 515–521; 39). During the two years 1961 and 1962 the United Kingdom produced 57 percent and Mexico furnished 30 percent of the total reported world production of strontium minerals. Italy supplied 7 percent and Pakistan furnished 3 percent.

England has long been the chief producer of celestite, which comes from marls of Triassic age, chiefly in Gloucestershire but also in Somersetshire. The celestite of Mexico has come from veins in Durango and from replacements of limestone in Coahuila. Celestite occurs at several places in the United States, but limited production has come chiefly from California, Texas, and Arizona. No production was reported in 1962.

Diatomite

Diatomite or diatomaceous earth (4, pp. 360–369; 9; 16; 18, pp. 181–188; 22, pp. 185–193; 32), which consists chiefly of siliceous remains of diatoms, is rather widespread throughout the world, but the known major production before 1958 was from the United States, France, and West Germany. Large production has been reported from the Soviet Union, starting in 1958, although actual unreported production probably took place before then. The important production, as percent of the world total, follows.

Production in the United States comes chiefly from the Lompoc area near Santa Barbara, California; other production comes mainly from Nevada, Washington, and Oregon. The deposits of the Lompoc area are large and remarkably pure. They occur in the lower (Miocene) part of the Sisquoc formation, which contains about 1,000 ft of diatomite and overlies the Monterey formation. The commercial deposits have been preserved in synclinal structures.

	Average of 1953–1957, percent	1962, percent
United States	36.0	33.0
France	7.0	7.6
West Germany	6.0	5.0
Soviet Union	(?)	23.0

Vermiculite

Commercial vermiculite is a mineral of micaceous habit that expands when it is heated and thus is suitable for use in lightweight aggregates. Such material is widespread throughout the world, but the major reported production comes from the United States—about 70 percent of the world total in 1962—and the Republic of South Africa—30 percent. The production in the United States in 1962 was chiefly from Montana and South Carolina. The Montana vermiculite is near Libby, in Lincoln County, and is associated with pyroxenite, into which it grades. The South Carolina vermiculite is in Greenville, Laurens, and Spartanburg Counties of the Piedmont. Lenses of pyroxenite that lie within gneisses and schists apparently were altered in part to biotite, which, in turn, was changed to vermiculite by groundwaters, as the vermiculite in weathered parts of the pyroxenite passes into biotite in depth (4, pp. 343–345; 22, pp. 574–580; 29; 31; 34).

Perlite

Commercial perlite (4, pp. 50–57; 18, pp. 445–450; 20, pp. 490–491; 22, pp. 375–379; 30; 33) is natural volcanic glass that will "pop" when heated sufficiently, because of contained water, and thus yield expanded material that can be used in lightweight aggregates. Perlite is produced chiefly in the United States. New Mexico supplied 76 percent of the domestic material in 1962 and was followed in order by Nevada, California, Arizona, Colorado, Idaho, and Utah. The deposit in New Mexico forms a large dome near Socorro.

Stone

Stone (6, 20, 40) occurs in every country of the world, and much of the material is used close to its source. In general stone is classed in two groups on the basis of use, dimension stone and crushed stone. Dimension stone is material that is cut into blocks or slabs and used for various types of construction, in both exterior and interior work, for paving work, as monumental stone, and other purposes for which large pieces are needed. Crushed stone is used in a great variety of operations, but more is used for concrete and road stone than for any other purpose. By far the greater amount of production is as crushed stone—about 250 times the tonnage of dimension stone in the United States in 1962—but the value of crushed stone is much less. Limestone (18, pp. 293–326) is commonly considered with discussions of cement and lime, because of its use in their manufacture. Actually, limestone has many uses, and a recent report (23) states that it is used for more than 70 purposes. Similarly, slate is considered separately in various discussions

(8; 20, pp. 800–803; 22, pp. 476–484) because of its particular characteristics and somewhat specialized uses. Slate in the United States is produced chiefly in the eastern part of the country; in 1962 Pennsylvania, Vermont, and Virginia supplied 89 percent of the total production.

Sand and Gravel

Every country in the world has at least some deposits of sand and gravel (19; 22, pp. 453–470; 24, 44) and much of this material is used near its source. Certainly sand and gravel make up the largest volume of any industrial mineral material, but at the same time the value of the material per ton is low. Nevertheless, in the aggregate the value of the production is large, 795 million dollars in the United States in 1962 from 777 million tons of sand and gravel. More than 97 percent of this material was used in construction work. The production from California was far greater than that of any other state, and was followed in turn by Michigan, Ohio, Illinois, Wisconsin, Texas, New York, Minnesota, Indiana, and Utah. Together these states supplied half of the total production of the United States.

Clays

Industrial materials termed *clay* (4, pp. 117–151; 11; 15; 18, pp. 69–75, 133–145; 22, pp. 141–164; 26; 28) vary considerably in their characteristics and properties, but they all contain essential amounts of the clay minerals, of which those of the kaolinite, montmorillonite, and illite groups are the most common. Much has been written about clays and clay deposits, and this material cannot so much as be summarized here. Some information, however, is presented in Table 15-5.

TABLE 15-5. The Six Chief Groups of Clays, Some of Their Characteristics, Uses, and Major Areas of Production, Especially in the United States*

Type of clay	General composition	Chief characteristics	Major uses	Major areas of production and reserves in the United States	Some other important world producers
Kaolin or china clay	Chiefly kaolinite, some halloysite	Nearly white color when dry and fired	Chinaware, various types of pottery and stoneware; filler in manufacture of paper, rubber, and some other products; making refractories	Georgia furnishes about 75 percent of total, South Carolina about 15 percent. These deposits of Cretaceous age. Tertiary deposits of Cretaceous age. These deposits of Cretaceous age. These deposits of Florida (Eocene age), Georgia, South Carolina, North Carolina	United Kingdom, especially Cornwall and Devon, England; production also from West Germany, Austria, India, Algeria, France, Spain, Italy, Australia
Ball clay	Chiefly kaolinite, but higher ratio silica to alumina than most kaolinites	High plasticity. Name originated in England because of softness and pliability	Manufacture of pottery, stoneware, floor tile, wall tile	Major production in Tennessee and Kentucky from large lenses in Tertiary formations; other important deposits in New Jersey (Cretaceous age). Major reserves in Kentucky and Tennessee	United Kingdom, in North and South Devon and in Dorset, England
Fire clay	Basically kaolinitic but usually contains halloysite, illite, perhaps others; inorganic and organic impurities	Withstands high temperatures; may be plastic, semiplastic, semiflint, flint (hard, nonplastic), nodular flint	Manufacture of refractories used by iron and steel industry and many others; manufacture of brick, sewer tile, and similar products	Major production in Ohio, Pennsylvania, and Missouri; important production also in Texas, Illinois, West Virginia, Indiana, Arkansas, Kentucky, Kansas, Colorado. Reserves well distributed but high-grade fire clays limited chiefly to Ohio, Pennsylvania, Missouri, possibly Kentucky and Tennessee	United Kingdom, especially in England—Northumberland, Durham, Yorkshire, Worcestershire, Staffordshire; also in Scotland and Wales; deposits in Mysore, India, and in West Germany

				Production in United States	Production elsewhere
Bentonite	Minerals of the montmorillonite group	Two types, swelling and nonswelling (metabentonites or subbentonites)	Oil refining; drilling mud; bonding of foundry and molding sands	Wyoming generally supplies more than 50 percent from beds of Cretaceous age; also production from Mississippi (Cretaceous) and Texas (Eocene)	Some production in Manitoba and Alberta, Canada. One of largest deposits outside of the United States on Isle of Ponza, Italy
Fuller's earth	Chiefly minerals of montmorillonite group; tendency for high magnesium content	Similar to nonswelling bentonites; absorb coloring matter; original use was by fullers for cleaning textiles, hence name	As adsorbents; in making insecticides and fungicides; as drilling muds; in mineral oil refining	Generally more than half of production comes from Florida; other production from Georgia and Tennessee. Chief reserves in Florida, Georgia, Texas	
Miscellaneous clays (includes shales)	Variable; illite generally predominates, but kaolinite and montmorillonite may be present	Variable	Making building brick, sewer pipe, drain tile, and similar products; making portland cement and lightweight aggregates	Major production in Ohio, Texas, North Carolina, California; important production in Pennsylvania, Michigan, Illinois, Alabama, Indiana, and New York; less important production in many other states	Produced in various countries, among which those of considerable importance are the United Kingdom, India, Austria, West Germany, and Denmark

* Compiled chiefly from *U.S. Bur. Mines Minerals Yearbook*, various years, and *U.S. Bur. Mines Bull. 556* and *Bull. 585.*

SELECTED REFERENCES

1. J. C. Arundale, 1956, Boron, *U.S. Bur. Mines Bull.* 556, pp. 137–141.
2. J. C. Arundale, 1956, Sodium compounds, *U.S. Bur. Mines Bull.* 556, pp. 795–800.
3. J. C. Arundale, 1956, Strontium, *U.S. Bur. Mines Bull.* 556, pp. 839–842.
4. R. L. Bates, 1960, "Geology of the Industrial Rocks and Minerals," Harper & Row, Publishers, Incorporated, New York, 441 pp.
5. J. R. Bogert, 1963, Production begins at Stauffer's Big Island trona mine, *Mining World,* vol. 25, no. 1, pp. 12–15.
6. O. Bowles, 1956, Stone, *U.S. Bur. Mines Bull.* 556, pp. 825–838.
7. O. Bowles, 1960, Dimension stone, in "Industrial Minerals and Rocks," 3d ed., pp. 321–337, American Institute of Mining, Metallurgical, and Petroleum Engineers, New York, 934 pp.
8. O. Bowles, 1960, Slate, in "Industrial Minerals and Rocks," 3d ed., pp. 791–798, American Institute of Mining, Metallurgical, and Petroleum Engineers, New York, 934 pp.
9. H. P. Chandler, 1956, Diatomite, *U.S. Bur. Mines Bull.* 556, pp. 259–262.
10. A. B. Cummins, 1960, Diatomite, in "Industrial Minerals and Rocks," 3d ed., pp. 303–319, American Institute of Mining, Metallurgical, and Petroleum Engineers, New York, 934 pp.
11. T. de Polo, 1960, Clays, *U.S. Bur. Mines Bull.* 585, pp. 199–212.
12. E. G. Enck, 1960, Strontium minerals, in "Industrial Minerals and Rocks," 3d ed., pp. 815–818, American Institute of Mining, Metallurgical, and Petroleum Engineers, New York, 934 pp.
13. D. E. Garrett and J. F. Phillips, 1960, Sodium carbonate from natural sources in the United States, in "Industrial Minerals and Rocks," 3d ed., pp. 799–808, American Institute of Mining, Metallurgical, and Petroleum Engineers, New York, 934 pp.
14. M. F. Goudge and R. V. Tomkins, 1960, Sodium sulfate from natural sources, in "Industrial Minerals and Rocks," 3d ed., pp. 809–814, American Institute of Mining, Metallurgical, and Petroleum Engineers, New York, 934 pp.
15. B. L. Gunsallus, 1956, Clays, *U.S. Bur. Mines Bull.* 556, pp. 185–201.
16. C. V. O. Hughes, Jr., 1953, Diatomaceous earth—non-metal of a thousand uses, *Mining Eng.,* Vol. 5, pp. 277–281.
17. H. W. Huse, 1960, Nitrogen compounds, in "Industrial Minerals and Rocks," 3d ed., pp. 639–648, American Institute of Mining, Metallurgical, and Petroleum Engineers, New York, 934 pp.
18. S. J. Johnstone and M. G. Johnstone, 1961, "Minerals for the Chemical and Allied Industries," 2d ed., John Wiley & Sons, Inc., New York, 788 pp.
19. W. W. Key, 1960, Sand and gravel, *U.S. Bur. Mines Bull.* 585, pp. 701–715.
20. W. W. Key, 1960, Stone, *U.S. Bur. Mines Bull.* 585, pp. 793–813.
21. T. A. Klinefelter, 1960, Lightweight aggregates, in "Industrial Minerals and Rocks," 3d ed., pp. 487–495, American Institute of Mining, Metallurgical, and Petroleum Engineers, New York, 934 pp.
22. R. B. Ladoo and W. M. Myers, 1951, "Nonmetallic Minerals," McGraw-Hill Book Company, New York, 605 pp.
23. J. E. Lamar, 1961, Uses of limestone and dolomite, *Illinois Geol. Survey Circ.* 321, 41 pp.

24. W. B. Lenhard, 1960, Sand and gravel, in "Industrial Minerals and Rocks," 3d ed., pp. 733–758, American Institute of Mining, Metallurgical, and Petroleum Engineers, New York, 934 pp.

25. R. T. MacMillan, 1960, Sodium and sodium compounds, *U.S. Bur. Mines Bull.* 585, pp. 745–765.

26. Helen Mark, 1963, High-alumina kaolinitic clay in the United States exclusive of Alaska and Hawaii, *U.S. Geol. Survey Mineral Inv. Resource Map* MR-37. Map and pamphlet of 22 pp.

27. B. L. Miller and J. T. Singewald, Jr., 1919, "The Mineral Deposits of South America," McGraw-Hill Book Company, New York, 598 pp.

28. H. H. Murray, 1960, Clay, in "Industrial Minerals and Rocks," 3d ed., pp. 259–284, American Institute of Mining, Metallurgical, and Petroleum Engineers, New York, 934 pp.

29. J. B. Myers, 1960, Vermiculite, in "Industrial Minerals and Rocks," 3d ed., pp. 889–895, American Institute of Mining, Metallurgical, and Petroleum Engineers, New York, 934 pp.

30. O. S. North, 1956, Perlite, *U.S. Bur. Mines Bull.* 556, pp. 595–600.

31. O. S. North, 1956, Vermiculite, *U.S. Bur. Mines Bull.* 556, pp. 961–966.

32. L. M. Otis, 1960, Diatomite, *U.S. Bur. Mines Bull.* 585, pp. 275–281.

33. L. M. Otis, 1960, Perlite, *U.S. Bur. Mines Bull.* 585, pp. 581–587.

34. L. M. Otis, 1960, Vermiculite, *U.S. Bur. Mines Bull.* 585, pp. 949–955.

35. J. L. Rich, 1941, The nitrate district of Taracapá, Chile, an aerial traverse, *Geog. Rev.,* vol. 31, pp. 1–22.

36. J. L. Rich, 1942, Physiographic setting of the nitrate deposits of Taracapá, Chile: its bearing on the problem of origin and concentration, *Econ. Geology,* vol. 37, pp. 188–214.

37. E. R. Ruhlman, 1956, Nitrogen compounds, *U.S. Bur. Mines Bull.* 556, pp. 569–576.

38. E. R. Ruhlman, 1960, Nitrogen compounds, *U.S. Bur. Mines Bull.* 585, pp. 565–572.

39. A. E. Schreck, 1960, Strontium, *U.S. Bur. Mines Bull.* 585, pp. 815–820.

40. N. Severinghaus, 1960, Crushed stone, in "Industrial Minerals and Rocks," 3d ed., pp. 285–302, American Institute of Mining, Metallurgical, and Petroleum Engineers, New York, 934 pp.

41. W. C. Smith, 1960, Borax and borates, in "Industrial Minerals and Rocks," 3d ed., pp. 103–118, American Institute of Mining, Metallurgical, and Petroleum Engineers, New York, 934 pp.

42. W. C. Smith, 1962, Borates in the United States, exclusive of Alaska and Hawaii, *U.S. Geol. Survey Mineral Inv. Resource Map* MR-14. Map and pamphlet of 4 pp.

43. H. E. Stipp, 1960, Boron, *U.S. Bur. Mines Bull.* 585, pp. 141–147.

44. J. R. Thoenen and O. Bowles, 1956, Sand and gravel, *U.S. Bur. Mines Bull.* 556, pp. 761–772.

45. Various authors, 1960, "Industrial Minerals and Rocks," 3d ed., American Institute of Mining, Metallurgical, and Petroleum Engineers, New York, 934 pp.

46. Anon., 1958, Where to look for boron, *Mining World,* vol. 20, no. 6, p. 59.

47. Anon, 1960, Wyoming's trona trend, *Mining World,* vol. 22, no. 4, p. 26.

chapter **16**

Diamond
and Other Gem
and Abrasive Minerals

The most desirable gem stones are those that are rare, beautiful, and durable. Diamond, above all other minerals, is characterized by these qualities. Other minerals that are highly prized as gems include emerald, ruby, and sapphire, whereas some others that furnish gem stones of varying degrees of value are turquois, aquamarine (beryl), amethyst, chrysoberyl, garnet, spinel, zircon, spodumene, and tourmaline. Many other materials pleasing to the eye are used as gems and ornamental stones, even though they lack great beauty, lack durability, and are not especially rare. Indeed, "more than 60 gem minerals, mostly semiprecious, have been produced commercially from domestic sources, and at least one variety of gem stone occurs in each State." (24, p. 326).

The durability of many minerals, which is in part a function of hardness, also makes them desirable abrasive materials, although some are not sufficiently abundant to be used for this purpose. Diamond, the hardest of all minerals, although an outstanding gem stone, is also the best abrasive material for many purposes and is widely used in industry. Corundum, garnet, and a few other minerals are used as abrasives, but the place of such minerals has been usurped to a large extent by artificial abrasives made from alumina.

Durability also enables certain minerals to be used as jewel bearings. Formerly sapphire and ruby were extensively used for this purpose, but such use also has been largely displaced by artificial products. Garnet and agate are used for making large bearings.

Some interesting gems

The world's largest diamond, the Cullinan, 3,106 carats in the rough, 10 by 6.5 by 5 cm, was found at the Premier mine, South Africa, on January 25, 1905. Other famous African diamonds are the Jonker, 726 carats uncut, found January

16, 1934, about 3 miles from the Premier mine and rated as the highest quality of any of the large famous diamonds; the Tiffany, 128.5 carats after cutting, a deep yellow stone from the Kimberley mine; and the Hope, 44.5 carats, the largest blue diamond. Famous diamonds of India include the Great Mogul, 787.5 carats uncut, the largest of the Indian stones, and the Kohinoor (mountain of light), 793 carats uncut, 186.1 carats first cutting, 106.1 carats second cutting, presented to Queen Victoria in 1850. The largest of the Brazilian diamonds is the Vargas, 726.6 carats uncut, found in Patrocinio, Minas Gerais, in July, 1938.

Important rubies and sapphires, both of which are varieties of corundum, have come from Asia. The best rubies are dark, somewhat purplish red (pigeon's blood red) and have come from the Mogok area, Burma. Blue sapphires, the true type, are Kashmir blue and have come from Kashmir, western India, for many years; also from Thailand and Ceylon.

Emerald, which is a variety of beryl, is a highly prized green gem. Many ancient emeralds came from upper Egypt, from "Cleopatra's Mines." Now the best emeralds come from Colombia, the Muzo mine, and some come from the Ural Mountains, Soviet Union. Aquamarine, which is a transparent, blue to sea-green stone, is also highly prized when of good quality. An outstanding aquamarine from Morambaya, Minas Gerais, Brazil, weighed 243 lb, was transparent from one end to the other, and sold for $25,000 uncut.

Considerable interest centers about other gem materials, such as alexandrite (which is a variety of chrysoberyl), jade, turquois, and lapis lazuli, but such discussion is beyond the scope of this book (See references 25, 29, and *Encyclopaedia Britannica.*)

Diamond

Most of the diamonds of the world are produced in Africa (Table 16-1). Some are produced in South America (Brazil, Venezuela, British Guiana), India, the Soviet Union, and a few other places. Recent developments in India and the Soviet Union indicate that production from those countries may be more important in the future. The Republic of the Congo is the outstanding producer of industrial diamonds and supplied about two-thirds of the world total in the five-year period 1958 to 1962, whereas the Republic of South Africa is the major producer of gem diamonds—about one-fifth of the world total.

Republic of South Africa

Diamond production of the world was dominated by the Union of South Africa (now the Republic of South Africa) for many years. Much has been written about the deposits there, and only recent summaries are cited (1; 7; 10; 12; 13; 20; 21, pp. 43–76; 22; 27, pp. 173–181; 32, pp. 176–180).

The diamond industry of South Africa began when, in 1866 or 1867, a fifteen-year-old boy, E. S. Jacobs, picked up a stone that weighed 21.75 carats[1] on the Orange River near Hopetown (1, p. 1093; 21, p. 43). A second diamond, the

_____
[1] A cut diamond of 1 carat weight measures about ¼ in. across.

TABLE 16-1.   World Production of Diamond, Average for 1958–1962*

Country	Total diamonds Thousand carats†	Percent of world total	Gem diamonds Thousand carats	Percent of world total	Industrial diamonds Thousand carats	Percent of world total
Congo, Republic of the	16,247	54.2	532	8.5	15,715	66.3
South Africa, Republic of	3,227	10.8	1,307	20.7	1,920	8.1
Ghana	3,120	10.4	832	13.2	2,288	9.6
Sierra Leone	1,771	5.9	712	11.3	1,059	4.5
South-West Africa	1,140	3.8	837	13.3	303	1.3
Angola	1,060	3.5	632	10.0	428	1.8
Liberia	954	3.1	438	6.9	516	2.1
All other African countries	1,688	5.6	712	11.3	976	4.1
Total Africa	29,207	97.3	6,002	95.2	23,205	97.8
Total South America (Brazil, British Guiana, Venezuela)	534	1.8	255	4.0	279	1.1
All others (Soviet Union, India, and others)	289	0.9	45	0.8	244	1.1

* Compiled from *U.S. Bur. Mines Minerals Yearbook,* 1958 through 1962.
† The standard unit of weight is the metric carat, equivalent to 0.2 g.

Star of South Africa, that weighed 82.5 carats was found in 1869, also in the vicinity of Hopetown. These finds started a diamond rush along the Orange and Vaal Rivers, but in 1870 and subsequently the search for diamonds led to their discovery on the surface of volcanic pipes—the sites of future mines. Among the pipes discovered in the first few years were the Dutoitspan, Bultfontein, Jagersfontein, DeBeers, Kimberley, and Koffiefontein (Fig. 16-1). The Wesselton pipe was discovered in 1890 and the Premier some time before 1902 (21, p. 43). The capital realized from the early diamond mining is stated to have been the means of opening up most of the important gold mining of South Africa and subsequently, along with the returns from gold, most of the other important mining.

The original source of the diamonds of South Africa is the kimberlite[1] pipes and "fissures," which are dikelike bodies that have filled fissures. The pipes are remarkable structures, most of which are somewhat oval in plan, are cylindrical or funnel-shaped with steep sides, and are vertical or nearly so. Some are compound and consist of two or more plugs of kimberlite that occur in the same orifice; many pipes are connected by narrow dikes of kimberlite, and some pass into dikes; some are relatively short, others, as at the Kimberley mine, have been worked to a

[1] Kimberlite is a variety of mica peridotite that was named from its occurrence at Kimberley, South Africa; the same type of rock occurs elsewhere.

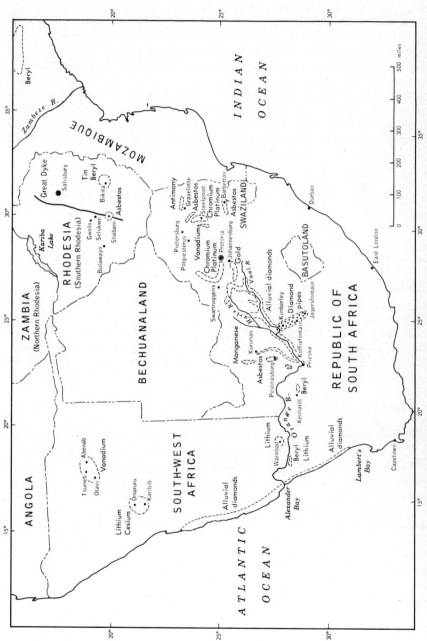

Fig. 16-1 Map showing approximate areas of production of various metallic and industrial deposits in southern Africa. Deposits of South Africa generalized from map in "The Mineral Resources of the Union of South Africa," 4th ed. Extracts from official publications of the Republic of South Africa reproduced under Government Printer's Copyright Authority No. 3390 of 16/9/1964.

depth of 3,520 ft and continue to depths as yet unknown. The pipes and dikes of kimberlite are as young as Lower Cretaceous.

The material within the pipes consists chiefly of kimberlite but in part of fragments of various rocks through which the magma passed in its ascent toward the surface, including materials now eroded from the surface and others nowhere known to be exposed and thought to exist at great depth. The kimberlite near the present surface, designated *yellow ground,* commonly is hydrated and oxidized; beneath this is *blue ground,* which is less decomposed material that may contain solid lumps or pass downward into *hardebank*—solid material that does not disintegrate when exposed. The pipes are regarded as products of successive volcanic eruptions of material that rose through fissures and solidified there. Serpentinization and other alteration took place during the postmagmatic stage and finally, near the present surface, weathering and circulation of water formed the yellow ground.

The diamonds of the pipes range in size from minute ones to those the size of the famous Cullinan diamond. Diamonds from different pipes show individual characteristics so distinctive that an expert can identify the source. Some diamonds occur as well-formed crystals, some as broken crystals and fragments, and some as impure stones. The amount of diamonds in even the most productive pipes is exceedingly small, only 5 to 30 carats in about 160,000 lb of rock.

Outstanding mines of the region are the Premier, of the Bronkhorstspruit district near Pretoria, noted as the largest pipe in South Africa, which now produces about half of the diamonds; the Dutoitspan, noted for large yellow stones, and the Bultfontein and Wesselton mines, all near Kimberley; the Jagersfontein mine near Jagersfontein, noted for high-quality blue-white stones. Two famous mines, the Kimberley and the DeBeers, were closed many years ago, but the DeBeers mine was reopened in 1963. It was closed in 1908 after producing 23,209,719 carats of diamonds, but it is known to contain a considerable amount of workable blue ground (46).

Early work on the "fissures" indicated that they contained very few diamonds or were barren, but later work showed that some of them are rich enough in diamonds to be mined. The chief ones being exploited are in the Swartruggens area, west of Rustenburg (18), and in the Barkly West district, northwest of Kimberley.

Alluvial (placer) deposits of South Africa occur chiefly between Prieska and Potchefstrom in the valleys of the Vaal and Harts Rivers; near Lichtenburg and Venterdorp in the high plains of western Transvaal; and on the coast of Namaqualand southward from the mouth of the Orange River for some 200 miles, from Lambert's Bay to Alexander Bay (Fig. 16-1). The Namaqualand region has recently become the site of considerable activity.

A new type of diamond activity, offshore mining from the seabed, has come into existence along the coast of South Africa and elsewhere in Africa. Work has been concentrated chiefly along the coast of South-West Africa, where it has been so successful that it is being applied in South Africa also.

Republic of the Congo

Diamonds were discovered in the Belgian Congo (now the Republic of the Congo) in 1909 in the Tshikapa district and in 1918 in the Bakwanga district, both in Kasai Province but about 200 miles apart by air and 300 miles by road (12, 13, 14). The major production comes from the Bakwanga district.

The Tshikapa deposits are in an alluvial gravel layer generally not more than 2 m thick, which is covered by as much as 40 m of eolian material except in the stream channels. This ancient gravel sheet is known throughout an area about 25 miles wide and 140 miles long, which extends southward into Angola. Recovery is made from about 55 mines.

The deposits of the Bakwanga district include both kimberlite and alluvial deposits derived from the kimberlite. The alluvial deposits are in the valley of the Kanshi River, a tributary of the Bushimaie. Both the alluvial deposits and the kimberlite are mined. This is a large, modern operation that produces most of the Congolese diamonds. Proven reserves have been estimated to be at least 500 million carats, but most of this material is the lowest-valued grade of industrial diamond.

South-West Africa

South-West Africa is becoming one of the most important producers of gem diamonds in Africa. During the five-year period 1958 to 1962 it ranked second to South Africa and, although during that period its production was about the same as that of Ghana, its production in 1960, 1961, and 1962 was considerably greater than that of Ghana. Most of the production comes from The Consolidated Diamond Mines of South-West Africa, Ltd., which has reserves estimated at 74,700,000 cu m of gravel containing 23,280,000 carats of diamonds (44). The average weight of the diamonds recovered is 0.86 carat, and about 95 percent of them are of gem quality (1). The deposits are in a strip of land along the coast that extends northward from the mouth of the Orange River. They consist of elevated marine terraces covered with barren sand. Thirteen washing and screening plants were in operation in 1959.

The most recent development in South-West Africa is underwater mining along the coast. Several announcements of this activity appeared in 1961 (47). Exploration was to extend north from the mouth of the Orange River and from the low water mark to the 3-mile limit, where the water is 70 ft deep. In 1962 reports appeared (48) that more than 2,100 diamonds had been recovered. The weight of the entire group was reported as 1,038 carats, the weight of the largest stone as 3.4 carats. In 1963 (50) enough information had been obtained from dredging and drilling to indicate that a diamond-bearing area of very considerable extent and importance is present beneath the sea for several hundred miles along the coast. Prospecting rights have been granted to a number of companies over more than 1,000 miles of the coastal waters of South-West Africa and South Africa, and much capital has been invested. Reserves cannot be accurately determined at this early stage of exploitation, but estimates of the order of tens of millions of carats have been made for only a part of the area. Most of the diamonds recovered have been of gem quality, but the average weight was only 0.41 carat.

Other African Deposits

Other gem diamond deposits of considerable importance in Africa occur in Ghana, Sierra Leone, and Angola. Diamonds were discovered in gravel of the Birim River, Ghana, in 1919 and subsequently were found to be widely distributed in stream gravels in the upper part of the Birim drainage basin (12). The gravel beds in which the diamonds occur generally are not more than 10 ft thick, are present on hillside slopes and in the valley floors, and are covered with but little overburden. The diamonds, which are small, have been eroded from several steeply dipping beds of graywacke of Precambrian age. The diamond deposits of Sierra Leone are in two areas about 50 miles apart in the eastern part of the country. Mining has been from alluvial deposits, but kimberlite pipes have been discovered recently. About 68 percent of the diamonds are reported to be of gem quality. The deposits of Angola are of the same character as those of the Congo and apparently are a southern extension of those deposits. About 60 percent of the diamonds are of gem quality.

South America

Diamonds in South America have been produced in Brazil, Venezuela, and British Guiana. Among these countries Brazil has been by far the most important.

Diamonds were discovered in Brazil sometime around 1727 and for many years, until the production from Africa became important, Brazil supplied most of the diamonds of the world (17, pp. 176–182; 32, p. 176; 34, pp. 207–214). The original discovery was at Diamantina (then known as Tejuco) in Minas Gerais. Later other districts were discovered in Minas Gerais and also in the states of Bahia, Goias, Mato Grosso, São Paulo, Paraná, and Piauí and are known to exist in the present river beds of 12 states (33). A new discovery was reported in 1962 in Mato Grosso (49).

The diamonds of Minas Gerais (34, 35) come from an area in the northern part of the state that extends from the headwaters of the Jequitinhona River (south of Diamantina) northward to Grão Mogol and Serra Nova. Diamonds have been found along the Jequitinhonha River for about 250 miles in a zone generally 6 to 8 miles wide but in places as much as 20 miles wide. They have been found, also, along tributary streams and in the tablelands. Further, diamonds occur along the Paraúna River and its tributaries, south of Diamantina. The diamond-bearing deposits were classed by Moraes and Guimaraes (35) as (1) sericitic rock that forms the matrix of the diamonds, (2) ancient conglomerate beds, (3) detrital material derived from older conglomerates or other diamond-bearing rock, and (4) alluvial deposits in the beds and along the banks of rivers. Mieretz (33) stated that active mining generally is done (1) along the present river channels and (2) on buried ancient river or lake deposits. The second type is considered to be the more profitable. He further classed the deposits as true placers, soft conglomeratic placers, and hard conglomeratic placers, both of the conglomeratic types being composed of the material of ancient river channels or lake beds. The soft conglomeratic deposits are actually weathered phases of the hard conglomeratic deposits and usually occur at higher elevations. In general the diamonds are of gem quality.

The diamonds of Bahia are chiefly carbonado or black diamond and are the main source of industrial diamonds in Brazil although some diamonds of gem quality are recovered also. The diamonds of Bahia occur throughout an extensive area in the central part of the state east of the São Francisco River, and in the southeastern part of the state along the Pardo and Salobro Rivers. They are found in alluvial deposits that apparently have been derived from the disintegration of the Lavras sandstone (Carboniferous age?).

The origin of the diamonds of Brazil is a matter of disagreement and will not be discussed here.

The diamond deposits of British Guiana (34, pp. 422–423; 38) occur along numerous streams flowing from the Pakaraima Mountains, and were first found in 1887 or 1888 along the Puruni River in the Mazaruni-Puruni gold district. Early recovery was almost wholly from gold placers. Later an independent diamond industry developed, and recently a study has been made of the diamond resources of the country (38).

The Pakaraima Mountains, which are in the west-central part of British Guiana and lie also in eastern Venezuela and northern Brazil, are composed chiefly of the Roraima formation, a nearly horizontal succession of sandstones and conglomerates 2,000 to 3,000 ft thick. The mountainous area consists of plateaus cut by streams, and is characterized around its margins by steep escarpments and precipitous gorges, and in the interior by mesas that rise above the general level of the plateaus, one of which, Mount Roraima, 8,600 ft in altitude, marks the joint boundaries of British Guiana, Brazil, and Venezuela. All of the diamond workings occur within about 15 miles outward from the escarpments or actually on the plateau. Some, however, occur at greater distances, but they are associated with outliers of the Roraima formation or with the White Sands series, believed to have been derived from the Roraima formation. The diamonds are generally considered to have come from erosion of the Roraima formation.

The diamonds of British Guiana occur in alluvial deposits of four types: (1) high alluvial deposits, generally associated with the White Sand series, and possibly representing dissected remnants of a former terrace system; (2) terrace deposits, which are extensive and lie between the elevation of the high alluvial deposits and the beds of the present streams; (3) river flat deposits along the present streams; (4) deposits directly associated with the plateau, especially in gorges at the plateau escarpment.

The most productive area is the Mazaruni diamond field along the Mazaruni River to the north of the plateau margin. All four types of deposits occur in this field.

Production of diamonds in Venezuela has been small, although diamonds are known to be rather widely distributed (15). The chief area from which production has occurred is eastern Venezuela south of Ciudad Bolivar, especially along the Rio Caroni and the Rio Paragua in the state of Bolivar. Diamonds have been reported also in Amazonas Territory. The diamond deposits occur in (1) old river terraces, some as much as 80 ft above present river levels, generally covered by overburden not more than 5 ft thick; (2) along present stream beds; and (3) along the slopes and floodplains between deposits of the first two types. The source of the diamonds is uncertain but is thought to be conglomerates of the Roraima series.

Soviet Union

Diamonds were known to be present in the Soviet Union as early as 1898, but concentrated geological work has been done and the size, possibilities, and some features of the diamond deposits have been described only recently (2, 16, 28, 36, 39, 42). Concentrated work that started in 1948 led to the discovery of a pipe of kimberlite in the headwaters of the Markha River, just about on the Arctic Circle, in 1954. Further exploration disclosed the existence of more than 20 kimberlite pipes by the end of the 1956 season. Most of these are in the basin of the Vilyui River, into which the Markha and other diamond-bearing tributaries flow. This area, which covers some 300,000 sq km and is centered about latitude 64°N, longitude 114°E, has been designated the Vilyui diamond district. Other fields containing alluvial deposits of commercial tenor occur farther north, in the basins of the Olenk, Muna, and Tyung Rivers.

The Vilyui district is situated in the Siberian platform. Siberian volcanic rocks of Late Carboniferous to Triassic age, which closely resemble the Karoo dolerites of South Africa, cover large areas to the west of the district. Pipes or diatremes of diamond-bearing kimberlite are associated with nepheline basalts and melilite basalts. Two principal fields occur within the district, a northern one in the basin of the Daaldyn River and a southern one several hundred kilometers away in the basin of the Batuobiya River, a tributary of the Vilyui. Kimberlite pipes are present in both regions. Those of the northern region are generally 40 to 50 m in diameter, but one pipe (Zarnista or Zarnitza) is 600 m across (16). The largest and best-known pipe of the southern area is the Mir, in which was discovered the largest diamond so far recorded in the area, weight 32.5 carats. Large stones generally are rare in the district. The average weight seems to be 10 to 15 mg in the pipes and 10 to 25 mg in alluvial deposits. Both terrace gravels and riverbed gravels are present. Usually the content of diamonds is greater in the riverbed gravels than it is in the terrace gravels, and greater in the lower (younger) terraces than it is in the higher ones.

Russian reports have indicated that the Siberian fields may be as important as the South African ones, and that the known kimberlite pipes would not be inferior to the Premier and other pipes of South Africa. It was stated that millions of carats of diamonds probably are present. Too little is yet known to judge the correctness of these statements, but it seems not unlikely that the field is somewhat overrated (16). The average size and quality of the stones so far reported are certainly less than they are in South Africa. Further, the climate of Russia is not favorable for recovery of the diamonds.

India

Diamonds have been recovered from several localities of India for many years. The first diamond apparently was collected from a stream bed in 800 B.C. Early production possibly occurred 2,000 years ago (25, 32), and some of the famous diamonds of history have come from India (4, pp. 576–586). All stones were recovered from placers, some of which were associated with a conglomerate.

Diamonds occur in three principal areas: (1) a southern area (Anantapur, Bellary, Cuddapah, Kurnool, Krishna, Godavari); (2) an eastern area (valley of Mahanadi in Orissa, also westward into Madhya Pradesh); (3) a central area (Vindhya Pradesh). Diamonds have been reported from a number of other places (25, p. 13). Production has been very small recently, but systematic exploitation of the diamond deposits in the Panna district of Madhya Pradesh is planned, and the Geological Survey of India has made a preliminary survey of some of the diamond-bearing area. Further, detailed mapping and exploration are planned (51).

United States

Diamonds have been found in stream and glacial deposits in various places in the United States (29, 32), especially in an area extending from Ohio to

Minnesota, another one in the Southern Appalachian region, and along the Pacific coast area, especially in California. Generally they have been small: the largest one of the Ohio-Minnesota area was reported to weigh something over 21 carats; one found at Manchester, Virginia, 23 carats; one, the largest, found at Petersburg, West Virginia, 34.46 carats; the largest of the California diamonds, a bit more than 7 carats.

Rock similar to the kimberlite of South Africa is present near Murfreesboro, Pike County, Arkansas, about a hundred miles southwest of Little Rock. Diamonds were discovered in this rock in 1906 and since then have been recovered from time to time. It has been stated that more than 10,000 diamonds have been obtained, the largest of which weighed slightly more than 40 carats. In 1962 the reported production was 200 carats valued at $8,850, including one frosty white diamond of 4.39 carats valued at $3,000.

Corundum and emery

Corundum and emery (6; 9; 19; 21, pp. 395–402; 23; 30; 31, pp. 6–9; 32, pp. 166–172), formerly important natural abrasives, have been largely displaced by artificial aluminum oxide abrasives. The chief production of corundum in the Free World is from the Republic of South Africa, Rhodesia (Southern Rhodesia), and India (Table 16-2). The corundum of Rhodesia (Southern Rhodesia), however, occurs with sillimanite, is not suitable for abrasives, and is used in the production of refractories. The Soviet Union is known to be a producer of corundum, but statistics are unavailable. Canada formerly furnished a considerable amount of corundum from deposits in Ontario, between 1900 and 1915, and the states of North Carolina and Georgia furnished the corundum of the United States from about 1870 to 1900.

TABLE 16-2.   World Production of Corundum, Average of 1945–1959 and in 1962*

	Percent of world total	
Country	Average of 1945–1959	1962 only
Republic of South Africa	25.8	3.8
Rhodesia (Southern Rhodesia)	14.4	37.2
India	4.8	3.7
Others	55.0	55.3
	Short tons	
	Average of 1945–1959	1962 only
World total	9,833	9,000

* Compiled from *U.S. Bur. Mines Minerals Yearbook,* various years.

The deposits of the northern and northeastern Transvaal, Republic of South Africa, are associated with basic rocks that are infolded isoclinal masses within gneiss. Both the basic rocks and the gneiss are cut by pegmatites, and apparently the corundum deposits always occur where there are both basic rocks and pegmatites. The corundum occurs as "reefs," which are veins or dikes a few feet wide that cut the basic rocks, and as eluvial deposits—loose crystals of corundum and boulders of corundum-bearing rock. The commercial reefs contain 20 to 40 percent corundum. Production of the Transvaal is about 60 percent from eluvial deposits. The origin of the reef deposits has been a matter of considerable discussion. This has been summarized by the Geological Survey of South Africa (21), by Hall (23), and by Kupferburger (30) and discussed by others. The discussions relate to the possible interaction of pegmatite magma with the basic rocks, accompanied by desilication of the pegmatite magma and the production of an alumina-rich magma from which corundum crystallized, and also to the possible role of hydrothermal solutions and metamorphism.

The deposits of Renfrew County, Ontario, Canada (5), form lenses in nepheline syenite and nepheline gneiss of Precambrian age and occur in three discontinuous belts, the major one of which is 103 miles long. The corundum generally composes 5 to 15 percent of the lenses, rarely as much as 40 percent. The deposits are large and contain a considerable amount of reserve material.

Emery generally is a mixture of corundum and magnetite (true emery), but may be a mixture of spinel, corundum, and magnetite (spinel emery), or of those minerals and 30 to 50 percent plagioclase feldspar (feldspathic emery). Important production has come from Greece, Turkey, the United States, and the Soviet Union and has not been affected as much as corundum by the manufacture of artificial abrasives. The deposits of Greece are on several islands, the most important ones being on Naxos. The deposits occur as lenses up to 300 ft long and 150 ft wide in crystalline limestone, and as huge boulders in red soil. The deposits of Turkey, in the province of Aidin, are somewhat similar to those of Greece, since they occur as masses in limestone and also as detrital deposits. Generally the grade of this emery is not as high as is that of Naxos. Emery has been produced in the United States at Chester, Massachusetts, near Whittles, Virginia, and near Peekskill, New York. The production for many years has come only from the New York deposits, which consist of spinel emery that makes replacement bodies in basic rock and schist.

Garnet

Garnet (8; 11; 31, pp. 9–13; 32, pp. 241–249) is widespread throughout the world but is mined for abrasives in but a few countries, and production comes chiefly from the United States. The most important deposits are in Warren County, New York, near the top of Gore Mountain, about 4.5 miles south of North River Village. The Gore Mountain garnets occur as nodules that vary in diameter from 2 in. to 3 ft, and are surrounded by shells of hornblende in a rock that consists chiefly of hornblende and feldspar, which is enclosed in gneiss.

Other abrasives

Numerous other natural materials are used as abrasives. Among these quartz, as loose grains, crushed, or in stones, is perhaps the one most widely used (31, pp. 2–5; 32, pp. 5–12). The stones include the buhrstone of France, the grindstones and pulpstones of the United States (northern Ohio, southern Michigan, and West Virginia), various sharpening stones, and grinding pebbles.

SELECTED REFERENCES

1. R. J. Adamson, 1961, Diamond recovery in southern Africa, Commonwealth Min. Metall. Cong., 7th, South Africa, *Trans.*, vol. 3, pp. 1093–1101.
2. E. A. Alexandrov, 1958, Review of "The Diamonds of Siberia," by A. P. Bobrievich and others, 158 pp., (Russian), *Econ. Geology,* vol. 53, pp. 220–221.
3. O. Bowles, 1952, Diamond, in W. Van Royan and O. Bowles, "Mineral Resources of the World," vol. 2, pp. 162–165, Prentice-Hall, Inc., Englewood Cliffs, N.J., 181 pp.
4. J. C. Brown and A. K. Dey, 1955, "India's Mineral Wealth," 3d ed., Oxford University Press, London, 761 pp. Precious stones, pp. 575–608; uncommon gem stones, pp. 609–620; semiprecious and decorative stones, pp. 625–643.
5. Canadian Institute of Mining and Metallurgy, 1957, Corundum, in "The Geology of Canadian Industrial Mineral Deposits," p. 73, Canadian Institute of Mining and Metallurgy, Montreal.
6. H. P. Chandler, 1956, Corundum, *U.S. Bur. Mines Bull.* 556, pp. 247–250.
7. H. P. Chandler, 1956, Diamonds, *U.S. Bur. Mines Bull.* 556, pp. 251–258.
8. H. P. Chandler, 1956, Garnet, *U.S. Bur. Mines Bull.* 556, pp. 295–297.
9. H. P. Chandler, 1960, Corundum and emery, *U.S. Bur. Mines Bull.* 585, pp. 261–266.
10. H. P. Chandler, 1960, Diamond—industrial, *U.S. Bur. Mines Bull.* 585, pp. 267–274.
11. H. P. Chandler, 1960, Garnet, *U.S. Bur. Mines Bull.* 585, pp. 321–324.
12. A. F. Daily, 1960, Africa's key role in diamond mining, *Mining World,* vol. 22, no. 10, pp. 44–49.
13. A. F. Daily, 1960, How diamonds are found and mined in Africa, *Mining World,* vol. 22, no. 11, pp. 36–41.
14. A. F. Daily, 1961, Economic aspects of interruption of diamond production in Congo Republic, *Mining Eng.,* vol. 13, pp. 475–479.
15. J. C. Davey, 1948, The diamond fields of Venezuela, *Eng. Mining Jour.,* vol. 149, no. 4, pp. 75–78.
16. C. F. Davidson, 1957, The diamond fields of Yakutia, *Mining Mag. (London),* vol. 97, no. 6, pp. 329–338.
17. J. B. DeMille, 1947, "Strategic Minerals," McGraw-Hill Book Company, New York, 626 pp. Diamond (industrial), pp. 176–182; emery (Turkish), pp. 183–185.
18. G. P. Fourie, 1958, The diamond occurrences near Swartruggens, Transvaal, *South Africa Geol. Survey Bull.* 26, 16 pp. English summary, pp. 14–16.

19. G. M. Friedman, 1956, The origin of spinel-emery deposits with particular reference to those of the Cortlandt complex, New York, *New York State Mus. Bull.* 351, 68 pp.
20. W. S. Gallagher, 1960, Mining blueground at DeBeers, *Eng. Mining Jour.*, vol. 161, no. 6, pp. 92–102.
21. South Africa Department of Mines, and Geological Survey, 1959, Diamonds [and] Corundum, in "The Mineral Resources of the Union of South Africa," 4th ed., Pretoria, South Africa, pp. 43–76, pp. 395–402.
22. G. S. Giles, 1961, Diamond mining practice in South Africa, Commonwealth Min. Metall. Cong., 7th, South Africa, *Trans.*, vol. 2, pp. 836–850.
23. A. L. Hall, 1920, Corundum in the Northern and Eastern Transvaal, *Geol. Survey Union of South Africa, Mem. No.* 15, 223 pp.
24. J. W. Hartwell, 1960, Gem stones, *U.S. Bur. Mines Bull.* 585, pp. 325–340.
25. L. A. N. Iyer, 1961, Indian precious stones, *Geol. Survey India Bull.*, ser. A, no. 18, 105 pp.
26. R. H. Jahns, 1960, Gem stones and allied materials, in "Industrial Minerals and Rocks," 3d ed., pp. 383–441, American Institute of Mining, Metallurgical, and Petroleum Engineers, New York, 934 pp.
27. S. J. Johnstone and M. G. Johnstone, 1961, "Minerals for the Chemical and Allied Industries," 2d ed., John Wiley & Sons, Inc., New York, 788 pp. Corundum and emery, pp. 164–172; Diamond, pp. 173–181; Garnet, pp. 209–211.
28. K. Kondakov and G. Naumov, 1959, Diamond deposits in Yakutia await development (Condensation and translation), *Internat. Geol. Review*, vol. 1, no. 5, pp. 73–76.
29. E. H. Kraus and C. B. Slawson, 1947, "Gems and Gem Materials," 5th ed., McGraw-Hill Book Company, New York, 332 pp.
30. W. Kupferburger, 1935, Corundum in the Union of South Africa, *Geol. Survey South Africa Bull.* 6, 81 pp.
31. R. B. Ladoo, 1960, Abrasives, in "Industrial Minerals and Rocks," 3d ed., pp. 1–21, American Institute of Mining, Metallurgical, and Petroleum Engineers, New York, 934 pp.
32. R. B. Ladoo and W. M. Meyers, 1951, "Nonmetallic Minerals," 2d ed., McGraw-Hill Book Company, New York, 605 pp.
33. R. E. Mieritz, 1959, Brazil: an untapped diamond source? *Mining World*, vol. 21, no. 1, pp. 41–43.
34. B. L. Miller and J. T. Singewald, Jr., 1919, "The Mineral Deposits of South America," McGraw-Hill Book Company, New York, 598 pp.
35. L. J. Moraes and D. Guimarães, 1931, The diamond-bearing region of northern Minas Geraes, Brazil, *Econ. Geology*, vol. 26, pp. 502–530.
36. M. M. Odintsov, 1959, Basic features of the geology of the Siberian diamond province, *Internat. Geol. Review*, vol. 1, no. 9, pp. 42–46.
37. V. Oppenheim, 1948, The Muzo emerald zone, Colombia, S.A. *Econ. Geology*, vol. 43, pp. 31–38.
38. E. R. Pollard, C. G. Dixon, and R. A. Dujardin, 1957, Diamond resources of British Guiana, *British Guiana Geol. Survey Bull. No.* 28, 45 pp.

39. G. I. Smirnov, 1959, Mineralogy of Siberian kimberlites, *Internat. Geol. Review,* vol. 1, no. 12, pp. 21–39.

40. R. D. Thomson, 1956, Gemstones, *U.S. Bur. Mines Bull.* 556, pp. 299–307.

41. R. D. Thomson, 1956, Jewel bearings, *U.S. Bur. Mines Bull.* 556, pp. 399–407.

42. S. I. Tomkeieff, 1958, Diamonds in Siberia, *Nature,* vol. 181, no. 4605, pp. 323–324.

43. R. D. Trietus, 1962, Diamonds—in their modern role, *Mining Jour. (London),* vol. 259, no. 6624, pp. 101–102; vol. 259, no. 6627, p. 167.

44. Anon., 1963, Diamonds in South-West Africa, *Mining Jour. (London),* vol. 260, no. 6667, p. 553.

45. Anon., 1963, Diamonds, *Mining Jour. (London),* vol. 261, no. 6695, pp. 559, 561.

MISCELLANEOUS REFERENCES CITED

46. Anon., 1963, DeBeers mine re-opens, *Mining Jour. (London),* vol. 261, no. 6683, p. 261. See also Famous diamond mine comes to life again, *Mining Eng.,* vol. 15, no. 9, pp. 44–45.

47. Anon., 1961, Submerged deposits may yield diamonds, *Eng. Mining Jour.,* vol. 162, no. 9, p. 128. See also Offshore diamond recovery, *Mining Jour. (London),* vol. 257, no. 6586, p. 485.

48. Anon., 1962, Dredge searches seabed, brings up diamonds, *Eng. Mining Jour.,* vol. 163, no. 6, p. 291. See also the following: Mining diamonds from the sea bed, *Mining Jour. (London),* vol. 258, no. 6616, p. 593; Offshore diamond dredge succeeds, second planned, *Eng. Mining Jour.,* vol. 163, no. 10, pp. 128–129.

49. Anon., 1962, *Mining Jour. (London),* vol. 259, no. 6624, p. 110. Notice of a new discovery of diamonds in Mato Grosso, Brazil.

50. Anon., 1963, Diamond strike off coast of S. W. Africa, *Mining Jour. (London),* vol. 260, no. 6648, p. 65. See also High grade diamond deposit is claimed offshore S.W. Africa, *Eng. Mining Jour.,* vol. 164, no. 1, p. 111.

51. Anon., 1961, Development of Panna diamond mines, *Mining Jour. (London),* vol. 257, no. 6588, p. 537.

chapter **17**

Mica, Feldspar, Nepheline Syenite, and Quartz

Commercial muscovite mica and feldspar commonly occur together in granitic pegmatites, and recoverable quartz may be present also. Many pegmatites, however, do not contain mica of sufficient size and so localized that the pegmatite may be mined for mica alone, and thus much mica is obtained as a by-product of feldspar mining. A minor amount of mica is obtained from schists and from the washing of kaolin. Commercial phlogopite mica generally is much more restricted in occurrence than is muscovite and is obtained chiefly from pyroxenites. Feldspar is obtained not only from pegmatites but also from aplites. Further, it is obtained from nepheline syenites, along with the nepheline, which is used as a substitute for feldspar.

Separate discussions of mica, feldspar, and quartz are not feasible for all deposits because two or more of these minerals may be obtained at a single deposit. Consequently the discussions are organized so as to consider the major material recovered or, in some cases, two or more materials.

Mica

Mica has developed into an outstanding strategic mineral about which much has been written, but several good general summaries have been published recently (2, 16, 18, 28, 34, 49). The outstanding use of mica is in the electrical industry, but ground mica is used in making roofing, paints, and some other products. Large sheets of mica from "books" are the most highly prized material. Sheet mica may be sold as blocks (trimmed books from which impurities and imperfections have been removed); as splittings (generally pieces of irregular shape from 0.0007 to 0.001 in. thick); as various small pieces suitable for electrical work, which have been given several designations, such as *punch, circle,* and *washer mica;* and as scrap mica. Scrap mica is material not suitable for any of the types stated and waste from trimming and preparation of sheet mica; it is generally ground. Sheet mica may be muscovite or phlogopite, but most of it is muscovite.

Production

Published statistics of mica production state separate types, such as block, splitting, and scrap, for some countries but do not subdivide production for others. Thus when one considers only total production the true importance of some countries is not realized. Table 17-1 shows the most important production of the world for the period 1945 to 1959 and also for 1962. It should be noted that, although the United States is far ahead of other countries in total production, almost the entire product is scrap mica, and much of it is recovered along with feldspar. India has for many years been the leading world producer of block and splitting muscovite mica. Other countries that now produce block and splitting mica of some importance are the Republic of South Africa (since 1951) and Brazil, but formerly a fair amount of such material was furnished by Canada and the Malagasy Republic (Madagascar). The mica from Canada and the Malagasy Republic is chiefly phlogopite.

TABLE 17-1. World Production of Mica, Average of 1945–1959 and in 1962*

| | Percent of world total | |
Country	Average of 1945–1959	1962 only
United States†	57.2	53.9
India‡	14.6	17.2
Republic of South Africa	1.6	1.2
Canada	1.5	0.4
Brazil	1.4	2.7
Norway	0.9	0.5
Malagasy Republic (Madagascar)	0.6	0.7
Argentina	0.5	0.01
Total of above	78.3	76.61

| | Thousand pounds | |
	Average of 1945–1959	1962 only
World total	250,000	400,000

* Compiled from *U.S. Bureau of Mines Yearbook,* various years.
† Almost entirely scrap mica.
‡ Much more block and splitting mica than scrap mica.

India

The deposits of India (2; 6, pp 539–549; 18; 34; 43; 45; 46) occur in at least seven states, but the most important ones are in the Hazaribagh district of Bihar and the Nellore district of Andhra, and several districts in Rajasthan.

The Bihar mica belt extends from Gaya easterly to Bhagalpur, a distance of about 90 miles, and is 12 to 16 miles wide. Production comes from a number of mining centers and from about 600 separate mines, about one-sixth of which are mechanized. The mines are of varying depth up to as much as 700 ft. The most important mining center has been within the Kodarma reserved forest, at the western end of the belt, but production there has been gradually declining and activity has increased to the northeast and farther east near Gawan. The Bihar area probably is the oldest producer of mica in India and it furnishes the greater part of the large-sized high-quality block mica of the world.

Within the Bihar mica belt, veins, tongues, and masses of mica-bearing pegmatite cut mica schists and gneisses. In parts of the area there are a great many small veins, some only a fraction of an inch thick; other veins attain a thickness of more than 100 ft. The thicker veins are characterized by a core of quartz, on either side of which are feldspar and mica. The thickness of mica zones generally is not more than 3 ft. Most veins are not over 100 ft long, but payable mica occurs along the strike of some pegmatites for as much as 1,000 ft. The books of mica average 6 to 12 in. across but exceptionally may attain a width of 3 ft or more. The average thickness is 3 to 4 in. but may attain 2 ft or more.

The Nellore mica belt (45) is 8 to 10 miles wide and extends for about 60 miles from Gudur to Sangham in the southern part of India between latitudes 14 and 15°N. At one time this area furnished about 20 percent of the Indian production, but lately it has supplied only 10 or 11 percent. Within the area there are about 1,000 mines, both open-pit and underground, but most of the production comes from mines on the eastern and western sides of the belt. The commercial mica-bearing pegmatites are from one hundred to several hundred feet long and a few feet to 50 ft wide, rarely as much as 200 ft. The average size of the mica books is 6 in. to 1 ft across and 2 to 6 in. thick, but books 6 ft across have been found.

The Rajasthan area extends from Jaipur to Udaipur (Mewar), a distance of about 200 miles, and is about 60 miles wide. Production began in this area in 1904, and between 1943 and 1946 the amount furnished was 19 percent of the mica of India. Sixty-seven mines were operating in 1953, and the area is rated as the second largest mica producer in India. One report (43) stated that the quality of mica is superior to that produced in the Bihar area but that it lacks flatness; other reports (6, 18) state that much of the mica is stained, warped, and cracked and that only a small proportion of good block mica is obtained. The occurrence of the pegmatites is about the same as it is in Bihar.

Brazil

Brazil (2; 18, p. 57; 34, p. 553; 41; 42; 50) probably is second only to India in the total production of sheet mica. About 95 percent of the Brazilian mica has come from four areas in southeastern and eastern Minas Gerais. One of the oldest and best producers of the area is the Golconda pegmatite, which was discovered in 1908 (41). Brazil exported about 8,500 metric tons of muscovite between 1900 and 1945, about 60 percent of which was after 1940 (42). The bulk of the mica has come from zoned pegmatite, in which it occurs in shoots, some of which contain as much as 40 percent mica, but commonly from 5 to 20 percent.

In addition to mica, the pegmatites of eastern Minas Gerais are important sources of feldspar, kaolin from the decomposition of feldspar, beryl, columbite-tantalite, gem-stone minerals, and lithium-bearing minerals.

Other production in Brazil has come from mines in the state of São Paulo, which at one time furnished a very considerable amount of sheet mica, chiefly as a by-product of kaolin mining. Other known areas of mica deposits follow more or less the general boundary of the main zone in Minas Gerais and have been investigated in the states of Espirito Santo, Rio de Janeiro, and Bahia, but in general production from them has been unimportant. Mica has been reported also from eastern Goias, Ceará, Pernambuco, Rio Grande do Sul, Mato Grosso, and Amazonas, but production is negligible (50).

Republic of South Africa

Moderate amounts of mica are produced in the Republic of South Africa (10, pp. 166–169; 20, 512–514). The deposits of commercial importance are in the Letaba district of northeastern Transvaal, just north of the Olifants River. In that district mica occurs in an extensive belt of pegmatites, the individual bodies of which vary in width up to as much as 60 ft and in length to possibly 1,000 ft or more. The mica is present in books of considerable size but sporadically distributed. Generally the mines are small and work is done in open pits.

Malagasy Republic (Madagascar)

Both muscovite and phlogopite occur in the Malagasy Republic (2; 5, pp. 83–88; 10, pp. 164–166; 18, p. 59; 30; 34; 47; 65), but the muscovite is not of good quality and commercial production has been chiefly phlogopite. Practically no production took place from the phlogopite deposits until 1919, but it has increased steadily since then until now most of the world production of phlogopite comes from these deposits.

The phlogopite deposits occur chiefly in the south part of the island in the vicinity of Fort Dauphin and Betroka. The mica is characteristically free from imperfections and available in large sizes, some with a surface area as much as 25 sq in.

The mica is associated with pyroxenites, metamorphosed limestone, and other rocks of Precambrian age. Numerous granite and syenite masses intrude the rocks, and a series of pegmatites is present. Some apatite, sphene, and scapolite are present with the deposits, which occur as seams and pockets always associated with pyroxenite. In many respects the deposits are similar to the phlogopite deposits of Canada.

The origin of the deposits is obscure. They may be of contact-metamorphic or of pegmatitic origin. Lacroix early described the deposits briefly, and later (30) in more detail. He noted that they always occur with a pegmatitic type of rock for which he proposed the name *dissogénite*. The *dissogénites* are composed of feldspar, diopside generally associated with sphene, and may contain grossularite, wollastonite, or scapolite, and more or less quartz. They always form veins or seams in association with granite masses, but they also always occur with metamorphosed calcareous sediments or crystalline limestones and silicate rocks such as amphibolite, pyroxenite, garnetite, and scapolite gneiss, derived from limestone. Lacroix ascribed the origin of the *dissogénites* to metamorphism of the limestone in which they are enclosed, along with assimilation of the limestone by magma. He proposed the term *dissogénite*, from the Greek *dissos* (different), to indicate that an origin different from either metamorphism or normal magma solidification is ascribed to the ingredients of which they are composed. Savornin (47) thought that the deposits were formed where magma, contemporaneous with metamorphism, invaded pyroxenites and gave rise to a complex magma which later crystallized into vein and pocket deposits. These veins he referred to as a type of offshoot dike (*dissogénite*).

Canada

The phlogopite deposits of Canada occur chiefly in Frontenac, Leeds, and Lanark Counties, Ontario, and in Gatineau, Papineau, and Argenteuil Counties, Quebec. They are in Precambrian metamorphic rocks of Grenville age, and the phlogopite is associated especially with pyroxenite, apatite, calcite, and some metamorphic minerals such as tremolite and scapolite. The deposits were mined from about 1870 to 1953, and both phlogopite and apatite were marketed. Wilson (59) gave the value of the phlogopite produced as $10,101,328, and the value of the apatite, nearly all of which was produced before 1894, as $4,744,388.

Descriptions of the deposits do not always agree in detail, but are similar. Landes (32, p. 362) stated that the phlogopite and apatite occur in dikes and irregular-shaped bodies of pyroxenite, which are definitely intrusive into the country rock. The largest pyroxenite in the area studied by him was about 400 yd across, but others are known to be ¾ mile long and ¼ mile wide. Microcline, mainly in pegmatite, accompanies the pyroxenite. Calcite, phlogopite, and apatite occur in veins which cut both the pyroxenite and microcline pegmatite. Spence (51) stated that all bodies show sharp, not gradational, contacts; that apatite and phlogopite, as well as augite, hornblende, and scapolite, commonly occur in "giant" crystals; also, that the mica bodies may pinch, swell, and fork, and occur as veins or shoots, not as single crystals. Wilson (59) stated that the most common type of occurrence is irregular zones or aggregates of mica and apatite in metamorphic pyroxenite, but veins or veinlike zones in pyroxenite and other rocks occur, especially where crystalline limestone, granite, and pegmatite are abundant.

The two major ideas advanced regarding the origin of these deposits are similar to those postulated for the deposits of Malagasy: (1) a metamorphic origin and (2) a pegmatitic origin, or some combination of these. The more recent discussions have been by Currie (11), Hoadley (25), and Wilson (59).

The outstanding muscovite producer of Canada has been the Purdy mica mine (23, 25), about two miles north of the village of Eau Claire, on highway No. 17 between North Bay and Mattawa, Ontario. The deposits, discovered in 1941, consist of 10 workable dikes in a zone 400 to 500 ft wide and 1,600 ft long. The dikes average 10 to 15 ft wide, 200 to 400 ft long. Most of the production has been from three dikes of the group. The mica obtained has been in unusually large blocks, many of them more than 1 ft in diameter. One block measured 9.5 by 7 ft, was nearly 3 ft thick, and yielded about 7 tons of trimmed sheet mica. The mine was active from 1941 to 1945 and again from 1949 to 1953.

Mica and feldspar in the United States

Pegmatites are rather widely distributed throughout the United States, as indicated in the discussion of deposits of beryllium and lithium. Both feldspar and mica are produced from many of the pegmatites, but the chief production of mica has come from New Hampshire, North Carolina, Virginia, South Dakota, New Mexico, and Idaho (52), and the chief production of feldspar has come from North Carolina, Colorado, New Hampshire, Maine, Virginia, Connecticut, Georgia, and Arizona (22). Mica and feldspar deposits will be discussed more or less together because many places furnish both materials.

Mica production began in the New England region about 1803 at the Ruggles mine, in Grafton County, New Hampshire. From then until about the middle of the nineteenth century New Hampshire furnished practically all the mica produced in the United States, after which North Carolina became the principal producer. At present North Carolina furnishes 50 percent or more of the total production of both mica and feldspar and has produced feldspar since 1744. The Spruce Pine district in Avery, Mitchell, and Yancey Counties is the most important one in North Carolina.

The deposits of the New England area have been rather extensively investigated by the United States Geological Survey (4, 8, 37, 55), and investigation by Bannerman and Cameron (3) covered the New England mica industry during the war-time period 1942 to 1944. The consensus of these studies indicates that the region can maintain satisfactory production of mica and feldspar for a considerable number of years.

The feldspar and mica deposits of the entire southeastern region were summarized by Parker (40), and various areas were covered in detail by others (1, 7, 19, 21, 26, 33, 38, 58). These investigations show that more than 1,600 mica deposits have been mined and that at least 595 yielded clear sheet mica of high quality during World War II. The main mica-producing areas have been the Amelia district, Virginia; the Ridgeway–Sandy Ridge district, North Carolina–Virginia; the Shelby-Hickory district, North Carolina; the Hartwell district, South Carolina–Georgia; the Thomaston-Barnesville district, Georgia; and three areas in east-central Alabama. The most important single mica deposit appears to be on the R. B. Phillips property in the Spruce Pine area, Mitchell County, North Carolina (1, 33). Feldspar has been mined chiefly in Virginia, especially in Bedford County, since about 1923, and the outlook for continued large production of feldspar appears to be good. Some kaolin has been obtained as a by-product (7). Parker's summary of some of the important possibilities appears in Table 17-2.

The Spruce Pine district of the Blue Ridge province is about 50 miles west of Asheville, North Carolina, and covers an area about 20 miles long and 12 miles wide. Schist and gneiss of Precambrian age are cut by many pegmatites and some larger bodies of alaskite—essentially a granitic rock that is about two-thirds

**TABLE 17-2.   Some Important Feldspar and Mica
Producers of Southeastern United States***

| District | Feldspar | | Mica | | | |
| | Actual production | Potential production | Actual production | | Potential production | |
			Sheet	Scrap	Sheet	Scrap
Bedford County, Virginia	Moderate	Moderate	—	—	—	Small
Spruce Pine, North Carolina	Large	Large	Large	Large	Large	Large
Bryson City, North Carolina	Moderate	Moderate	Negligible	Negligible	Small	Small
Franklin-Sylva, North Carolina	Small	Moderate	Large	Large	Large	Large
Jasper County, Georgia	Moderate	Large	Negligible	Negligible	—	Small

* From "Symposium on Mineral Resources of the Southeastern United States," pp. 42–48, Feldspar and Mica Deposits of Southeastern United States, by J. M. Parker, III. Copyright 1950. University of Tennessee Press. Used by permission.

feldspar, one-fourth quartz, and the rest chiefly muscovite, with dark minerals practically absent. Originally production in this area was only from the pegmatites, but since the successful development of flotation for the recovery of feldspar the production is chiefly from the alaskite, which is crushed, ground, and concentrated by flotation. Several masses of alaskite and associated pegmatites are present, the largest of which is about 2 miles long and 4,000 ft wide.

Perhaps the most outstanding production of the western United States has been that of the Black Hills, South Dakota. Two main pegmatite areas are present: (1) around the Harney Peak uplift in the southern Black Hills, and (2) around the Nigger Hill uplift in the northern Black Hills. The Harney Peak area includes three mining districts, all of which have produced feldspar and mica as well as other materials, but the Custer district of Custer County has been the main source of sheet mica. Since 1879 the entire Harney Peak area has furnished material valued at more than 6 million dollars as shown by Table 17-3.

Some recent investigations (44, 48, 54, 61) indicate the possibility of development of deposits in Idaho, Montana, Colorado, New Mexico, Arizona, and California. The largest muscovite mine in the western United States is stated to be owned by the Idaho Beryllium and Mica Corporation and to be located about 10 miles northwest of Deary in Latah County, Idaho (61).

Feldspar and nepheline syenite

Production of feldspar is more widespread than is that of mica, since many pegmatites contain commercial feldspar but not commercial mica. The major countries producing feldspar are shown in Table 17-4. Information in 1962 (62) indicated that the Soviet Union ranked second or third in world production.

TABLE 17-3. **Value of Material Produced from Pegmatites in the Harney Peak Area, South Dakota, 1879–1953***

Mica	$2,558,200
Feldspar	2,053,969
Spodumene	669,425
Amblygonite	261,984
Lepidolite	226,318
Beryl	126,466
Tin (1884–1936)	95,730
Columbite-tantalite	62,717
Rose quartz	40,444
Total	$6,095,253

* From *U.S. Geol. Survey Prof. Paper* 247, p. 3, 1953.

In a general way throughout the world feldspar is produced in those regions that contain pegmatites. Thus in the United States it comes from the pegmatite areas of the eastern and western parts of the country; in Canada it comes from the pegmatite areas of Ontario and Quebec; in the Soviet Union, from pegmatites of Karelia and the Kola Peninsula, and in the future may come from the Kansk pegmatite district of central Siberia; in Sweden, from pegmatites near Margretelund, a short distance northeast of Stockholm, and from a number of other pegmatite districts.

TABLE 17-4. **World Production of Feldspar, Average of 1945–1959 and in 1962***

	Percent of world total	
Country	Average of 1945–1959	1962 only
United States	55.9	32.8
West Germany	11.8	17.3
France	6.6	6.1
Sweden	4.6	3.7
Norway	3.8	3.6
Italy	3.4	6.0
Japan	3.1	3.2
Canada	2.9	0.6
Soviet Union†	?	13.3
All others	7.9	13.4
	Long tons	
	Average of 1945–1959	1962 only
World total	864,666	1,500,000

* Compiled from *U.S. Bur. Mines Minerals Yearbook,* various years.
† The production of the Soviet Union for the period 1945–1959 is unavailable.

Nepheline syenite is becoming of increasing importance as industrial material in glass manufacturing (63), and in Canada, where deposits are numerous, it has largely displaced feldspar for that purpose. Nepheline deposits are being worked also in Norway and the Soviet Union.

The Canadian deposits (24, 63) are exploited chiefly in Ontario. There, in the Haliburton-Bancroft district of Renfrew, Hastings, and Peterborough Counties, nepheline-bearing rocks extend throughout a length of about 80 miles but are best developed east of Bancroft Village. Many of the rocks are gneissic, but some are pegmatitic. The largest nepheline-bearing body, at Blue Mountain, Peterborough County, 18 miles northeast of Lakefield, is 5¼ miles long and as much as 1½ miles wide. It has been an important producer for more than 10 years. The assured or reasonably certain reserves are stated to be 4,640,000 tons of nepheline-bearing rock, with additional indicated reserves of 1,500,000 tons (14). Some question apparently exists regarding the origin of the deposits, whether it was by intrusion of igneous magma accompanied by some nephelinization or replacement of sedimentary rock, or completely by replacement of sedimentary rocks by nephelinizing solutions.

Recent exploration in Hastings County (35) has outlined a tabular nepheline-bearing body that is exposed for 1,200 ft along the crest of a ridge adjacent to York River in Monteagle Township. The average thickness of the body is 90 ft, the dip about 70°, and the depth is unknown but has been shown by drilling to be as much as 600 ft. The rock is Precambrian gneiss composed of nepheline, scapolite, feldspar, mica, and corundum. The material mined is to be beneficiated so as to recover the following products: (1) a mixture of nepheline, feldspar, and scapolite, acceptable for glass manufacture, 55 percent; (2) muscovite mica, acceptable in making paint and as a filler, 8 percent; (3) corundum, acceptable as an abrasive in the optical industry, 2.4 percent.

The important deposit in Norway, a large mass of nepheline syenite, is on the island of Stjernoy, some distance north of the Arctic Circle. Utilization of this deposit began late in 1960.

Deposits of nepheline rock in the Soviet Union are used not only as a source of ceramic material but also, and chiefly, as an ore of aluminum. Apatite-nepheline deposits at Khibiny in the Kola Peninsula near Kirovsk (67°30′N, 34°E), are important (36, p. 16; 63). Also, much nepheline is contained in other deposits of the Kola Peninsula. Recently two large nepheline syenite deposits were discovered in the interior of Kazakhstan.

Quartz

Quartz sands, sandstones, quartzites, massive quartz, and crystal quartz are used for making abrasives, glass, and pottery and for a number of other purposes. Quartz crystal of high quality that is clear and free from imperfections is in great demand for electronic work. Such crystals should contain at least one cu in. of flawless material. Other high-quality quartz crystals that do not meet the stringent specifications for electronic-grade material may be used in optical work. Here we will be concerned only with high-quality crystal quartz (15; 17; 28, pp. 503–522; 31, pp. 419–431; 53; 57). Almost all important production of quartz crystal of electronic grade has come from Brazil, but small amounts have been furnished by Malagasy Republic (Madagascar), and a limited amount by the United States, chiefly from Arkansas. Such quartz crystals also occur in Uganda, Australia, and possibly in the Soviet Union. Synthetic electronic-grade quartz crystals have been made and may eventually compete with natural crystals.

Quartz crystals in Brazil occur chiefly in Minas Gerais, Goiaz, and Bahia; minor amounts are present in Espirito Santo and northeastern Brazil (27, 29). The crystals occur in primary veins, pegmatites, eluvial, colluvial, and alluvial deposits. Clear crystals form but a small part of masses of milky quartz, and enormous quantities of milky quartz must be mined to obtain a few kilograms of usable clear quartz. The veins are composed chiefly of milky comb quartz, usually completely intergrown. Vugs are usually abundant and many are lined with well-formed crystals, occasional ones of which are clear, transparent, and of electronic grade. Small, clear crystals are common whereas large ones are relatively rare, yet crystals that weigh from 1 to 5 metric tons do occur in many deposits, and some may weigh as much as 40 metric tons. The veins yield far more quartz than do the pegmatites, but the distinction between pegmatitic quartz and vein quartz is difficult and, indeed, both may have the same origin.

SELECTED REFERENCES

1. D. H. Amos, 1959, D.M.E.A. project blossoms into best U.S. mica mine, *Mining World,* vol. 21, no. 10, pp. 30–34.

2. J. C. Arundale, 1952, Mica, in W. Van Royan and O. Bowles, "Mineral Resources of the World," vol. 2, pp. 178–181, Prentice-Hall, Inc., Englewood Cliffs, N.J., 181 pp.

3. H. M. Bannerman and E. N. Cameron, 1946, The New England mica industry, *Am. Inst. Mining Metall. Engineers Trans.,* vol. 173, pp. 524–531.

4. E. S. Bastin, 1911, Geology of the pegmatites and associated rocks of Maine, including feldspar, quartz, mica, and gem deposits, *U.S. Geol. Survey Bull.* 445, 152 pp.

5. H. Besairie, 1961, Les Ressources minérales de Madagascar, *Annales Géologiques Madagascar,* pt. 30, 116 pp. Mica, pp. 83–88.

6. J. C. Brown and A. K. Dey, 1955, "India's Mineral Wealth," 3d ed., Oxford University Press, London, 761 pp. Mica, pp. 539–549.

7. W. R. Brown, 1962, Mica and feldspar deposits of Virginia, *Virginia Div. Mineral Resources, Mineral Resources Rept.* 3, 195 pp.

8. E. N. Cameron and others, 1945, Structural and economic characteristics of New England mica deposits, *Econ. Geology,* vol. 40, pp. 369–393.

9. J. E. Castle and J. L. Gillson, 1960, Feldspar, nepheline syenite, and aplite, in "Industrial Minerals and Rocks," 3d ed., pp. 339–362, American Institute of Mining, Metallurgical, and Petroleum Engineers, New York, 934 pp.

10. R. R. Chowdhury, 1941, "Handbook of Mica," Chemical Publishing Company, Inc., New York, 340 pp.

11. J. B. Currie, 1951, The occurrence and relationships of some mica and apatite deposits in southeastern Ontario, *Econ. Geology,* vol. 46, pp. 765–781.

12. J. P. DeMille, 1947, "Strategic Minerals," McGraw-Hill Book Company, New York, 626 pp. Mica, pp. 326–338.

13. T. dePolo, 1960, Feldspar, *U.S. Bur. Mines Bull.* 585 pp. 283–289.

14. D. R. Derry and C. V. G. Phipps, 1957, Nepheline syenite deposit, Blue Mountain, Ontario, in "The Geology of Canadian Industrial Mineral Deposits," pp. 190–195, Canadian Institute of Mining and Metallurgy, Montreal.

15. W. F. Dietrich, 1956, Quartz crystal, *U.S. Bur. Mines Bull.* 556, pp. 719–728.

16. W. F. Dietrich and R. D. Thomson, 1956, Mica, *U.S. Bur. Mines Bull.* 556, pp. 521–542.

17. W. F. Dietrich and C. R. Gwinn, 1960, Quartz crystal, *U.S. Bur. Mines Bull.* 585, pp. 665–671.

18. J. A. Dunn, 1962, Mica, *Geol. Survey India Bull.,* ser. A, no. 19, 141 pp.

19. A. S. Furcron and K. H. Teague, 1943, Mica-bearing pegmatites of Georgia, *Georgia Geol. Survey, Bull.* 48, 192 pp.

20. South Africa Department of Mines, and Geological Survey, 1959, Mica, in "The Mineral Resources of the Union of South Africa," 4th ed., Pretoria, South Africa, pp. 512–514.

21. W. R. Griffitts and others, 1953, Mica deposits of the southeastern Piedmont, *U.S. Geol. Survey Prof. Paper* 248, 462 pp.

22. B. L. Gunsallus, 1956, Feldspar, *U.S. Bur. Mines Bull.* 556, pp. 263–271.

23. D. F. Hewitt, 1957, The Purdy mica mine, in "The Geology of Canadian Industrial Mineral Deposits," pp. 181–185, Canadian Institute of Mining and Metallurgy, Montreal.

24. D. F. Hewitt, 1957, Nepheline syenite, in "The Geology of Canadian Industrial Mineral Deposits," pp. 186–190, Canadian Institute of Mining and Metallurgy, Montreal.

25. J. W. Hoadley, 1960, Mica deposits of Canada, *Geol. Survey Canada, Econ. Geology Ser., No.* 19, 141 pp.

26. R. H. Jahns and F. W. Lancaster, 1950, Physical characteristics of commercial sheet muscovite in the southeastern United States, *U.S. Geol. Survey Prof. Paper* 225, 110 pp.

27. W. D. Johnston, Jr., and R. D. Butler, 1946, Quartz crystal in Brazil, *Geol. Soc. America Bull.,* vol. 57, pp. 601–650.

28. S. J. Johnstone and M. G. Johnstone, 1961, "Minerals for the Chemical and Allied Industries," John Wiley & Sons, Inc., New York, 788 pp.

29. F. L. Knouse, 1947, Deposits of quartz crystal in Espirito Santo and eastern Minas Gerais, Brazil, *Am. Inst. Mining Metall. Engineers Trans.,* vol. 173, pp. 173–184.

30. A. Lacroix, 1941, Les gisements de phlogopite de Madagascar et les pyroxénites qui les renferment, *Annales Géologiques Service Mines,* pt. 11, 119 pp.

31. R. B. Ladoo and W. M. Myers, 1951, "Nonmetallic Minerals," 2d ed., McGraw-Hill Book Company, New York, 605 pp.

32. K. K. Landes, 1938, Origin of Quebec phlogopite-apatite deposits, *Am. Mineralogist,* vol. 23, pp. 359–390.

33. C. S. Maurice, 1940, The pegmatites of the Spruce Pine district, North Carolina, *Econ. Geology,* vol. 35, pp. 49–78, 158–187.

34. S. A. Montague, 1960, Mica in "Industrial Minerals and Rocks," 3d ed., pp. 551–566, American Institute of Mining, Metallurgical, and Petroleum Engineers, New York, 934 pp.

35. L. Moyd, P. Moyd, and H. L. Noblitt, 1962, The Monteagle nepheline-corundum-mica deposit, Hastings County, Ontario. Case history of the exploration and evaluation of an industrial mineral deposit, *Canadian Mining Metall. Bull.*, vol. 55, no. 604, pp. 563–570.

36. D. V. Nalivkin, 1960, "The Geology of the U.S.S.R., a Short Outline," Pergamon Press, New York, 170 pp., geologic map of U.S.S.R.

37. J. C. Olson, 1942, Mica-bearing pegmatites in New Hampshire, *U.S. Geol. Survey Bull.* 931-P, pp. 363–403.

38. J. C. Olson and others, 1946, Mica deposits of the Franklin-Sylva district, North Carolina, *North Carolina Dept. Cons. and Development, Div. Min. Resources, Bull.* 49, 56 pp.

39. L. R. Page and others, 1953, Pegmatite investigations, 1942–1945, Black Hills, South Dakota, *U.S. Geol. Survey Prof. Paper* 247, 228 pp.

40. J. M. Parker, III, 1950, Feldspar and mica deposits of southeastern United States, in "Symposium on Mineral Resources of the Southeastern States," pp. 42–48, University of Tennessee Press, Knoxville, Tenn.

41. W. T. Pecora and others, 1950, Structure and mineralogy of the Golconda Pegmatite, Minas Gerais, Brazil, *Am. Mineralogist*, vol. 35, pp. 889–901.

42. W. T. Pecora and others, 1950, Mica deposits in Minas Gerais, Brazil, *U.S. Geol. Survey Bull.* 964-C, pp. 205–305.

43. C. M. Rajgarhia, 1951, "Mining, Processing, and Uses of Indian Mica," McGraw-Hill Book Company, New York, 388 pp.

44. D. E. Redmon, 1961, Reconnaissance of selected pegmatite districts in north-central New Mexico, *U.S. Bur. Mines Inf. Circ.* 8013, 79 pp.

45. B. C. Roy, 1956, The Nellore mica belt, *Geol. Survey India Bull.*, ser. A, no. 11, 156 pp.

46. S. K. Roy, N. L. Sharma, and G. C. Chattapadhyay, 1939, The mica-pegmatites of Kodarma, India, *Geol. Mag.*, vol. 76, pp. 145–164.

47. A. Savornin, 1937, Les gisements de mica phlogopite du sud de Madagascar, *Annales Géologiques Service Mines*, pt. 7, pp. 9–23.

48. T. I. Sharps, 1962, Colorado mica, *Colorado School Mines, Min. Ind. Bull.*, vol. 5, no. 1, pp. 1–15.

49. M. L. Skow, 1960, Mica, *U.S. Bur. Mines Bull.* 585, pp. 521–535.

50. D. D. Smythe, 1947, Muscovite mica in Brazil, *Am. Inst. Mining Metall. Engineers Trans.*, vol. 173, pp. 500–523.

51. H. S. Spence, 1929, Mica, *Canada Dept. Mines and Geol., Mines Branch, Report* 701, 142 pp.

52. D. B. Sterrett, 1923, Mica deposits of the United States, *U.S. Geol. Survey Bull.* 740, 342 pp.

53. R. E. Stoiber, C. Tolman, and R. D. Butler, 1945, Geology of quartz crystal deposits, *Am. Mineralogist*, vol. 30, pp. 245–268.

54. W. C. Stoll, 1950, Mica and beryl pegmatites in Idaho and Montana, *U.S. Geol. Survey Prof. Paper* 229, 64 pp.

55. F. Stugard, Jr., 1959, Pegmatites of the Middletown area, Connecticut, *U.S. Geol. Survey Bull.* 1042-Q, pp. 613–683.

56. P. M. Tyler, 1953, Economics of pegmatites, *Mining Eng.,* vol. 5, pp. 894–898.

57. H. H. Waesche, 1960, Quartz crystal and optical calcite, in "Industrial Minerals and Rocks," 3d ed., pp. 687–698, American Institute of Mining, Metallurgical, and Petroleum Engineers, New York, 934 pp.

58. L. P. Warriner and B. C. Burgess, 1949, The pegmatites of Jasper County, Georgia, *Mining Eng.,* vol. 1, pp. 376–380.

59. M. E. Wilson, 1957, The phlogopite-apatite deposits of eastern Ontario and the southern Laurentian Highlands, Quebec, in "The Geology of Canadian Industrial Mineral Deposits," pp. 175–181, Candian Institute of Mining and Metallurgy, Montreal.

60. M. E. Wilson and R. M. Buchanan, 1957, Feldspar, in "The Geology of Canadian Industrial Mineral Deposits," pp. 85–89, Canadian Institute of Mining and Metallurgy, Montreal.

61. Anon., 1953, Strategic mica from Idaho, *Mining World,* vol. 15, no. 1, pp. 31–32.

62. Anon, 1962, Feldspar in the U.S.S.R., *Mining Jour. (London),* vol. 258, no. 6610, p. 416.

63. Anon, 1962, Survey of nepheline syenite, *Mining Jour. (London),* vol. 259, no. 6638, p. 437.

64. Anon, 1962, Mica in India, *Mining Jour. (London),* vol. 259, no. 6639, p. 463.

65. Anon, 1963, Mining in Madagascar, *Mining Jour. (London),* vol. 261, no. 6693, pp. 517–519.

chapter *18*

Kyanite, Sillimanite, Andalusite; Pyrophyllite; Talc and Soapstone

The industrial minerals grouped together in this chapter are related to some extent in origin and use, yet marked differences exist among them in both of these features. Some authors discuss pyrophyllite separately, others discuss it with talc and soapstone, and still others discuss it with kyanite, sillimanite, and andalusite. Based on chemical composition, pyrophyllite, $A_2O_3 \cdot 4SiO_2 \cdot H_2O$, belongs with kyanite, sillimanite, and andalusite, all $Al_2O_3 \cdot SiO_2$, and because of this it is used in the manufacture of ceramic articles that must withstand high temperatures and abrupt changes of temperature, the major use for the three other minerals of similar chemical composition. Based on physical properties, pyrophyllite belongs with talc and soapstone because the softness of these minerals makes them valuable for use in making insecticides, paints, rubber products, roofing, paper, and other materials where a filler is needed. Chemically, talc, $3MgO \cdot 4SiO_2 \cdot H_2O$, and soapstone, which is impure massive talc, are both used in the ceramic industry also, in making magnesium refractories. Based on origin, kyanite generally is formed by moderate regional metamorphism and sillimanite by high regional metamorphism; andalusite generally is formed by contact metamorphism, but sillimanite also may be formed in this manner; pyrophyllite may be formed by low-grade regional metamorphism or by hydrothermal activity, as may talc and soapstone, although talc is formed by regional metamorphism also. Field associations of these minerals, however, differ because of the differing chemical compositions. Thus pyrophyllite, kyanite, sillimanite, and andalusite originate from aluminous rocks, such as some shales and granitic igneous rocks, and commonly occur with schists, gneisses, and quartzose rocks. Talc and soapstone, however, commonly originate from basic to ultrabasic igneous rocks, especially serpentinites, and from dolomitic rocks, and consequently are associated with such rocks in regions of relatively low-grade metamorphism.

Kyanite, sillimanite, andalusite

Kyanite and sillimanite are the chief aluminum silicate minerals that are used commercially and are more abundant than andalusite. Two other minerals are used to a minor extent, also. These are dumortierite, $8Al_2O_3,Be_2O_3 \cdot 6SiO_2 \cdot H_2O$, and topaz, $2Al_2O_3 \cdot 2Al(F,OH)_3 \cdot 3SiO_2$ (6; 8, pp. 245–249, 431–434; 11; 13; 16; 19, pp. 522–529; 21). Detailed information is not available about many deposits throughout the world, but statistics indicate that the most important production of both kyanite and sillimanite has come from India, and that important production of kyanite and some production of sillimanite and andalusite has come from Africa. A considerable amount of kyanite is present in the United States, but it is the disseminated rather than the massive type. The massive type is much preferred for industrial use because it yields strong, large aggregates of mullite[1] crystals, whereas the concentrate from the disseminated type yields only fine-grained aggregates.

India

The important deposits of India (1, pp. 414–422; 21, Chap. 3) consist of (1) kyanite and (2) sillimanite and sillimanite-corundum. In the Singhbhum district, state of Bihar, massive kyanite and kyanite-quartz rock occur at various places throughout a curved zone about 70 miles long. At the western end of this zone is the principal deposit, Lapsa Bura, which is rated as the largest of its kind in the world and has been the chief producer of kyanite in India since 1924. Kyanite is present as almost the only constituent of a massive, medium-grained to coarse-grained rock and also as segregations of large, bladed crystals in beds of kyanite-quartz rock. Much kyanite is recovered as pebbles and boulders as much as 6 ft across from surface deposits at a depth of 5 or 6 ft. On the Khasi Plateau of Assan numerous sillimanite-corundum deposits are known, the most important of which are around the village of Sona Pahar. Most of the deposits consist of massive sillimanite, with which occurs a small amount of corundum, but some are almost entirely sillimanite and others are almost entirely corundum, for which the deposits were originally worked. Associated rocks are gneiss and schist. Large boulders of sillimanite are present, some of which weigh as much as 40 tons. At Pipra (Rewa), state of Vindhya Pradesh, are corundum-sillimanite deposits that have been worked chiefly for corundum.

Africa

The important deposits of Africa are in Kenya, in the Republic of South Africa, and in South-West Africa (7; 15, pp. 352–356; 21, Chap. 3).

Throughout Kenya kyanite is present in crystalline rocks of Precambrian age at various places, but the most important kyanite deposits are at Murka Hill and vicinity, a few miles northeast of Taveta. Lenticular segregations of massive kyanite as much as 100 ft long and 15 ft thick are present in biotite gneiss.

---

[1] Mullite, $3Al_2O_3 \cdot 2SiO_2$, which is rare in nature, is the material to which the various minerals under discussion are converted on heating to temperatures of 1595°C or somewhat below. These minerals yield silica also during the process. The three minerals kyanite, sillimanite, and andalusite yield 88 percent mullite and 12 percent silica (13, p. 775).

In the Republic of South Africa there are important deposits of andalusite. At the western end of the Bushveld complex, the sedimentary rocks have been metamorphosed by the intrusion of the magma of the complex and changed in (1) an inner zone to a mass of minerals that includes both andalusite and sillimanite so closely intergrown with other minerals that economic separation is difficult and (2) an outer zone into andalusite in softer micaceous minerals but not intergrown with them. In many places the andalusite is in well-formed crystals which are separated from the matrix during weathering and erosion and are concentrated in sands along stream courses. The best deposits are in the Marico district, western Transvaal, along the upper parts of the Little Marico River and its tributaries. Some deposits occur along stretches of 4 to 5 miles and contain grains of andalusite generally 0.75 to 1.75 mm across but in places as much as 5 mm across. Some samples contain at least 50 percent recoverable andalusite.

Deposits of sillimanite and corundum occur in Namaqualand, South-West Africa, the chief deposits being those in the general vicinity of Pella West. The deposits of Pella West, now practically worked out, originally consisted of boulders and blocks scattered over an area of about 200,000 sq yds. These varied from less than a foot to more than 10 ft across. The Swartkoppies deposit some distance east of Pella West, an important producer, covers an area of about 200,000 sq ft and consists of hills composed almost wholly of corundum-sillimanite rock that weathers black. Heavily jointed outcrops of this rock make up the eastern and central part of the district, whereas loose blocks make up the western part. The average mineral composition for the whole deposit was calculated as 53 percent sillimanite, 41 percent corundum, 2.77 percent ilmenite, and 1.19 percent rutile (total 97.96 percent). The ore consists of (1) coarse, veinlike material composed almost wholly of sillimanite; (2) mottled material composed of sillimanite and corundum, with sillimanite in excess; (3) fine-grained material consisting of sillimanite and corundum but with corundum in excess. The three types are rather well segregated within the deposit. Another deposit of this general area, about 18 miles southwest of Pella West, known as Deposit No. 1, consists of a lenticular sillimanite body about 260 ft long and 140 ft wide in quartz-mica schist. A number of other deposits are known, but the major reserves are in Deposit No. 1 (190,000 short tons) and in the Swartkoppies deposit (255,000 short tons).

United States

The producing deposits of the United States (11, 12, 24) consist of (1) pyrophyllite, chiefly in California and North Carolina; (2) kyanite, chiefly in Virginia and North Carolina; and (3) andalusite, along with pyrophyllite, in North Carolina. In the past dumortierite was mined in Nevada, and topaz in South Carolina. Production of the sillimanite group minerals—sillimanite, kyanite, andalusite—has been confined to the southeastern United States since 1950 and has consisted chiefly of kyanite concentrates from mines at Baker Mountain and Willis Mountain, Virginia, and Henry Knob, South Carolina, and andalusite contained in pyrophyllite ore in North Carolina. Measured and indicated reserves of the southeastern region have been estimated to be 100 million tons containing 10 to 30 percent kyanite.

The kyanite-sillimanite-andalusite deposits of the southeastern United States (12, 24) occur throughout a very large area of metamorphic rocks of the Blue Ridge and Piedmont provinces. They consist of four major types: (1) quartzose rocks; (2) micaceous schists and gneisses; (3) quartz veins and pegmatites; and (4) residual soils and placers. Among these the quartzose deposits are the most important, and the quartz veins and pegmatites have not been worked successfully. Among the quartzose deposits, which consist of rocks composed chiefly of quartz-kyanite, quartz-sillimanite, and quartz-andalusite-pyrophyllite, quartz-kyanite deposits are the largest, most numerous, and most productive. In these deposits kyanite makes up 10 to 30 percent of the rock. The largest and most persistent deposits are in the Farmville district and the Kings Mountain district, North Carolina—South Carolina (Fig. 8-1), and the Graves Mountain district, Georgia. The larger deposits form ridges a mile or more long and several hundred feet high. Kyanite in micaceous schist and gneiss occurs throughout extensive areas of the Blue Ridge province in western North Carolina and northern Georgia. The kyanite content generally ranges from a few percent to about 10 percent, rarely 15 percent or more. Only a few of these deposits have been mined in North Carolina and Georgia. Sillimanite schist is rather widespread in two belts, one in the western Piedmont, the other in the Blue Ridge. The sillimanite is in very fine crystals intimately intergrown with mica and generally constitutes less than 10 percent of the rock. These deposits have not been mined although they have been extensively explored.

Pyrophyllite

Pyrophyllite is included with talc and soapstone in most production figures. Consequently information about the amount of pyrophyllite mined and also about the most important deposits of pyrophyllite is meager. According to the statistics for the production of talc, soapstone, and pyrophyllite combined, Japan furnished 20.2 percent of the world total for the period 1945 to 1959 and 22.5 percent for 1962, but most of this material was pyrophyllite, as there is very little talc in Japan (14, p. 236; 22). Japan probably holds first place in the world as a producer of pyrophyllite. Reserves of pyrophyllite in Japan have been given in one estimate as about 59 million tons (14, p. 230) and in another one as more than 70 million tons (18, p. 857). The deposits apparently resulted from hydrothermal alteration of rhyolite or quartz porphyry.

Formerly China was accredited with important production of talc, but much of the Chinese material was pyrophyllite (19, p. 466).

The pyrophyllite of North Carolina (23) occurs in a zone of metamorphosed volcanic and sedimentary rocks in the Piedmont province. The pyrophyllite bodies are irregular, oval, or lenticular, and generally consist of pyrophyllite, quartz, and some sericite and other minerals, the pyrophyllite being present as foliated masses, as radial fibrous aggregates, or as massive granular bodies. The lenses of the larger bodies are 500 to 2,000 ft long, 100 to 500 ft wide, but the workable pyrophyllite occurs near the centers of lenses throughout a width of 25 to 200 ft. Drilling has shown that some of the larger deposits persist in depth more than 500 ft vertically.

If a depth of 400 to 500 ft is assumed, the reserves of the North Carolina area are 12 to 14 million tons. The deposits are associated with acidic tuffs and breccias and apparently were formed by hydrothermal replacement of such rocks.

Talc and soapstone

Talc and soapstone are rather widespread throughout the world, but major production has been chiefly from the United States (Table 18-1), since the material produced by Japan is chiefly pyrophyllite. China at one time held second rank in the production of talc and soapstone, but there, also, much of the material was pyrophyllite. Production in the Soviet Union is unknown for a number of years, but statistics indicate that in 1962 it was nearly 12 percent of the world total.

TABLE 18-1.   World Production of Talc, Soapstone, and Pyrophyllite, Average of 1945–1959 and in 1962*

	Percent of world total	
Country	*Average of 1945–1959*	*1962 only*
United States	37.8	26.3
Japan†	20.2	22.5
France	8.3	6.8
Italy	5.2	4.7
Norway	4.5	4.8
India	2.6	3.8
Soviet Union	?	11.6
China†	?	5.6
Others	21.4	13.9
	Short tons	
	Average of 1945–1959	*1962 only*
World total	1,609,333	2,930,000

* Compiled from *U.S. Bur. Mines Minerals Yearbook,* various years.
† Chiefly pyrophyllite.

United States

Talc and soapstone deposits are known in many places in the United States (4, 5) and production came from 16 states in 1962, but major production came only from New York, California, and North Carolina. Many deposits occur in Georgia and Vermont, and at one time Vermont was the leading producer.

The commercially important deposits of New York (5; 9; 10; 22, p. 532) are in St. Lawrence County, near Gouverneur, and in Lewis County near Natural Bridge, which is in Jefferson County. Most of the production comes from St. Lawrence County, especially in the Talcville area. The talc occurs in the folded Grenville (Precambrian) marble belt of the area as elongate zones interlayered with impure marble. These zones persist for as much as 6 miles along the strike throughout a width of as much as 400 ft, and probably extend more than 2,000 ft down the dip, which averages about 45° but is exceedingly variable. Although classed as talc commercially, and satisfactorily used as talc for certain products, the mineral talc constitutes less than one-fourth of the rock that is mined and ground. Much of the rock is tremolite schist or tremolite-anthophyllite schist partly altered to serpentine and talc. The deposit near Natural Bridge is somewhat different. A pluglike mass of brecciated and altered marble has been partly replaced by serpentine, talc, and a few other minerals. Massive serpentine generally is more abundant than talc.

The California deposits (4, 5, 10, 25) that have been most productive are near the California-Nevada border in a zone about 200 miles long. Three districts are present within this zone. These, from north to south, are (1) the Inyo Range–Northern Panamint Range district in Inyo County; (2) the Southern Death Valley–Kingston Range district in Inyo and San Bernardino Counties; and (3) the Silver Lake–Yucca Grove district in San Bernardino County. In 1962 talc was shipped from 12 mines in Inyo County and also from 12 mines in San Bernardino County. The deposits of the Inyo Range–Panamint Range district are chiefly alterations of relatively unmetamorphosed dolomite of Ordovician age, in which talc has formed by replacement along sheared and fractured zones. The deposits of the Southern Death Valley–Kingstone Range district are chiefly in the Crystal Spring formation of Precambrian age. This formation, about 4,000 ft thick, contains a carbonate member that is commonly chiefly dolomite, and has been intruded by diabase sills, most of which are immediately below or immediately above this carbonate member. Many of the talc bodies are above the diabase sill, in contact with it and replacing the lower part of the carbonate member; some lie above the sill but are separated from it by noncarbonate strata; some are below the sill; and others form septa within the sill. The deposits of the Silver Lake–Yucca Grove district are talc replacements of earlier-formed tremolite.

The talc and soapstone deposits of North Carolina occur in (1) a northwestern belt, (2) the Murphy marble belt, which extends into northern Georgia, and (3) a southwestern belt. Other deposits in Georgia occur in the Chatsworth district, Murray County. The deposits of the northwestern and the southwestern belts of North Carolina are associated chiefly with and derived from ultramafic and some mafic igneous rocks. The deposits of the Murphy marble belt are talc rock formed from dolomitic rocks. The deposits of the Chatsworth district, Georgia, are talc-carbonate and talc rock that probably came from alteration of dolomitic parts of a schist but possibly came from the alteration of ultramafic rocks.

The commercial talc deposits of Vermont (3, 4, 5, 11) occur in a zone of ultramafic igneous rocks that extends through the central part of the state and is about 25 miles wide at the north but only about 5 miles wide at the south. Individual ultramafic bodies range in length from at least 3.5 miles to less than 100 ft and in width from about a mile to only a few feet. Most of these ultramafic bodies, and all that contain workable deposits of talc, are completely serpentinized and more or less altered to talc. A typical body of this type contains a core of serpentinite and an outer border of talc commonly a few inches to a few feet thick. At the margins there may be a mixture of talc and carbonate, locally designated *grit*.

The most important deposits have been in the northern part of the ultramafic zone, especially in Lamoille County, and toward the southern part of the zone in Windsor and Windham Counties. Production in 1962 came from Lamoille County near Johnson and from Windsor County near Reading.

Other Countries

Japan, France, Italy, Norway, and India are the most important sources of talc and soapstone except for the Communist countries the Soviet Union and China. The deposits of Japan are chiefly pyrophyllite and have been discussed briefly. The deposits of France and Italy (10; 22, p. 534) are of high quality. The important ones in France occur chiefly in Ariège along the north slopes of the Pyrenees Mountains, and those in Italy occur chiefly near Perosa and west of Pineroo in the Alps. Both the French and the Italian deposits appear to be a result of low-grade metamorphism or hydrothermal replacement of dolomitic limestones and dolomites. The Norwegian deposits, of lower grade than the French and Italian ones, were also formed by metamorphism and alteration of dolomitic rocks. The deposits of India (1, pp. 524–531) of commercial importance are chiefly in Rajasthan, Andhra, and Madhya Pradesh, and apparently all were formed by replacement of dolomite. The deposits of China (20, pp. 466–467) are in Liaoning, Chekiang, and Fukien (Fig. 9-5). The deposits of Chekiang and Fukien are not talc but are pyrophyllite that occurs as veins in rhyolite and tuff from which it was formed by alteration. The deposits of Liaoning are good-quality talc associated with deposits of magnesite. Little specific information seems to be available regarding the talc deposits of the Soviet Union. The ultramafic rocks of the Ural Mountains should be a source of deposits of talc and soapstone. Recently it was reported that a large deposit of talc had been found in the foothills of the Ural Mountains near Miass. The deposit was estimated to contain 30 million metric tons (28).

SELECTED REFERENCES

1. J. C. Brown and A. K. Dey, 1955, "India's Mineral Wealth," 3d ed., Oxford University Press, London, 761 pp.
2. F. Chappell, 1960, Pyrophyllite, in "Industrial Minerals and Rocks," 3d ed., pp. 681–686, American Institute of Mining, Metallurgical, and Petroleum Engineers, New York, 934 pp.
3. A. H. Chidester, M. P. Billings, and W. M. Cady, 1951, Talc investigations in Vermont, preliminary report, *U.S. Geol. Survey Circ.* 95, 33 pp.
4. A. H. Chidester and H. W. Worthington, 1962, Talc and soapstone in the United States, exclusive of Alaska and Hawaii, *U.S. Geol. Survey Mineral Inv. Resource Map* MR-31. Map and pamphlet of 9 pp.
5. A. H. Chidester, A. E. J. Engel, and L. A. Wright, 1964, Talc resources of the United States, *U.S. Geol. Survey Bull.* 1167, 61 pp.
6. J. D. Cooper, 1960, Kyanite and related minerals, *U.S. Bur. Mines Bull.* 585, pp. 423–428.
7. D. H. de Jager and J. W. von Backström, 1961, The sillimanite deposits in Namaqualand near Pofadder, *South Africa Geol. Survey Bull.* 33, 49 pp.
8. J. B. DeMille, 1947, "Strategic Minerals," McGraw-Hill Book Company, New York, 626 pp.

9. A. E. J. Engel, 1949, New York talcs, their geological features, mining, milling, and uses, *Am. Inst. Mining Metall. Engineers Trans.,* vol. 184, pp. 345–348.

10. A. E. J. Engel and L. A. Wright, 1960, Talc and soapstone, in "Industrial Minerals and Rocks," pp. 835–850, American Institute of Mining, Metallurgical, and Petroleum Engineers, New York, 934 pp.

11. G. H. Espenshade, 1962, Pyrophyllite, and kyanite and related minerals in the United States, exclusive of Alaska and Hawaii, *U.S. Geol. Survey Mineral Inv. Resource Map* MR-18.

12. G. H. Espenshade and D. B. Potter, 1960, Kyanite, sillimanite, and andalusite deposits of the southeastern states, *U.S. Geol. Survey Prof. Paper* 336, 121 pp.

13. W. R. Foster, 1960, The sillimanite group—kyanite, andalusite, sillimanite, dumortierite, topaz, in "Industrial Minerals and Rocks," 3d ed., pp. 773–789, American Institute of Mining, Metallurgical, and Petroleum Engineers, New York, 934 pp.

14. Geological Survey of Japan, 1960, Geology and mineral resources of Japan, 2d ed., Hisamoto-cho, Kawasaki-shi, Japan, 304 pp.

15. South Africa Department of Mines, and Geological Survey, 1959, Andalusite, sillimanite, kyanite, in "The Mineral Resources of the Union of South Africa," 4th ed., Pretoria, South Africa, pp. 352–356.

16. B. L. Gunsallus, 1956, Kyanite and related minerals, *U.S. Bur. Mines Bull.* 556, pp. 411–415.

17. R. D. Irving, 1956, Talc, soapstone, and pyrophyllite, *U.S. Bur. Mines Bull.* 556, pp. 853–866.

18. R. D. Irving, 1960, Talc, soapstone, and pyrophyllite, *U.S. Bur. Mines Bull.* 585, pp. 835–844.

19. S. J. Johnstone and M. G. Johnstone, 1961, "Minerals for the Chemical and Allied Industries," 2d ed., John Wiley & Sons, Inc., New York, 788 pp.

20. V. C. Juan, 1946, Mineral resources of China, *Econ. Geology,* vol. 41, pt. 2, pp. 399–474.

21. T. A. Klinefelter and J. D. Cooper, 1961, Kyanite, a materials survey, *U.S. Bur. Mines Inf. Circ.* 8040, 55 pp. Chapter 3, pp. 8–19, by G. H. Espenshade, discusses deposits.

22. R. B. Ladoo and W. M. Myers, 1951, "Nonmetallic Minerals," 2d ed., McGraw-Hill Book Company, New York, 605 pp.

23. J. L. Stuckey, 1958, Resources and utilization of North Carolina pyrophyllite, *Mining. Eng.,* vol. 10, pp. 97–99.

24. K. H. Teague, 1950, Sillimanite in the southeast, *Am. Inst. Mining Metall. Engineers, Trans.* vol. 187, pp. 785–789.

25. L. S. Wright, 1957, Talc and soapstone, *California Div. Mines Bull.* 176, pp. 623–634.

26. Anon., 1961, Sillimanite in Namaqualand, *Mining Jour. (London),* vol. 257, no. 6584, p. 425.

27. Anon, 1962, Kyanite—a survey, *Mining Jour. (London),* vol. 259, no. 6624, p. 103.

28. Anon, 1961, *Mining Jour. (London),* vol. 257, no. 6583, p. 405. Notice of a newly discovered deposit of talc in the Ural Mountains, U.S.S.R.

chapter 19

Asbestos

The commercial deposits of asbestos are characterized by the presence of certain fibrous minerals, the fibers being generally strong, flexible, and resistant to heat. These asbestos minerals belong in two groups, chrysotile and amphibole. Chrysotile, also designated serpentine asbestos, $H_4Mg_3Si_2O_9$, is the major asbestos mineral, and 95 percent of the world production consists of this type. Five minerals occur in deposits of amphibole asbestos: amosite, $(Fe, Mg)_7Si_8O_{22}(OH)_2$; anthophyllite, $Mg_7Si_8O_{22}(OH)_2$; tremolite, $Ca_2Mg_5Si_8O_{22}(OH)_2$; actinolite, $Ca_2(Mg, Fe)_5Si_8O_{22}(OH)_2$; and crocidolite (blue asbestos), $Na_2Fe_5Si_8O_{22}(OH)_2$. The compositions of these minerals vary considerably, especially with iron substituting in part for magnesium. Among the amphibole minerals, amosite is mined only in Africa, crocidolite is mined chiefly in Africa, and deposits of these two types make up most of the world production not accounted for by chrysotile asbestos. Anthophyllite and tremolite make up the small remainder, as actinolite is of little commercial importance. Most chrysotile asbestos deposits occur in ultrabasic igneous rocks, especially serpentinized peridotite and pyroxenite, but a few deposits are in serpentinized limestone. Deposits of amosite and crocidolite generally occur in metamorphosed sedimentary rocks that are ferruginous and siliceous.

Various characteristics determine the usefulness and hence the importance of the types of asbestos. The uses of anthophyllite, tremolite, and actinolite asbestos are limited because the fibers are weak and brittle as contrasted with other types. Amosite commonly consists of long and fairly strong fibers, shows good flexibility and good resistance to acids and alkalies, and is used chiefly as insulating material. Crocidolite fibers generally are not as long as are those of amosite but they are strong and their resistance to acids and alkalies is good, but their resistance to heat is poor. Crocidolite makes good fabric for acid-resistant packings and similar uses. Chrysotile is usually highly fibrous; the fibers may be separated almost indefinitely and vary from short to long. Chrysotile shows good resistance to heat, poor resistance to acids and alkalies.

Based on use, asbestos may be separated into spinning and non-spinning grades. The types of chrysotile and crocidolite with the longer fibers are spinning grades; the types with shorter fibers, and generally all types of the other amphibole members, are nonspinning grades. The spinning grades are spun and woven into various fabrics, whereas the nonspinning grades are used for various types of packings, for heat insulation, and for making building materials, such as roofing shingles.

Production and reserves

Commercial production of asbestos began in Italy in the early 1800s, but large production did not take place until the extensive development of highly important deposits in Canada, mining of which began in 1878. These are the most important deposits of chrysotile asbestos in the world. World production for the period 1945 to 1959 and for 1962 is shown in Table 19-1. Recent estimates of the most important reserves of asbestos are shown in Table 19-2.

Canada

The asbestos deposits of Canada (1; 3; 5, pp. 29–34; 7, Chap. 2; 10; 22; 23; 24; 28) are chiefly in Quebec, but deposits of some importance are being developed

TABLE 19-1. World Production of Asbestos, Average of 1945–1959 and in 1962*

	Percent of world total	
Country	Average of 1945–1959	1962 only
Canada	55.2	40.0
Soviet Union	20.9	36.0
Republic of South Africa	6.5	7.2
Rhodesia (Southern Rhodesia)	5.7	4.6
All others	11.7	12.2
	Short tons	
	Average of 1945–1959	1962 only
World total	1,532,000	3,055,000

* Compiled from *U.S. Bur. Mines Minerals Yearbook,* various years.

TABLE 19-2. The Most Important Reserves of Asbestos in the World*

	Thousand short tons		
Country	Chrysotile	Amosite	Crocidolite
Canada	50,000	—	—
Soviet Union	25,000	—	—
Rhodesia (Southern Rhodesia)	10,000	—	—
United States	1,500	—	—
Republic of South Africa	250	1,500	1,000

* From *U.S. Bur. Mines Inf. Circ.* 7880, p. 35, 1959.

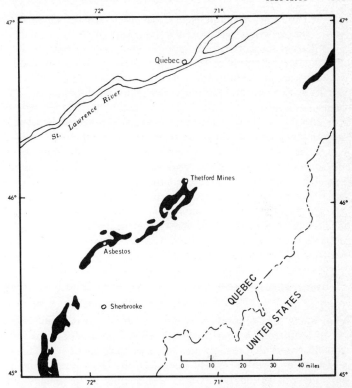

Fig. 19-1 Map showing areas of serpentinized rock (solid black) in the asbestos region of Quebec, Canada. Generalized after Oliver Bowles, *U.S. Bur. Mines Bull.* 552, p. 28.

also in Ontario, British Columbia, and Newfoundland. Production has been continuous in Quebec since 1878, and that province still furnishes 95 percent of the asbestos of Canada.

Quebec

The asbestos area of Quebec is in the eastern part of the province in a zone 5 to 6 miles mide and about 70 miles long (Fig. 19-1), but the productive part of it lies between Asbestos on the south and East Broughton on the north, with the most extensive mining between the Thetford mines and Black Lake, and around Asbestos. The Jeffrey mine of Canadian Johns-Manville Corporation Limited, at Asbestos, produces about 57 percent of the Canadian total and is rated as the world's largest asbestos mine. The most recent development in the area has been the draining of Black Lake and the operation of an open-pit mine on the former bed of the lake (29).

Throughout the asbestos area, peridotite and some dunite, pyroxenite, and gabbro were intruded into sedimentary and volcanic rocks generally of Paleozoic age but possibly including some of Precambrian age. A large part of the ultrabasic rock is peridotite, and this is the rock in which the chrysotile asbestos deposits occur. The peridotite was extensively serpentinized and contains an average of 30 to 40 percent serpentine, although in some places it consists almost wholly of serpentine, especially around acidic intrusions, faults, broken zones, and joints.

The asbestos occurs in veins of cross-fiber type[1] that are a fraction of an inch to 4 or 5 in. thick although the majority of them are less than $\frac{1}{4}$ in. thick. They may intersect, branch, and make lenses that are exceedingly small or more than 20 to 30 ft. long, and some veins are characterized by one or two partings. At both the Thetford mines and the Jeffrey mine the ore bodies consist of a stockwork of cross-fiber asbestos veins, although some slip-fiber veins are present also and account for some of the production. Generally the amount of asbestos in a deposit is from 5 to 7 percent, although locally it may reach 25 percent. The deposits of the area are mined by open-pit and also by underground methods.

The exact mode of formation of the asbestos veins is uncertain. Both growth of fibers in an open fracture and replacement have been suggested, and it seems probable that both have occurred. The genesis of asbestos in ultrabasic rocks was discussed in 1955 by Riordon (21).

Other Deposits of Canada

The most important asbestos deposit of Ontario, at the Munro mine of the Johns-Mansville Corporation, is about 55 miles east of Timmins and about 12 miles east of Matheson. A nearly vertical serpentinized sill 500 to 900 ft wide contains cross-fiber veins that reach a maximum width of about 1 in. The asbestos apparently persists to a depth of several hundred feet. Open-pit mining began in 1950. In 1961 the production of Ontario was slightly more than 25,000 short tons of asbestos compared with a total Canadian production of 1,173,695 short tons (28, p. 2).

The major production of asbestos in British Columbia, which furnished slightly more than 45,000 short tons in 1961, comes from the Cassiar Asbestos Corporation, Ltd. The deposit (10), which is in serpentinite, is on a spur of McDame Mountain at an elevation of 6,000 ft and is in the northern part of the province, about 70 miles from the Alaska Highway and about 60 miles south of the Yukon Territory border. Low-iron chrysotile with an exceptionally high proportion of spinning fibers and large reserves are very favorable features of the deposit, but offsetting these to some extent are severe climate and difficult transportation.

Several deposits of asbestos occur in Newfoundland, but the one now known to have commercial importance is at Baie Verte (3), where production began in 1963 and is expected to reach 60,000 tons per year (35). The explored deposit is about 5,200 ft long, averages 800 to 1,000 ft wide, and is bowl-shaped. Proven reserves are slightly more than 41 million tons, but much of the property has not been tested.

---

[1] Asbestos veins are classed as cross-fiber and slip-fiber. Cross-fiber veins are those in which the fibers of asbestos lie approximately at right angles to the walls. Slip-fiber veins are those in which the fibers are parallel to the walls; such veins not uncommonly show effects of movement along a fault surface.

Soviet Union

The Soviet Union is next to Canada in production and reserves of asbestos. The more important deposits are in the Ural Mountains (5; 7; 20, p. 57–58) about 90 km northeast of Sverdlovsk in the Bazhenov area (also designated Bazhenovo and Bajenova). A new town in the area, Asbest, is the center of the industry. The deposits are in a serpentinized intrusion that is 21 km long and 20 to 1,200 m wide and consist of ellipsoidal masses that attain a maximum length of 3,500 ft, a width of 1,000 ft, and contain veins composed chiefly of cross-fiber chrysotile, the highest percentage of which is in the central parts of these masses. The percentage of commercial fiber in the Russian deposits is about the same as it is in the Quebec deposits of Canada, although the percentage of spinning fiber is a bit higher.

Africa

Rhodesia (Southern Rhodesia)

The reserves of asbestos in Rhodesia (Southern Rhodesia) are the third largest in the world, and the production, although smaller than that of Russia, is highly important in international trade, since the material of Russia is consumed at home to a large extent. The deposits (5, 7) occur in the ultramafic rocks of the country and are more or less centered about Shabani, which is about 80 miles south of Gwelo and 120 miles east of Bulawago (Fig. 16-1). The Shabani group of mines, the most important in the area, is in the central part of a mass of serpentine that is 1 to 3 miles wide and about 10.5 miles long. The chrysotile veins, which may be as much as 6 in. thick, are concentrated in a zone 20 to 200 ft thick and 2.5 to 3 miles long. Extension of the deposits down the dip, which is 5 to 45°, is unknown but considerable. Other deposits of importance are in the Mashaba district, about 40 miles east of Shabani, and in the Filabusi district, about 60 miles west of Shabani. The proportion of long fiber of spinning grade in the deposits of Rhodesia (Southern Rhodesia) is exceptionally high, 25 to 30 percent as compared with 4 to 6 percent in the deposits of Canada.

Republic of South Africa

Asbestos produced in the Republic of South Africa is chiefly amosite, crocidolite, and chrysotile, along with small amounts of anthophyllite and tremolite (7, Chap. 2; 9; 11, pp. 357–367; 16; 34). Crocidolite (blue asbestos) is mined in Cape Province and the Transvaal; amosite in the Transvaal; chrysotile in the Transvaal and to a small extent in Natal; and small amounts of anthophyllite in the Transvaal, and of tremolite in Natal. Most of the production comes from the Transvaal and the Cape of Good Hope.

The crocidolite and amosite deposits of the Transvaal occur in an east-west zone some 60 miles long and several miles wide near the Olifants River. The zone is south of Pietersburg and north of the chromium and platinum deposits of the Steelpoort area (Fig. 16-1). The deposits are in the Dolomite series of the Transvaal system, which has been subdivided, from the bottom upward, into the Main

Dolomite stage, the Banded Ironstone stage, and the Upper Dolomite stage. The asbestos is confined to various stratigraphic positions within the banded ironstone of the Banded Ironstone stage (11, pp. 360, 362). The commercial amosite deposits, which are the only ones known in the world, are best developed near Penge, which is about 30 miles north of Lydenburg. These deposits are associated with diabase sills, which certainly contributed to the metamorphism of the ironstone and the formation of the amosite. The fibers are 3 to 6 in. long; much material contains fibers 6 in. or more long and is in strong demand. Farther west, in the area known as the Pietersburg field, both amosite and crocidolite are mined.

Chrysotile of the Transvaal has come from the Carolina district and the Barberton district. Only a small amount has come from the Carolina district, which is about 150 miles east of Johannesburg and 60 miles south of Lydenburg. Those deposits occur in altered dolomite above a diabase sill. Production from the Barberton district, to the east of the Carolina district, has furnished most of the chrysotile of South Africa, but production became very low for some time and then increased somewhat. The deposits occur in a serpentine zone, the asbestos-bearing part of which has been outlined over a length of 3 miles and a width of 2 miles.

The crocidolite deposits of the Cape of Good Hope, discovered between 1803 and 1806, are the most extensive ones known in the world. They occur throughout a zone 250 miles or more long and as much as 30 miles wide in Griqualand West that extends northward from the vicinity of Prieska to the Bechauanaland border, in the same general region as some of the manganese deposits of South Africa (Fig. 16-1). The largest mine of the area, and also the largest producer of crocidolite in South Africa, is the Koegas (9), about 45 miles northwest of Prieska.

Throughout the crocidolite zone the deposits occur in the banded ironstones of the Lower Griquatown stage of the Pretoria series, which is the uppermost division of the Transvaal system. Three horizons[1] of crocidolite occur in the Koegas-Prieska area, a lower one between 50 and 250 ft above the base of the banded ironstone beds, an intermediate one approximately 700 ft above the lower horizon, and an upper one (Westerberg asbestos horizon) about 900 ft above the intermediate one. All three of these horizons are mined at some place in the area but it is the upper one, about 150 ft thick, that furnishes the production of the Koegas asbestos mine. Most of the production is from the Westerberg syncline, where material has been mined for nearly 70 years and reserves are enough to maintain production at the present rate for many years.

Swaziland

Important production of chrysotile has come from Swaziland (5, 7) from a deposit close to the border of Transvaal and only a short distance east of the Barberton district (Fig. 16-1). A zone of chrysotile-bearing veins of cross-fiber type, about 110 ft wide and dipping 40 to 60°, lies within a mass of serpentine. The Havelock mine, the most important one in the area, transports material by

---

[1] Locally material is designated as *seams* or *bands, reefs,* and *horizons.* A seam or band is a single layer of crocilodite; a reef is a group of seams that can be mined together as a unit; a horizon is a group of reefs separated by layers of barren rock of varying thickness (9, p. 1426).

means of an overhead cableway 12.6 miles long to the nearest railroad, which is at Barberton.

United States

The United States is not an outstanding producer of asbestos. It furnished but 1.7 percent of the world total in 1962 and 2.6 percent for the period 1945 to 1959. The more important deposits (5, 7, 8) occur in two zones of serpentinized peridotite, one in the eastern United States extending from Maine to Alabama, the other in the western United States extending from Washington to California. The most important production has been from Vermont, especially in Lamoille and Orleans Counties, where chrysotile occurs in serpentine, possibly an extension of the Canadian zone. The asbestos is similar to that occurring at Thetford mines, Quebec. The most recent interest in asbestos centers about deposits in California and in Arizona. The most important development in California is that of the Jefferson Lake Asbestos Corporation at Copperopolis, Calaveros County, near Stockton; a second development is at Coalinga, between San Francisco and Los Angeles, on the western edge of the San Joaquin Valley. These two deposits are expected to produce 100,000 tons annually. Other deposits in California are being investigated. The development in Arizona is near Globe.

The Jefferson Lake deposit, California (2, 12, 18), formerly designated the Vorhees deposit, has been known for many years, but it was not until 1959 that results of drilling were highly favorable. In 1961 enough information was available to bring about a decision to mine the ore body by open-pit methods and to construct a mill at a cost of 5 million dollars. In 1962 the production and milling of ore had reached 2,500 tons a day. Reserves are not definitely known but seem to be about 23 million tons. The chrysotile zone in serpentine is boat-shaped, averages 340 ft wide, and is 2,000 ft long. The best material is in the central part of the zone at a depth of 150 to 450 ft. The ore body is cut by numerous fractures, many of which contain veins of cross-fiber chrysotile.

The Coalinga deposit, California (19), occurs in a serpentinized peridotite—the New Idria intrusive—on the northern flank of which is the New Idria mercury mine. The asbestos occurrence here is unusual, since chrysotile is present as white to greenish-white flakes that vary in size from very minute ones to those that are over an inch across and are flexible. Detailed study has shown that these flakes are made up of short fibers and that the material is chrysotile. These fibers differ from those of the Quebec, Canada, deposits, however, as they are randomly oriented and matted in flakes rather than occurring in veins. Production, consequently, will be only short-fiber material. Research by Johns-Manville has shown that this particular type of chrysotile can be used very advantageously in several operations and thus can have a ready market. Material is easily mined by open-pit methods and processed at the Coalinga mill, which was designed to produce 12,000 tons of fiber annually.

The chrysotile deposits of Arizona (5, 7, 13, 25, 26, 30) are present at numerous places throughout an area about 60 miles long and 25 miles wide in the Salt River and Cherry Creek basins of Gila County. The nearest railway station is

Globe, and the center of milling is near there although some of the mines are about 40 miles away and material must be transported to the mills by truck. Operations are not large compared with other places, but in 1960 four companies were producing a total of 150 tons of finished fiber a week. The cost of mining is generally high, as veins pinch and swell irregularly and are very erratic in size and direction. The occurrence is different from that of California, Vermont, Quebec, and many other places. The deposits of Arizona occur in veins parallel to the bedding of the Mascal limestone, tentatively designated as Cambrian in age, near intrusions of diabase. The veins are generally confined to more or less horizontal zones several inches thick. The asbestos is a low-iron type.

Other countries

Mention should be made of the asbestos deposits of Bolivia (5, 7) and Australia (5, 31). Crocidolite has been mined on a small scale in Bolivia for many years, and this has been the only production of crocidolite in the Western Hemisphere. The deposits, about 200 km northeast of Cochabamba, province of Chapere, department of Cochabamba, are veins 0.5 to 30 cm wide that fill fractures and bedding planes in sandstone. Numerous deposits of asbestos are known throughout Australia, but approximately 86 percent of the production comes from Wittenoom Gorge (33). These deposits are in an arid and isolated area about 200 miles by road inland from Roebourne on the north coast of Western Australia. Earliest production was in 1937, large-scale operations were started in 1943, production was 12,400 short tons in 1957 and had increased to 14,400 short tons in 1960. Reserves are estimated to be at least 3 million tons. The deposits are seams of crocidolite in nearly horizontal beds of ferruginous quartzite.

SELECTED REFERENCES

1. C. C. Allen, J. C. Gill, and J. S. Koski, 1957, The Jeffrey Mine of Canadian Johns-Manville Company Limited, in "The Geology of Canadian Industrial Mineral Deposits," pp. 27–36, Canadian Institute of Mining and Metallurgy, Montreal.
2. G. O. Argall, Jr., 1960, Asbestos awakens Mother Lode at Jefferson Lake's new mine and mill, *Mining World*, vol. 24, no. 9, pp. 22–27.
3. W. J. Bichan, 1960, Commercial chrysotile deposits, Baie Verta, Notre Dame district, Newfoundland; preliminary notes, *Econ. Geology,* vol. 55, pp. 399–401.
4. O. Bowles, 1952, Asbestos, in W. Van Royan and O. Bowles, "The Mineral Resources of the World," vol. 2, pp. 166–169, Prentice-Hall, Inc., Englewood Cliffs, N.J., 181 pp.
5. O. Bowles, 1955, The asbestos industry, *U.S. Bur. Mines Bull.* 552, 122 pp.
6. O. Bowles, 1956, Asbestos, *U.S. Bur. Mines Bull.* 556, pp. 77–85.
7. O. Bowles, 1959, Asbestos, a materials survey, *U.S. Bur. Mines Inf. Circ.* 7880, 94 pp.
8. A. H. Chidester and A. F. Shride, 1962, Asbestos in the United States, *U.S. Geol. Survey Mineral Inv. Resource Map* MR-17. Map and pamphlet of 11 pp.

9. J. J. LeR. Cilliers and others, 1961, Crocidolite from the Koegas-Westerberg area, South Africa, *Econ. Geology*, vol. 56, pp. 1421–1437.

10. H. Gabrielse, 1960, The genesis of chrysotile asbestos in the Cassiar asbestos deposit, northern British Columbia, *Econ. Geology*, vol. 55, pp. 327–337.

11. South Africa Department of Mines, and Geological Survey, 1959, Asbestos, in "The Mineral Resources of the Union of South Africa," 4th ed., Pretoria, South Africa, pp. 357–367.

12. J. Huttl, 1962, Jefferson Lake—California's premier asbestos producer, *Eng. Mining Jour.*, vol. 163, no. 10, pp. 84–89.

13. D. W. Jaquays and A. W. Gerhardt, 1957, How low-iron chrysotile asbestos is mined and milled in central Arizona, *Mining World*, vol. 19, no. 7, pp. 54–58.

14. G. F. Jenkins, 1960, Asbestos, in "Industrial Minerals and Rocks," 3d ed., pp. 23–53, American Institute of Mining, Metallurgical, and Petroleum Engineers, New York, 934 pp.

15. S. J. Johnstone and M. G. Johnstone, 1961, "Minerals for the Chemical and Allied Industries," 2d ed., John Wiley & Sons, Inc., New York, 788 pp.

16. F. E. Keep, 1961, Amphibole asbestos in the Union of South Africa, Commonwealth Min. Metall. Cong., 7th, South Africa, *Trans.* vol. 1, pp. 91–121.

17. D. O. Kennedy, 1960, Asbestos, *U.S. Bur. Mines Bull.* 585, pp. 77–84.

18. P. C. Merritt, 1962, California asbestos goes to market, *Mining Eng.*, vol. 14, no. 9, pp. 57–60.

19. R. C. Munro and K. M. Reim, 1962, Coalinga—newcomer to the asbestos industry, *Mining Eng.*, vol. 14, no. 9, pp. 60–62.

20. D. V. Nalivkin, 1960, "The Geology of the U.S.S.R., a Short Outline," Pergamon Press, New York, 170 pp., geological map of the U.S.S.R.

21. P. H. Riordon, 1955, The genesis of asbestos in ultrabasic rocks, *Econ. Geology*, vol. 50, pp. 67–81.

22. P. H. Riordon, 1957, The asbestos belt of southeastern Quebec, in "The Geology of Canadian Industrial Mineral Deposits," pp. 3–8, Canadian Institute of Mining and Metallurgy, Montreal.

23. P. H. Riordon, 1957, The asbestos deposits of Thetford Mines, Quebec, in "The Geology of Canadian Industrial Mineral Deposits," pp. 9–17, Canadian Institute of Mining and Metallurgy, Montreal.

24. P. H. Riordon, 1962, Geology of the asbestos belt in southeastern Quebec, *Canadian Mining Metall. Bull.*, vol. 55, no. 601, pp. 311–313.

25. L. A. Steward, 1955, Chrysotile-asbestos of Arizona, *U.S. Bur. Mines Inf. Circ.* 7706, 124 pp.

26. L. A. Steward, 1956, Chrysotile-asbestos deposits of Arizona (supplement to Inf. Circ. 7706), *U.S. Bur. Mines Circ.* 7745, 41 pp.

27. F. J. Wiebelt and M. C. Smith, 1959, A reconnaissance of asbestos deposits in the serpentine belt of northern California, *U.S. Bur. Mines Inf. Circ.* 7860, 52 pp.

28. H. M. Woodrooffe, 1961, Asbestos chapter (9 pp.), in The Canadian mineral industry 1961, *Geol. Survey of Canada, Mineral Resources Div., Dept. Mines. Tech. Surveys.*

29. Anon., 1956, Canada moves a lake to mine asbestos, *Eng. Mining Jour.*, vol. 157, no. 1, pp. 91–93. See also 1959, Asbestos production underway at Black Lake, *Mining Eng.*, vol. 8, pp. 55–57.

30. Anon., 1960, Arizona asbestos industry is growing steadily around Globe, *Mining World, vol. 22*, no. 9, pp. 44–45.

31. Anon., 1961, More blue asbestos from Australia, *Mining Jour. (London)*, vol. 257, no. 6580, pp. 312–313.

32. Anon., 1962, *Mining Jour. (London)*, vol. 258, no. 6600, p. 173. Relates to new deposits of asbestos at the Pangani mine, Southern Rhodesia.

33. Anon., 1962, Mineral resources of Australia's Northern Territory, *Mining Jour. (London)*, vol. 258, no. 6617, pp. 620–621.

34. Anon., 1963, The Cape's blue asbestos, *Mining Jour. (London)*, vol. 260, no. 6663, p. 417.

35. Anon., 1963, Advocate to treat 5,000 tpd of asbestos ore, *Eng. Mining Jour.*, vol. 164, no. 10, p. 114. Relates to operation at Baie Verte, Newfoundland.

36. Anon., 1964, Development of Pangani asbestos mine is underway, *Eng. Mining Jour.*, vol. 165, no. 4, pp. 174, 176.

chapter 20

Graphite

Graphite is used for a variety of purposes, among the most important of which are (1) in foundry facings, where graphite provides a smooth surface and permits castings to be removed easily; (2) in steelmaking; (3) in the manufacture of lubricants; (4) in the manufacture of dry cells and batteries; (5) in the manufacture of crucibles for melting nonferrous metals, as in the aluminum and brass industries; (6) in the manufacture of lead pencils. The largest tonnage of graphite is used for foundry facings; smaller tonnages, in the order of quantity, are used in making lubricants, steel, crucibles, pencils, and batteries. The tonnages used, however, do not necessarily reflect the importance of material in the graphite industry, because only certain types of graphite can be used for some manufactured materials.

Commercial graphite is classed as (1) amorphous (actually cryptocrystalline material), (2) flake or crystalline flake, (3) lump, and (4) chip. Industrial usage of these terms is a bit confusing, since some lump graphite is further subdivided into amorphous and crystalline and again divided into grades based on size, carbon content, and degree of consolidation. Some information regarding size types of graphite follows:

Size type	Approximate size	Approximate percent carbon
Lump	Walnut to pea	90 or more
Chip	Pea to slightly less than grain of wheat	90–92
Dust	About 40-mesh to 60-mesh screen	55–90

Details of the use of different types of graphite cannot be given here, but in general much amorphous graphite is used in making foundry facings and in the steel industry, where there is no need for a high carbon content of the graphite. Graphite of all types is used extensively in making lubricants if it can be freed from any grit that may be present, but flake graphite is needed for certain lubricants. Flake graphite is needed also in making crucibles and related refractory materials. Lump and chip graphite are used for making crucibles, carbon brushes, and other products that require graphite of high purity (4; 10; 11, pp. 225–234; 15).

The various commercial types of graphite occur in deposits of different kinds. Lump and chip graphite are obtained from veins in regions of igneous and metamorphic rocks; flake graphite is obtained from metamorphic rocks, particularly schists but also marbles; amorphous graphite is obtained from coal seams that have undergone contact metamorphism, and also from various other sources, since the finest sizes of flake and lump graphite are classed as amorphous graphite in commercial operations.

Production

Graphite is widely distributed throughout the world and hence is produced in many countries. Important commercial production, however, comes from a relatively few places (Table 20-1).

TABLE 20-1. World Production of Graphite, Average of 1945–1959 and in 1962*

	Percent of world total	
Country	*Average of 1945–1959*	*1962 only*
Republic of Korea	—	35.8
Korea	22.0	—
North Korea	—	9.6
Austria	8.0	17.2
Soviet Union	?	10.5
China	?	7.9
Mexico	12.4	5.6
Malagasy Republic (Madagascar)	5.7	2.9
West Germany (Germany)	4.0	2.4
Ceylon	4.4	1.7
All others	43.5	6.4

* Compiled from *U.S. Bur. Mines Minerals Yearbook,* various years.

Among the outstanding producers in 1962—Korea, Austria, and the Soviet Union—only Austria was an important early producer, furnishing 41.2 percent of the world total in 1900, 37.9 percent in 1905, 34.8 percent in 1910, and 28.7 percent in 1915, after which production declined considerably but was high again in 1960, 20.9 percent of the world total. Korean production had not started in 1910, was slightly less than 5 percent of the world total in 1915, and thereafter rose rapidly until the Korean conflict, but after a decline it again increased. Information regarding Russian production is not available for many years. Production reported in 1930 was 3.9 percent of the world total and in 1935 it was 7.7 percent. Among countries that were not important in 1962, Italy furnished about 12 to 13 percent of the world total around 1900 to 1910, then declined to only

0.9 percent in 1960; Czechoslovakia supplied nearly 20 percent of the world total in 1920 and considerable amounts thereafter through 1945, since which time accurate information has not been available. Most other producing countries furnished but small amounts, those that generally contributed from 1 to 4 percent of the world total throughout most of the period between 1900 and 1960 being Canada, the United States, and India.

Europe

The most important deposits of Europe are in Austria, West Germany on the Austrian border, Czechoslovakia, Italy, and the Soviet Union (3, 15).

The deposits of Austria are of two types: (1) those that are associated with mica gneiss and marble and (2) those associated with schists and quartzites. Deposits of the first type are between 39 and 55 miles west of Vienna in the Muehldorf area and consist of a layer of amorphous to finely crystalline graphite 2 to 40 m thick in marble and gneiss. Deposits of the second type occur in folded metasediments in the Styrian Alps throughout a zone 30 miles long that extends from Leoben to Rottenmann. The principal mine is at Kaisersberg, where the aggregate thickness of a series of graphitic beds is 40 ft. The crude ore averages 50 to 60 percent amorphous graphite.

The deposits of West Germany are east of Passau, Bavaria, near the junction of Germany, Austria, and Czechoslovakia (3, p. 457; 21). These are among the oldest known deposits of Europe and they have been worked since at least the fifteenth century, although the first recorded production was in 1868. Seams and lenses of crystalline graphite occur in micaceous gneiss and crystalline limestone. Graphite is also disseminated in gneisses and schists and in limestone, but the seams and lenses furnish the important production. Mining is done only at Kropfmuehl, about 10 miles east of Passau, where crystalline graphite constitutes 10 to 30 percent of the fresh minable lenses, which are 1 to 5 ft thick and as much as several hundred feet long.

The graphite deposits of Czechoslovakia, at one time worked to a considerable extent, have lately been inactive so far as is known. The deposits are lenses of graphite schist interlayered with other metasedimentary rocks. Some lenses reach a thickness of 65 ft, whereas the thickness of others is but a few feet.

The graphite of Italy consists of finely crystalline material in layers or lenses in phlyllite, schist, and gneiss; locally the seams are around masses of anthracite. The principal deposits are near Turin (Pinerolo).

Information about the deposits of the Soviet Union is meager. Nalivkin (16, pp. 36, 58, 149) stated in 1960 that graphite occurs in nepheline syenite at Botogolsk (53°N, 97°E) and with metamorphosed coal seams in the valley of the Kureika River (66°30'N, 87°E), both in Siberia. Also, that large reserves of graphite associated with coal beds are present in the Boevskoe deposit near Kamensk (56°30'N, 62°E) on the eastern flanks of the middle Ural Mountains, and that a large deposit of scaly graphite in graphitic schists and gneisses is present in the Lesser Khingan Range of the Far East. Pinkow (18) stated in 1944 that more than 90 percent of the reserves of massive graphite of Russia are east of the Yenessei River near Turuchansk (Turukhansk) (66°N, 88°E). Other deposits have been reported from various places (3, p. 460; 15, p. 331).

Asia

The largest deposits in Asia are those of Korea, but the deposits of Ceylon are highly important because they are suitable for uses for which other deposits are not satisfactory.

Graphite is widely distributed throughout Korea (17), but production has come to a large extent from southern Korea, south of latitude 38°N, especially from the provinces of Ch'ungch'ong-pukto and Kyonsang-pukto. Nearly 90 percent of the material is amorphous graphite, large reserves of which are present. Most of the amorphous graphite occurs as irregular lenses in schists and phyllite originally of sedimentary origin, but some impure amorphous graphite is associated with coal beds. Deposits of crystalline graphite that are worked consist of flakes generally disseminated in granitic rock that cuts calcareous schists. These deposits generally contain 3 to 4 percent graphite.

Graphite deposits in Ceylon (1; 3, pp. 459–460; 14, p. 252; 15, p. 331) have been known from very early times but were not exploited until between 1820 and 1830, and commercial production was not important until 1834. The deposits are located in the mountainous southwestern part of the island, and the principal deposits are in Southern, Western, and Sabaragamuwa Provinces. The rocks of the region are gneisses of various types, with which are interlayered crystalline limestone, and all of these are cut by igneous rocks, chiefly granites and granite pegmatites. Noncommercial graphite is disseminated throughout various rocks, but the commercial graphite occurs in veins that cut the gneisses and limestone where fractures were formed. The width of the veins varies from less than ⅛ in. to several feet, but veins may branch and form stockworks. Usually the veins are composed almost entirely of graphite but contain some other minerals in subordinate amounts, especially pyrite and quartz. The graphite in the smaller veins occurs in aggregates of platy needles and in the larger veins mostly as coarse plates. The needle lump graphite and the platy lump graphite are highly prized for making crucibles and other refractory products. Reserves are thought to be very large.

Bastin (1) considered the deposits to be true fissure veins related in origin to the invasion of the granitic magma. Clark (5) favored formation by interaction of magmatic gases and calcareous sediments.

Little specific information is available about the graphite deposits of China. Juan (12) stated that graphite deposits occur in Suiyuan (central Inner Mongolia), Honan, Kiangsu, and Hunan (Fig. 9-5), but that they do not appear to be large. Wang (20) stated that the graphite deposits of China are important. He stated that the graphite is present chiefly in Manchuria, Suiyuan, and Shansi, and that the leading producing areas include Liu-mao (or Chining) and P'an-shih (Kirin Province) in Manchuria; Hsing-ho and Wuyuan in Suiyuan; and Tatung (Tat'ung) in Shansi. Wei (22) also stated that the graphite deposits of China are important, that graphite is produced chiefly in the northeastern part of the country, and that a large part of that mined in 1961 came from the Liu-mao deposit in Heilung-kiang, Manchuria. At least some of the deposits of Manchuria are in gneisses and schists.

Africa

Graphite deposits are known in Kenya and Tanganyika, but the important deposits of Africa are those in the Malagasy Republic (Madagascar).

The deposits of the Malagasy Republic (2, 3, 13) have been worked commercially since 1907. Although deposits are known throughout almost the eastern half of the island, the chief producing area extends along the east-central coast from Tamatave to Marovintsy, a distince of about 90 miles, and an area of subordinate importance is inland in the uplands, extending southward from Tananarive. The deposits occur in schists and gneisses of Precambrian age and make layers that vary in thickness from 10 to more than 100 ft, some of which are thousands of feet long. The average size of the flakes of graphite is perhaps 1 to 3 mm, but flakes larger than 1 cm across have been reported and the deposits are noted for their high proportion of coarse flakes. The world standard for high-quality flake graphite is set by these deposits. The average grade of deposits worked recently is 4 to 11 percent graphite.

An interesting feature of these deposits, and one of practical importance as well, is the deep lateritic weathering to which they have been subjected. In general most of the minerals except graphite and quartz have been converted into clay, and weathering is so deep that operations do not enter unweathered material. Water for concentrating the deposits is abundant along the coast, and because of that, as well as because of greater accessibility, deposits of somewhat lower grade can be mined there than can be mined in the uplands.

Mexico

Graphite is known from several places in Mexico, but the deposits of major importance are in the state of Sonora, district of Guayamas, in the Sierra de Moradillas between 28°30′ and 28°40′N latitude. The deposits are about 50 miles southeast of Hermosillo and about 14 miles east of Escalante. Gonzáles Reyna (8) gives information about 10 mines in the district, of which Moradillas is of first importance and San Francisco ranks second. More than half of the production comes from the Moradillas mine (3, p. 458). High-grade deposits contain an average of 85 percent carbon, and some contain as much as 95 percent.

The deposits have been described as beds of amorphous graphite intercalated between layers of sandy and quartzitic black shale, the thickness of the graphite layers varying from 30 cm to 2.7 m. Gonzáles Reyna (8) classed the deposits as metamorphic ones formed at high temperature by the intrusion of diorite, and others (3, 15) have classed them as metamorphosed coal beds. Recent detailed work at the Moradillas mine indicates that they are hydrothermal veins, which occupy fissures that cross the bedding at low angles (3).

Canada

Canada has produced practically no graphite since 1954, when the Black Donald mine of Ontario was closed after producing almost continuously since 1896 (9). At the Black Donald mine, in Renfrew County, the graphite occurs in a bed of marble within a sequence of marble, quartzite, and gneisses of Precambrian (Grenville) age. The average width of the deposit was 20 ft, the maximum width about 70 ft. Underground mining was done along a length of about 1,000 ft.

A 1962 report (23) indicates that a large deposit of graphite has been discovered in the Rideau Lake district of Ontario, northwest of Portland. Two veins have been located, but only one had been developed in 1962 when about 5,000 tons of ore had been mined. The vein that had been developed is about 200 ft wide and 2,000 ft long. The deposit is easily accessible and is stated to be ideal for open-pit mining, with the average thickness of overburden 1 to 4 ft, but overburden is almost entirely absent in many places.

United States

Large reserves of flake graphite in the United States are present in in central Texas, northeastern Alabama, eastern Pennsylvania, and the eastern Adirondack Mountains of New York, and deposits of less extent occur in at least 21 other states (3, 4, 15). Production began between 1644 and 1648 in Massachusetts and expanded into other states, so that by 1900 a domestic graphite industry was established. Peak production was attained in the period 1916 to 1920, under wartime conditions, when the United States supplied 51,603 short tons of graphite, which was 5.8 percent of the world production for the five-year period. Since then production has declined markedly—2.9 percent of the world total in 1950. A small amount of flake graphite was mined in Rhode Island, Alabama, and Texas in 1953, but production had stopped entirely by the end of 1954, although some amorphous graphite was then mined in Rhode Island. Since then production of flake graphite has been resumed in Texas. During recent years statistics have not been released, since the small amount of production has been almost wholly from one area in Texas.

Graphite deposits of the United States occur as (1) graphite disseminated in metamorphosed siliceous sediments, (2) graphite disseminated in marble, (3) graphite formed by metamorphism of coal beds or other highly carbonaceous sedimentary rocks, (4) vein deposits, and (5) contact-metasomatic deposits in marble. Most of the production and reserves of flake graphite are of type 1, and such deposits are of particular importance in Texas, Alabama, Pennsylvania, and New York. Deposits of type 2 occur in rocks of the Precambrian Grenville series of New York but generally appear to be of little economic importance. Deposits of type 3, metamorphosed coal beds, occur in the Raton coal field of New Mexico and in the Narragansett basin of Rhode Island, both of which have furnished some amorphous graphite. Vein deposits, type 4, occur chiefly in the Ticonderoga district of New York and near Dillon, Montana. Contact-metasomatic deposits in marble, type 5, occur west and northwest of Ticonderoga, New York, near the eastern edge of the Adirondack Mountains.

The graphite deposits of Texas occur in Llano and Burnet Counties as graphitic zones in the Precambrian Packsaddle schist, which has been intruded by granite. The graphite varies from exceedingly fine-grained to medium-grained, commercial flake coming only from the medium-grained material. The entire thickness of rocks in which graphite occurs is 130 to 160 ft.

The graphite deposits of Alabama occur throughout a length of about 55 miles in Chilton, Coosa, and Clay Counties, in two zones that are generally parallel, are 1 to 2 miles wide, and trend northeast. The southernmost zone, which extends from Verbena in Chilton County nearly to Goodwater in Coosa County, is the larger one, but the northern zone, which lies in Clay County, contains the majority of the deposits that have been mined. Graphitic rock of the northern zone occurs as lenses in the Ashland mica schist of Precambrian age. These lenses, locally designated *leads,* contain 1 to 5 percent of disseminated flake graphite. Most of the leads mined contain several layers of graphitic rock 1 in. to 10 ft thick, the aggregate thickness of which may be as much as 100 ft. Some leads have been mined for more than 500 ft along the strike, and some have been traced for about 4,000 ft. The deposits of the southern zone appear to be similar to those of the northern zone, but are not well known.

The Piedmont Plateau area in the southeastern part of Pennsylvania contains the graphite deposits of that state. The deposits occur as graphitic layers in the Pickering gneiss of Precambrian age and are largest and best known in the valley of Pickering Creek west of Phoenixville, Chester County. Some of the graphitic layers or zones are known to be present throughout a length of 8 miles, and in places graphite makes up as much as 10 percent of the rock. The rocks have been cut by pegmatites, some of which contained enough graphite to be mined. The graphite flakes contained in the pegmatites are generally larger than those in the gneiss, some reaching a diameter of 25 mm.

The eastern Adirondack Mountain area of New York, in Essex, Warren, Saratoga, and Washington Counties, contains the principal deposits of that state, although graphite occurs also in Rockland and St. Lawrence Counties. The graphite deposits occur in rocks of the Precambrian Grenville series, which consists of schists, gneisses, quartzites, amphibolites, and marbles, all cut by igneous intrusions.

The most important deposits consist of disseminations in schist composed of graphite and quartz along with some mica and feldspar. The content of graphite is 5 to 7 percent in some of the deposits, in flakes that generally are not more than 3 mm in diameter. The graphitic layers, which are less than 1 ft to more than 12 ft thick, persist for hundreds to thousands of feet along the strike. The best example of a mine in deposits of this type probably is the mine of the Dixon-American Graphite Co., 4.5 miles west of Lake George in Warren County, which was operated continuously for more than 30 years and probably was the largest and most productive mine in New York.

Two other types of graphite deposits occur in New York, vein deposits along Lake Champlain between Ticonderoga and Split Rock Point, and contact-metasomatic deposits in marble, chiefly in southern Essex County.

Vein deposits of considerable interest occur at the Crystal Graphite mine, Beaverhead County, Montana, 13 miles southeast of Dillon (4, 7), at the southern end of the Ruby Range. Production began in 1902 and mining was conducted intermittently until at least 1948, with the greatest production in 1918 to 1919 during World War I. The mine furnished lump graphite similar to that of Ceylon, but never in large amounts. The richest deposits are veins that occur in fractures in gneiss and pegmatite, vary in thickness from less than ⅛ in. to more than 2 ft but chiefly between ¼ in. and 4 in. Few are over 50 ft long, but in places they are numerous and intersecting and thus resemble stockworks. The graphite of most of the veins is in coarse interlocking plates, some as much as 1 in. long, but some of it shows comb or needle structure. The deposits were worked underground on four levels. Graphite in the area also occurs disseminated in pegmatites and gneisses, but apparently not in commercial amounts.

SELECTED REFERENCES

1. E. S. Bastin, 1912, Graphite deposits of Ceylon, *Econ. Geology,* vol. 7, pp. 419–443.
2. H. Beasirie, 1961, Les Ressources minérales de Madagascar, *Annales géologiques de Madagascar,* pt. 30, 116 pp.
3. E. N. Cameron, 1960, Graphite, in "Industrial Minerals and Rocks," 3d ed., pp. 455–469, American Institute of Mining, Metallurgical, and Petroleum Engineers, New York, 934 pp.
4. E. N. Cameron and P. L. Weis, 1960, Strategic graphite: a survey, *U.S. Geol. Survey Bull.* 1082-E, pp. 201–321.
5. T. H. Clark, 1921, The origin of graphite, *Econ. Geology,* vol. 16, pp. 167–183.
6. J. B. DeMille, 1947, "Strategic Minerals," McGraw-Hill Book Company, New York, 626 pp.
7. R. B. Ford, 1954, Occurrence and origin of the graphite deposits near Dillon, Montana, *Econ. Geology,* vol. 49, pp. 31–43.
8. J. González Reyna, 1956, "Riqueza Minera y Yacimientos Minerales de México," 3d ed., 497 pp., Internat. Geol. Cong., 20th, Mexico 1956.
9. D. F. Hewitt, 1957, Graphite, in "The Geology of Canadian Industrial Mineral Deposits," pp. 104–108, Canadian Institute of Mining and Metallurgy, Montreal.
10. D. R. Irving, 1960, Graphite, *U.S. Bur. Mines Bull.* 585, pp. 357–366.
11. S. J. Johnstone and M. G. Johnstone, 1961, "Minerals for the Chemical and Allied Industries," 2d ed., John Wiley & Sons, Inc., New York, 788 pp.
12. V. C. Juan, 1946, Mineral resources of China, *Econ. Geology,* vol. 41, pt. 2, pp. 399–474.
13. A. Lacroix, 1922, "Minéralogie de Madagascar," vol. 2, pp. 148–155, A. Challamel, Paris, 3 vols.
14. R. B. Ladoo and W. M. Myers, 1951, "Nonmetallic Minerals," 2d ed., McGraw-Hill Book Company, New York, 605 pp.
15. F. D. Lamb and D. R. Irving, 1956, Graphite, *U.S. Bur. Mines Bull.* 556, pp. 327–337.
16. D. V. Nalivkin, 1960, "The Geology of the U.S.S.R., a Short Outline," Pergamon Press, New York, 170 pp., geologic map of U.S.S.R.
17. W. C. Overstreet, 1947, Graphite deposits in southern Korea [abs.], *Econ. Geology,* vol. 42, pp. 424–425.

18. Hans-Heinz Pinkow, 1944, Die russichen Graphitvorkommen und ihre Bedeutung für die Graphitindustrie der Sowjetunion (the Russian graphite occurrences and their importance for the graphite industry of the Soviet Union), *Zeitsch. prakt. Geologie.*, vol. 52, no. 8, pp. 73–82.

19. D. G. Runner, 1952, Graphite, in W. Van Royan and O. Bowles, "The Mineral Resources of the World," vol. 2, pp. 174–177, Prentice-Hall, Inc., Englewood Cliffs, N.J., 181 pp.

20. K. P. Wang, 1960, Mineral wealth and industrial power, *Mining Eng.*, vol. 12, pp. 901–912.

21. R. G. Wayland, 1951, The graphite of the Passau area, Bavaria, *Am. Inst. Mining Metall. Engineers Trans.*, vol. 190, pp. 166–172.

22. A. W. T. Wei, 1962, Minerals in China in 1961—III, *Mining Jour (London)*, vol. 258, no. 6610, pp. 409–411.

23. Anon, 1962, Ontario firm uncovers large graphite deposit, *Eng. Mining. Jour.*, vol. 163, no. 6, pp. 300, 302.

chapter 21

Barite and Witherite, Fluorite, Magnesite, and Brucite

Barite and fluorite occur together or separately in many hydrothermal veins or other cavity-filled deposits and in replacements, generally formed at low temperature, in which chalcopyrite, galena, or sphalerite may be present either in very small amount as accessory minerals or in sufficient abundance to warrant mining for their recovery. Barite also is present as commercial deposits interbedded with sedimentary rocks and as important residual deposits derived either from veins and similar deposits or from bedded deposits. Further, some of the bedded deposits appear to have been formed by hydrothermal replacement. Commercial deposits of witherite, $BaCO_3$, are confined almost entirely to lead deposits of England, which supply barite and fluorite also. Deposits of magnesite resemble the bedded deposits of barite to some extent, as they appear to have been formed either as normal sedimentary beds or as replacements of sedimentary beds. The magnesite deposits, however, generally are associated with dolomites whereas the barite deposits are associated with shales, limestones, and cherts. Further, some magnesite deposits were formed from serpentinites. Brucite, $Mg(OH)_2$, occurs with some deposits of magnesite or alone but is relatively uncommon and is not an important mineral except in a few places.

Barite and fluorite are common minerals that are widely distributed and that make commercial deposits in a number of countries, yet the major production during the 15-year period 1945 to 1959 came from the United States, West Germany, and Mexico, which collectively furnished 59.8 percent of the barite of the world and 46.3 percent of the fluorite. Three other countries, Canada, the Soviet Union, and the United Kingdom, together supplied 17.3 percent of the barite and 18.8 percent of the fluorite, so that six countries produced 77.1 percent of the barite and 65.1 percent of the fluorite of the world during that period.

Commercial deposits of magnesite show a somewhat different distribution throughout the world, emphasizing differences in the formation of those deposits. Thus the outstanding deposits of magnesite are in the Soviet Union, Austria, and

China—67.3 percent of the world production in 1962—and important but smaller deposits are in Czechoslovakia, the United States, and Yugoslavia—18.0 percent of the world total in 1962—with 85.3 percent of the world production supplied by these six countries.

Differences in distribution and occurrence of the commercial deposits of barite, fluorite, and magnesite are distinctive enough to warrant separate discussion of those three materials. Only the outstanding occurrences can be considered briefly.

Barite and witherite

The major world production of barite during the period 1945 to 1959 and for the year 1962 is shown by Table 21-1.

TABLE 21-1. World Production of Barite, Average of 1945–1959 and in 1962*

	Percent of total world production	
Country	Average of 1945–1959	1962 only
United States	41.3	26.8
West Germany	14.5	14.2
Canada	7.9	6.9
Soviet Union	4.9	6.0
United Kingdom	4.5	2.5
Mexico	4.1	10.6
Italy	3.6	4.0
Yugoslavia	2.9	4.3
Peru	1.3	3.1
Morocco	0.5	2.9
China	?	2.6
All others	14.5	16.1
	Short tons	
	Average of 1945–1959	1962 only
World total	2,153,333	3,310,000

* Compiled from *U.S. Bur. Mines Minerals Yearbook,* various years.

United States

The large production of barite in the United States is stimulated by the extensive use of barite as drilling mud in the petroleum industry. More than 90 percent of the consumption in the United States is for that purpose (5, 35, 55). The glass, paint, rubber, and chemical industries use much of the barite not consumed by the oil industry in this country, but in some countries most of the consumption of barite is by those other industries.

The major production of barite in the United States has come from Arkansas, Missouri, Nevada, and Georgia, although some production has come from 10 or more other states. Missouri was the leading producer in 1962, with slightly more than 35 percent of the total, followed by Arkansas, about 30 percent. These two states have furnished much of the total during recent years, with Arkansas generally in the lead—40 percent during the 10 years 1950 to 1959 compared with 31 percent for Missouri.

Deposits in the United States (5, 6, 35) are of three types: (1) vein and cavity filling, (2) bedded, and (3) residual. The vein and cavity filling deposits occur chiefly in fault zones and breccia zones. Common associated minerals are dolomite, quartz, pyrite, chalcopyrite, galena, and sphalerite. Generally such deposits in the western United States are considered to be of low-temperature (epithermal) origin, whereas those in the midwestern and eastern United States may be of very low-temperature (telethermal) origin, although their relation to igneous activity may be obscure. The bedded deposits generally occur in shale and limestone sequences of late Paleozoic age, in which the barite is fine grained, not uncommonly gray and fetid, and may be accompanied by chert, pyrite, secondary iron oxides, and various carbonates and clay minerals. Individual beds vary in thickness from a few inches to about 10 ft and are essentially barite-rich zones within the sedimentary sequence. The total thickness of such a barite-rich sequence may be as much as 200 ft. The residual deposits have accumulated to a considerable extent from previous sedimentary rocks of Cambrian to Ordovician age in which the barite may have been contained in veins and breccia zones, in replacement deposits, or in bedded deposits, although the origin of the barite in the original beds is not always clear. The barite of the residual deposits commonly occurs in clays derived from weathering of the original sedimentary rocks and may be in particles considerably less than an inch across up to irregular masses that weigh hundreds of pounds, although most material is 1 in. to 6 in. across.

Detailed description of the important deposits cannot be undertaken here, but the major features of some of them are summarized (5).

Major Features of Some Important Deposits of Barite in the United States

Arkansas-Missouri Area

Magnet Cove, Hot Springs County, Arkansas.   Bedded deposits intercalated in Stanley shale of Mississippian and Pennsylvanian age. Deposit is at least 0.75 mile long, about 60 ft thick, one of largest in the United States. May be of replacement origin with source of solutions the magma that formed the Magnet Cove intrusive.

Southeastern District, Missouri, chiefly Washington County.   Barite occurs chiefly in residual clay derived from underlying Potosi and Eminence dolomites of Late Cambrian age. Many deposits as large as 100 acres and 4 to 30 ft thick. Some deposits contain as much as 20 to 25 percent barite. One of most important districts in the United States.

Central District, Missouri, chiefly in Cole, Miller, Montineau, and Morgan Counties.   Residual and vein (circle) deposits. Conical or bell-shaped deposits, circular in ground plan, are designated *circle deposits.* Barite and accessory galena, sphalerite, chalcopyrite, calcite, fill cavities in collapse breccia in Gasconade, Roubidoux and Jefferson City formations of Ordovician age. Deposits generally small but have been persistent producers and have supplied material for some valuable residual deposits. Galena and sphalerite are recovered from some deposits.

Georgia-Tennessee Area

Cartersville, Bartow County, Georgia. Productive part of district is about 4.5 miles long, 2 miles wide. Residual deposits in clays associated with Weisner, Shady, and Rome formations of Cambrian age. Barite fragments up to 4 ft across but average diameter probably less than 6 in. Some of the richest deposits, now chiefly mined out, yielded 1 long ton of barite for 2.9 to 4.6 cu yd of material.

Sweetwater district, Tennessee, chiefly in McMinn, Monroe, and Loudon Counties. Residual deposits in clay overlying rocks of the Knox group of Cambrian and Ordovician age. Deposits occur in three narrow belts, the easternmost of which is the most productive. Some masses of barite weighed as much as 1 ton. Original material from which the residual deposits were derived consists of veins containing barite, fluorite, and pyrite in shattered zones of Knox group.

Nevada-California Area

Battle Mountain–Carlin area, Nevada, in Lander, Eureka, and Elko Counties. The "Barite belt" of Nevada extends northeastward through the central part of the state, but most mining has been done in the northern part of the belt. The productive areas are as long as $\frac{1}{2}$ mile, as wide as $\frac{1}{4}$ to a $\frac{1}{2}$ mile. Barite generally interlayered with chert and minor amounts of limestone in Paleozoic bedded chert units. Baritized microfossils similar to those in the chert are rather common in the barite. Some beds contain up to 97 percent barite. Deposits classed either as bedded or replacement type. Some people (23, 28) suggest replacements unrelated to metallic sulfide deposits in the area. Numerous veins of barite present throughout the belt.

Mountain Pass, San Bernardino County, California. Barite is associated with carbonates and the rare-earth mineral bastnaesite (about 20 percent barite, 10 percent bastnaesite) in veins around a large shonkinite body. Reported to be one of largest single concentrations of barite in the world. Highly important as a potential by-product of rare-earth mineral recovery.

Canada

Barite occurs at various places throughout Canada but generally is mined only in Nova Scotia and British Columbia. Normally about 90 percent of the production comes from the Walton deposits of Nova Scotia, operated by the Magnet Cove Barium Corporation (3, 25, 34).

The Walton deposit is 3 miles southwest of Walton and 1.5 miles inland from the Bay of Fundy. It consists of a large, irregular mass of barite overlying limestones, shales, and quartzites of the Horton and Windsor series of Mississippian age. This mass lies against a prominent vertical fault. The deposit as originally discovered contained considerable amounts of ferruginous and calcareous material, but mining and exploration have shown material of much better grade, 98 to 99 percent barium sulfate throughout sections 50 ft thick. From 1941 through 1960 the production from this deposit was 2,710,422 tons. The reserves are not published but were estimated to be 2,500,000 tons.

Operations were conducted at four deposits in British Columbia during 1961. Two of these were on vein deposits, one consisted of recovery from mine tailings, and one was recovery of by-product barite during the mining of lead-zinc ores.

Europe

The most important barite deposit in Europe is the one at Meggen, Westphalia, now a part of West Germany, which has been mined since 1845. Prior to World War II Germany produced half or more of the barite of the world from this deposit and from veins in Hessen, Thuringia, Bavaria, Baden, and Silesia, the latter five areas yielding only about a fourth or a third of the total German production (6, p. 58; 29, p. 69). The Meggen deposit, about 60 miles east of Cologne, consists of pyrite and barite replacements of folded limestone of Devonian age; some galena, sphalerite, and other sulfides are present in subordinate amounts. The deposit is about 2.5 miles long, 6 to 21 ft thick, and is known to extend to a vertical depth of at least 6,000 ft. The central part of the deposit consists of pyrite, whereas the ends, each about half a mile long, consist of barite and are separated from the major pyrite body by transition material that contains considerable amounts of both minerals (58).

Barite and witherite, and also fluorite, have been produced from lead deposits of England that have been mined ever since the Roman occupation (14, 48). The deposits occur in the general vicinity of the Pennine Chain in central and northern England. They consist of veins and replacements in limestone of Carboniferous age. Although galena and some sphalerite were the major minerals mined during early development, recent attention has been centered on barite, witherite, and fluorite. Most of the world's production of witherite comes from Northumberland County.

Information about the barite deposits of the Soviet Union is scanty. Nalivkin (31, pp. 59, 97, 122) stated that the largest deposit is a vein of barite 4 to 8.5 m thick in shales and limestones of Upper Proterozoic age at Medvedevskoe near Zlatoust in the Ural Mountains (55°N, 59°30′E). Other veins of considerable importance are north of Kutaisi (42°N, 43°E) in the Caucasus. The veins, in tuff and apparently associated with Jurassic igneous activity, average 0.1 m to 2 m wide but are as much as 4 to 6 m wide in some places. Numerous small deposits of barite and witherite, associated with rocks of Cretaceous age, are present at Kopet-Dagh (39°N, 56°E to 37°N, 60°E) in Central Asia.

Fluorite

Fluorspar[1] is produced in considerable amounts in many countries that produce barite, but the relative importance of the production is not the same (Table

---

[1] The term *fluorite* is used for the mineral, and the term *fluorspar* is used for an aggregate or mass containing enough fluorite to meet commercial qualifications.

21-2). Further, although both fluorite and barite are produced from some deposits, generally the major deposits of fluorspar occur in different areas from the deposits of barite.

TABLE 21-2.   World Production of Fluorspar,
Average of 1945–1959 and in 1962*

	Percent of total world production	
Country	Average of 1945–1959	1962 only
United States	23.6	8.5
Mexico	13.4	23.0
West Germany	9.3	4.5
United Kingdom	6.4	3.2
Canada	6.2	3.1
Soviet Union	6.2	9.5
Italy	6.1	7.1
France	5.1	9.1
East Germany	4.9	3.3
Spain	4.7	6.5
China	?	9.1
Republic of South Africa	—	4.6
All others	14.1	8.5
	Short tons	
	Average of 1945–1959	1962 only
World total	1,264,000	2,405,000

* Compiled from *U.S. Bur. Mines Minerals Yearbook,* various years.

United States

Illinois-Kentucky field.   The most important commercial fluorspar deposits of the United States, by far, are those of the Illinois-Kentucky field. The production there in 1962 was slightly more than 80 percent of all fluorite produced in the United States, about four-fifths of which came from Illinois.

The Illinois-Kentucky field (15, 20, 46, 52, 56) occupies about 700 sq miles in southeastern Illinois and northwestern Kentucky, in which the chief mineralized areas lie in Hardin County, Illinois, and in Crittenden, Livingston, and Caldwell Counties, Kentucky. The rocks throughout the area consist chiefly of limestones, sandstones, and shales of late Paleozoic age that occupy two anticlinal uplifts, one in Illinois, the other in Kentucky. The rocks have been broken by many faults and locally intruded by dikes, sills, and irregularly shaped masses of lamprophyre and mica peridotite. Generally the sedimentary rocks are nearly horizontal except close to faults.

The deposits have been classed as (1) vein deposits, (2) bedding-replacement deposits, and (3) residual or gravel-spar deposits. The main production in Illinois has been from vein deposits of the Rosiclare district and from bedding-replacement deposits of the Cave in Rock area, both close to the Ohio River. Most of the production in Kentucky has come from vein deposits, although some has come from the bedding-replacement type associated with the veins. Locally some production has come from residual deposits above noncommercial veins.

Most of the larger veins are along faults that trend northeast to east, but some are in cross faults. The most abundant mineral of the veins is calcite, which may be almost the sole mineral or may be associated with fluorite, sphalerite, galena, quartz, barite, or a number of other minerals in minor amounts. Fluorite generally occurs in somewhat lenticular bodies in the calcite and shows much diversity in abundance, from very little to almost the entire vein. Sphalerite and galena commonly are definitely in subordinate amounts, but may be in sufficient quantities to be mined for those minerals alone or to make important by-products of fluorspar mining. Veins are 1 in. to as much as 25 ft thick. Some ore shoots have been 1,000 ft long and were mined to a depth of 500 ft, but the average length and minable depth are much less.

The bedding-replacement deposits of the Cave in Rock area generally are more or less parallel to the beds of limestone in which they occur, but are lenticular and are irregular in detail. Typical deposits are 200 to more than 2,500 ft long, 50 to 300 ft wide, and 4 to 15 ft thick. The deposits are concentrated in three stratigraphic positions within limestone formations, at each of which the limestone is overlain by a sandstone formation or member. The mineral composition of the bedding-replacement deposits is similar to that of the veins, but generally calcite is less abundant.

Some investigators have thought that the veins were formed chiefly by the filling of cavities, others have thought that replacement was of considerable importance, as it was in the bedding deposits. Generally the fluorite and associated minerals are thought to have been deposited by hydrothermal solutions, possibly at low temperature—epithermal to telethermal type—but perhaps at moderate temperature—mesothermal type. The genesis of the Cave in Rock deposits was discussed in some detail in 1962 (4).

Western states.   Fluorite is rather widespread throughout the western United States, both as a constituent of metal-bearing deposits and as separate deposits of sufficient size to justify mining. Production in 1962 was chiefly from Colorado, Montana, Nevada, and Utah. The Northgate district of Colorado (43) on the western flank of the Medicine Bow Mountains, in Jackson County near the Wyoming border, is rated among the largest fluorspar deposits in the western United States. Fluorite occurs as encrustations along faults on Pinkham Mountain and as finely granular aggregates that cement and replace fault gouge and tuffaceous silts and clays. The Crystal Mountain deposit, Montana (36, 44), has been the important producer of that state. The deposit, east of Darby in Ravalli County, consists of lenticular bodies of fluorspar of exceptionally high grade (96 percent CaF_2) in granite. Numerous fluorspar deposits are scattered throughout western Nevada (22), but production has come chiefly from extreme northeastern Mineral County from veins in andesite, northwestern Nye County near Gabbs from veins of several types, and southwestern Nye County from veins and pipelike deposits. The Thomas Range district, Juab County, Utah (41, 45), has been an important producer of fluorspar since 1950. Pipes, veins, and disseminated deposits are present, but almost all the production has come from the pipes in dolomite of Paleozoic age. All the fluorspar deposits show abnormal radioactivity, as stated in the discussion of uranium deposits. The geologic characteristics of fluorspar deposits in the western United States were discussed in 1958 by Peters (33).

Canada

The most important production of fluorite in Canada comes from the St. Lawrence district of the Burin Peninsula of southeastern Newfoundland (54). Numerous fluorite-bearing veins occur in granite, some of which attain a length of several thousand feet. Generally the veins are considerably shorter, are 5 ft or less thick if they are of high grade but 15 to 35 ft thick if they are of lower grade. Some have been followed for as much as 900 ft down the dip, which generally is between 70 and 90°. Quartz is the most common gangue mineral and its presence, along with inclusions of the granite wallrock, causes the silica content of the fluorspar deposits to be higher than the amount desirable for metallurgical use, but the fluorspar is an excellent source of acid-grade material.

Mexico

Mexico was the world's largest producer of fluorspar in 1962, a position held since 1956. The deposits (16, 18, 47, 53) to a large extent occur in limestone of Early Cretaceous age, and some occur in shales or volcanic rocks that overlie the limestone and are within or adjacent to rhyolite of Tertiary age. They are nearly flat mantos, steeply inclined veins, and variously shaped replacement bodies, and also they fill sinkholes and spaces in collapse breccias. Generally the mantos are smaller than some of the other bodies but contain higher-grade fluorspar. The major minerals present in addition to fluorite are calcite and quartz or chalcedony. The average grade of the fluorspar is estimated to be about 65 percent CaF_2. Deposits are widespread, but the most important ones appear to be in the states of Chihuahua, Coahuila, Durango, Guanajuato, Guerrero, Mexico, San Luis Potosí, Sonora, and Zacatecas. The more recent developments, which have been largely instrumental in placing Mexico at the top of world production, are in San Luis Potosí and Guanajuato (16).

Europe

The reserves of fluorspar in Europe were estimated in 1955 (24) to be 17 million metric tons, about 83 percent of which are in five countries, as follows: Soviet Union, 23 percent; United Kingdom, 18 percent; Spain, 18 percent; Italy, 12 percent; West Germany, 12 percent. France and East Germany each contain nearly 6 percent of the reserves, bringing the total of seven countries to about 95 percent.

The deposits of the Soviet Union (24; 31, pp. 79, 97, 149) that seem to yield the largest production are in eastern Transbaikal, where hydrothermal veins are associated with granitic intrusions into sandstones and shales of Jurassic age. Other vein deposits of considerable importance are in Turkeystan associated with lead-zinc-silver mineralization around Aurakhmat (about 41°N, 69°E) and Takob

(39°N, 69°E), and in the northern Ural Mountains, within the Arctic Circle, associated with granite.

The chief deposits of the United Kingdom (13, 24) from which production is obtained are in the northern part of Derbyshire and the western part of Durham. In West Durham, fluorspar veins occur in Lower Carboniferous limestones. These veins contain, also, some galena, sphalerite, and a few other minerals. Barite and witherite are abundant in the area, but because of zoning they are not associated with workable fluorspar deposits. In North Derbyshire the deposits are of several types: veins that are fracture fillings but in some places pass into replacements of the limestone walls, mineralized joints in limestone, blanket replacements, and pipelike replacements. Associates of fluorite in these deposits are calcite, barite, galena, some sphalerite, and quartz and chalcedony.

Fluorspar deposits in Spain (24) occur in Gerona Province, northeastern Spain, in Oviedo Province, northern Spain, and in Córdoba Province, southern Spain. The largest deposits are those near Osar, Gerona Province, where fluorite occurs in veins along with galena and sphalerite. The veins traverse gneiss, schists, and limestones. One vein is 2.5 km long, 2 to 8 m wide, and has been developed throughout a length of 500 m and a depth of 110 m. The deposits of Oviedo Province are veins and pockets in limestone of Liassic age, whereas at least some of the veins of Córdoba Province are in granite.

The fluorspar deposits of Italy are in the northern and central parts of the country. In northern Italy the veins are in limestones and schists of Triassic age, whereas in central Italy they are in limestones and shales of Eocene age. By-product fluorspar is recovered from lead-zinc deposits in Lombardia and in Sardinia.

The most active fluorspar district of West Germany in 1954 was the Schwandorf area of Bavaria, where the veins are in granite of Permian age that intrudes gneiss of Archean age. The chief minerals that accompany the fluorite are quartz, barite, calcite, and locally galena, pyrite, and chalcopyrite. Fluorspar veins in granite are present in Baden, also.

The deposits of France, which formerly were very productive, are veins in the Central massif and in the Esterel massif. The deposits of East Germany are veins in Saxony, Thuringia, and Anhalt.

Other Countries

Little information is available about the fluorspar deposits of China. Juan (27, pp. 459–460) stated that the most important deposits are irregular veins in rhyolite in Chekiang. The length of the veins generally is about 300 m, but a few attain a length of 3,500 m. Fluorspar deposits are present in South-West Africa and some large deposits were discovered about 1962 in the Republic of South Africa (57).

Magnesite and brucite

Statistical information regarding the production of magnesite has not been available consistently during the 15-year period 1945 to 1959, but from material that has been published it seems likely that the important producers during that time were the same as those in 1962 (Table 21-3).

TABLE 21-3.   Some Important World Production of Magnesite, Average of 1945–1959 and in 1962*

	Percent of world production	
Country	*Average of 1945–1959*	*1962 only*
Austria	19.7	21.6
United States	11.9	6.0
Yugoslavia	3.1	5.0
India	2.2	2.9
Greece	1.5	2.0
Australia	1.3	0.7
Soviet Union	?	33.5
China	?	12.2
Czechoslovakia	?	7.0
Not given above	60.3	9.1

	Short tons	
	Average of 1945–1959	*1962 only*
World total	3,690,000	8,200,000

* Compiled from *U.S. Bur. Mines Minerals Yearbook,* various years.

United States

The principal deposits of magnesite in the United States (7, 8, 17, 50, 51) are in Stevens County, Washington, and in Nye County, Nevada. Early production (1886 to 1916) was entirely from California.

The principal magnesite deposits of Washington occur in a narrow zone in the Huckleberry Mountains, Stevens County. The main shipping point for the processed magnesite is Chewelah. The commercial deposits are confined to the Stensgar dolomite of Precambrian (Belt) age, which is part of a group of metasedimentary rocks composed chiefly of fine-grained clastics that have been intruded by greenstones, subsequently closely folded, and considerably metamorphosed in places by later emplacement of a granitic batholith of Cretaceous age. The magnesite deposits seem to have resulted from replacement of dolomite by hydrothermal solutions that may have had their origin in the granitic magma or have been related to its emplacement. The larger bodies of magnesite are as much as 200 ft thick and 1,000 ft long. The material as mined is high in lime from inclusions of dolomite, and also high in silica, but the quality is much improved by beneficiation.

The magnesite and brucite deposits of Nevada (50, 51) are in the extreme northwest corner of Nye County, east of Gabbs, and lie on the western side of Paradise Mountain. They were formed by replacement of the Luning dolomite of Triassic age, probably by solutions that emanated from granodiorite magma that was intruded during Jurassic time or later. The commercial deposits of magnesite are north and northwest of a mass of granodiorite, and the largest body, about 3,000 ft wide and 5,000 ft long, is associated with a prong of granodiorite that extends northwest from the main mass. Magnesite is known to continue to a depth of at least 600 ft. Brucite is associated with magnesite in places, the major deposits being on the east and west sides of the prong of granodiorite. Reserves of the deposits are difficult to calculate because of the gradational nature of the boundaries of various types of material and the irregularities of replacement, but apparently they are considerable.

Most of the magnesite deposits of California are replacements of serpentine in the Coast Ranges and the western foothills of the Sierra Nevada (49). At Red Mountain, in Santa Clara and Stanislaus Counties, the largest known and most recently worked deposit, magnesite occurs as replacements of serpentine in shear zones and tension fractures; locally it occurs as fissure fillings. Some masses of magnesite were as much as 30 ft thick, but 5 to 10 ft was more usual, and those less than 2 or 3 ft thick could not be mined profitably.

Europe

Important deposits in former Austria-Hungary furnished most of the world's supply of magnesite before World War I. Now these deposits are in Austria and Czechoslovakia. The most important ones are in the province of Styria, Austria, at Veitsch and Radenthein (9, p. 489; 29, p. 300). Throughout the region magnesite occurs as lenses in dolomite, which seem to be replacement deposits formed by hydrothermal emanations. The deposit at Veitsch is about a mile long, 500 ft wide, and extends to a depth of at least 700 ft.

The magnesite deposits of the Soviet Union (29, p. 300; 31, p. 57) may be among the largest in the world. The most important mining area is in the western Ural Mountains near Satka (55°N, 59°E). The magnesite is interbedded with dolomites of Upper Proterozoic age, according to Nalivkin (31, p. 57).

China

The magnesite deposits of China are in Manchuria (27, pp. 465–466; 29, p. 300; 32), in Liaoning Province (Fig. 9-5), and may be among the largest in the world. The magnesite is coarsely crystalline, interbedded with dolomite of Precambrian age, contains 44 to 48 percent MgO, and is in deposits that are suitable for open-pit mining. Juan (27, p. 465) stated that the deposits are replacements, whereas Nishihara (32) suggested that the magnesite was deposited directly from waters of Precambrian seas.

SELECTED REFERENCES

1. A. L. Anderson, 1954, Fluorspar deposits near Myers Cove, Lemhi County, Idaho, *Idaho Bur. Mines and Geol., Pamphlet No.* 98, 34 pp.
2. J. C. Arundale, 1956, Barium, *U.S. Bur. Mines Bull.* 556, pp. 87–93.
3. R. W. Boyle, 1963, Geology of the barite, gypsum, manganese, and lead-zinc-copper-silver deposits of the Walton-Cheverie area, Nova Scotia, *Geol. Survey Canada Paper* 62-25, 36 pp.

4. E. A. Brecke, 1962, Ore genesis of the Cave-in-Rock fluorspar district, Hardin County, Illinois, *Econ. Geology,* vol. 57, pp. 499–535.

5. D. A. Brobst, 1958, Barite resources of the United States, *U.S. Geol. Survey Bull.* 1072-B, pp. 67–130.

6. D. A. Brobst, 1960, Barium minerals, in "Industrial Minerals and Rocks," 3d ed., pp. 55–64, American Institute of Mining, Metallurgical, and Petroleum Engineers, New York, 934 pp.

7. I. Campbell and J. S. Loofbourow, Jr., 1957, Preliminary geologic map and sections of the magnesite belt, Stevens County, Washington, *U.S. Geol. Survey Mineral Inv. Field Studies Map* MF 117.

8. I. Campbell, 1962, Geology of the magnesite belt of Stevens County, Washington, *U.S. Geol. Survey Bull.* 1142-F, 53 pp.

9. H. B. Comstock, 1956, Magnesium, *U.S. Bur. Mines Bull.* 556, pp. 471–492.

10. H. B. Comstock, 1960, Magnesium and magnesium compounds, *U.S. Bur. Mines Bull.* 585, pp. 481–492.

11. H. W. Davis, 1952, Fluorspar, in W. Van Royan and O. Bowles, "The Mineral Resources of the World," vol. 2, pp. 170–173, Prentice-Hall, Inc., Englewood Cliffs, N.J., 181 pp.

12. J. B. DeMille, 1947, "Strategic Minerals," McGraw-Hill Book Company, New York, 626 pp.

13. K. C. Dunham and others, 1952, Fluorspar, *Great Britain Geol. Survey, Mem., Spec. Rept. Min. Resources,* vol. 4, 143 pp.

14. K. C. Dunham, 1959, Non-ferrous mining potentialities of the Northern Pennines, in "Future of Non-ferrous Mining in Great Britain and Ireland," pp. 115–117, Institute of Mining and Metallurgy, London.

15. G. C. Finger, H. E. Risser, and J. C. Bradbury, 1960, Illinois fluorspar, *Illinois Geol. Survey Circ.* 296, 36 pp.

16. M. H. Frohberg, 1962, Geological features of some fluorite deposits in the state of San Luis Potosi, Mexico, *Geol. Assoc. Canada, Proc.,* vol. 14, pp. 9–19.

17. B. Gildersleeve, 1962, Magnesite and brucite in the United States, *U.S. Geol. Survey Mineral Inv. Resource Map* MR-27. Map and pamphlet of 4 pp.

18. J. González Reyna, 1956, Los yacimientos de fluorita de México, "Riqueza Minera y Yacimientos Minerales de México," 3d ed., pp. 397–414, Internat. Geol. Cong., 20th, Mexico 1956.

19. R. M. Grogan, 1960, Fluorspar and cryolite, in "Industrial Minerals and Rocks," 3d ed., pp. 363–382, American Institute of Mining, Metallurgical, and Petroleum Engineers, New York, 934 pp.

20. G. C. Hardin, Jr., and R. D. Trace, 1959, Geology and fluorspar deposits, Big Four fault system, Crittenden County, Kentucky, *U.S. Geol. Survey Bull.* 1042-S, pp. 699–724.

21. J. E. Holtzinger, 1956, Fluorine, *U.S. Bur. Mines Bull.* 556, pp. 279–290.

22. R. C. Horton, 1961, An inventory of fluorspar occurrences in Nevada, *Nevada Bur. Mines, Report* 1, 31 pp.

23. R. C. Horton, 1963, An inventory of barite occurrences in Nevada, *Nevada Bur. Mines, Report* 4, 18 pp.

24. H. R. Hose, 1955, European fluorspar supplies, *Mining Eng.*, vol. 7, pp. 383–390.

25. G. A. Jewett, 1957, The Walton, N. S., barite deposit, in "The Geology of Canadian Industrial Mineral Deposits," pp. 54–58, Canadian Institute of Mining and Metallurgy, Montreal.

26. S. J. Johnstone and M. G. Johnstone, 1961, "Minerals for the Chemical and Allied Industries," 2d ed., John Wiley & Sons, Inc., New York, 788 pp.

27. V. C. Juan, 1946, Mineral resources of China, *Econ. Geology*, vol. 41, pt. 2, pp. 399–474.

28. K. B. Ketner, 1963, Bedded barite deposits of the Shoshone Range, Nevada, *U.S. Geol. Survey Prof. Paper* 475-B, pp. 38–41.

29. R. B. Ladoo and W. M. Myers, 1951, "Nonmetallic Minerals," 2d ed., McGraw-Hill Book Company, New York, 605 pp.

30. R. B. McDougal, 1960, Fluorine, *U.S. Bur. Mines Bull.* 585, pp. 305–315.

31. D. V. Nalivkin, 1960, "The Geology of the U.S.S.R., a Short Outline," Pergamon Press, New York, 170 pp., geologic map of U.S.S.R.

32. H. Nishihara, 1956, Origin of the bedded magnesite deposits of Manchuria, *Econ. Geology*, vol. 51, pp. 698–711.

33. W. C. Peters, 1958, Geologic characteristics of fluorspar deposits in the western United States, *Econ. Geology*, vol. 53, pp. 663–688.

34. J. S. Ross, 1961, Barite chapter (6 pp.), in The Canadian mineral industry 1961, *Geological Survey Canada, Mineral Resources Div., Dept. Mines and Tech. Surveys.*

35. E. L. H. Sackett, 1962, Barite, little-known industry that means "mud" to oil men, *Mining Eng.*, vol. 14, no. 5, pp. 46–49.

36. U. M. Sahinen, 1962, Fluorspar deposits in Montana, *Montana Bur. Mines and Geol. Bull.* 28, 38 pp.

37. G. A. Schnellmann, 1962, Occurrence and distribution of fluorspar, *Mining Jour. (London)*, vol. 258, no. 6611, pp. 442–443.

38. G. A. Schnellmann, 1962, Production and consumption of fluorspar, *Mining Jour. (London)*, vol. 258, no. 6614, pp. 526–527.

39. A. E. Schreck, 1960, Barium, *U.S. Bur. Mines Bull.* 585, pp. 85–93.

40. B. J. Scull, 1958, Origin and occurrence of barite in Arkansas, *Arkansas Geol. and Cons. Comm., Inf. Circ.* 18, 101 pp.

41. M. H. Staatz and F. W. Osterwald, 1959, Geology of the Thomas Range fluorspar district, Juab County, Utah, *U.S. Geol. Survey Bull.* 1069, 97 pp.

42. T. A. Steven, 1954, Geology of the Northgate fluorspar district, Colorado, *U.S. Geol. Survey Mineral Inv. Field Studies Map* MF 13.

43. T. A. Steven, 1960, Geology and fluorspar deposits, Northgate district, Colorado, *U.S. Geol. Survey Bull.* 1082-F, pp. 323–422.

44. J. W. Taber, 1953, Montana's Crystal Mt. fluorite deposit is big and high grade, *Mining World*, vol. 15, no. 6, pp. 43–46.

45. W. R. Thurston, M. H. Staatz, D. C. Cox, and others, 1954, Fluorspar deposits of Utah, *U.S. Geol. Survey Bull.* 1005, 53 pp.

46. R. D. Trace, 1962, Geology and fluorspar deposits of the Levias-Keystone and Dike-Eaton areas, Crittenden County, Kentucky, *U.S. Geol. Survey Bull.* 1122-E, 26 pp.

47. R. E. Van Alstine, 1961, Investigation of the principal fluorspar districts of Mexico, *U.S. Geol Survey Prof. Paper* 424-C, pp. 212–215.
48. W. W. Varvill, 1959, The future of lead-zinc and fluorspar mining in Derbyshire, in "Future of Non-ferrous Mining in Great Britain and Ireland," pp. 175–203, Institute of Mining and Metallurgy, London.
49. W. E. Ver Planck, 1957, Magnesium and magnesium compounds, *California Dept. Nat. Res., Dept. Mines, Bull* 176, pp. 313–323.
50. C. J. Vitaliano and E. Callaghan, 1956, Geologic map of the Gabbs magnesite and brucite deposits, Nye County, Nevada, *U.S. Geol. Survey Mineral Inv. Field Study Map MF* 35.
51. C. J. Vitaliano, E. Callaghan, and N. L. Silberling, 1957, Geology of Gabbs and vicinity, Nye County, Nevada, *U.S. Geol. Survey Mineral Inv. Field Studies* Map MF 52.
52. J. S. Williams and others, 1954, Fluorspar deposits in western Kentucky, *U.S. Geol. Survey* Bull. 1012, 127 pp.
53. D. R. Williamson, 1961, Mexican fluorite deposits, *Colorado School Mines, Min. Ind. Bull.,* vol. 4, no. 2, 15 pp.
54. D. H. Williamson, R. F. Jooste, and D. M. Baird, 1957, St. Lawrence fluorite district, in "The Geology of Canadian Industrial Mineral Deposits," pp. 90–97, Canadian Institute of Mining and Metallurgy, Montreal.
55. T. Wilson, 1954, Magcobar—mud is their business, *Mining Eng.,* vol. 6, pp. 494–496.
56. Anon., 1958, Fluorspar mining in Hardin County, Illinois, *Mining Eng.,* vol. 10, pp. 65–67.
57. Anon., 1962, Fluorspar discovery in South Africa, *Mining Jour. (London),* vol. 259, no. 6633, p. 321.
58. Anon., 1962, West Germany's largest pyrites mine, *Mining Jour. (London),* vol. 259, no. 6644, p. 589. Much barite is associated with pyrite.

chapter **22**

Salt, Potash,
Gypsum and Anhydrite,
Sulfur, and Pyrite

Salt, potash, and gypsum all belong to the evaporite group and hence may occur in the same general region. Deposits of salt and gypsum are much more widespread and abundant than are those of potash, since potassium minerals form late in the cycle of evaporation of a body of water and the original concentration of potassium in the water is much less than is that of sodium and calcium. Sulfur is commonly associated with salt domes, which furnish the major production, but sulfur deposits of volcanic origin are unrelated to salt, potash, and gypsum. Pyrite may occur with volcanic sulfur or in deposits unrelated to any of the other materials being considered.

Salt

Salt is produced to some extent in most countries, but 51 percent of the total production in the 15-year period 1945 to 1959 and also in 1962 came from the United States, the Soviet Union, and China (Table 22-1).

Nearly 75 percent of the total production of salt is used as chemical raw material and for indirect chemical operations (29). Among other important uses are the preparation of food stuff, to preserve and to add flavor; the curing and tanning of hides and skins; the manufacture of textiles; and ice and snow removal. These are but a few of 14,000 uses that are credited to salt.

Salt is recovered in commercial amounts from (1) deposits of rock salt, which occur in sedimentary beds, in salt domes or diapiric folds, and in playas, and (2) salt in solution in seawater, lake water, and groundwater (33). The recovery of salt from seawater has attained great importance, and, in 1958, although it was estimated to be but 5 percent of the total production of salt in the United States, it was about 45 percent of the production outside the United States (49).

Bedded salt deposits throughout the world accumulated during various

**TABLE 22.1.   World Production of Salt,
Average of 1945–1959 and in 1962***

Country	*Percent of total world production*	
	Average of 1945–1959	*1962 only*
United States	31.7	28.8
Soviet Union	10.2	9.0
China	9.1	13.2
United Kingdom	7.6	6.7
France	5.1	4.2
India	5.0	4.2
West Germany	4.6	5.4
Italy	3.0	3.0
Canada	2.0	3.6
Spain	2.0	1.7
All others	19.7	20.2
	Thousand short tons	
	Average of 1945–1959	*1962 only*
World total	62,084	100,500

* Compiled from *U.S. Bur. Mines Minerals Yearbook,* various years.

geologic times from the Cambrian to the Tertiary, but the greatest period of world-wide accumulation was the Permian.

United States

Rock salt deposits are known in 24 states in the United States, and range in age from Silurian to Pliocene (?) (33, 44), but the shorter range, Pennsylvanian to Jurassic, is more general. Salt is recovered not only from rock salt deposits but also from natural brines and from solar evaporation of seawater. The leading producing states in 1962, and the percent of the total production of the United States, were Texas (19), Louisiana (18), New York (15), Michigan (15), Ohio (15), and California (6).

The rock salt deposits and natural brines of the United States (33, 44) occur in three major basins: (1) the eastern or northeastern, in which the salt is chiefly of Silurian age; (2) the Gulf Coast embayment, in which the bedded deposits may be of Jurassic or Permian age but the important salt domes occur in younger rocks; (3) the southwest or Permian basin, where the deposits are in rocks of Permian age. Salt is recovered by evaporation of seawater and lake water in California and Utah.

The northeastern basin has been the most productive one in the United States. It includes parts of New York, Pennsylvania, West Virginia, Ohio, and Michigan, and also extends into southwestern Ontario (6, 21, 31, 33, 44). The area underlain by salt is approximately 100,000 sq miles, and the amount of salt in the basin is estimated to be 2,800 cu miles, or 2.7×10^{13} tons. This major basin may be separated into two basins that, at times, were connected, an eastern one in New

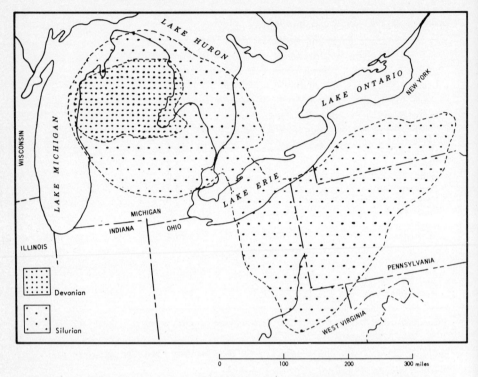

Fig. 22-1 Map showing Northeastern salt basin of the United States. Generalized after W. G. Pierce and E. I. Rich, *U.S. Geol. Survey Bull.* 1148, plate 1.

York and Pennsylvania, a western one in Michigan. Throughout most of the area the salt is in the Salina formation of Silurian age (Fig. 22-1), but in Michigan salt occurs also in the Detroit River formation of Devonian age. The greatest development of the Salina formation is in Michigan, where over 3,000 ft of sediment was deposited in the center of the basin. The Salina formation consists of shale, dolomite, limestone, salt, anhydrite, and gypsum. It ranges in thickness from nearly nothing to 3,000 ft or more and is thickest in the central parts of the eastern and western basins of New York and Michigan. The maximum thickness of salt, more than 2,000 ft, is in Michigan.

At present salt is recovered in Michigan from one underground mine and from 13 brine operations. The underground mine, at Detroit, obtains salt from a bed of salt about 20 ft thick at a depth of 1,020 ft. A much thicker bed of salt underlies this one. Three centers of brine operations are active: southeastern Michigan, where the brine is obtained by leaching salt beds of the Salina formation; western Michigan, near Manistee, where brine is obtained by circulating water through salt beds of the Detroit River formation; and central Michigan, in Midland County, where brine is obtained by leaching both the Salina and the Detroit River formations.

Salt occurrences in Ohio are similar to those in Michigan but are confined to the eastern part of the state (21, 43). Salt has been obtained commercially since

1889 by pumping water into the Salina salt beds and repumping the brine to the surface. Operations are chiefly in Summit, Wayne, and Lake Counties. Two underground mines are operating in Lake County, one of which is to recover salt from beneath Lake Erie by means of a shaft 2,000 ft deep on Whiskey Island in Cleveland.

New York was one of the early salt producers of the United States. The salt all comes from the Salina formation, and is recovered by underground mining and from artificial brine wells. Rock salt is mined by underground workings in Livingston and Tompkins Counties from a bed 5 to 60 ft thick at a depth of about 2,175 ft. Brine salt plants are operating in Wyoming, Schuyler, Tompkins, and Onondaga Counties.

Salt of the Gulf Coast embayment (Fig. 22-2) occurs in parts of Arkansas, Mississippi, Alabama, Texas, and Louisiana and extends southward beneath the Gulf of Mexico. Within this large area, bedded salt salt deposits are known to be

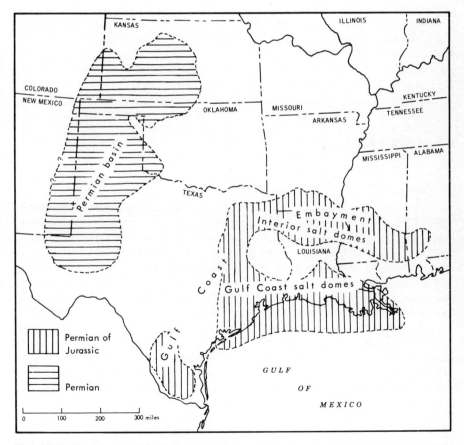

Fig. 22-2 Map showing Gulf Coast Embayment, Permian basin, and general location of salt domes in the United States. After W. G. Pierce and E. I. Rich, *U.S. Geol. Survey Bull.* 1148, plate 1.

present in the Eagle Mills formation of Jurassic age in southern Arkansas, northern Louisiana, and northeastern Texas; in interior salt domes in northeast Texas, northern Louisiana, and Mississippi; and in coastal salt domes in Texas and Louisiana. Major production comes from the interesting salt domes, both from underground and brine operations. In Texas during 1958 more than 90 percent of the production was from brine, whereas in Louisiana much of the production was from underground mines, especially those in Iberia Parish, southern part of the state—the Five Islands deposits, which are five mounds rising from the surrounding marshes. Salt occurs in a huge vertical column that is several hundred feet across and of unknown depth. Shafts are slightly more than 500 ft to a bit more than 800 ft deep.

Kansas furnishes most of the salt produced in the Permian basin, the source of the salt being the Wellington formation. The aggregate thickness of salt in the area of underground mining is 200 ft, but only a small part of this is exploited by shafts 600 to 1,000 ft deep.

The salt production of California (49, 54) is chiefly from solar evaporation of seawater. The center of the industry is on the southeast shore of San Francisco Bay, where one of the world's largest solar evaporation plants is located at Mount Eden. Seawater is passed through a series of concentrating ponds, in which sun and wind cause evaporation of some of the water, bring the solution to the point of saturation for sodium chloride, and cause precipitation of slightly soluble carbonates and gypsum. The resulting brine, which has only about one-tenth of the volume of the original seawater, is passed into a group of crystallizing ponds, where salt is precipitated, after which the remaining liquid (bittern) is sold to chemical plants for the recovery of some of the remaining constituents. Salt is also recovered in California from various dry lakes in San Bernardino County, especially Britol Lake and Searles Lake.

Europe

The bedded salt deposits of the Stassfurt area in the salt basin of north Germany have been mined by underground methods for more than a hundred years. They are of Permian age and are associated with the much more important potash deposits. They will be discussed in the section relating to potash.

Information about the salt deposits of the Soviet Union is scanty. According to Nalivkin (41, pp. 30, 37, 58, 98, 122) a large zone of salt-bearing rocks of Lower Cambrian age extends for many miles along the upper course of the Lena River to the basin of the Vilui River. The most important deposits are at Usolsk near Irkutsk, at Peledui on the Lena River, and at Kempendyaisk on the Vilui River. Salt domes occur at Kempendyaisk (roughly 62°N, 118 to 120°E) and much farther to the north on the east shore of the Khatanga estuary at Nordvic (74°N, 112°E) and south from there. Further, salt domes in which the salt is of Jurassic age occur near Kulyab (38°N, 70°E), and enormous salt deposits of Jurassic age, with which potash deposits are associated, are present around Kugitang (38°N, 66°E). Rock salt and potash salts occur in large deposits in the Carpathian Mountains, and other deposits of rock salt are present in the southern part of the Lesser Caucasus. In addition, rock salt occurs with potash in deposits of Permian age around Solikamsk (56°30′N, 56°30′E) in the Kama River Basin. According to

Nalivkan (41, p. 58) these deposits contain reserves of potash many times greater than those of Stassfurt, Germany. Some other sources of salt in the Soviet Union are many playas and solar salt from operations on the shores of the Caspian Sea.

Asia

More than three-fourths of the salt of China comes from the evaporation of seawater, operations being concentrated in the coastal provinces of Liaoning, Kiangsu, Shantung, and Hopeh (Fig. 9-5) (28, pp. 469–470). Salt is also obtained from brine wells in Szechwan and Yunnan Provinces, some of which are as much as 1,000 m deep. Red beds in Hupeh, Hunan, Sinkiang and western Yunnan Provinces contain rock salt in thin beds and disseminations, and this salt is recovered by leaching. Also, some salt is obtained from inland lakes.

Most of the salt of India (7, pp. 499–512) is produced by solar evaporation of seawater, especially in the coastal states of Bombay, Madras, and Andhra, but to some extent in Orissa. Lake salt is obtained from some lakes, the largest of which, Sambhar, is in Jaipur, Rajasthan State. Some salt is obtained from rock salt deposits of Mandi, state of Himachal Pradesh. Formerly rock salt was obtained from northwestern areas, especially the Punjab Salt Range, now in West Pakistan.

Potash

The major production of potash comes from the United States, Germany, France, and the Soviet Union (Table 22-2). The combined production from West Germany and East Germany generally is about 40 percent of the world total. The

**TABLE 22-2.   World Production of Potash,
Average of 1945–1959 and in 1962***

Country	Percent of world total	
	Average of 1945–1959	*1962 only*
United States	26.7	22.9
West Germany	22.3	20.0
East Germany	20.5	17.7
France	17.4	17.7
Soviet Union†	10.0	15.4
All others	3.1	6.3
	Thousand short tons K_2O *equivalent*	
	Average of 1945–1959	*1962 only*
World total	6,083	10,700

* Compiled from *U.S. Bur. Mines Minerals Yearbook*, various years.
† Based on the production of the Soviet Union for 1950–1959 compared with the world production during the same time, as some data unavailable for 1945-1959.

production picture of the future could be changed considerably because of the development of the Esterhazy potash mine of Saskatchewan, Canada, which has been rated as the largest potash mine in the world, and other mines to be developed there.

United States

About 90 percent of the potash produced in the United States in 1962 came from deposits in New Mexico, and most of the remainder came from California and Utah.

The rich potash deposits of Carlsbad, New Mexico (10; 32, pp. 401–402; 46; 47) are on the western flank of the Permian basin that extends through parts of Colorado, Kansas, Oklahoma, Texas, and New Mexico. The deposits occur in the Permian Salado formation at depths of 650 to 1,500 ft. The Salado formation consists of numerous beds of rock salt, gypsum, anhydrite, and polyhalite, $K_2Ca_2Mg(SO_4)_4 \cdot 2H_2O$. The potash deposits, which lie about in the middle of the Salado formation, occur in a thickness of about 250 ft and consist chiefly of sylvite, KCl, along with some langbeinite, $K_2Mg_2(SO_4)_3$, and carnallite, $KMgCl_3 \cdot 6H_2O$. The production from these potash deposits is expected to increase considerably in the future, as new shafts were being sunk in 1963 and a new flotation process permits the mining and treatment of mixed ores not formerly of commercial value because of difficulty of recovering the potash (62).

The potash produced by California comes from brines of Searles Lake in San Bernardino County (53). Commercial operations recover, in addition to potash, borax, lithium carbonate, and a number of other products.

Some potash is recovered from the Salduro Marsh about 100 miles west of Salt Lake, Utah. Important future production is expected from deposits of the Crane Creek area about 9 miles southwest of Moab, Utah. Reserves stated at 10 million tons of proved and probable ore are to be mined from a depth of nearly 3,000 ft. The deposit is in the Paradox formation of Pennsylvanian age (13).

Canada

One of the outstanding recent undertakings is the development of important potash deposits in Saskatchewan, Canada (4, 18, 52, 61). Salt beds 400 to 600 ft thick that extend over an area 350 miles long and 120 miles wide were cut during the drilling of numerous oil wells, in many of which potash was reported at depths varying from 2,760 ft to more than 2,900 ft. Active drilling for potash disclosed important deposits near the top of the Prairie Evaporites formation of Devonian age, composed chiefly of rock salt. Shaft sinking was undertaken in 1957 and proved to be exceptionally difficult (51) because of a surface covering of 289 ft of water-bearing glacial till and because of water under high pressure at various depths. The important potash-bearing bed, up to 10 ft thick, was finally reached at a depth of 3,132 ft in June, 1962, and the shaft bottomed at 3,378 ft at a cost of 10.5 million dollars which is $3,150 per foot! The mine and accompanying mill represent an investment of 40 million dollars.

The ore consists chiefly of sylvite, carnallite, and rock salt. It assays about 25 percent K_2O, which is stated to be double the grade of the Carlsbad, New Mexico, ore. The potash is recovered by leaching and flotation. The Esterhazy mine, reported to be the largest and most automated potash mine in the world, was officially opened in September, 1963, and was expected soon to reach full capacity of 1,200,000 tons a year.

A second important project is planned east of Saskatoon by Alwinsal Potash of Canada Limited. Sinking of a 3,200 ft shaft was to start early in 1964 and production at a rate of 1 million tons a year was expected by 1968 (63).

Europe

The major potash deposits of Europe (32, pp. 396–397; 45; 46; 47; 50) are in Germany. Deposits of lesser importance are in France and Spain, and probably in the Soviet Union, although the Russian deposits may be of major importance.

Germany. The potash deposits of north-central Germany (33, pp. 47–49; 50) have furnished much of the world's potash for many years. Within the German salt basin, salt and potash occur in beds of Upper Permian age throughout an area of some 10,000 sq miles. The region may be divided into a main basin, in which production comes from the districts of Stassfurt, Sud Hartz, Halle, and Hannover-Braunschweig, and a subsidiary basin, with production from the Werra-Fulda district. Somewhat less than half of the production comes from the main basin. Throughout the region there are two major groups of salt beds, designated the *older* and the *younger salt*. One potash source, the Stassfurt bed, lies near the top of the older salt throughout the entire main basin. Two other sources, designated the *Ronnenberg* and the *Reidel beds*, are in the younger salt and lie 200 m and 300 m respectively above the Stassfurt bed. These two beds are present only in the Hannover area. Two potash beds that are known only in the subsidiary or Werra basin lie 150 m and 210 m respectively below the position of the Stassfurt bed.

The commercial beds are mixtures of common salt (halite) and various potash salts, of which sylvite, carnallite, and polyhalite are the most common. Some associated minerals are kieserite, $MgSO_4 \cdot H_2O$, and anhydrite.

France. The principal potash deposits of France are in the Alsace basin, north and east of Mulhouse, in the upper valley of the Rhine River. They consist of two beds of sylvite, 1.5 m and 4 m thick, in rocks of Lower Oligocene age that consist of shale and rock salt and are overlain by Middle Oligocene beds containing rock salt, anhydrite, dolomite, and gypsum (50). The area in which production occurs is an oval covering approximately 75 sq miles. In addition to these principal deposits, some potash salts in rocks of Triassic age in southwestern France near Bayonne and Dax formerly yielded some production. Modernization of plants in Alsace in the early 1960s is expected to increase the French production (64).

Spain. Spain has furnished only a small amount of potash from deposits in Barcelona Province. These deposits occur near the top of a thick series of salt beds of Oligocene age.

Israel and Jordan. Potash has been recovered from the Dead Sea by evaporating the seawater. Early production came to an end in 1948 because of the conflict between Israel and Jordan, and the plant at the north end of the Dead Sea was destroyed. Since then there has been production at the south end by Dead Sea Works Ltd., of Israel, and large expansion is planned. Also, Arab Potash Co. Ltd., of Jordan, plans to extract potash from sea brines (60).

Soviet Union.   The potash deposits of the Soviet Union (32; 41, pp. 30, 37, 58, 98, 122; 46) were mentioned in connection with the discussion of salt deposits. The major potash region is along the western slope of the Ural Mountains in the vicinity of Solikamsk (56°30′N, 56°30′E). The deposits are of Permian age, are 600 to 750 ft beneath the surface, and consist of a carnallite zone that contains a maximum of 17 percent K_2O and a sylvite-halite zone that contains as much as 20 percent K_2O. The overlying rocks consist of salt, gypsum, and clay. The Galician deposits of the Carpathian Mountains, which were originally controlled by Austria, then by Poland, and now are in the Soviet Union, occur throughout a distance of slightly more than 40 miles in a zone 2 to 3 miles wide in the vicinity of Kalusz and Stebnik, south of Lwow. The deposits consist of lenticular beds that contain kainite, $KMg(SO_4)Cl \cdot 3H_2O$, along with some sylvite, langbeinite, and carnallite.

Gypsum and anhydrite

Gypsum and anhydrite are widely distributed and commonly occur with or near deposits of rock salt. Many, perhaps most, large bodies of gypsum were deposited originally as anhydrite and subsequently converted into gypsum near the earth's surface. Consequently commercial deposits of gypsum commonly occur within a distance of a few hundred feet of the surface and generally less than one hundred feet.

Gypsum and anhydrite, although of very similar chemical composition, are not interchangeable for many uses. Gypsum is used to a large extent in making plaster, and a few percent of anhydrite will render gypsum useless for this purpose. Some anhydrite mixed with gypsum can be used as a retarder in making portland cement, and anhydrite is preferred for agricultural use. Anhydrite is used as a source of sulfur for the making of sulfuric acid and ammonium sulfate in Great Britain, France, Germany, Poland, and India (25, p. 246).

Production statistics generally do not separate gypsum from anhydrite but merely express everything as gypsum, which is by far the greater part of the material mined. On this basis five countries—the United States, Canada, France, the United Kingdom, and the Soviet Union—furnished 70 percent of the total production of the world during the period 1945 to 1959 and 65 percent in 1962 (Table 22-3).

United States

Nearly 54 percent of the gypsum produced in the United States during 1962 came from four states: California, 17.5 percent; Michigan, 12.7 percent; Iowa, 12.5 percent; Texas, 11.2 percent. Other important production came from Nevada, 8.1 percent, New York, 6.0 percent, and Oklahoma, 5.0 percent.

Commercial gypsum deposits of California (55, 57) consist of (1) rock gypsum of Tertiary and pre-Tertiary age and (2) gypsite deposits, which consist

TABLE 22-3. World Production of Gypsum, Average of 1945–1959 and in 1962*

Country	Percent of world total	
	Average of 1945–1959	*1962 only*
United States	29.5	20.0
Canada	13.3	10.3
France	10.5	9.5
United Kingdom	10.5	8.9
Soviet Union	7.2?	16.5
Spain	5.6	5.5
Italy	2.5	4.4
West Germany	2.6	2.3
All others	18.3	22.6

	Thousand short tons	
	Average of 1945–1959	*1962 only*
World total	27,678	49,965

* Compiled from *U.S. Bur. Mines Minerals Yearbook,* various years.

of gypsum mixed with sand or clay. Rock gypsum has been produced recently in the Little Maria Mountains, Riverside County, in the Fish Creek Mountains, Imperial County, and near Cuyama Wash, Ventura County. The largest rock gypsum mine of the state is the one operated by the United States Gypsum Company in the Fish Creek Mountains, where gypsum beds occur in a shallow synclinal basin 3 miles long and ½ mile wide—a remnant of a much more extensive deposit removed by erosion. The deposit occurs in rocks of Miocene age. Gypsite deposits may (1) form cappings on upturned gypsiferous beds (cap gypsum), (2) occur along the margins of periodic lakes, or (3) occur in the beds of dry washes (channel gypsum). Of these, the cap gypsum type is the most abundant. The largest gypsite deposits of California are those of the Lost Hills, Kern County. Flat-lying lenses of gypsite, which reach a maximum thickness of 20 ft, occur in silty sand and clay for about 5 miles along the southwest side of the Lost Hills. They attain their best development where present-day washes cut through the hills. A notable former gypsite producer is Bristol Lake, San Bernardino County, where selenite crystals occur in mud near the lake margins outside of the salt-bearing area. A large plaster mill at nearby Amboy was supplied with gypsum from Bristol Lake for nearly 20 years—1906 to 1925.

The production of gypsum in Michigan comes from the Lower Grand Rapids formation of Mississippian age, which crops out at various places away from the central salt-producing part of the Michigan basin. Operations of major importance are conducted in Kent County in the Grand Rapids area on the southwestern side of the basin and in Iosco County in the vicinity of Tawas City on the eastern side of the basin.

The Iowa gypsum is produced in the area around Fort Dodge from beds 10 to 30 ft thick that are probably of Permian age. The production of Texas and

Oklahoma is from gypsum beds of the Permian basin, some of which attain a thickness of 60 ft. Some production of gypsum and anhydrite in Texas comes also from the cap rock of salt domes. The important deposits of Nevada are in the vicinity of Las Vegas, where some of the gypsum may be of Permian age and some may be of Triassic age. The deposits of New York are in the Salina formation of the Silurian basin. For many years New York was the leading producer of gypsum in the United States.

Canada

Nova Scotia is the most important gypsum-producing province of Canada and supplied 84 percent of that country's output in 1961. The remaining production came, in order, from Ontario, British Columbia, Manitoba, New Brunswick, and Newfoundland (11). The operations of the Canadian Gypsum Company Limited, in Nova Scotia, have been the principal ones in that area, but the operations of that company were taken over in 1961 by the Fundy Gypsum Company Limited.

The major deposits of Nova Scotia (58) are in Hants and Halifax Counties with the center of the industry located at Wentworth. The deposits form two beds in the Windsor group of Upper Mississippian age, a white one and a dark one. The white bed is underlain by great thicknesses of anhydrite. The dark bed consists of clay, marl, and selenitic gypsum, and in places little or no anhydrite. The dark bed furnishes much the greater amount of gypsum.

The deposits in other parts of Canada, except in Newfoundland, are somewhat different. The deposits of southwestern Newfoundland (2) are very similar to those of Nova Scotia and are of the same age. Those of Ontario occur in the Salina formation of Silurian age and are similar to the deposits of New York. The major deposits of British Columbia occur in the Burnais formation of Middle Devonian age in the Stanford Mountains of the southern part of the province, and are quarried especially near the villages of Windemere and Canal Flats. Other deposits of British Columbia occur around Falkland, 40 miles east of Kamloops, in large shear zones. These deposits apparently were formed by hydrothermal replacement of argillites and tuffs along the shear zones. Some anhydrite is present and this may have been the original mineral from which the gypsum was formed, but the evidence seems to be indefinite (16). Gypsum deposits of Manitoba (8, 9) occur at Amaranth, 90 miles northwest of Winnipeg, and near Gypsumville, 150 miles north of Winnipeg. The Amaranth deposit, thought to be of Jurassic age, consists of about 43 ft of gypsum separated into two layers by 4 ft of anhydrite and underlain by red shale. The larger deposits around Gypsumville consist of 20 to 50 ft of gypsum that passes into anhydrite in depth, the total thickness of the material being 150 ft. They seem to be part of the Ashern formation that lies at the contact between definite Silurian and Devonian rocks and has not yet been assigned to either age group.

Europe

The United Kingdom and France together furnished 21 percent of the gypsum and anhydrite of the world in the period 1945 to 1959, and slightly more than 18 percent in 1962, the production of each country being in about equal amounts.

The production of the United Kingdom (19, pp. 46–55) during recent years has consisted of about equal amounts of gypsum and anhydrite, the anhydrite being used in the manufacture of sulfuric acid and ammonium sulfate. The gypsum and anhydrite deposits occur in strata of Triassic and Jurassic age and are present in four areas: (1) northwestern England, in Cumberland, Westmorland, and the Furness district of Lancashire; (2) northeastern England; (3) the Midlands; and (4) southern England. The production of anhydrite at Billingham, Durham, in northeastern England, has been especially important. The most extensive workable deposits of gypsum are in the East Midlands. The most important gypsum deposits of France (19, pp. 85–86) are of Tertiary age, although known deposits include the range from Permian to Tertiary. The extensive Tertiary deposits extend over 8,000 sq km in the Paris basin and the reserves are considered to be almost unlimited. There the gypsum occurs in four beds or masses throughout a total thickness of 55 m. The uppermost mass is the one most extensively worked.

The two other countries of Europe that have supplied relatively large amounts of gypsum are Spain and the Soviet Union. Production comes from at least 34 provinces of Spain, but Madrid furnishes much more than any other province. Information regarding deposits of gypsum and anhydrite in the Soviet Union is sparse. Both are known to occur in the same general areas as some of the deposits of salt and potash, and reserves apparently are extensive.

The production of Germany comes from deposits of gypsum and anhydrite of Permian age in the Harz Mountains, whereas the production of Italy comes from numerous places and to a considerable extent is associated with deposits of sulfur. One feature of some of the Italian deposits is the presence of pure white alabaster, which is highly desirable for interior architectural work and for the manufacture of many ornamental objects (19, p. 87).

Sulfur and pyrites

Sulfur for various commercial operations is obtained from deposits of elemental (native) sulfur; from various sulfides grouped together as pyrites, especially pyrite, marcasite, pyrrhotite; from sulfates, especially anhydrite; as a by-product of various mining and metallurgical operations that involve sulfide ores; and as a by-product, as hydrogen sulfide, from natural gas and petroleum refinery gases. Here we will be concerned with elemental sulfur and pyrites, especially the former.

The major part of the world's elemental sulfur for more than 20 years has come from the United States, with appreciable and increasing amounts coming from Mexico (Table 22-4). Elemental sulfur occurs (1) with salt, gypsum, and anhydrite of salt domes, (2) as sedimentary deposits, generally associated with gypsum, and (3) as volcanic sulfur. The largest production now comes from the cap rock of salt domes.

Sulfur

United States. The elemental sulfur produced in the United States comes chiefly from salt domes of Louisiana and Texas (14). Not all salt domes contain sulfur in the cap rock, but 20 of the domes in the Gulf Coast area have produced

**TABLE 22-4.   World Production of Elemental Sulfur,
Average of 1945–1959 and in 1962***

	Percent of total world production	
Country	Average of 1945–1959	1962 only
United States	84.5	62.2
Mexico	5.4	16.7
Italy	3.3	0.6
Japan	2.3	2.7
Soviet Union	?†	11.7
All others	4.5	6.1
	Thousand long tons	
	Average of 1945–1959	1962 only
World total	5,934	8,070

* Compiled from *U.S. Bur. Mines Minerals Yearbook,* various years.
† Data unavailable. The production for 1955–1959 was 3.4 percent of
the world total for that period.

sulfur. Typically the sulfur occurs in limestone of the cap rock, which is generally underlain by gypsum and anhydrite. The sulfur may be in masses, veinlets, or disseminated as small crystals, and in places crystals line cavities. The amount and disposition of the sulfur in the cap rock is variable, in places erratic, and shows no necessary relationship to the size of the salt dome.

The more recent exploration in the Gulf Coast area has been in the Gulf of Mexico, using geophysical surveys to locate the domes. One producing mine so located is the Grand Island mine off the coast of Grand Island, Jefferson Parish, Louisiana (37). Drilling in 1949 cut more than 200 ft of sulfur-bearing limestone between 1,813 and 2,075 ft below sea level, and subsequent drilling proved the existence of a large deposit of sulfur. Production began in 1960 after the expenditure of more than 22 million dollars. This is rated as the third largest sulfur deposit in the United States. Rock containing less than 5 percent sulfur is not classed as ore, and generally the sulfur content of the commercial material is 15 to 30 percent.

Sulfur is recovered from the salt-dome deposits by the Frasch process, whereby water heated to about 325°F is forced into the deposit, the enclosing rock and sulfur are heated until the sulfur melts and separates from the rock and finally from the water, since the specific gravity of the sulfur is about twice that of water. It then collects in a reservoir and is brought to the surface.

Minor amounts of sulfur in the United States come from several western states. The most important single deposit is the Leviathan, 9 miles east of Markleeville, Alpine County, California (38). The sulfur occurs in masses, stringers, and veins in andesite throughout a maximum thickness of 135 ft. The deposit is of volcanic origin.

Mexico. Sulfur production by the Frasch process began in Mexico in 1953 from salt domes on the Isthmus of Tehuantepec, state of Vera Cruz (14, 30, 48). The deposits are similar to those of Texas and Louisiana and have aided materially in advancing Mexico to second rank as a producer of sulfur.

Italy. Italy was the chief source of sulfur of the world before production began from the salt-dome deposits of Texas and Louisiana (14, pp. 822–823; 32, pp. 524–525). The deposits occur both on the mainland and on Sicily (12), and production comes from many small deposits and a few large ones. The deposits are present in rocks designated the gypsum-sulfur series—evaporites of Upper Miocene age. The beds of the series consist of gypsum, limestone, and shale, along with some thick lenses of salt and the accumulations of sulfur, which commonly are veins, seams, and linings of cavities in limestone. Generally the sulfur occurs in several beds or lenses throughout a thickness of 5 ft to more than 100 ft. Many of the small deposits cover only a few acres, but the larger deposits may extend along the strike of the beds for a mile or more and down the dip, which is 15 to 90°, for several thousand feet. Generally the material mined contains 12 to 30 percent sulfur. The sulfur cannot be recovered by the Frasch process but is mined by underground methods.

The deposits are classed as sedimentary in origin. Apparently they were deposited in a vast series of lagoons, then folded and eroded. The present deposits are remnants that were preserved in synclines or structural basins.

Japan. The sulfur deposits of Japan (3; 15; 17, pp. 224–226, 233–236) furnish an outstanding example of the volcanic type of sulfur deposit. They occur in andesitic tuffs, breccias, agglomerates, and in sediments.

The Matsuo deposits, in the Iwate Prefecture in northeastern Honshu Island, occur in andesitic lavas and pyroclastics that dip 8 to 12°. They are the largest known deposits in Japan and occur throughout an area with a diameter of 3,300 ft and extend to a depth of 500 ft. The sulfur occurs as disseminations, replacements, veins, and as a sedimentary type formed by deposition from hot springs in hot crater lakes. Four flattened ore bodies are present, one of which is 1,000 m long, 800 m wide, and 80 m thick. Material of the ore bodies varies from rich deposits of free sulfur to mixed deposits of sulfur and pyrite and marcasite, to deposits composed chiefly of pyrite. At the Matsuo mine, material containing more than 25 percent free sulfur is designated sulfur ore, whereas material containing very little free sulfur but with a total sulfur content of more than 36 percent is designated pyrite ore.

A number of other deposits of volcanic sulfur occur in Japan, but the major mines are in Hokkaido and Honshu.

Other countries. The largest reserves of volcanic sulfur are stated to be in the Andes Mountains of Chile (3, 20). Estimates indicate that as much as 100 million tons of sulfur may be contained in over 100 deposits throughout a length of 2,000 miles.

Little is known about the sulfur deposits of the Soviet Union. Several deposits are stated to be associated with gypsum of Lower Cretaceous, Miocene, and Paleocene age in various places in Central Asia, and volcanic sulfur is stated to occur in Kamchatka (41, p. 149).

Pyrites

Ordinary pyrite is one of the most common minerals and deposits are widespread. Spain was for many years the leading world producer, but first place has been held by Japan for some years (Table 22-5).

TABLE 22-5.   World Production of Pyrites,
Average of 1945–1959 and in 1962*

| | Percent of total world production | |
Country	Average of 1945–1959	1962 only
Japan	15.2	19.2
Spain	13.2	10.5
Italy	7.5	7.8
Cyprus	7.2	4.0
United States	7.1	4.5
Norway	5.2	3.9
Portugal	4.3	3.1
Canada	3.9	2.3
Soviet Union	?	15.9
China	?	5.9
All others	36.4	22.5

| | Thousand long tons | |
	Average of 1945–1959	1962 only
World total	13,353	20,100

* Compiled from *U.S. Bur. Mines Minerals Yearbook,* various years.

The largest pyrite mine of Japan is the Matsuo, discussed with the deposits of sulfur. The second largest is the Yanahara, Okayama Prefecture. Together these furnish nearly one-third of the Japanese production. The Yanahara deposits consist of elongated masses of pyrite in slates and sandstones of Paleozoic age and also in some igneous rocks. In some places where diorite intrusions occur a considerable amount of pyrrhotite is present.

The deposits of Spain and Cyprus have been productive since the very early mining by the Phoenicians and the Romans. The principal deposits in Spain are those of Huelva, mentioned with pyritic gold and silver deposits. They are part of a pyritic zone some 100 miles long and 12 to 18 miles wide that extends throughout Spain into Portugal and yields the production of the latter country also.

Much of the pyrite production of Canada and the United States is obtained as a by-product of the mining of nonferrous base-metal deposits. An outstanding example in Canada is the Noranda gold-copper area of Quebec, and one in the United States is the Ducktown copper deposit of Tennessee.

SELECTED REFERENCES

1. J. C. Arundale, 1956, Salt (including sodium), *U.S. Bur. Mines Bull.* 556, pp. 755–759.
2. D. M. Baird, 1957, Gypsum deposits of southwestern Newfoundland, in "The Geology of Canadian Industrial Mineral Deposits," pp. 124–130, Canadian Institute of Mining and Metallurgy, Montreal.
3. A. F. Banfield, 1954, Volcanic deposits of elemental sulphur, *Canadian Mining Metall. Bull.,* vol. 47, no. 511, pp. 769–775.
4. C. M. Bartley, 1961, Potash chapter, in The Canadian mineral industry 1961, *Geological Survey Canada, Mineral Resources Div., Dept. Mines and Tech. Surveys.*
5. A. C. Bersticker, editor, 1963, "Symposium on Salt," The Northern Ohio Geological Society, Cleveland, Ohio, 661 pp.
6. W. C. Bleimeister, 1961, Rock salt mining operations in Michigan, Ohio, and Ontario, *Mining Eng.,* vol. 13, pp. 467–471.
7. J. C. Brown and A. K. Dey, 1955, "India's Mineral Wealth," Oxford University Press, London, 761 pp.
8. G. M. Brownell, 1957, The Amaranth deposit of Western Gypsum Products Limited, in "The Geology of Canadian Industrial Mineral Deposits," pp. 130–131, Canadian Institute of Mining and Metallurgy, Montreal.
9. G. M. Brownell, 1957, The Gypsumville, Manitoba, deposit of Gypsum, Lime and Alabastine, Canada, Limited, in "The Geology of Canadian Industrial Mineral Deposits," pp. 131–132, Canadian Institute of Mining and Metallurgy, Montreal.
10. M. F. Byrd, 1955, Potash occurrences in the United States, *U.S. Geol. Survey Mineral Inv. Resource Appraisals Map* MR 3.
11. Collings, 1961, Gypsum and anhydrite chapter, in The Canadian mineral industry 1961, *Geological Survey Canada, Mineral Resources Div., Dept. Mines and Tech. Surveys.*
12. G. Dessau, M. L. Jensen, and N. Nakai, 1962, Geology and isotopic studies of Sicilian sulfur deposits, *Econ. Geology,* vol. 57, pp. 410–438.
13. B. W. Dyer, 1947, Discoveries of potash in eastern Utah, *Am. Inst. Mining Metall. Engineers Trans.,* vol. 173, pp. 56–61.
14. C. F. Fogarty and R. D. Mollison, 1960, Sulfur and pyrites, in "Industrial Minerals and Rocks," 3d ed., pp. 819–833, American Institute of Mining, Metallurgical, and Petroleum Engineers, New York, 934 pp.
15. Y. Fujita, 1956, Japan's Matsuo sulphur mine is big underground producer, *Eng. Mining Jour.,* vol. 157, no. 8, pp. 98–103.
16. Geologic Staff, 1957, Gypsum deposits of Gypsum, Lime and Alabastine, Canada, Limited, at Falkland, British Columbia, in "The Geology of Canadian Industrial Mineral Deposits," pp. 133–137, Canadian Institute of Mining and Metallurgy, Montreal.
17. Geological Survey of Japan, 1960, Geology and mineral resources of Japan, 2d ed., Hisamoto-cho, Kawasaki-shi, Japan, 304 pp.
18. M. A. Goudie, 1961, Saskatchewan potash deposits, *Am. Inst. Mining, Metall. Petroleum Engineers Trans.,* vol. 220, pp. 169–173.
19. A. W. Groves, 1958, Gypsum and anhydrite, London, Overseas Geological Surveys, 108 pp.

20. C. S. Haley, 1952, Sulphur—from Alaska to Chile, *Mining World,* vol. 14, no. 10, pp. 34–36.

21. J. F. Hall, 1963, Distribution of salt in Ohio, in "Symposium on Salt," pp. 27–30, The Northern Ohio Geological Society, Cleveland, Ohio, 661 pp.

22. Florence E. Harris, 1952, Salt, in W. Van Royan and O. Bowles, "The Mineral Resources of the World," vol. 2, pp. 157–161, Prentice-Hall, Inc., Englewood Cliffs, N.J., 181 pp.

23. Florence E. Harris, 1960, Economics of the salt industry, in D. W. Kaufmann (ed.), Sodium chloride; the production and properties of salt and brine, pp. 627–661, *American Chem. Soc. Mon. Ser. No.* 145, 743 pp.

24. B. L. Johnson, 1952, Potash, in W. Van Royan and O. Bowles, "The Mineral Resources of the World," vol. 2, pp. 147–152, Prentice-Hall, Inc., Englewood Cliffs, N.J., 181 pp.

25. S. J. Johnstone and M. G. Johnstone, 1961, "Minerals for the Chemical and Allied Industries," John Wiley & Sons, Inc., New York, 788 pp.

26. G. W. Josephson, 1952, Sulfur and pyrites, in W. Van Royan and O. Bowles, "The Mineral Resources of the World," vol. 2, pp. 153–156, Prentice-Hall, Inc., Englewood Cliffs, N.J., 181 pp.

27. G. W. Josephson, 1956, Sulfur, *U.S. Bur. Mines Bull.* 556, pp. 843–851.

28. V. C. Juan, 1946, Mineral resources of China, *Econ. Geology,* vol. 41, pt. 2, pp. 399–474.

29. D. W. Kaufmann, 1960, Uses of salt and brine, in D. W. Kaufmann (ed.), Sodium chloride; the production and properties of salt and brine, pp. 662–685, *American Chem. Soc. Mon. Ser. No.* 145, 743 pp.

30. J. H. Kearney, 1955, A new empire of Frasch process sulphur is rising from the jungles of Mexico, *Eng. Mining. Jour.,* vol. 156, no. 1, pp. 72–77.

31. W. L. Kreidler, 1963, Silurian salt of New York State, in "Symposium on Salt," pp. 10–18, The Northern Ohio Geological Society, Cleveland, Ohio, 661 pp.

32. R. B. Ladoo and W. M. Myers, 1951, "Nonmetallic Minerals," 2d ed., McGraw-Hill Book Company, New York, 605 pp.

33. K. K. Landes, 1960, The geology of salt deposits, in D. W. Kaufmann (ed.), Sodium chloride; the production and properties of salt and brine, pp. 28–69; Salt deposits of the United States, *idem.,* pp. 70–95; *American Chem. Soc. Mon. Ser. No.* 145, 743 pp.

34. K. K. Landes, 1963, Origin of salt deposits, in "Symposium on Salt," pp. 3–9, The Northern Ohio Geological Society, Cleveland, Ohio, 661 pp.

35. L. B. Larson, 1960, Gypsum, *U.S. Bur. Mines Bull.* 585, pp. 367–375.

36. L. B. Larson, 1960, Sulfur and pyrites, *U.S. Bur. Mines Bull.* 585, pp. 821–834.

37. C. O. Lee, Z. W. Bartless, and R. H. Feierabend, 1960, Freeport Sulphur's offshore venture, the Grand Island Mine, *Mining Eng.,* vol. 12, pp. 578–590.

38. P. A. Lydon, 1957, Sulfur and sulfuric acid, *California Dept. Nat. Res. Div. Mines, Bull.* 176, pp. 613–622.

39. R. T. MacMillan, 1960, Salt, in "Industrial Minerals and Rocks," 3d ed., pp. 713–731, American Institute of Mining, Metallurgical, and Petroleum Engineers, New York, 934 pp.

40. R. T. MacMillan, 1960, Sodium and sodium compounds, *U.S. Bur. Mines Bull.* 585, pp. 745–765.

41. D. V. Nalivkin, 1960, "The Geology of the U.S.S.R., a Short Outline," Pergamon Press, New York, 170 pp., geologic map of the U.S.S.R.

42. O. S. North, 1956, Gypsum, *U.S. Bur. Mines Bull.* 556, pp. 339–345.

43. J. F. Pepper, 1947, Areal extent and thickness of the salt deposits of Ohio, *Ohio Geol. Survey Rept. Inv. No.* 3, 15 pp.

44. W. G. Pierce and E. I. Rich, 1962, Summary of rock salt deposits in the United States, *U.S. Geol. Survey Bull.* 1148, 91 pp.

45. E. R. Ruhlman, 1956, Potash, *U.S. Bur. Mines Bull.* 556, pp. 703–713.

46. E. R. Ruhlman, 1960, Potash, in "Industrial Minerals and Rocks," 3d ed., pp. 669–680, American Institute of Mining, Metallurgical, and Petroleum Engineers, New York, 934 pp.

47. E. R. Ruhlman, 1960, Potassium compounds, *U.S. Bur. Mines Bull.* 585, pp. 651–658.

48. S. Schwartz, 1953, Mexico challenges sulphur shortage, *Eng. Mining. Jour.,* vol. 154, no. 6, pp. 77–79.

49. D. S. See, 1960, Solar salt, in D. W. Kaufmann (ed.), Sodium chloride; the production and properties of salt and brine, pp. 96–108, *Am. Chem. Soc. Mon. Ser. No.* 145, 743 pp.

50. J. P. Smith, 1950, Notes on the geology of the potash deposits of Germany, France, and Spain, *Am. Inst. Mining Metall. Engineers Trans.,* vol. 187, pp. 117–121.

51. A. B. Thut, 1962, International Minerals conquers high-pressure water and quicksand to sink Canadian potash shaft, *Mining World,* vol. 24, no. 11, pp. 20–24.

52. R. V. Tomkins, 1957, Potash, in "The Geology of Canadian Industrial Mineral Deposits," pp. 198–202, Canadian Institute of Mining and Metallurgy, Montreal.

53. W. E. Ver Planck, 1957, Salt, *California Dept. Nat. Res., Div. Mines, Bull.* 176, pp. 483–494.

54. W. E. Ver Planck, 1957, Salines, *California Dept. Nat. Res., Div. Mines, Bull.* 176, pp. 475–482.

55. W. E. Ver Planck, 1957, Gypsum, *California Dept. Nat. Res., Div. Mines, Bull.* 176, pp. 231–240.

56. W. H. Voskuil, 1955, "Minerals in World Industry," McGraw-Hill Book Company, New York, 324 pp.

57. C. W. Withington, 1962, Gypsum and anhydrite in the United States, exclusive of Alaska and Hawaii, *U.S. Geol. Survey Mineral Inv. Resources Map* MR-33. Map and pamphlet of 18 pp.

58. M. F. Zaskalicky, 1957, The gypsum deposits of Canadian Gypsum Company Limited in Nova Scotia, New Brunswick and Ontario, in "The Geology of Canadian Industrial Mineral Deposits," pp. 119–123, Canadian Institute of Mining and Metallurgy, Montreal.

59. Anon., 1958, Potash producer looks to California reserves, *Mining Eng.,* vol. 10, pp. 570–572.

60. Anon., 1962, Israeli potash firm plans $50-million expansion, *Eng. Mining*

Jour., vol. 163, no. 3, p. 116. See also the following: 1962, Dead Sea works gets dike system, *Mining Eng.*, vol. 14, no. 11, p. 21; 1962, Arab potash completes feasibility studies, contracts for plant construction at Sofi, *Eng. Mining Jour.*, vol. 163, no. 10, p. 110; 1964, Dead Sea expansion, *Mining Jour. (London)*, vol. 262, no. 6717, p. 377.

61. Anon., 1963, World's biggest potash mine Esterhazy gets into gear, *Mining Jour. (London)*, vol. 260, no. 6653, pp. 173–174.

62. Anon., 1963, Duval's Nash Draw potash mine will be operated in 1964, *Mining World*, vol. 25, no. 9, p. 65; 1963, Langbeinite flotation doubles potash reserves for I.M.C., *Mining World*, vol. 25, no. 5, p. 44.

63. Anon., 1963, *Mining Jour. (London)*, vol. 261, no. 6686, p. 341. Notice of new potash mine in Saskatchewan, Canada.

64. Anon., 1963, Potash mining in Alsace, *Mining Jour. (London)*, vol. 260, no. 6658, pp. 294–295.

chapter 23

Phosphate Rock

Commercial phosphate rock is essentially calcium phosphate plus various impurities —carbonates of calcium and magnesium, iron oxides, clay, silica, fluorine, and perhaps traces of uranium. Several grades of phosphate rock are utilized, acid grade (high grade, shipping grade, or fertilizer grade), containing more than 30 percent P_2O_5; furnace grade, 24 to 40 percent P_2O_5; low grade, less than 24 percent P_2O_5; low grade for direct application, at least 20 percent P_2O_5.

Phosphorous is one of the three important elements of commercial fertilizers, the other two being potassium and nitrogen. The major use of phosphate rock is in the manufacture of fertilizers. Other important uses are in the chemical industry, especially in making detergents. Some by-products, especially fluorine, uranium, vanadium, and rare earths, may be recovered from some phosphate deposits.

Generally something more than 40 percent of the phosphate rock of the world is produced in the United States, about 25 percent in Africa (Morocco, Tunisia, Algeria), and possibly 15 percent in the Soviet Union (Table 23-1).

TABLE 23-1. World Production of Phosphate Rock, Average of 1945–1959 and in 1962*

	Percent of world total	
Country	Average of 1945–1959	1962 only
United States	45.6	42.1
Morocco	17.1	17.4
Soviet Union†	14.2	18.6
Tunisia	7.0	4.5
Nauru Island	3.5	3.3
All others	12.6	14.1
	Thousand long tons	
	Average of 1945-1959	1962 only
World total	25,361	46,040

* Compiled from *U.S. Bur. Mines Minerals Yearbook,* various years.
† Includes apatite.

United States

Phosphate rock in the United States (2; 8; 9; 10; 12; 13; 15, pp. 379–393; 16; 17; 18; 21; 22; 23) is produced chiefly in Florida, Tennessee, and four western states—Idaho, Montana, Utah, and Wyoming—designated the *Western Field*. Florida furnished 72 percent of the total production in 1962, Tennessee supplied 12 percent, and the Western Field 16 percent, almost two-thirds of which came from Idaho. Early production in the United States, starting in 1867, was chiefly from South Carolina until 1894, as production in Florida did not begin until 1888. The reserves were given in 1960 by Ruhlman (22, p. 636) as shown in Table 23-2.

In addition to the reserves given by Table 23-2, a distinct possibility exists that economic mining of phosphatic nodules on the sea floor off the coast of California may be feasible (19). Economic recovery may be possible between latitude 38°N and 32°30′N, about from San Francisco to somewhat south of Los Angeles. The estimated reserve of material that it may be possible to recover economically is enough to maintain production for 200 years at an annual rate of 500,000 tons.

TABLE 23-2.   Reserves and Potential Resources of Phosphate Rock in the United States*

Area	Total reserves, million long tons		Potential resources, million long tons	
	Marketable product	P_2O_5 content	Marketable product	P_2O_5 content
Arkansas	—	—	20	5
Florida	2,040	660	23,350	4,932
South Carolina	—	—	9	2.4
Tennessee	85	15	5,398	1,129
Western Field	3,000	870	20,000	5,800
Total (rounded)	5,100	1,500	49,000	12,000

* From *U.S. Bur. Mines Bull.* 585, p. 636, 1960.

Florida

Phosphate-bearing rock is rather widespread throughout western Florida in a zone about 30 to 40 miles wide and 200 miles long that lies 20 to 50 miles inland from the coast and extends northward from about 27°30′N latitude (Fig. 23-1). Within this zone are two general areas of commercial phosphate deposits, a southern somewhat oval one, the land-pebble district, which now furnishes most of the production of Florida, and a more northern elongate one, the hard-rock district, which now furnishes very little of the production but which supplied most of the material mined between 1888 and 1906. The Florida deposits were investigated in some detail in the early 1960s (5, 6, 7, 11) in order to furnish additional information that may be of value in prospecting or in development.

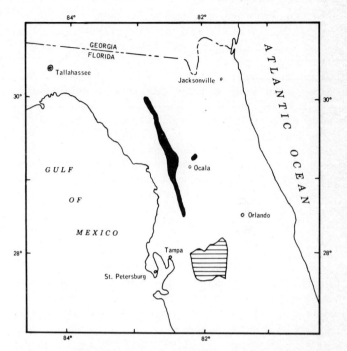

Fig. 23-1   Generalized map of northern Florida showing approximate areas of hard-rock phosphate mining district (solid black) and land-pebble phosphate mining district (horizontal ruling). After G. H. Espenshade and C. W. Spencer, *U.S. Geol. Survey Bull.* 1118, p. 4.

The phosphate-bearing rocks at or near the surface in this part of Florida are the Hawthorne formation of Miocene age, the Bone Valley formation of Pliocene age, and the Alachua formation, which may be of Pliocene age (16) or of Miocene age, possibly in part equivalent to the Hawthorne formation and in part younger (24, pp. 181–208); or the Bone Valley formation may be in part the equivalent of the Alachua formation and in part somewhat younger than the Alachua (9, pp. 221, 234). The Hawthorne formation is composed of (1) lower phosphatic marls and hard limestone that contains clay and quartz sand, (2) overlying clastic phosphatic sands, clays, and sandy clays, and (3) upper dolomites and dolomitic limestone. The Bone Valley formation in the important producing part of the land-pebble district is composed of (1) a lower unit of clay, silt, sand, and gravel and (2) an upper unit composed of quartz sand, aluminum phosphate minerals, and kaolinite. The gravel and much of the sand of the lower unit consists of phosphate, identified as carbonate-fluorapatite. The Alachua formation is composed of clays, fine sands, phosphate rock, and fragments of some older formations. The matrix of much of the sand and the fragments is a phosphate mineral.

Throughout the phosphate-bearing areas (2, 3, 8, 9, 11, 13, 16, 24) the phosphates may be classed in three general types—rock phosphate or hard-rock phosphate, soft phosphate, and pebble phosphate (land-pebble and river-pebble). The hard-rock phosphate is material of the Alachua formation, the land-pebble phosphate is material of the Bone Valley formation, the soft phosphate is fine material that forms much of the matrix of the hard-rock phosphate and also the land-pabbe phosphate. River-pebble phosphate consists of phosphatic fragments along certain streams and was utilized only during the very early history of mining in the region.

The hard-rock phosphate occurs at the base of the Alachua formation, which rests unconformably on older rocks. The deposits consist of phosphate masses and particles—pellets, pebbles, boulders—in a matrix of fuller's earth, phosphatic clay, and quartz sand. They are of irregular form and generally occur in depressions in the underlying limestone. Present mining occurs only in Marion County, near Dunellon, on the Ocala uplift.

Various concepts of the origin of the hard-rock phosphate deposits have been advanced, and these have been reviewed by Vernon (24, pp. 195–200). The origin that has been most generally accepted is solution of older materials, chiefly the Hawthorne formation, by groundwater, and subsequent precipitation. Vernon, however, suggested that phosphoric acid, derived possibly from guano deposits on islands that were bird rookeries, reacted with limestone bedrock and thus formed the deposits in place.

The land-pebble phosphate occurs in the lower unit of the Bone Valley formation and is locally designated *matrix*. It consists of particles from sand size to as much as 2 ft across composed chiefly of phosphate, along with phosphatic clay-sized material. The lower unit of the Bone Valley formation generally is about 20 ft thick, although it varies in thickness from less than 1 ft to more than 50 ft. The upper unit generally is about 6 ft thick but also varies in thickness, from almost nothing to 60 ft. The phosphate of this unit consists of wavellite and crandallite, both aluminum phosphates, and thus this unit constitutes an aluminum phosphate zone, thought to have been formed by leaching of material similar to that of the lower unit. In the southern part of the district the two units cannot be separated. Throughout the area these phosphate zones are not fixed, and aluminum phosphate minerals have been formed in the Bone Valley formation and in the Hawthorne formation, and the calcium phosphate zone may include the bottom part of the Bone Valley formation and the weathered top of the Hawthorne formation.

The phosphorite nodules of the land-pebbe phosphate deposits contain 30 to 36 percent P_2O_5, the higher percentages generally occurring in the smaller nodules. The amount of recoverable phosphate obtained by open-pit mining is about 5,000 tons per acre, but may be as low as 500 tons and as high as 35,000 tons.

Some interesting and efficient methods of benefication and recovery of phosphate have been developed. These include hydraulic transportation through pipes of as much as 20 in. in diameter and pipelines more than 5 miles long in some cases, disintegration of material, removal of clay slimes, screening, and flotation (26).

Tennessee

Most of the commercial phosphate deposits of Tennessee (1, 2, 23) are in the Central basin, a depressed area about 100 miles long and 50 to 60 miles wide, with Murfreesboro a few miles south of its center. Most of the important deposits occur along the western side of this area or in adjoining Highland Rim areas. Three types of phosphate rock are present in the state—brown, blue, and white—but production comes almost entirely from the brown type.

The commercial brown phosphate deposits are practically confined to the Bigby-Cannon limestone of Ordovician age, but low-grade deposits are present in the underlying Hermitage formation. Deposits of some importance occur, however, in the Leipers formation, also of Ordovician age but stratigraphically about 80 ft above the Bigby-Cannon limestone. The major producing area is in Maury, Giles, and Williamson Counties on the west and southwest flanks of the Nashville dome. This is a result of the character and distribution of the Bigby-Cannon limestone, which, according to Alexander (1, pp. 5–6), consists of three facies, the Bigby, the Cannon, and the Dove-colored. The Bigby facies is a highly phosphatic limestone, whereas the other two facies are, in general, nonphosphatic limestones. The part of the formation on the extreme western side of the Central basin consists entirely of the phosphatic Bigby facies; the part on the extreme eastern side, almost entirely of Cannon facies. These facies interfinger and may be replaced by lentils of Dove-colored facies near the axis of the Nashville dome.

The brown phosphate deposits form (1) residual blankets, which have extensions downward into solution-enlarged joints of the limestone, such as extension being locally designated a *cutter*; and (2) rim accumulations. The blanket deposits form where the Bigby facies underlies a flat or only gently rolling surface, whereas the rim deposits form where the Bigby facies crops out along a slope and is overlain by younger formations. The deposits may be exposed at the surface or they may be overlain by soil and clay generally less than 20 ft thick. The phosphatic material of the deposits commonly consists of loose grains of calcium phosphate mixed with fine-grained phosphate and clay. Such deposits are locally designated *muck*.

Western Field

The commercial phosphate deposits of the Western Field occur within the area of the Permian Phosphoria formation in Montana, Idaho, Wyoming, Utah, and Nevada (2, 10, 12, 17, 18). Within this region there are two chief mining areas, one near Helena, Montana, the other near Pocatello, Idaho, in the southeastern corner of that state and the northeast corner of Utah (Fig. 23-2). About two-thirds of the production in 1962 came from Idaho.

The Phosphoria formation (18) at the type locality in southeastern Idaho is composed mainly of chert, carbonaceous mudstone, and phosphorite. These rocks give away, at least in part, to sandstone and carbonate rocks around the margin of the phosphate field. At the type locality the Meade Peak phosphatic shale member, which is the lowermost member of the formation, is about 200 ft thick. In general the Meade Peak member consists of phosphatic and carbonaceous mudstone, phosphorite, and minor amounts of dark carbonate rocks. The phosphorites

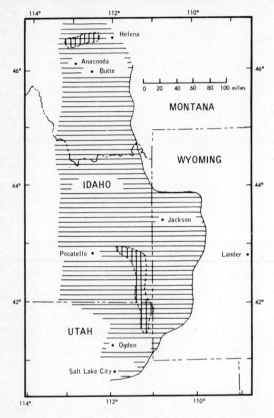

Fig. 23-2 Map showing distribution of mudstone, chert, and phosphorite of the Phosphoria formation (horizontal ruling) and approximate location of the two major phosphate-mining areas (cross-hatching) in part of the Western Phosphate Field. After V. E. McKelvey and others, *U.S. Geol. Survey Prof. Paper* 313-A, p. 4.

are composed of carbonate-fluorapatite along with varying amounts of detrital minerals, calcite, dolomite, chert, carbonaceous matter, and glauconite.

Commercial phosphate deposits generally occur in two zones, but at a few places in three zones. The main phosphatic zone, near the base of the Meade Peak member and designated the *lower phosphate zone* at the type locality, where it is 7 ft thick, may contain as much as 90 percent phosphate pellets and is high-grade phosphate rock. A second zone, the *upper phosphate zone* of the type locality, where it starts about 25 ft below the top of the Meade Peak member and is about 15 ft thick, contains pisolitic and oolitic phosphorite and some slightly phosphatic mudstone.

Mining in the Western Field is by underground and open-pit methods. Before 1946 all mining was of the underground type, and by about 1960 this method produced only 20 to 30 percent of the material mined. The current underground mining is chiefly in Montana. Much of the phosphate rock is beneficiated before it is processed.

North Carolina

Intensive exploration of phosphate deposits in Beaufort County, eastern North Carolina, has been in progress since 1961. The activity has been in the bottom land areas of the Pamlico and Pungo Rivers. Texas Gulf Sulphur had invested more than 2 million dollars in exploration and mining through 1963, and several other companies were conducting exploration. The phosphatic beds are 150 to 200 ft below the surface and contain up to 21 percent P_2O_5 (29, 30).

Northern Africa

The important phosphate deposits of Morocco, Algeria, and Tunisia supply about one-fourth of the world production, a little more than two-thirds of which comes from Morocco. A considerable amount of expansion of the Moroccan phosphate industry has taken place since 1960 (27, 28).

Morocco

The principal deposits of Morocco (2, pp. 659–660; 4, vol. 4, pp. 1–34; 15, pp. 384–385) occur throughout a distance of several hundred miles inland from Safi and north of the High Atlas in the region known as the Phosphate Plateau, which is composed of Upper Cretaceous and Eocene sedimentary rocks. Within this region there are a number of phosphate deposits, but two are of special importance, Khouribga, productive since 1922, and Louis Gentil (Kachkate), productive since 1932 (Fig. 11-8).

The deposit at Khouribga is on the Ouled Abduon Plateau which, in this area, is underlain by phosphate throughout about 150 sq km. The phosphatic formation, of Eocene age, about 20 to 25 m thick around Khourigba, consists of marls, sandy phosphate, and phosphatic limestone. Reserves of the Ouled Abduon Plateau area have been estimated to be about 1,000 million tons of high-grade phosphatic material. The material at Khourigba is known as Moroccan phosphate and is of high grade, 72.5 to 75 percent BPL.[1]

The Louis Gentil (Kachkate) deposit, on the northeastern slopes of the Gantour Hills, consists of phosphate beds of Eocene age similar to those at Khourigba. The grade of phosphate is more variable and generally not as high as at the Khouribga deposit, averaging about 70 percent BPL. The material is exported from Safi and is known as Safi phosphate. Reserves have not been estimated but are stated to be very important.

Algeria and Tunisia

The deposits of Algeria and Tunisia (2, pp. 660–661; 4, vol. 4; 15, p. 385; 25) occupy a zone south of the Mediterranean Sea that extends south for about 250 miles and east-west for about 500 miles. Deposits are mined in three principal districts: Setif in Algeria (deposits of M'Zaita and Tocqueville); Tebessa, in Algeria and Tunisia (deposits of Djebel Kouif and Kalaa-Djerda); and Gafsa, in Tunisia and Algeria (deposits of Metlaoui, Redeyef, Ain Moulares, and Djeb M'Dilla in Tunisia and Djebel Onk in Algeria). Among these, the deposits of the Gafsa area are by far the most important and contain the largest reserves. The

---

[1] BPL is the trade designation for bone phosphate of lime, and one percent BPL is equvalent to 0.458 percent P_2O_5. The designation TPL, tricalcium phosphate of lime, is used in some countries.

deposits are similar to those of Morocco and are also of Eocene age, but the content of P_2O_5 is somewhat lower, the deposits generally ranging from 58 to 64 percent of BPL. The reserves, especially those of Tunisia, are large.

Soviet Union

Phosphate is obtained both from apatite and phosphorite deposits in the Soviet Union (2; 4, vol. 4, pp. 47–61; 15, pp. 385–386). The apatite deposit, Kola, is in the Kola Peninsula near Kirovsk (67°30'N, 34°E) associated with a nepheline-syenite complex and was discussed briefly with nepheline-syenite deposits. The reserves of apatite are very great.

Seven phosphate-bearing basins are present in the Soviet Union (4, vol. 4, pp. 47–53), in which the age of the phosphatic rocks ranges from Cambrian to Tertiary, although the principal phosphatic formations are of Upper Jurassic and Lower Cretaceous age. Most of the known deposits are in the area of the Russian platform.

The largest phosphorite deposit, according to Nalivkin (20, p. 98), is on the southeastern slope of the Karatau Mountains (45°N, 67°E, to 42°30'N, 71°E) in Kazakhstan, near the base of marine limestones of Middle Cambrian age. The main phosphate horizon is about 31 ft thick and averages slightly less than 29 percent P_2O_5. This phosphorite deposit and the apatite deposit at Kola provide most of the phosphate of the Soviet Union (4, vol. 4, p. 54).

Other deposits of the Soviet Union incllude phosphorite nodules in marine deposits of Upper Cretaceous age in the western part of Mangyshlak (44°30'N, 50°E, to 43°N, 53°E) and along the shore of the Aral Sea, and similar deposits along the western slopes of the Karatau Mountains near Tashkent (41°N, 69°E), and in the Fergana Valley. Also, a large phosphorite mine is stated to be located on the western flanks of the Mugodzshar Mountains at Aktyubinsh (50°N, 57°E).

Oceanic Islands

Nauru Island in the Gilbert group of islands, just south of the equator at about 167°E longitude, has furnished slightly more than 3 percent of the phosphate of the world for the last 20 years. It is typical of a number of islands on which large deposits of guano have furnished solutions of phosphoric acid that reacted with and gradually replaced the calcium carbonate of coral limestone.

SELECTED REFERENCES

1. F. M. Alexander, 1953, Field trip to central Tennessee phosphate district, *Southeastern Section Geol. Soc. America Guidebook,* Nashville, Tenn., pp. 3–8.
2. J. A. Barr, 1960, Phosphate rock, in "Industrial Minerals and Rocks," 3d ed., pp. 649–668, American Institute of Mining, Metallurgical, and Petroleum Engineers, New York, 934 pp.

3. R. L. Bates, 1960, "Geology of the Industrial Minerals and Rocks," Harper & Row, Publishers, Incorporated, New York, 441 pp.

4. British Sulphur Corporation, 1961, "World Survey of Phosphate Deposits;" vol. 1, Near and Middle East, 48 pp; vol. 2, Canada and eastern U.S.A., 70 pp; vol. 3, West, south and east Africa, 97 pp; vol. 4, North Africa, with U.S.S.R., China, India, Antarctica and western Europe, 69 pp.; vol. 5, Western states of America and Pacific Ocean, 69 pp.; vol. 6, Central America, the Caribbean, and South America, 63 pp; London.

5. J. B. Cathcart, 1963, Economic geology of the Keysville quadrangle, Florida, *U.S. Geol. Survey Bull.* 1128, 82 pp.

6. J. B. Cathcart, 1963, Economic geology of the Chicora quadrangle, Florida, *U.S. Geol. Survey Bull.* 1162-A, 66 pp.

7. J. B. Cathcart, 1963, Economic geology of the Plant City quadrangle, Florida, *U.S. Geol. Survey Bull.* 1142-D, 56 pp.

8. J. B. Cathcart, L. V. Blade, D. F. Davidson, and K. B. Ketner, 1953, The geology of the Florida land-pebble phosphate deposits, Internat. Geol. Cong., 19th, Algiers 1953, Comptes rendus, sec. 11, pt. 11, pp. 77–91.

9. J. B. Cathcart and L. J. McGreevy, 1959, Results of geologic exploration by core drilling, 1953, land-pebble phosphate district, Florida, *U.S. Geol. Survey Bull.* 1046-K, pp. 221–298.

10. F. A. Crowley, 1962, Phosphate rock, *Montana Bur. Mines and Geol. Spec. Pub.* 25 (*Mineral Resource data sheet no. 1*), 8 pp.

11. G. H. Espenshade and C. W. Spencer, 1963, Geology and phosphate deposits of northern peninsular Florida, *U.S. Geol. Survey Bull.* 1118, 115 pp.

12. F. C. Jaffé, 1961, Phosphate rock of the western United States, *Colorado School Mines, Min. Ind. Bull.,* vol. 4, no. 5, 11 pp.

13. B. L. Johnson, 1952, Phosphate rock, in W. Van Royan and O. Bowles, "The Mineral Resources of the World," vol. 2, pp. 141–146, Prentice-Hall, Inc., Englewood Cliffs, N.J., 181 pp.

14. S. J. Johnstone and M. G. Johnstone, 1961, "Minerals for the Chemical and Allied Industries," John Wiley & Sons Inc., New York, 788 pp.

15. R. B. Ladoo and W. M. Myers, 1951, "Nonmetallic Minerals," 2d ed., McGraw-Hill Book Company, New York, 605 pp.

16. G. C. Matson, 1915, The phosphate deposits of Florida, *U.S. Geol. Survey Bull.* 604, 101 pp.

17. V. E. McKelvey, R. W. Swanson, and R. P. Sheldon, 1953, The Permian phosphorite deposits of western United States, Internat. Geol. Cong., 19th, Algiers 1953, Comptes rendus, sec. 11, pt. 11, pp. 45–64.

18. V. E. McKelvey and others, 1959, The Phosphoria, Park City and Shedhorn formations in the Western Phosphate Field, *U.S. Geol. Survey Prof. Paper* 313-A, 47 pp.

19. J. L. Mero, 1961, Sea floor phosphate, *California Div. Mines, Mineral Inf. Service,* vol. 14, no. 11, pp. 1–12.

20. D. V. Nalivkin, 1960, "The Geology of the U.S.S.R., a Short Outline," Pergamon Press, New York, 170 pp., geologic map of U.S.S.R.

21. E. R. Ruhlman, 1956, Phosphate rock, *U.S. Bur. Mines Bull.* 556, pp. 681–693.

22. E. R. Ruhlman, 1960, Phosphate rock, *U.S. Bur. Mines Bull.* 585, pp. 631–641.

23. R. W. Smith and G. S. Whitlatch, 1940, The phosphate resources of Tennessee, *Tennessee Div. Geol., Bull.* 48, 444 pp.

24. R. C. Vernon, 1951, Geology of Citrus and Levy Counties, Florida, *Florida Geol. Survey Bull.* 33, 256 pp.

25. L. D. Visse, 1952, Genèse des gites phosphatés du sud-est algero-tunisien, Internat. Geol. Cong., 19th, Algeria, Mon. regionales ser. 1, Algéria, no. 27, 58 pp.

MISCELLANEOUS REFERENCES CITED

26. H. B. Hardy and S. A. Canariis, 1961, Hydraulic transportation of Florida phosphate matrix, *Mining Eng.,* vol. 13, pp. 274–281. See also W. M. Houston and W. A. LaVenue, 1962, Current benefication practices for pebble phosphate in Florida, *Mining Eng.,* vol. 14, no. 11, pp. 45–49.

27. Anon., 1961, Morocco's phosphate industry expands, *Mining Jour., (London),* vol. 257, no. 6586, p. 477.

28. Anon., 1963, Mining in Morocco, *Mining Jour. (London),* vol. 261, no. 6678, p. 151.

29. B. Barnes, 1964, U.S.G.S. report gives boost to mining plans in North Carolina's phosphate fields, *Eng. Mining Jour.,* vol. 165, no. 3, pp. 92–95.

30. Anon., 1964, North Carolina phosphate industry soon a reality?, *Metal Mining and Processing,* vol. 1, March, p. 19.

Index

Rhodesia, metallic ores production of, 110
platinum deposits of, 120
Rhodochrosite, sedimentary formation of, and
environment, 68–69, 75
Riddle, Oregon, residual nickel deposit, 85,
126
Rio Tinto mines, Spain, 281
Robinson copper district, Nevada (*see* Ely
copper district)
Rock character and localization of ore (*see*
Mineral deposits, localization of)
Rubidium, association of, with cesium and
lithium, 148
Rubidium deposits, 148
Rubidium minerals (*see* Cesium and rubidium
minerals)
Rubidium production, distribution among
countries, 108
Rubidium reserves, 148
Rubies, 441
Russia (*see* Soviet Union)
Rutile, occurrence of, 319
Rutile deposits, of Australia, 345–346
of the United States, 347
Rutile production, of Australia, 344, 346
of the United States, 347

Saddle reefs of the Bendigo gold field, Australia, 22, 25, 199–201
nature of, 15, 25
Saline lakes, composition of, 71–72
Salinity and chemical deposition, 65–66
Salmo area, British Colombia, silver-lead-
zinc-cadmium deposits of, 218–219
production of, 218
tungsten deposits of, 170
Salt, early importance of, 3
uses of, 508
Salt deposits, 508–513
of China, 513
of Germany, 512, 515
of India, 513
of the Soviet Union, 512–513
types of, 508
of the United States, 509–512
in the Gulf Coast embayment, 511–512
map of, 511
in the northeastern basin, 509–511
map of, 510
in the Permian basin, 512
map of, 511
Salt domes, characteristics of, 99
sulfur deposits in, 99–100
Salt production, 429–431, 509, 512–513
of Canada, 509
of China, 429, 509
by solar evaporation, 513
distribution among countries, 509
of France, 430, 509
of India, 429, 509
by solar evaporation, 513
of Italy, 509
of the Soviet Union, 430, 509
of Spain, 509

Salt production, of the United Kingdom, 430,
509
of the United States, 431, 509, 512
by solar evaporation in California, 512
of West Germany, 509
of the world, 509
Salt reserves of the United States in the
northeastern basin, 509
San Francisco del Oro silver-lead-zinc-copper
area, Chihuahua, Mexico, 207
San José antimony mines, Mexico, 300
San Luis Potosí, Mexico, production of, 204
San Manuel copper mine, Arizona, 257, 259–
260
Sand and gravel, 435
Sangdong tungsten deposit, Republic of Korea,
166–167
Santa Barbara silver-lead-zinc-copper area,
Chihuahua, Mexico, 207
Santa Eulalia silver-lead-zinc-copper area,
Chihuahua, Mexico, 207–209
Santa Rita district, New Mexico (*see* Central
district)
Sapphires, 441
Scheelite, occurrence of, 153
Schürmann, E., replacement series of sulfides,
92–93
Searles Lake, California, borate deposits of,
431
lithium in, 134–135
potash production from, 514
salt production from, 512
sodium carbonate deposits of, 432
sodium sulphate deposits of, 433
tungsten in, 169
Sedimentary bed deposits, characteristics of,
12, 60–66
selected examples of, containing iron, 360–
362, 368–369, 372–375, 378–381, 385,
399
containing manganese, 390–391, 396, 399
containing phosphate rock, 529–532
containing potash minerals, 514–515
containing salt, 509–512
Sedimentary beds, formation of, by chemical
and biochemical activity, 61–66
by mechanical deposition, 60–61
Selukwe chromite deposit, Rhodesia, 116–117
Semiprecious stones of Brazil, 137
Sericitization, 41
Serpentinization, 42
Shinkolobwe mine, Katanga, Republic of the
Congo, 337–338
Siderite, sedimentary formation of, and environment, 68–69, 75
Sierra Leone, diamond deposits of, 445–446
diamond production of, 429, 442
industrial minerals and rocks production of,
429
Silesian region, lead-zinc deposits of, 252–253
Silica, precipitation of, 63
separation of, from aluminum, 64
from iron, 64
solution of, effect of pH on, 64
and transportation of, 63